D1273270

SELECTED TOPICS IN GRAPH THEORY

SELECTED TOPICS IN GRAPH THEORY

Edited by

LOWELL W. BEINEKE

Department of Mathematics
Purdue University at Fort Wayne
Indiana, U.S.A.

and

ROBIN J. WILSON

Faculty of Mathematics
The Open University
England

1978

ACADEMIC PRESS
London New York San Francisco
A Subsidiary of Harcourt Brace Jovanovich, Publishers

ACADEMIC PRESS INC. (LONDON) LTD.
24/28 Oval Road,
London NW1 7DX

United States Edition published by
ACADEMIC PRESS INC.
111 Fifth Avenue
New York, New York 10003

Library of Congress Catalog Card Number: 78-18676
ISBN 0-12-086250-6

Typeset by Eta Services (Typesetters) Ltd., Beccles, Suffolk
Printed in Great Britain by Whitstable Litho Ltd., Kent

Preface

One of the consequences of the recent rapid expansion in graph theory is that it has become increasingly difficult to ascertain what is currently known about any particular topic in the field. In view of this, we felt that it would be worth while to collect together a series of expository surveys written by a distinguished group of authors, and covering various areas of graph theory, in the hope that such a collection might prove useful to professional graph theorists, to newcomers to the field, and to experts in other fields who may want to learn about specific topics.

The selection of topics chosen is entirely ours, and we are well aware that many important areas have had to be omitted. Our choices were made on the basis of several criteria, including a particular need for surveys on certain topics, the timeliness of particular areas, various suggestions from colleagues and friends, and, of course, our own particular preferences. We are hoping to follow this book with a sequel which includes some of the areas we have omitted, and we shall be happy to receive any suggestions as to which subjects should be covered. In both of these books, the emphasis is on "pure" graph theory. The reader primarily interested in the applications of graph theory in the Sciences and Social Sciences should refer to our companion volume *Applications of Graph Theory*. This latter book contains expository chapters covering the applications of graph theory to a wide variety of subjects, ranging from Communications Networks and Chemistry to Geography and Architecture.

An important feature of this book is that we have attempted, as far as possible, to impose a uniform terminology and notation throughout, in the belief that this will aid the reader in going from one chapter to another. It should also make the book more accessible to groups using it for an advanced course or seminar. In order to give the chapters a fairly consistent style, we asked our contributors to undergo the ordeal of having the early versions of their chapters subjected to severe criticism. We believe that this resulted in a considerable improvement in the final drafts, and we should like to express our thanks and appreciation to all our contributors for their co-operation in this, and in particular for their tolerance and their willingness to put up with our idiosyncracies.

We should also like to thank the many reviewers of the early drafts of chapters – and in particular, the seminar group at Western Michigan University – for their helpful and pertinent comments; the secretarial staff of the Open University, for typing the manuscript; the Mathematics Departments of the Open University, Oxford University, and Purdue University, for financial support and the use of facilities; Academic Press for encouragement, support and co-operation; (and finally) our wives and children, who have had to put up with us during the writing of this book.

August, 1978.

L.W.B.

R.J.W.

Notes on Contributors

Lowell Beineke is Professor of Mathematics at the Fort Wayne campus of Purdue University, where he moved after receiving his Ph.D. in topological graph theory from the University of Michigan. He has contributed to a wide variety of areas in graph theory, including topological graph theory, line graphs and tournaments, and is an Associate Editor of the *Journal of Graph Theory*. *Present address: Department of Mathematics, Purdue University at Fort Wayne, Indiana 46805, U.S.A.*

Jean-Claude Bermond is a Research Fellow at the Centre National de la Recherche Scientifique, where he earlier obtained his doctorate. He has worked in several areas of graph theory—in particular, Hamiltonian graphs and tournaments—as well as in the theory of designs and hypergraphs. He was co-organizer of the 1976 Paris-Orsay Conference on Combinatorial Mathematics. *Present address: Université Paris XI, Informatique, bât 490, 91405-Orsay, France.*

Peter Cameron is Fellow and Tutor in Mathematics at Merton College, Oxford University. He received his undergraduate degree in Australia, and his D.Phil. in Oxford, where he has remained apart from brief appointments at the University of Michigan and Bedford College, London. His research interests include permutation groups, finite groups, finite geometries, designs, and graphs. He is the author of *Parallelisms of Complete Designs* (Cambridge University Press, 1976), co-author (with J. H. van Lint) of *Graph Theory, Coding Theory and Block Designs* (Cambridge University Press, 1975), and editor of *Combinatorial Surveys: Proceedings of the Sixth British Combinatorial Conference* (Academic Press, 1977). *Present address: Merton College, Oxford, England.*

Stanley Fiorini is Lecturer in Mathematics at the University of Malta, where he went after receiving his Ph.D. from the Open University. His research interests include edge-colorings of graphs and the reconstruction problem, and he is co-author (with R. J. Wilson) of *Edge-Colourings of Graphs* (Pitman Publishing, 1977). *Present address: Department of Mathematics, The University of Malta, Msida, Malta.*

Robert Hemminger is Professor of Mathematics at Vanderbilt University, where he went after receiving his Ph.D. from Michigan State University. His main research interests concern the reconstruction problem, line graphs and digraphs, and automorphism groups. *Present address: Department of Mathematics, Vanderbilt University, Nashville, Tennessee 37235, U.S.A.*

Crispin Nash-Williams is Professor of Pure Mathematics at the University of Reading, and also holds a part-time appointment at the University of Waterloo. He is on the editorial boards of the *Journal* and the *Proceedings of the London Mathematical Society*, the *Journal of Combinatorial Theory* (*B*), and *Aequationes Mathematicae*. His main research interests in combinatorics include the reconstruction problem, traversability, infinite graphs and transversal theory. He is co-editor (with J. Sheehan) of the *Proceedings of the Fifth British Combinatorial Conference (1975)*. *Present address: Department of Mathematics, The University of Reading, Reading RG6 2AX, England.*

Edgar Palmer is Professor of Mathematics at Michigan State University. His main research interest lies in the field of enumerative graph theory. He has written many research papers in this area, and is co-author (with F. Harary) of *Graphical Enumeration* (Academic Press, 1973). *Present address: Department of Mathematics, Michigan State University, East Lansing, Michigan 48824, U.S.A.*

Torrence Parsons is Associate Professor of Mathematics at Pennsylvania State University, to which he moved after several years at Princeton University. He has contributed to several areas of graph theory, and in particular to Ramsey graph theory, algebraic graph theory, and topological graph theory. *Present address: Department of Mathematics, Pennsylvania State University, University Park, Pennsylvania 16802, U.S.A.*

Ronald Read is Professor in the Department of Combinatorics and Optimization at the University of Waterloo. He has contributed to several areas of graph theory—in particular, to the role of algorithms and computing—and is also interested in enumeration problems in organic chemistry. He is the editor of *Graph Theory and Computing* (Academic Press, 1972). *Present address: Department of Combinatorics and Optimization, The University of Waterloo, Waterloo, Ontario N2L 3G1, Canada.*

Brooks Reid is Associate Professor of Mathematics at Louisiana State University. His primary interest is in tournaments, and he has written many papers on this topic. He has also been involved with the organization of several of the Southeastern Conferences on Combinatorics, Graph Theory and Computing. *Present address: Department of Mathematics, Louisiana State University, Baton Rouge, Louisiana 70803, U.S.A.*

Allen Schwenk is Assistant Professor of Mathematics at the U.S. Naval Academy in Annapolis. He received his Ph.D. from the University of Michigan, and spent brief periods at Michigan State University and Oxford University. His research interests include spectral graph theory, Ramsey graph theory, and the enumeration of graphs and is an Associate Editor of the *Journal of Graph Theory*. *Present address: Department of Mathematics, U.S. Naval Academy, Annapolis, Maryland 21402, U.S.A.*

Arthur White is Associate Professor of Mathematics at Western Michigan University. He received his Ph.D. from Michigan State University, to which he went after a period as a Communications-Electronics Officer with the U.S. Air Force. His main research interest is topological graph theory. He is the author of *Graphs, Groups and Surfaces* (North-Holland, 1973), co-author (with N. L. Biggs) of *Permutation Groups and Combinatorial Structures* (Cambridge University Press, 1979), and is a co-editor of *Graph Theory and Applications* (Springer-Verlag, 1972), and an Associate Editor of the *Journal of Graph Theory*. *Present address: Department of Mathematics, Western Michigan University, Kalamazoo, Michigan 49008, U.S.A.*

Robin Wilson is Lecturer in Mathematics at the Open University. He received his Ph.D. in number theory from the University of Pennsylvania, and changed to graph theory shortly afterwards. His main interests are in edge-colorings of graphs, spectral graph theory, and the history of graph theory and combinatorics. He is the author of *Introduction to Graph Theory* (Academic Press, 1972), co-author (with N. L. Biggs and E. K. Lloyd) of *Graph Theory 1736–1936* (Oxford University Press, 1976), and co-author (with S. Fiorini) of *Edge-Colourings of Graphs* (Pitman Publishing, 1977). *Present address: Faculty of Mathematics, The Open University, Milton Keynes MK7 6AA, England.*

Douglas Woodall is Reader in Pure Mathematics at the University of Nottingham where he earlier received his Ph.D. for work in block designs and enumerative combinatorics. He has contributed to many areas of graph theory and combinatorics, in particular to the theories of circuits in graphs and minimax theorems. He was the founder, and for four years Editor, of the British Combinatorial Information Sheet (now the British Combinatorial Bulletin), and was a co-editor of *Combinatorics* (Institute of Mathematics and its Applications, 1972). He is also the present Mathematical Consultant to the Oxford English Dictionary. *Present address: Department of Mathematics, University Park, Nottingham NG7 2RD, England.*

Contents

1 Introduction

In this introductory chapter we shall present those definitions and theorems in graph theory which will be assumed throughout the rest of this book. Further explanation of these terms, together with the proofs of stated results, will be found in the standard texts in the subject (see, for example, [**2**], [**3**], [**4**], [7] and [**11**]), although not all of the terminology is completely standardized. Definitions and results not included here will be introduced later on as they are needed, or may be found in the above-mentioned texts.

Graphs

A **graph** G is a pair $(V(G), E(G))$, where $V(G)$ is a finite non-empty set of elements called **vertices**, and $E(G)$ is a finite set of distinct unordered pairs of distinct elements of $V(G)$ called **edges** (see Fig. 1). We call $V(G)$ the **vertex-set**

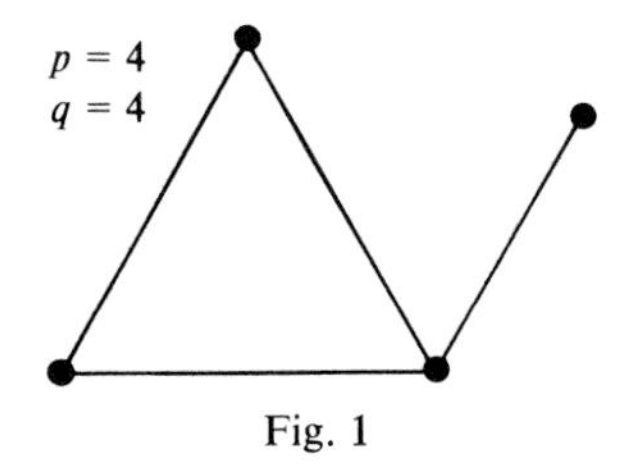

Fig. 1

of G, and $E(G)$ the **edge-set** of G; when there is no possibility of confusion, these are sometimes abbreviated to V and E, respectively. The number of vertices of G will be called the **order** of G, and will usually be denoted by p; the number of edges of G will generally be denoted by q. For convenience, we shall usually denote the edge $\{v, w\}$ (where v and w are vertices of G) by vw.

If $e = vw$ is an edge of G, then e is said to **join** the vertices v and w, and these vertices are then said to be **adjacent**. In this case, we also say that e is

incident to v and w, and that w is a **neighbor** of v; the **neighborhood** of v, denoted by $N(v)$, is the set of all vertices of G adjacent to v. Two edges of G incident to the same vertex will be called **adjacent edges**. An **independent set of vertices** in G is a set of vertices of G no two of which are adjacent, and the size of the largest such set is called the **independence number** of G. Similarly, an **independent set of edges**, or **matching**, in G is a set of edges of G no two of which are adjacent, and the size of the largest such set is called the **edge-independence number** of G. An independent set of edges which includes every vertex of G is called a **1-factor**, or **complete matching**, in G.

Two graphs G and H are said to be **isomorphic** (written $G \cong H$) if there is a one-to-one correspondence between their vertex-sets which preserves the adjacency of vertices. An **automorphism** of G is a one-to-one mapping ϕ of $V(G)$ onto itself with the property that $\phi(v)$ and $\phi(w)$ are adjacent if and only if v and w are. The automorphisms of G form a group $\Gamma(G)$ under composition, called the **automorphism group** of G; $\Gamma(G)$ is said to be **transitive** if it contains transformations mapping each vertex of G to every other vertex.

Some Variations

If, in the definition of a graph, we remove the restriction that the edges must be distinct, then the resulting object is called a **multigraph** (see Fig. 2); two or more edges joining the same pair of vertices are then called **multiple edges**. If M is a multigraph, its **underlying graph** is the graph obtained by replacing each set of multiple edges by a single edge; for example, the underlying graph of the multigraph in Fig. 2 is the graph in Fig. 1.

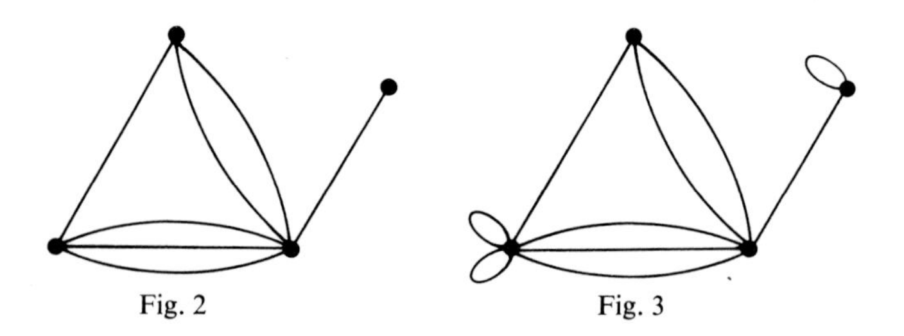

Fig. 2 Fig. 3

If we also remove the restriction that the edges must join distinct vertices, and allow the existence of **loops**, then the resulting object is called a **general graph**, or **pseudograph** (see Fig. 3). When concentrating our attention on graphs, as opposed to multigraphs or general graphs, we shall sometimes use the term "simple graph" to emphasize the fact that we are excluding loops and multiple edges.

A graph in which one vertex is distinguished from the rest is called a **rooted graph**. The distinguished vertex is called the **root-vertex**, or simply the **root**, and is often indicated by a small square (see Fig. 4). A **labeled graph** of order p is a graph whose vertices have been assigned the numbers 1, 2, . . ., p in such a way that no two vertices are assigned the same number (see Fig. 5).

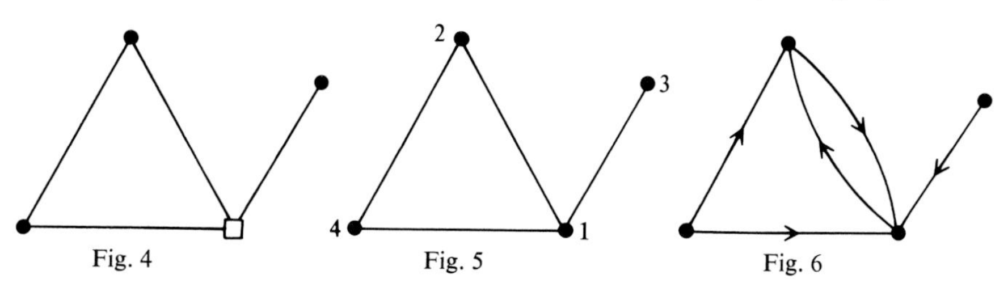

Fig. 4 Fig. 5 Fig. 6

We can also consider directed graphs, in which the word "unordered" in the definition of a graph is replaced by "ordered". More formally, we define a **digraph** D to be a pair $(V(D), A(D))$, where $V(D)$ is a finite non-empty set of elements called **vertices**, and $A(D)$ is a finite set of distinct ordered pairs of distinct elements of $V(D)$ called **arcs** (see Fig. 6); we shall usually denote the arc (v, w) (where v and w are vertices of D) by vw. If $e = vw$ is an arc of D, then we say that v and w are **adjacent**, and that e is **incident from *v*** and **incident to *w***; we also say that v **dominates** w. A pair of arcs of the form vw and wv is called a **symmetric pair**. If D is a digraph, its **converse** D' is the digraph obtained by replacing each arc vw by its "opposite" wv, and the **underlying graph** of D is the graph or multigraph obtained from D by replacing each arc by an (undirected) edge joining the same pair of vertices. A **complete symmetric digraph** is a digraph in which every two vertices are joined by exactly two arcs, one in each direction, and a **tournament** is a digraph in which every two vertices are joined by exactly one arc; tournaments are discussed at length in Chapter 7.

Finally, we can consider **infinite graphs**, in which we drop the restriction that $V(G)$ and/or $E(G)$ are finite. If the number of edges incident with each vertex remains finite, then the infinite graph is said to be **locally finite**.

Valencies

For each vertex v in a graph G, the number of edges incident to v is called the **valency** of v, denoted by $\rho(v)$ (or by $\rho_G(v)$, if there is any possibility of confusion). The maximum and minimum valencies in G will be denoted by $\rho_{\max}(G)$ and $\rho_{\min}(G)$, or simply by $\rho_{\max}$ and $\rho_{\min}$. A vertex of valency 0 is called an **isolated vertex**, and a vertex of valency 1 is called an **end-vertex**.

The **valency-sequence** of G is the set of valencies of the vertices of G, usually arranged in non-decreasing order—for emphasis, this is sometimes called the **non-decreasing valency-sequence**; for example, the non-decreasing valency-sequence of the graph in Fig. 1 is (1, 2, 2, 3).

If all of the vertices of G have the same valency (ρ, say), then G is said to be a **regular graph**, or **ρ-valent graph**. A 0-valent graph (that is, one with no edges) is called a **null graph**, and a 3-valent graph is usually called a **trivalent graph**.

Analogous concepts can also be defined for digraphs. If v is a vertex of a digraph D, then its **in-valency** $\rho_{\text{in}}(v)$ (or $\rho^-(v)$) is the number of arcs in D of the form wv, and its **out-valency** or **score** $\rho_{\text{out}}(v)$ (or $\rho^+(v)$) is the number of arcs in D of the form vw. The **score-list** of D is the sequence of scores of D, usually arranged in non-decreasing order.

Subgraphs

A **subgraph** of a graph $G = (V(G), E(G))$ is a graph $H = (V(H), E(H))$ such that $V(H) \subseteq V(G)$ and $E(H) \subseteq E(G)$. If $V(H) = V(G)$, then H is called a **spanning subgraph** of G. If W is any set of vertices in G, then the **subgraph induced by W** is the subgraph of G obtained by taking the vertices in W and joining those pairs of vertices in W which are joined in G. An **induced subgraph** of G is a subgraph which is induced by some subset W of $V(G)$. Similar definitions may be given for digraphs and multigraphs.

If e is an edge of G, then the **edge-deleted subgraph** $G-e$ is the graph obtained from G by removing the edge e; more generally, we write $G-\{e_1, \ldots, e_k\}$ for the graph obtained from G by removing the edges $e_1, \ldots, e_k$. Similarly, if v is a vertex of G, then the **vertex-deleted subgraph** $G-v$ is the graph obtained from G by removing the vertex v together with all the edges incident to v; more generally, we write $G-\{v_1, \ldots, v_k\}$ for the graph obtained from G by removing the vertices $v_1, \ldots, v_k$ and all edges incident to any of them. These concepts are illustrated in Fig. 7.

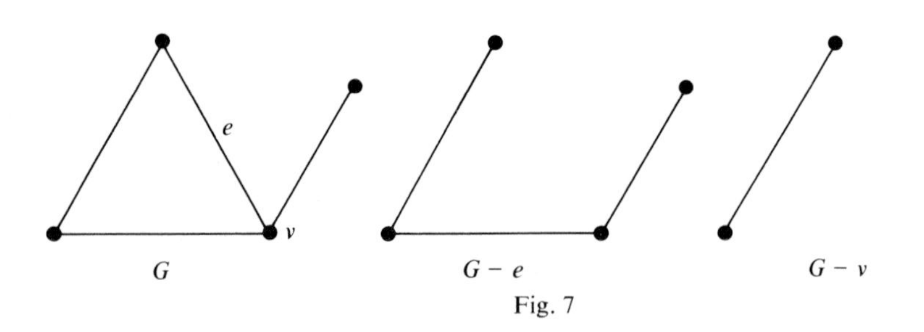

Fig. 7

New Graphs from Old

If G is a graph, then there are several graphs that we can obtain from G. For example, the **complement** of G (denoted by $\bar{G}$) is the graph with the same vertex-set as G, but where two vertices are adjacent if and only if they are not adjacent in G. The **line graph** $L(G)$ of G is the graph whose vertices correspond to the edges of G, and where two vertices are joined if and only if the corresponding edges of G are adjacent; line graphs are discussed in Chapter 10. The **total graph** $T(G)$ of G is the graph whose vertices correspond to the vertices *and* edges of G, and where two vertices are joined if and only if the corresponding vertices or edges of G are adjacent or incident.

If $e = vw$ is an edge of G, then we can obtain a new graph by replacing e by two new edges vz and zw, where z is a new vertex; this is called **inserting a vertex into an edge** (see Fig. 8). If two graphs can be obtained from the same

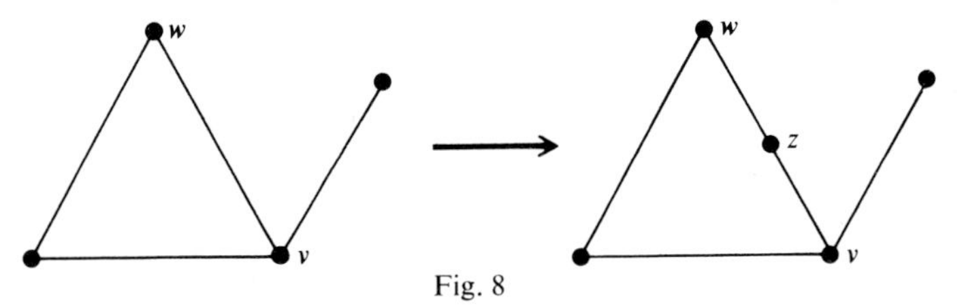

Fig. 8

graph by inserting vertices into its edges, then these two graphs are called **homeomorphic**. We can also obtain a new graph from G by removing the edge $e = vw$ and identifying v and w in such a way that the resulting vertex is incident to all those edges (other than e) which were originally incident to v or to w; this is called **contracting the edge e** (see Fig. 9), and the resulting graph is denoted by $G\backslash e$. If the graph H can be obtained from G by a succession of "edge-contractions" such as this, we say that G is **contractible** to H.

If G and G' are graphs with the same vertex-set, then their **intersection** $G \cap G'$ is the graph with edge-set $E(G) \cap E(G')$, and their **union** $G \cup G'$ is the graph with edge-set $E(G) \cup E(G')$. If G and G' are disjoint graphs, then $G \cup G'$ is the graph with vertex-set $V(G) \cup V(G')$ and edge-set $E(G) \cup E(G')$;

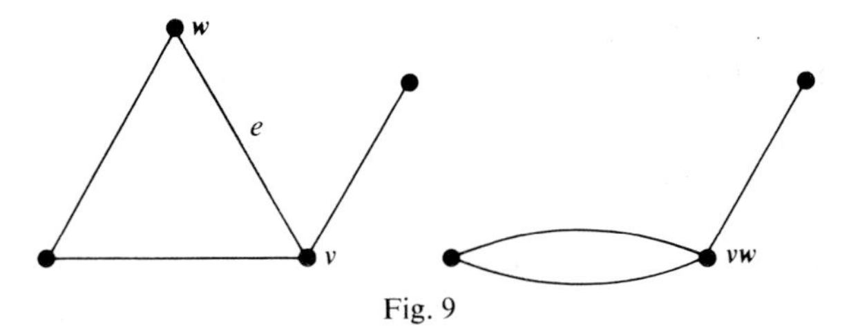

Fig. 9

it is sometimes called the **disjoint union** of G and G'. The disjoint union of k copies of G is often written kG. The **join** of G and G', denoted by $G+G'$, is obtained from their disjoint union by adding edges joining each vertex of G to each vertex of G'. Finally, the **Cartesian product** of G and G' is the graph with vertex-set $V(G)\times V(G')$ in which the vertex (v, w) is adjacent to the vertex (v', w') whenever $v = v'$ and w is adjacent to w', or $w = w'$ and v is adjacent to v'.

Examples of Graphs

A graph in which every two vertices are adjacent is called a **complete graph**; the complete graph with p vertices and $\frac{1}{2}p(p-1)$ edges is denoted by K_p. The **circuit graph** of order p, denoted by C_p, consists of the vertices and edges of a p-gon. The **wheel** W_p is the graph $C_{p-1}+K_1$, and the path graph P_p is obtained by removing an edge from C_p. The **null graph** N_p of order p is the graph with p vertices and no edges. The graphs K_5, C_5, W_5, P_5 and N_5 are shown in Fig. 10. It is also occasionally useful to introduce the **empty graph** (not strictly speaking a graph at all), which consists of no vertices or edges. A **clique** in a graph G is a subgraph of G which is complete.

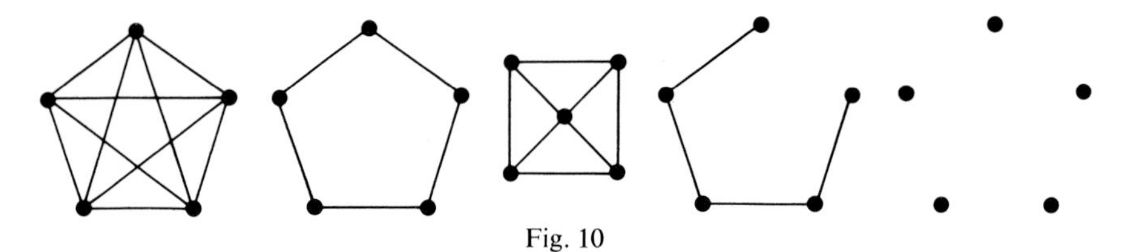

Fig. 10

A **bipartite graph** is one whose vertex-set can be partitioned into two sets (called **partite sets**) in such a way that each edge joins a vertex of the first set to a vertex of the second set. A **complete bipartite graph** is a bipartite graph in which every vertex in the first set is adjacent to every vertex in the second set; if the two partite sets contain r and s vertices, respectively, then the complete bipartite graph is denoted by $K_{r,s}$. Any complete bipartite graph of the form $K_{1,s}$ is called a **star graph**. A **complete r-partite graph** is obtained by partitioning the vertex-set into r sets, and joining two vertices if and only if they lie in different sets; if all of these sets have size k, then the resulting graph is denoted by $K_{r(k)}$; note that $K_{r(k)}$ is the complement of rK_k. The graphs $K_{3,3}$ and $K_{3(3)}$ are shown in Fig. 11.

The **Petersen graph** is the graph shown in Fig. 12; it is the complement of the line graph of K_5. The **Platonic graphs** are the graphs corresponding to the

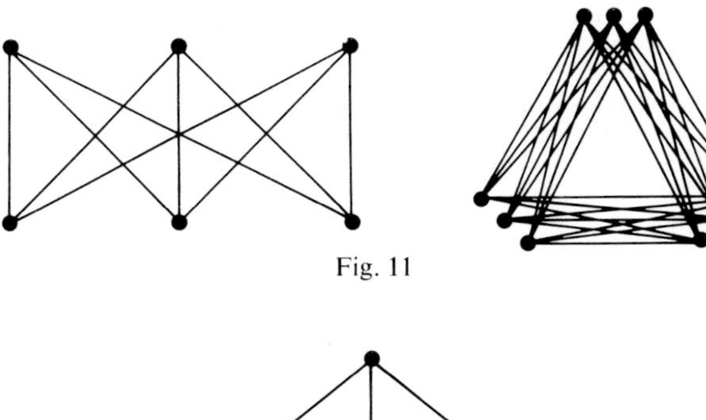

Fig. 11

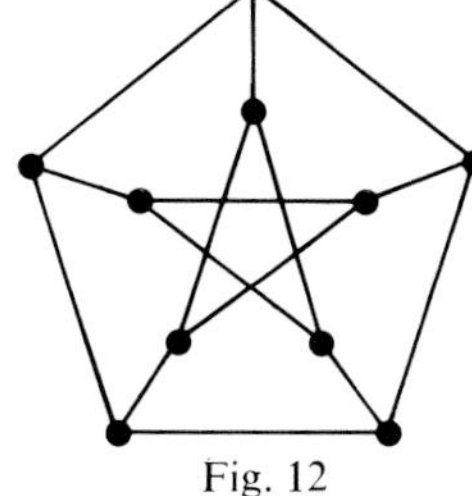

Fig. 12

vertices and edges of the five regular solids, the tetrahedron, cube, octahedron, dodecahedron and icosahedron (see Fig. 13). The ***n*-cube** Q_n is the graph whose vertices correspond to the sequences $(a_1, a_2, \ldots, a_n)$, where each $a_i = 0$ or 1, and whose edges join those pairs of vertices which correspond to sequences differing in just one place; thus $Q_2 = C_4$, and Q_3 is the graph of the cube. The ***n*-dimensional octahedron** is the graph $K_{n(2)}$, and is the complement of the graph nK_2. Finally, a **self-complementary graph** is a graph which is isomorphic to its complement.

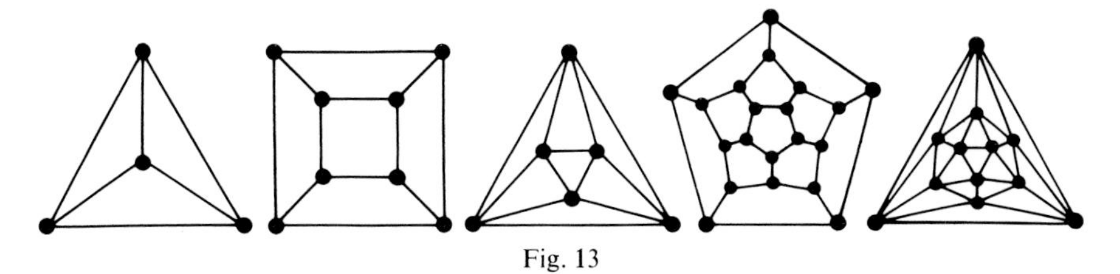

Fig. 13

Paths and Circuits

A sequence of edges of the form $v_0v_1, v_1v_2, \ldots, v_{r-1}v_r$ (sometimes abbreviated to $v_0, v_1, \ldots, v_r$) is called a **walk of length *r*** from v_0 to v_r; v_0 is called the **initial vertex** of the walk, and v_r is called the **terminal vertex**. If these edges are all distinct, then the walk is called a **trail**, and if the vertices $v_0, v_1, \ldots, v_r$

are also distinct, then the walk is called a **path** (or **open path**). Two paths in a graph are said to be **edge-disjoint** if they share no common edges; they are also said to be **vertex-disjoint** if they share no common vertices, although one frequently relaxes this condition to allow the initial vertices of the path to coincide, and also the terminal vertices. A walk or trail is said to be **closed** if $v_0 = v_r$, and a path in which the vertices $v_0, v_1, \ldots, v_r$ are all distinct except for v_0 and v_r (which coincide) is called a **circuit** (or **closed path**).

A circuit is said to be **even** if it has an even number of edges, and **odd** otherwise. A circuit of length 3 is called a **triangle**. The length of a shortest circuit in a graph G is called the **girth** of G, and the length of a longest circuit in G is called the **circumference** of G. If v and w are vertices in G, the length of any shortest path from v to w is called the **distance** between v and w, denoted by $d(v, w)$. The largest distance between two vertices in G is called the **diameter** of G; for example, the Petersen graph has diameter 2.

These definitions can be extended to directed graphs and infinite graphs. In particular, a **directed trail** in a digraph is a sequence of distinct arcs of the form $v_0v_1, v_1v_2, \ldots, v_{r-1}v_r$, a **directed path** is a sequence of arcs of the form $v_0v_1, v_1v_2, \ldots, v_{r-1}v_r$, where $v_0, v_1, \ldots, v_r$ are all distinct, and a **directed circuit** is a sequence of arcs of the form $v_0v_1, v_1v_2, \ldots, v_{r-1}v_0$, where $v_0, v_1, \ldots, v_{r-1}$ are all distinct. In an infinite graph, a **one-way infinite path** is a sequence of distinct edges of the form

$$v_0v_1, v_1v_2, \ldots, v_{r-1}v_r, \ldots \text{ or } \ldots, v_{-r}v_{-r+1}, \ldots, v_{-2}v_{-1}, v_{-1}v_0,$$

and a **two-way infinite path** is a sequence of distinct edges of the form

$$\ldots, v_{-r}v_{-r+1}, \ldots, v_{-1}v_0, v_0v_1, \ldots, v_{r-1}v_r, \ldots.$$

Connectivity

A graph G is **connected** if there is a path joining each pair of vertices of G (or, equivalently, if G cannot be expressed as the union of two disjoint graphs); a graph which is not connected is called **disconnected**. Clearly, every disconnected graph can be split up into a number of maximal connected subgraphs, and these subgraphs are called **components**. There are analogous definitions for digraphs; in particular, a digraph D is called **strongly connected** if, for each pair of vertices v and w, there is a directed path in D from v to w, **unilaterally connected** if there is either a directed path in D from v to w or a directed path from w to v, and **connected** if there is a path from v to w in the underlying graph of D.

If G is a connected graph, and if the graph $G-v$ is disconnected, for some vertex v, then v is called a **cut-vertex** of G. More generally, a **separating set of vertices** in G is a set of vertices whose removal disconnects G. We say that a graph G with at least $k+1$ vertices is ***k*-connected** if every two vertices v and w are connected by at least k paths which are pairwise disjoint except for the vertices v and w; a 2-connected graph is often called a **block** or a **non-separable graph**. The **connectivity** of G, denoted by $\kappa(G)$, is then defined to be the largest value of k for which G is k-connected.

If G is a connected graph, and if the graph $G-e$ is disconnected, for some edge e, then e is called a **bridge** of G. More generally, a **cutset** in G is a set of edges whose removal disconnects G. We say that a graph G is ***k*-edge-connected** if every two vertices v and w are connected by at least k edge-disjoint paths; the **edge-connectivity** of G, denoted by $\lambda(G)$, is then defined to be the largest value of k for which G is k-edge connected. (Note that $\kappa(G) \leqslant \lambda(G)$.) The most important theorem relating these concepts is **Menger's theorem**; this takes several forms, among which are the following:

Theorem 1.1 (Menger's Theorem). *Let G be a connected graph with at least* $k+1$ *vertices. Then*

(*i*) *G is k-connected if and only if G cannot be disconnected by the removal of* $k-1$ *or fewer vertices*;

(*ii*) *G is k-edge-connected if and only if G cannot be disconnected by the removal of* $k-1$ *or fewer edges.* ||

Further discussion of Menger's theorem and its many variations is given in Chapter 9, where the analogs for digraphs are also presented.

Traversability

A connected graph G is **Eulerian** if it has a trail which includes every edge of $E(G)$; such a trail is called an **Eulerian trail**. Similarly, a strongly connected digraph D is **Eulerian** if it has a directed trail which includes every arc of $A(D)$. Necessary and sufficient conditions for a graph or digraph to be Eulerian are given in the following theorem:

Theorem 1.2. (*i*) *A connected graph G is Eulerian if and only if every vertex of G has even valency*;

(*ii*) *a strongly connected digraph D is Eulerian if and only if the in-valency and out-valency of each vertex are equal.* ||

A graph G is **Hamiltonian** if it has a circuit which includes every vertex of $V(G)$; such a circuit is called a **Hamiltonian circuit**. More generally, a graph G

is **traceable** if it has a path which includes every vertex of $V(G)$; such a path is called a **Hamiltonian path**. Analogous definitions can be given for digraphs, and the theory of Hamiltonian graphs and digraphs is discussed in Chapter 6.

Trees

A connected graph which contains no circuits is called a **tree**, and a graph whose components are trees is called a **forest**, or **acyclic graph**. The trees of order 5 are shown in Fig. 14.

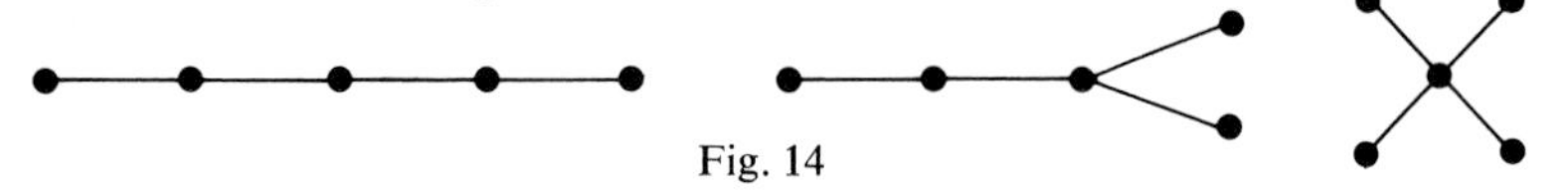

Fig. 14

The main properties of trees are summarized in the following theorem:

Theorem 1.3. *If T is a tree of order p, then*

(*i*) *T is a connected graph with $p-1$ edges;*

(*ii*) *every edge of T is a bridge;*

(*iii*) *if v is a vertex of T with $\rho(v) > 1$, then v is a cut-vertex;*

(*iv*) *if v and w are distinct vertices of T, then there is exactly one path from v to w.* ‖

If G is a connected graph, then a **spanning tree** in G is a connected spanning subgraph containing no circuits. A method for calculating the number of spanning trees in a given graph is described later in this chapter. The **arboricity** of G is the minimum number of forests whose union is G; results concerning the arboricity of a graph may be found in Section 9 of Chapter 2.

Finally, if T is any tree, we can obtain another tree by removing all the end-vertices (vertices of valency 1) from T. Repeating this procedure as often as necessary, we eventually obtain either a single vertex (the **center** of T), or two vertices joined by an edge (the **bicenter** of T); T is called **central** or **bicentral** according as T has a center or bicenter.

Planar Graphs

A **planar graph** is a graph which can be embedded in the plane in such a way that no two edges intersect geometrically except at a vertex to which they are both incident. A graph embedded in the plane in this way is called a **plane graph**; in this case, the points of the plane not on G are partitioned into open sets called **regions** (see Fig. 15), and the number r of such regions is given by **Euler's polyhedral formula**:

Theorem 1.4 (Euler's Polyhedral Formula). *Let G be a connected plane graph with p* ($\geqslant 3$) *vertices, q edges and r regions. Then*

$$p-q+r = 2. \parallel$$

Since G has no loops or multiple edges, every region must be bounded by at least three edges. It follows that $2q \leqslant 3r$, and hence that $q \leqslant 3p-6$; equality holds when every region is bounded by a triangle, and such a graph is called a **triangulation**. If G contains no triangles, then $2q \leqslant 4r$, and so $q \leqslant 2p-4$. It is a simple matter to check that every planar graph contains a vertex of valency 5 or less, and has girth at most 5. A connected plane graph which contains no bridges is sometimes called a **map**; and a planar graph which can

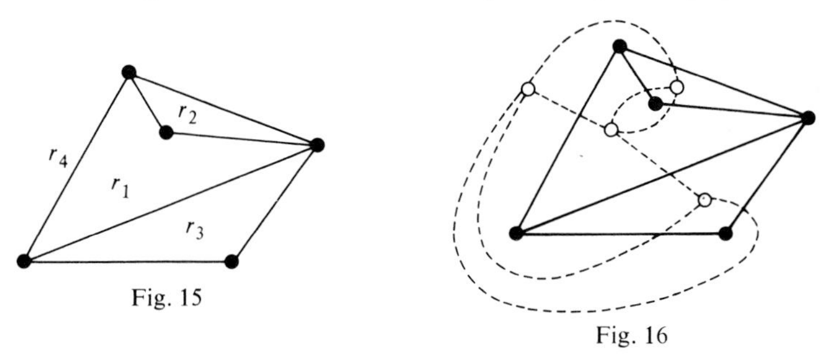

Fig. 15

Fig. 16

be embedded in the plane in such a way that every vertex lies on the boundary of the same region is called an **outerplanar graph**.

A necessary and sufficient condition for a graph to be planar has been given by Kuratowski [**8**]; we present two forms of this result:

Theorem 1.5 (Kuratowski's Theorem). *A graph G is planar if and only if*

either (*i*) *G has no subgraph homeomorphic to* K_5 *or* $K_{3,3}$;

or (*ii*) *G has no subgraph contractible to* K_5 *or* $K_{3,3}$. $\parallel$

If G is a connected plane graph, then its **dual graph** G^* is the (general) graph obtained by the following procedure: (*i*) place a point inside each region of G—these points correspond to the vertices of G^*; (*ii*) for each edge e of G, draw a line joining the vertices in the two regions bounded by e—these lines correspond to the edges of G^* (see Fig. 16). It is easy to see that G^* is a plane graph whose dual graph is isomorphic to G, and that if G has p vertices, q edges and r regions, then G^* has r vertices, q edges and p regions.

One can also consider the embedding of graphs on other surfaces—in particular, one can derive an analog of Euler's polyhedral formula for any orientable or non-orientable surface. Further details may be found in Chapters 2 and 3.

The Coloring of Graphs

If G is a graph, we define its **chromatic number** $\chi(G)$ to be the minimum number of colors needed to color the vertices of G in such a way that no two adjacent vertices are assigned the same color. If $\chi(G) = k$, we say that G is ***k*-chromatic**, and if $\chi(G) \leqslant k$, we say that G is ***k*-colorable**. For example, the complete graph K_p is p-chromatic, the path graph P_p ($p \geqslant 2$) is 2-chromatic, and the circuit graph C_p is 2-chromatic or 3-chromatic according as p is even or odd. Note that if G is a bipartite graph, then G is 2-colorable. The set of all vertices with the same color is called a **color class**.

An upper bound for the chromatic number of a graph G has been given by Brooks [**5**], and involves the maximum valency of G:

Theorem 1.6 (Brooks' Theorem). *Let G be a connected graph which is not a complete graph or a circuit of odd length, and let ρ be the largest valency in G. Then G is ρ-colorable.* ||

There are also bounds involving the chromatic numbers of a graph and its complement; the following theorem is due to Nordhaus and Gaddum [**9**].

Theorem 1.7. *Let G be a graph of order p, and let $\bar{G}$ be its complement. Then*

$$2\sqrt{p} \leqslant \chi(G)+\chi(\bar{G}) \leqslant p+1, \qquad \textit{and} \qquad p \leqslant \chi(G)\cdot\chi(\bar{G}) \leqslant \tfrac{1}{4}(p+1)^2. \; \|$$

For each graph G, let $P_G(k)$ be the number of ways of coloring the vertices of G in such a way that no two adjacent vertices are assigned the same color. For example, if $G = K_p$, then $P_G(k) = k(k-1)\ldots(k-p+1)$, and if $G = P_p$, then $P_G(k) = k(k-1)^{p-1}$. It is not difficult to show that if G has p vertices and q edges, then $P_G(k)$ is a monic polynomial in k of degree p, in which the coefficients alternate in sign, the constant coefficient is zero, and the coefficient of k^{p-1} is $-q$; $P_G(k)$ is called the **chromatic polynomial** of G. When working with chromatic polynomials, the following result is often very useful; it is proved by counting the number of colorings of G in which the vertices incident to e have, and have not, different colors:

Theorem 1.8. *Let G be a graph containing an edge e, and let $G-e$ and $G\backslash e$ be the graphs obtained from G by respectively deleting and contracting e. Then*

$$P_G(k) = P_{G-e}(k) - P_{G\backslash e}(k). \; \|$$

We end this section with the famous **four-color theorem** for planar graphs, discussed at some length in Chapters 4 and 15:

Theorem 1.9 (Four-Color Theorem). *Every planar graph is four-colorable.* ||

Matrices

If G is a graph with vertex-set $\{v_1, v_2, \ldots, v_p\}$, then the **adjacency matrix** of G is the $p \times p$ matrix $\mathbf{A}(G) = (a_{ij})$, where

$$a_{ij} = \begin{cases} 1, & \text{if } v_i \text{ and } v_j \text{ are adjacent,} \\ 0, & \text{if not.} \end{cases}$$

In Chapter 11 we study the eigenvalues of $\mathbf{A}(G)$, which are clearly independent of the way in which the vertices are labeled. For convenience, we refer to these eigenvalues as the **eigenvalues of G**, and in a similar way, the characteristic polynomial of $\mathbf{A}(G)$ is called the **characteristic polynomial of G**.

There are various other matrices associated with G. For example, if G has edge-set $\{e_1, e_2, \ldots, e_q\}$, then the **incidence matrix** of G is the $p \times q$ matrix $\mathbf{B}(G) = (b_{ij})$, where

$$b_{ij} = \begin{cases} 1, & \text{if } v_i \text{ is incident to } e_j, \\ 0, & \text{if not.} \end{cases}$$

Also the **distance matrix** of G is the $p \times p$ matrix $\mathbf{D}(G) = (d_{ij})$, where d_{ij} is the distance from v_i to v_j.

An example of the use of matrices in graph theory is provided by the **matrix-tree theorem** (see, for example, [**2**]):

Theorem 1.10 (Matrix-tree Theorem). *Let G be a connected labeled graph with adjacency matrix* $\mathbf{A}$, *and let* $\mathbf{M}$ *be the matrix obtained from* $-\mathbf{A}$ *by replacing each diagonal entry by the valency of the corresponding vertex. Then all of the cofactors of* $\mathbf{M}$ *are equal, and their common value is the number of spanning trees in G.* ‖

Block Designs

A **(v, b, r, k, λ)-balanced incomplete block design** (or simply, a **block design**) is an arrangement of v objects into b blocks so that

(*i*) each object appears in exactly r blocks;

(*ii*) each block contains exactly k $(<v)$ objects;

(*iii*) each pair of distinct objects appear together in exactly λ blocks.

The parameters v, b, r, k and λ are not independent; for example, simple counting arguments show that $vr = bk$, and $r(k-1) = \lambda(v-1)$. Also, an inequality of Fisher states that $b \geqslant v$.

If $k = 3$ and $\lambda = 1$, we have a **Steiner triple system**. It is easily checked that in this case, $b = \frac{1}{6}v(v-1)$ and $r = \frac{1}{2}(v-1)$, so that $v \equiv 1$ or 3 (modulo 6).

If $v = 7$ we get the **Fano plane**, the projective plane of order 2 in which the objects (points) are 1, 2, 3, 4, 5, 6, 7 and the blocks (lines) are 124, 156, 137, 235, 267, 346 and 457. More generally, the **projective plane of order *n*** has n^2+n+1 points and n^2+n+1 lines, where each point lies on $n+1$ lines and each line passes through $n+1$ points—that is, it is a block design with parameters $v = b = n^2+n+1$, $r = k = n+1$, $\lambda = 1$. These last examples are particular instances of **(v, k, λ)-designs** or **symmetric designs**—that is, designs in which $b = v$, and consequently, $r = k$.

Further information on block designs can be found in, for example, [**1**], [**6**] or [**10**].

And finally . . .

If S is a finite set, we denote the number of elements in S by $|S|$; the empty set will be denoted by $\varnothing$. We shall use $[x]$ for the largest integer not greater than x, and $\{x\}$ for the smallest integer not smaller than x (so that, for example, $[\pi] = 3$, $\{\pi\} = 4$). The set of real numbers and the set of integers will be denoted, respectively, by $\boldsymbol{R}$ and $\boldsymbol{Z}$. As in this chapter, the end or absence of a proof will be denoted by $\|$. In the references for each chapter, *Mathematical Reviews* numbers will be indicated by *MR*15–234 (page 234 of Volume **15**), or *MR*35#234 (review number 234 of Volume **35**).

References

1. I. Anderson, *A First Course in Combinatorial Mathematics*, Clarendon Press, Oxford, 1974; *MR*49#2402.
2. M. Behzad and G. Chartrand, *An Introduction to the Theory of Graphs*, Allyn and Bacon, Boston, Mass., 1971; *MR*55#5449.
3. C. Berge, *Graphs and Hypergraphs*, North-Holland, Amsterdam, 1973; *MR*50#9640.
4. J. A. Bondy and U. S. R. Murty, *Graph Theory with Applications*, American Elsevier, New York, and MacMillan, London, 1976; *MR*54#117.
5. R. L. Brooks, On colouring the nodes of a network, *Proc. Cambridge Phil. Soc.* **37** (1941), 194–197; *MR*6–281.
6. M. Hall, Jr., *Combinatorial Theory*, Blaisdell, Waltham, Mass., 1967; *MR*37#80.
7. F. Harary, *Graph Theory*, Addison-Wesley, Reading, Mass., 1969; *MR*41#1566.
8. K. Kuratowski, Sur le problème des courbes gauches en topologie, *Fund. Math.* **15** (1930), 271–283.
9. E. A. Nordhaus and J. W. Gaddum, On complementary graphs, *Amer. Math. Monthly* **63** (1956), 175–177; *MR*17–1231.
10. H. J. Ryser, *Combinatorial Mathematics*, Carus Mathematical Monographs **14**, Mathematical Association of America, Washington D.C., 1963; *MR*27#51.
11. R. J. Wilson, *Introduction to Graph Theory*, Academic Press, New York, 1972, and Longman Group, Harlow, Essex, 1975; *MR*50#9643.

2
Topological Graph Theory

ARTHUR T. WHITE AND LOWELL W. BEINEKE

1. Introduction

A graph exists in its own right as an abstract mathematical system, but if we wish to "visualize" it then we must try to realize it geometrically as a subset of some Euclidean space $\mathbf{R}^n$ (where n is no larger than necessary), with vertices and edges being modeled, respectively, by points and lines (or arcs).

It is well known that every graph can be realized in $\mathbf{R}^3$ using straight line segments; for example, the vertices can be represented by distinct points along the curve $C = \{(t, t^2, t^3): 0 \leqslant t \leqslant 1\}$, with the edges represented in the obvious manner. However, for ease of presentation, it is often desirable to model a graph on a locally 2-dimensional "drawing-board"—that is, on a surface. (If the surface is orientable, then it is a subset of $\mathbf{R}^3$; if not, then it is a subset of $\mathbf{R}^4$. If the graph is planar, then it can be represented either on a sphere in $\mathbf{R}^3$, or on $\mathbf{R}^2$.)

Topological graph theory is primarily concerned with representing graphs on surfaces. An "embedding" (and the more general "drawing") of a graph can be considered as an identification of the graph with a subset of a topological space in an appropriate fashion. Most of this chapter will be devoted to the study of embeddings of graphs on surfaces.

We begin by looking at some aspects of surfaces in Section 2. In Sections

3 and 4 we look at embeddings and at a systematic procedure for obtaining embeddings—the use of voltage graphs to generate rotation schemes, which in turn determine the embeddings. The main topic of Section 5 is finding the "genus" of a graph—that is, finding the "most efficient" orientable surface on which it can be embedded. In the following section we consider the corresponding concept for non-orientable surfaces. In Section 7 we discuss an application of embeddings—to the generation of block designs. In the final two sections, we consider topics for which the plane $\mathbf{R}^2$ is the main space of interest. These involve the minimum number of crossings in a drawing of a given graph (the "crossing number") and the minimum number of planar graphs needed to form a given graph (the "thickness"). In each case, extensions are discussed to surfaces of positive genus.

The authors wish to thank R. K. Guy and G. M. Haggard for their helpful comments and suggestions regarding this chapter.

2. Surfaces

The topological spaces with which we shall be working in this chapter are the "compact 2-manifolds", which we call **surfaces**. In topological terms, this means that (*i*) every point has a neighborhood which is homeomorphic to an open disk (or "2-cell"), and (*ii*) every covering with open disks has a finite subcovering. For our purposes, however, it will be convenient to have other, more intuitive, ways of looking at these structures. We note that spaces such as the plane and Möbius strip are not surfaces under this definition—the plane is not compact, the Möbius strip with boundary included does not have the 2-cell property, and with boundary excluded, it is not compact.

Handles and Crosscaps

One way to obtain our surfaces is by adding "handles" and "crosscaps" to a sphere. A sphere is "orientable", in that a positive sense of rotation can be defined consistently at all points; alternatively, the boundaries of the regions of any map on the sphere can be assigned orientations so that each edge is oriented in opposite directions on its two regions. We now give an intuitive procedure for obtaining other orientable surfaces.

In order to add a handle to the sphere, first take two disjoint closed disks on the sphere, orient their boundaries in the same direction, and remove their interiors; then take a (truncated) cylinder whose ends are similarly oriented and identify its boundary curves with those on the sphere, preserving orientation. The result is another orientable surface, equivalent to the torus (see

Fig. 1). This procedure can be repeated to obtain the "sphere with h handles", which we denote by S_h. (No matter how the handles are added, the result is effectively the same.) The number h is called the **genus of the surface**.

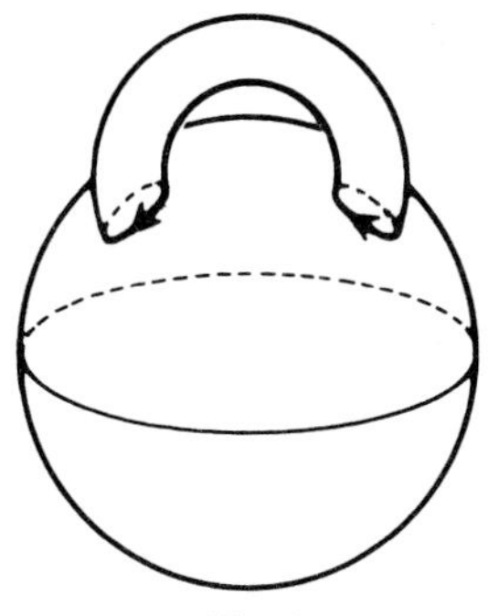

Fig. 1

A second important modification is the addition of a crosscap: this can be considered as the result of removing the interior of a disk and then identifying opposite points on the boundary (or by identifying the boundary curve with that of a Möbius strip). The addition of a crosscap results in a non-orientable surface; that is, one in which no positive sense of rotation can be assigned consistently at all points. If one crosscap is added to the sphere, we get the "projective plane"; if two crosscaps are added, we get the "Klein bottle". In general, we denote by N_k the non-orientable surface obtained by adding k crosscaps ($k > 0$) to the sphere.

A remarkable theorem due to Brahana **[15]** states that every surface (compact 2-manifold) is topologically equivalent either to S_h, for some $h \geqslant 0$, or to N_k, for some $k > 0$. This means that no matter how crosscaps and handles have been added to form a surface, it can be obtained by adding just crosscaps or just handles. Another consequence of this result is that any non-orientable surface can be obtained by adding at most two crosscaps—one if k is odd, and two is k is even. This implies the somewhat surprising further result that the addition of a handle to a surface S is equivalent to adding two crosscaps if and only if S is non-orientable.

Representations of Surfaces

In order to view a graph on a surface, it is frequently convenient to have a representation on a sheet of paper. Each of our surfaces can in fact be "assembled" from a polygon by appropriately identifying edges. For the torus, for example, this is done by identifying one pair of opposite sides of a rectangle to form a cylinder, and then identifying the two boundary circles of the cylinder. This is indicated in Fig. 2, and is given by the scheme α, β, α^{-1}, β^{-1}. The general orientable surface S_h can be formed from a $4h$-gon by identifying corresponding sides when the boundary labels are

$$\alpha_1, \beta_1, \alpha_1^{-1}, \beta_1^{-1}, \ldots, \alpha_h, \beta_h, \alpha_h^{-1}, \beta_h^{-1}.$$

In Fig. 3 we give a second example, that of the "double-torus".

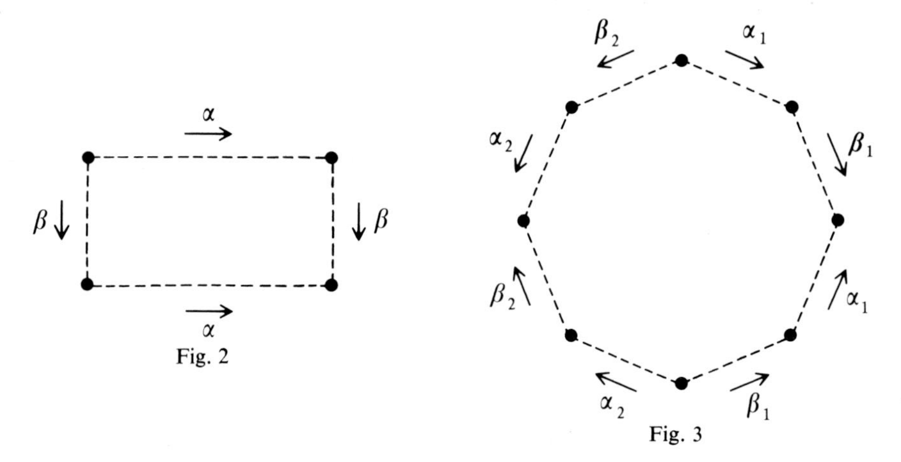

Fig. 2

Fig. 3

For the non-orientable surface N_k there is a similar construction, namely the identification of sides on a $2k$-gon with oriented boundary $\alpha_1, \alpha_1, \ldots, \alpha_k, \alpha_k$, but we should need four dimensions in order to carry it out physically. The representations for the projective plane and Klein bottle are indicated in Fig. 4. We note that each construction corresponds to appropriate cuts in the cor-

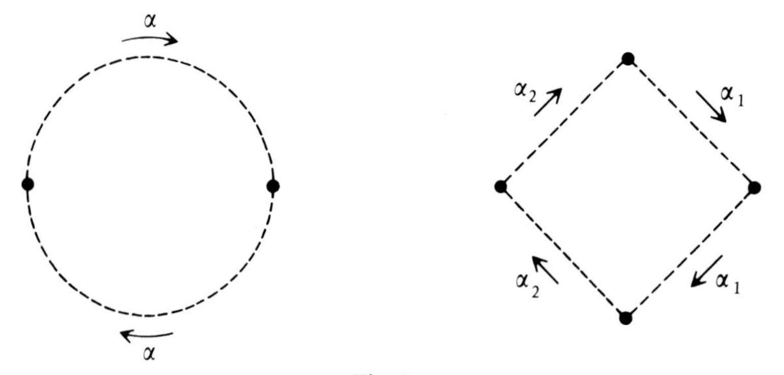

Fig. 4

responding surface, and that in each case the vertices of the polygon are the same point of the surface. Also, these polygonal representations are not unique (except for N_1); for example, S_2 can also be represented by α_1, β_1, α_2, β_2, α_1^{-1}, β_1^{-1}, α_2^{-1}, β_2^{-1}, and N_2 by α, β, α^{-1}, β.

3. Elementary Aspects of Embeddings

We define an **embedding** of a general graph G in a surface S to be a mapping of the vertices of G to distinct points of S, and the edges of G to disjoint open arcs of S, such that

(*i*) the image of no edge contains that of a vertex;

(*ii*) the image of an edge vw joins the points corresponding to v and w.

We frequently speak of the images of the vertices and edges of a graph embedded in a surface as the vertices and edges of the graph. In Figs. 5 and 6 we schematically show embeddings of K_6 and K_7 in the projective plane and torus, respectively; the shading in the embedding of K_7 will be useful in Section 7.

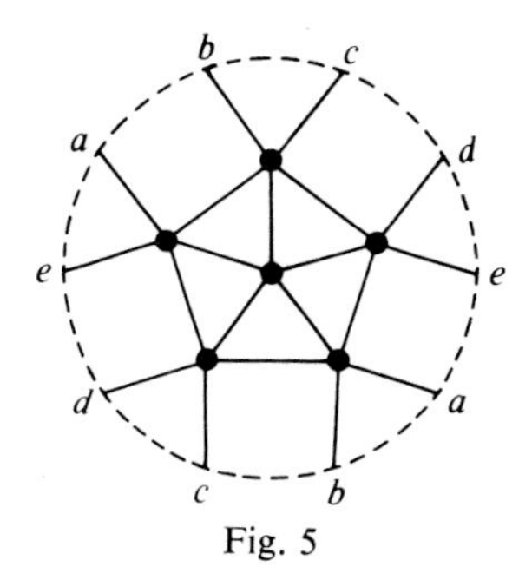

Fig. 5

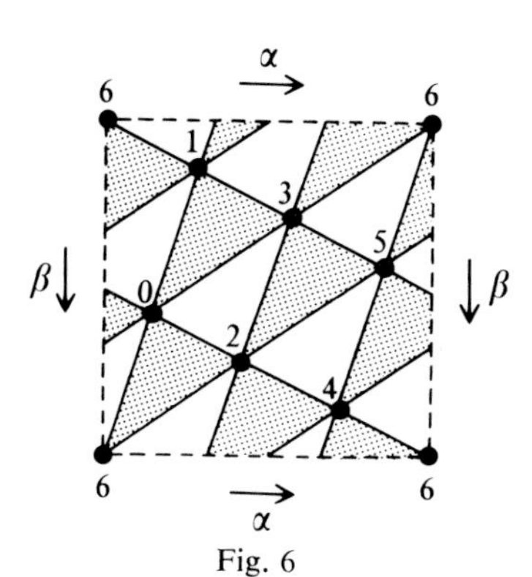

Fig. 6

If a graph G is embedded in a surface S, the complement of G relative to S is a collection of open sets, which we call **regions**. Some, but not necessarily all, of these regions may be topological disks (2-cells). If in fact *all* of them are 2-cells, we say that the embedding is a **2-cell embedding**. Figure 7 shows three embeddings of K_4 in the torus, but only the third (with just two regions) is a 2-cell embedding.

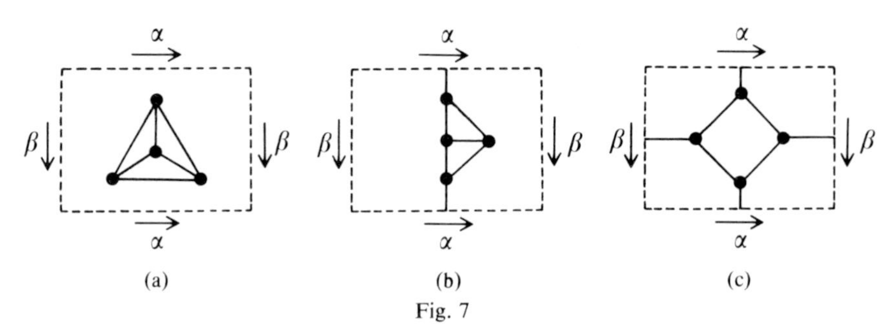

Fig. 7

The Euler Characteristic

We define the **Euler characteristic** $\chi(S)$ of a surface S by

$$\begin{aligned} \chi(S_h) &= 2-2h, && \text{for the orientable surface } S_h, \\ \chi(N_k) &= 2-k, && \text{for the non-orientable surface } N_k. \end{aligned}$$

In particular, the Euler characteristic is 2 for the sphere, 1 for the projective plane, 0 for the torus and Klein bottle, and negative for all other surfaces. The importance of this concept is that it gives the following relationship (a generalization of Euler's polyhedron formula for the plane) among the numbers of vertices, edges and regions in a 2-cell embedding; a proof of this result may be found in, for example, **[56]**:

Theorem 3.1. *For a 2-cell embedding of a connected (general) graph with p vertices, q edges, and r regions in a surface S, we have*

$$p-q+r = \chi(S). \quad \|$$

More explicitly, for our surfaces S_h and N_k we have $p-q+r = 2-2h$ if $S = S_h$, and $p-q+r = 2-k$ if $S = N_k$.

If an embedding of a connected graph G is not a 2-cell embedding, then Theorem 3.1 does not apply directly (see Fig. 7(a) and (b), for example). However, by a result of Kagno **[43]** (see also Youngs **[76]**), the embedding can be modified so that an embedding of G on a surface S' is obtained, where $\chi(S') > \chi(S)$. (This modification can leave r unaltered, as in the situation beginning with Fig. 7(a), or it can increase r, as with Fig. 7(b).) If necessary, this process is repeated until a 2-cell embedding of G is found, say in a surface S''. The procedure must terminate in a 2-cell embedding, since all surface characteristics are bounded above by $\chi(S_0) = 2$. Thus for any embedding of a connected graph G in a surface S, there exists a 2-cell embedding of G in a surface S'', where $\chi(S'') \geqslant \chi(S)$. (If the embedding in S is a 2-cell embedding, then $S'' = S$.) Just as for the sphere, one can derive from these considerations an upper bound for the number of edges in a graph embeddable on a surface S:

Theorem 3.2. *If G is a connected graph with p vertices, q edges, and girth g, and if G is embeddable in a surface S, then*

$$q \leqslant \frac{g}{g-2}(p-\chi(S)).$$

Proof. We first find a 2-cell embedding of G in a surface S'', where $\chi(S'') \geqslant \chi(S)$, as described above. Let r'' be the number of regions in this embedding. By counting the edges on each region boundary, and using the fact that each region boundary has at least g boundary edges, we find that $gr'' \leqslant 2q$. Now we use Theorem 3.1 to find that

$$\chi(S) \leqslant \chi(S'') = p-q+r'' \leqslant p-q+\frac{2q}{g},$$

and the result follows. ‖

Some special cases of this theorem, stated in the next two corollaries, are of particular interest; they will be used in later sections:

Corollary 3.3. *If G is a connected graph with p vertices and q edges, which is embeddable in S_h, then*

(*i*) $q \leqslant 3(p+2h-2)$;

(*ii*) $q \leqslant 2(p+2h-2)$, *if G has no triangles.* ‖

Corollary 3.4. *If G is a connected graph with p vertices and q edges, which is embeddable in N_k, then*

(*i*) $q \leqslant 3(p+k-2)$;

(*ii*) $q \leqslant 2(p+k-2)$, *if G has no triangles.* ‖

Rotation Schemes

Embeddings of small graphs on surfaces of low genus can actually be drawn as in Figs. 5 and 6, using the polygonal representation of the surface. Clearly this is impractical for embeddings in general, however. A major tool in the study of graph embeddings is a method for describing a 2-cell embedding algebraically, so that no actual drawings need be made. This device, called a "rotation scheme", was introduced by Heffter [**37**], and has been used extensively by Ringel (see [**52**], for example) and others. Rotation schemes were also discussed briefly by Edmonds [**18**], and a detailed justification of their use was provided by Youngs [**76**]. Here we outline the development of Youngs, but with a slightly different notation. This approach covers only orientable embeddings; for an extension to the non-orientable case, see Stahl [**62**].

Let G be a connected graph with vertex-set $\{1, 2, \ldots, p\}$. For each vertex i, let $E(i)$ denote the set of edges at i, considered as being oriented from i, and let π_i be a cyclic permutation of $E(i)$. (Each edge vw in G thus determines two oriented edges, (v, w) and (w, v); the former goes from v to w, and the latter goes from w to v.) Then there is a one–one correspondence between the 2-cell embeddings of G and the choices of the π_i, as stated in the following theorem:

Theorem 3.5. *Each choice of permutations* $(\pi_1, \pi_2, \ldots, \pi_p)$ *determines a 2-cell embedding of G in some orientable surface* S_h, *and* S_h *may be oriented so that, at each vertex i, the oriented edge* (i, k) *is followed by* $\pi_i(i, k)$. *Conversely, given any 2-cell embedding of G in* S_h, *there is a corresponding set of permutations which yields that embedding.*

Proof. Let G^* be the set of oriented edges in G, and define the mapping $\Pi^*: G^* \to G^*$ by $\Pi^*(v, w) = \pi_w(w, v)$. Then Π^* is a permutation of G^*, and each orbit determines a region. These regions may be "pasted together" (with (v, w) identified with (w, v) as shown in Fig. 8) to form a surface S_h in which G is 2-cell embedded. Since each π_i is cyclic, the result of this pasting is a 2-manifold. Since every edge (v, w) in the boundary of a given region is matched with the edge (w, v) in the boundary of another (possibly the same) region, the 2-manifold is without boundary, and hence is a surface. Since each (v, w) is matched with (w, v), and never with (v, w), the surface is orientable. The genus h of S_h can then be determined by Euler's formula, with r given by

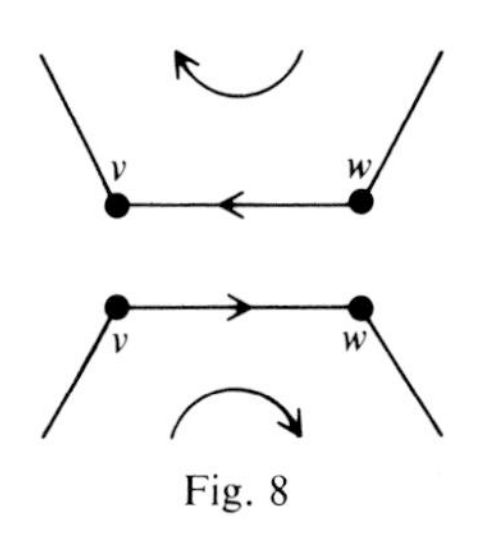

Fig. 8

the number of orbits under Π^*. The converse follows from similar considerations. ‖

We observe that if G is simple, each permutation π_i can also be regarded as a cyclic permutation of the vertices of G adjacent to i; we then define $\Pi^*(v, w) = (w, \pi_w(v))$.

For example, consider the toroidal embedding (necessarily 2-cell) of K_5 given in Fig. 9. Assuming a clockwise orientation on the torus, we find:

$$\begin{aligned}
\pi_1&: ((1, 2), (1, 3), (1, 5), (1, 4)) \sim (2, 3, 5, 4),\\
\pi_2&: ((2, 3), (2, 4), (2, 1), (2, 5)) \sim (3, 4, 1, 5),\\
\pi_3&: ((3, 4), (3, 5), (3, 2), (3, 1)) \sim (4, 5, 2, 1),\\
\pi_4&: ((4, 5), (4, 1), (4, 3), (4, 2)) \sim (5, 1, 3, 2),\\
\pi_5&: ((5, 1), (5, 2), (5, 4), (5, 3)) \sim (1, 2, 4, 3).
\end{aligned}$$

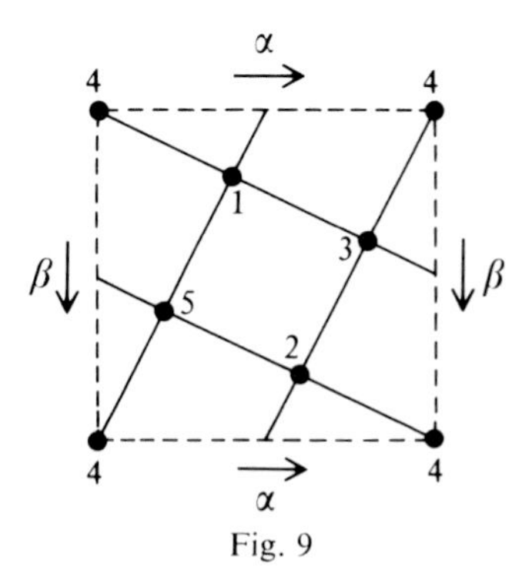

Fig. 9

The orbits under Π^* (corresponding to counter-clockwise region boundaries) are:

$$\begin{aligned}
(1)&: \{(1, 2), (2, 5), (5, 4), (4, 1)\},\\
(2)&: \{(1, 3), (3, 4), (4, 2), (2, 1)\},\\
(3)&: \{(1, 4), (4, 3), (3, 5), (5, 1)\},\\
(4)&: \{(1, 5), (5, 2), (2, 3), (3, 1)\},\\
(5)&: \{(2, 4), (4, 5), (5, 3), (3, 2)\}.
\end{aligned}$$

(Note that

$$\Pi^*(4, 1) = (1, 2), \qquad \Pi^*(2, 1) = (1, 3), \qquad \Pi^*(5, 1) = (1, 4),$$
$$\Pi^*(3, 1) = (1, 5), \qquad \text{and} \qquad \Pi^*(3, 2) = (2, 4),$$

so that these orbits are complete as indicated.) For notational convenience, we abbreviate an orbit such as (*1*) above by 1, 2, 5, 4; it is implicit that $\pi_4(5) = 1$ and $\pi_1(4) = 2$ if we regard π_i as permuting the vertices adjacent to i, or that $\pi_1(4, 5) = (4, 1)$ and $\pi_1(1, 4) = (1, 2)$ if π_i permutes $E(i)$.

Of course, in practice, rotation schemes are used not so much to describe existing embeddings, but more to produce new ones. Such schemes may be found by chance, by a flash of combinatorial insight, or perhaps by some more systematic approach. It is to the last of these possibilities that we turn our attention in the next section.

4. Covering Spaces and Voltage Graphs

An embedding of a graph in a surface can often be determined from a simpler embedding of another graph on another surface, by interpreting it as a covering space of the simpler embedding. In this section we look first at such coverings, and then at voltage graphs, which provide a systematic way of obtaining embeddings.

Covering Spaces

A map $\rho: \tilde{X} \to X$ from one topological space to another is a **covering projection** if every point x in X has a neighborhood U such that ρ maps each component of $\rho^{-1}(U)$ homeomorphically onto U. The space $\tilde{X}$ is then called a **covering space** for X. Some standard examples include the following, where S^1 denotes the unit circle in the complex plane:

(*i*) $\rho: \mathbf{R} \to S^1$, given by $\rho(t) = \cos t + i \sin t$;

(*ii*) $\rho: S^1 \to S^1$, given by $\rho(z) = z^n$ for some positive integer n;

(*iii*) $\rho: S_0 \to N_1$, where the map is antipodal identification.

Our interest lies in the case where $\tilde{X}$ and X are both surfaces, as in example (*iii*).

A basic result on covering projections is that $|\rho^{-1}(x)|$ is independent of the choice of x in X; if this value is n, then we call ρ an ***n*-fold covering projection**. The third example above is thus a 2-fold covering projection.

For an n-fold covering projection for surfaces, $\rho: \tilde{S} \to S$, it turns out that the surface characteristics are related by $\chi(\tilde{S}) = n\chi(S)$. In particular, if $\tilde{S} = S_k$ and $S = S_h$, then $k = n(h-1)+1$; if $\tilde{S} = N_k$ and $S = N_h$, then $k = n(h-2)+2$. Non-orientable surfaces can also have orientable covering surfaces (see (*iii*) above), but orientable surfaces cannot have non-orientable covers. In particular we see that (among the surfaces) only the sphere can cover the sphere, and then only for the trivial case $n = 1$, and that only the torus can cover the torus. For $h > 1$ (with S_m covering S_h), however, we see that m can increase without limit, depending upon n. We remark that the

Klein bottle can be covered (among the surfaces) only by itself and by the torus.

In this chapter and the next, a map $\rho: \tilde{X} \to X$ from one topological space to another is a **branched covering projection** (and $\tilde{X}$ is a **branched covering space** of X) if there exists a finite set B of points of X (called **branch points**) such that the restricted map $\rho: \tilde{X} - \rho^{-1}(B) \to X - B$ is a covering projection. If $b \in B$, if U is a sufficiently small neighborhood of b, and if $\tilde{U}$ is a component of $\rho^{-1}(U - \{b\})$, then the restricted map $\rho: \tilde{U} \to U - \{b\}$ is an n-fold covering projection for some integer n, which we call the **multiplicity** of branching at b. As examples here we give the following:

(*i*) $\rho: D \to D$ (the closed unit disk), given by $\rho(z) = z^n$, where n is a positive integer; here the origin is a branch point of multiplicity n.

(*ii*) $\rho: S_0 \to S_0$, given in spherical coordinates by $\rho(1, \theta, \phi) = (1, n\theta, \phi)$; here the poles are branch points of multiplicity n.

Again, our interest lies in the cases where X and $\tilde{X}$ are both surfaces, as in example (*ii*).

One additional item of terminology is that if Y and $\tilde{Y}$ are subsets of X and $\tilde{X}$, respectively, and if ρ maps $\tilde{Y}$ homeomorphically onto Y, then we say that Y **lifts** to $\tilde{Y}$. Thus, for a covering projection $\rho: \tilde{X} \to X$, every point $x \in X$ has a neighborhood U which lifts to each component of $\rho^{-1}(U)$.

Voltage Graphs

The theory of "current graphs" was introduced by Gustin [**30**], and developed by Youngs (see, for example, [77]), exclusively for the purpose of generating rotation schemes to embed the complete graphs, so as to prove the Heawood map-coloring theorem (see Chapter 3). Jacques [**38**] unified and extended this theory, and Gross and Alpert (see, for example, [**28**], [**29**]) extended the current graph theory to orientable 2-cell embeddings of arbitrary graphs, while setting the theory in the context of branched covering spaces. Gross [**26**] then dualized this to obtain the more natural theory of voltage graphs, which we discuss here. (The theory has been further extended to cover 2-cell embeddings into non-orientable surfaces, and into pseudosurfaces as well; see Stahl and White [**64**], and Garman [**22**], respectively. Here we treat only the orientable case.)

Let K be a general graph. With each edge uv of K associate the two oriented edges $e = (u, v)$ and $e^{-1} = (v, u)$, and let K^* be the set of oriented edges in K. A **voltage graph** is a triple (K, Γ, ϕ), where Γ is a group, and $\phi: K^* \to \Gamma$ is a function satisfying $\phi(e^{-1}) = (\phi(e))^{-1}$ for all e in K^*. The **covering graph** (or

derived graph) $K \times_\phi \Gamma$ for (K, Γ, ϕ) has vertex-set $V(K) \times \Gamma$, with each edge $e = uv$ of K determining the $|\Gamma|$ edges of $K \times_\phi \Gamma$ joining the vertices (u, g) and $(v, g\phi(e))$, for $g \in \Gamma$.

Now let (K, Γ, ϕ) have a 2-cell embedding in an orientable surface S, described algebraically by a rotation scheme $\Pi = (\pi_1, \pi_2, \ldots, \pi_p)$. We define the lift $\tilde{\Pi}$ of Π to $K \times_\phi \Gamma$ as follows: if $\pi_v(v, u) = (v, w)$ then $\tilde{\Pi}_{(v,g)}((v, g), (u, g\phi(v, u))) = ((v, g), (w, g\phi(v, w)))$ for each $g \in \Gamma$ (see Fig. 10). Then $\tilde{\Pi}$ is the collection of all such $\tilde{\Pi}_{(v,g)}$, one for each vertex of

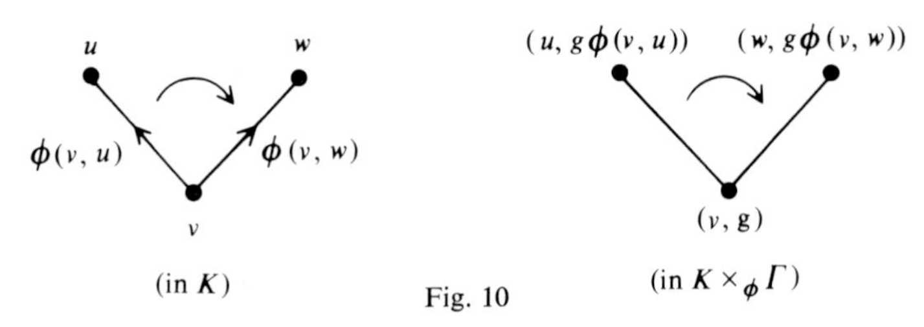

Fig. 10

$K \times_\phi \Gamma$, and thus $\tilde{\Pi}$ determines a 2-cell embedding of $K \times_\phi \Gamma$. The power of voltage graph theory is that the simpler embedding of K gives much information about the embedding of $K \times_\phi \Gamma$. To see this, let R be a region of the embedding of K on S induced by Π, and let $|R|_\phi$ be the order of $\phi(c)$ in Γ, where $c = e_1, e_2, \ldots, e_m$ is a closed walk in K consisting of the boundary of R, and $\phi(c) = \prod_{i=1}^{m} \phi(e_i)$. (Since $\phi(c)$ is unique up to inverses and conjugacy, $|R|_\phi$ is independent of the orientation of R and of the initial vertex of c.) We then have the following result:

Theorem 4.1. *Let (K, Γ, ϕ) be a voltage graph with rotation scheme Π, and let $\tilde{\Pi}$ be the lift of Π to $K \times_\phi \Gamma$. If Π and $\tilde{\Pi}$ determine 2-cell embeddings of K and $K \times_\phi \Gamma$ on the orientable surfaces S and $\tilde{S}$, respectively, then there exists a branched covering projection $\rho: \tilde{S} \to S$ such that*

(i) $\rho^{-1}(K) = K \times_\phi \Gamma$;

(ii) if b is a branch point of multiplicity n, then b is in the interior of a region R such that $|R|_\phi = n$;

(iii) if R is a region of the embedding of K which is a k-gon, then $\rho^{-1}(R)$ has $|\Gamma|/|R|_\phi$ components, each of which is a $k|R|_\phi$-gon region of $K \times_\phi \Gamma$. ||

We remark that, if $|R|_\phi = n$, then each component of $\rho^{-1}(R)$ is mapped by ρ onto R, essentially by $\rho: D \to D$, $\rho(z) = z^n$, as described above.

If $|R|_\phi = 1$, so that $\phi(c) = 1$ in Γ, then we say that R satisfies the **Kirchhoff voltage law** (KVL); if the KVL holds for all regions R of K on S, then we say that the embedding of (K, Γ, ϕ) satisfies the KVL.

Corollary 4.2. *If the embedding of* (K, Γ, ϕ) *satisfies the KVL, then* $\rho: \tilde{S} \to S$ *is a covering projection; that is, there is no branching.* ‖

In either case (branching or not), if K has order m, then the voltage graph (K, Γ, ϕ) is said to have **index *m*,** and the embedding of $K \times_\phi \Gamma$ is an **index *m* embedding.**

For two applications of this theory, consider the index 1 voltage graph of Fig. 11; K has one vertex and two loops, Γ is Abelian, and ϕ is as indicated in the figure. (Note that Fig. 11(a) displays the single region conveniently, whereas the single vertex rotation is more readily discerned from Fig. 11(b).)

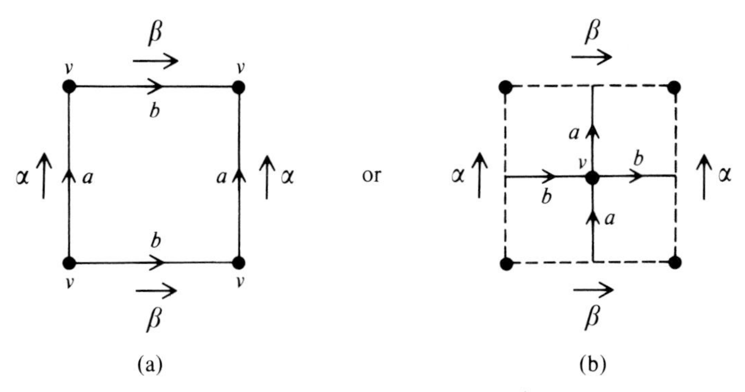

Fig. 11

Since Γ is Abelian, the KVL holds on the only region, so that Corollary 4.2 applies. The embedding of K is described by the single rotation

$$\pi_v: ((v, v)_a, (v, v)_b, (v, v)_{-a}, (v, v)_{-b}),$$

where the subscripts are used to distinguish between the two loops and between their orientations. If we take $a = 1$ and $b = 2$ in $\Gamma = \mathbf{Z}_5$, then $K \times_\phi \Gamma = K_5$, and $\tilde{\Pi}$, determined from Π, is as follows:

$$\tilde{\Pi}_{(v,1)}: (((v, 1), (v, 2)), ((v, 1), (v, 3)), ((v, 1), (v, 0)), ((v, 1), (v, 4))),$$
$$\tilde{\Pi}_{(v,2)}: (((v, 2), (v, 3)), ((v, 2), (v, 4)), ((v, 2), (v, 1)), ((v, 2), (v, 0))),$$
$$\tilde{\Pi}_{(v,3)}: (((v, 3), (v, 4)), ((v, 3), (v, 0)), ((v, 3), (v, 2)), ((v, 3), (v, 1))),$$
$$\tilde{\Pi}_{(v,4)}: (((v, 4), (v, 0)), ((v, 4), (v, 1)), ((v, 4), (v, 3)), ((v, 4), (v, 2))),$$
$$\tilde{\Pi}_{(v,0)}: (((v, 0), (v, 1)), ((v, 0), (v, 2)), ((v, 0), (v, 4)), ((v, 0), (v, 3))).$$

For an index 1 embedding such as this, since $|V(K)| = 1$, we lose no information by discarding the first coordinate in $V(K \times_\phi \Gamma) = V(K) \times \Gamma$; thus we can write $(v, g) = g$. Moreover, since K_5 has no loops or multiple edges, we can regard each $\tilde{\Pi}_g$ as permuting the neighbors of g rather than the edges in $E(g)$. Thus the above scheme can be shortened considerably to the following:

$$\tilde{\Pi}_1:(2, 3, 0, 4),$$
$$\tilde{\Pi}_2:(3, 4, 1, 0),$$
$$\tilde{\Pi}_3:(4, 0, 2, 1),$$
$$\tilde{\Pi}_4:(0, 1, 3, 2),$$
$$\tilde{\Pi}_0:(1, 2, 4, 3).$$

Now writing $5 = 0$ (in $\mathbf{Z}_5$), we see that this is exactly the rotation scheme for the embedding of Fig. 9. (In practice, this shortened form of the scheme is written down directly from the voltage graph.) The covering projection is five-fold; the single vertex, two edges, and one 4-gon for K lift respectively to five vertices, ten edges, and five 4-gons for $K \times_\phi \Gamma = K_5$.

For a second voltage graph construction using Fig. 11, take $a = (1, 0)$ and $b = (0, 1)$ in $\Gamma = \mathbf{Z}_m \times \mathbf{Z}_n$ (a direct product, with $m, n \geqslant 3$); then $K \times_\phi \Gamma = C_m \times C_n$ (a Cartesian product) has a toroidal embedding in which all regions are quadrilaterals. We illustrate the case $m = n = 3$ in Fig. 12.

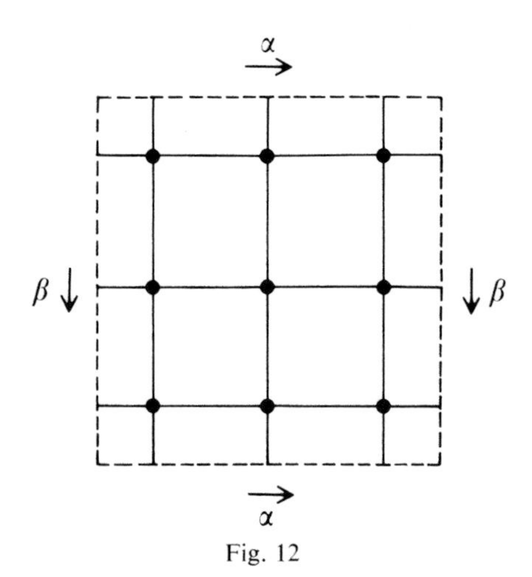

Fig. 12

The covering-space nature of this embedding is readily apparent; the single vertex, two edges, and one 4-gon for K now lift to nine vertices, eighteen edges, and nine 4-gons, under this nine-fold projection.

We reiterate the importance of the theory of voltage graphs for graph embeddings. Quite often a simple picture such as Fig. 11 above, or Fig. 13 in the next section, determines rotation schemes for all graphs in a given class simultaneously, and these rotation schemes in turn determine 2-cell embeddings for the graphs. Moreover, the genus of these derived embeddings, and other properties as well, are readily determined from the one initial picture—the embedded voltage graph.

5. The Genus of a Graph

Clearly, every graph is embeddable in some orientable surface—all we need to do is to draw the graph on the sphere and add enough handles to eliminate crossings. The **genus** $\gamma(G)$ of a graph G is defined to be the minimum genus of the orientable surfaces on which G is embeddable, In this section we present some of the very few general results known about the genus, and then discuss this parameter for certain families of graphs.

An embedding of G in a surface of genus $\gamma(G)$ is called a **minimal embedding**. The following theorem is a consequence of the result of Kagno mentioned earlier (see Youngs [**76**]):

Theorem 5.1. *A minimal embedding of a connected graph is a 2-cell embedding.* ‖

We lose no generality in restricting our attention to connected graphs; in fact, one need only consider 2-connected graphs, as the following result, due to Battle, Harary, Kodama and Youngs [**5**], states:

Theorem 5.2. *The genus of a graph is the sum of the genera of its blocks.* ‖

Lower bounds for the genus parameter are elementary consequences of Euler's formula. In particular, we have the following inequalities, which are directly obtainable from Corollary 3.3:

Theorem 5.3. *If G is a connected graph with p ($\geqslant 3$) vertices and q edges, then*

(*i*) $\gamma(G) \geqslant \{\frac{1}{6}(q-3p+6)\}$;

(*ii*) $\gamma(G) \geqslant \{\frac{1}{4}(q-2p+4)\}$, *if G has no triangles.* ‖

A survey of graphs for which strict inequality holds in these and similar expressions has been given by Ringel [**58**].

In theory, the genus is computable for any connected graph. In fact, there is an algorithm for doing this, presented by Youngs [**76**], and based upon rotation schemes for 2-cell embeddings. Let $V(G) = \{1, 2, \ldots, p\}$, and let ρ_i be the valency of vertex i. Then by Theorem 3.5 there is a one–one correspondence between orientable 2-cell embeddings of G and ordered p-tuples $(\pi_1, \pi_2, \ldots, \pi_p)$, where π_i is a cyclic permutation of the neighbors of i. For each such p-tuple, we compute r as the number of orbits under Π^*, the permutation of G^* induced by the π_i. The genus of the surface for that particular embedding is easily calculated using Theorem 3.1, and $\gamma(G)$ is the minimum genus thus determined. (Since minimal embeddings are 2-cell embeddings, by Theorem 5.1, we know that $\gamma(G)$ will be obtained by this procedure.)

Unfortunately, this algorithm for genus is very inefficient, as $\prod_{i=1}^{p} (\rho_i - 1)!$ such n-tuples must be surveyed, and the number of regions calculated for each. For $G = K_p$, for example, $(p-2)!^{p-1}$ such calculations must be made, since, by symmetry, π_1 can be fixed in this case.

In practice, the upper bound (to agree with a lower bound from Theorem 5.3, or some counterpart) is established by constructing an appropriate embedding. This can be done in a variety of ways, such as (*i*) by induction (see Theorem 5.9 below), (*ii*) by using a voltage graph (see Theorem 5.6 below) or a current graph (the dual structure), or (*iii*) by using some other means to generate an appropriate rotation scheme (see Theorem 5.5 below). Ringel (see **[52]**, for example) often seeks schemes which satisfy what he calls "Rule Δ^*": if $\pi_i(j) = k$, then $\pi_k(i) = j$. Such schemes always produce orientable triangular embeddings.

We now survey some of the more significant classes of graphs for which the genus has been found.

Theorem 5.4. *The genus of the complete graph K_p ($p \geqslant 3$) is*

$$\gamma(K_p) = \{\tfrac{1}{12}(p-3)(p-4)\}. \;\|$$

The proof (due primarily to Ringel and Youngs; see, for example, **[59]**) uses the lower bound of Theorem 5.3(*i*) and current graph embeddings in twelve cases depending upon the residue class of p (modulo 12). The construction of these embeddings is examined in some detail in Chapter 3, as this genus calculation provides a proof of the Heawood map-coloring theorem.

Theorem 5.5. *The genus of the complete bipartite graph $K_{r,s}$ ($r, s \geqslant 2$) is*

$$\gamma(K_{r,s}) = \{\tfrac{1}{4}(r-2)(s-2)\}. \;\|$$

The proof, due to Ringel **[53]**, uses the lower bound of Theorem 5.3(*ii*) and embeddings constructed by the generation of appropriate rotation schemes.

The general class of complete n-partite graphs has attracted considerable interest, and those that are regular have drawn particular attention. We denote by $K_{n(m)}$ the regular complete n-partite graph of order mn; it is the complement of nK_m:

Theorem 5.6. *The genus of the regular complete tripartite graph $K_{3(n)}$ is*

$$\gamma(K_{3(n)}) = \binom{n-1}{2}.$$

Remark. The first proof of this result was by Ringel and Youngs [**60**], and a second proof was by White [**68**], who found that $\gamma(K_{n,n,mn}) = \frac{1}{2}(n-1)(mn-2)$. We give a short proof here, found in [**64**], using voltage graphs.

Proof. It follows from Theorem 5.3(i) that $\gamma(K_{3(n)}) \geqslant \binom{n-1}{2}$, and it remains to prove the opposite inequality. The index 2 voltage graph of Fig. 13 uses $\Gamma = \mathbf{Z}_n$ to embed $K \times_\phi \Gamma = K_{2(n)}$ in S_h, where $h = \binom{n-1}{2}$, in such a way that

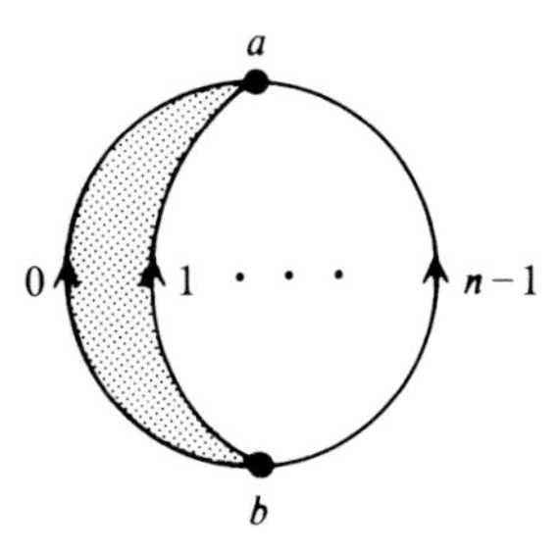

Fig. 13

there are n regions, each bounded by a Hamiltonian circuit. (Refer to Theorem 4.1: here there are n branch points, each of multiplicity n; each 2-gon region of K contains one branch point, and lifts to one component—a $2n$-gon which is readily seen to be bounded by a Hamiltonian circuit. For instance, the region covering the shaded region in Fig. 13 is bounded by $(a, 0)$, $(b, 0)$, $(a, 1)$, $(b, 1)$, . . ., $(a, n-1)$, $(b, n-1)$.) Now add one vertex in the interior of each $2n$-gon, and join it to all $2n$ vertices in the boundary. The result is a triangular (and hence genus) embedding of the join

$$K_{2(n)} + \overline{K}_n = K_{3(n)}. \quad \|$$

Theorem 5.7. *The genus of the regular complete quadripartite graph $K_{4(n)}$ is*

$$\gamma(K_{4(n)}) = (n-1)^2,$$

except that $\gamma(K_{4(3)}) = 5$. ‖

Again, the lower bound comes from Theorem 5.3(i), except when $n = 3$. Constructions have been provided by Garman [**22**] for n even, by Jungerman [**39**] for $n \neq 3$, and by White [**73**] for $n = 3$.

Theorem 5.8. *The genus of the n-dimensional octahedron $K_{n(2)}$ is*

$$\gamma(K_{n(2)}) = \tfrac{1}{3}(n-3)(n-1),$$

if $n \not\equiv 2$ (modulo 3). ||

For constructions, see [**23**], [**28**] and [**42**].

Bouchet [**13**] has developed a construction method for $K_{n(m)}$ using "generative m-valuations". He considered the residue classes of n(modulo 12), and of m (modulo 6); the 72 cases thus determined split into 30 cases for which Euler's formula does not allow an orientable triangulation, 7 cases for which the theory in use does not apply, 3 cases for which the theory is applicable but the issue has not yet been resolved, and 32 cases for which triangular embeddings have been found. Thus it could be said that $\gamma(K_{n(m)})$ has been found in four-ninths of the possible cases.

Other graphs for which the genus has been determined are the n-cubes. Several proofs of this are known: the first was given by Ringel [**51**], a second (which we give here) by Beineke and Harary [**7**], and a third (which uses the voltage graph of Fig. 13) by Gross and Alpert [**29**].

Theorem 5.9. *The genus of the n-cube Q_n ($n \geqslant 2$) is*

$$\gamma(Q_n) = (n-4)2^{n-3}+1.$$

Proof. For convenience, let γ_n denote the number $(n-4)2^{n-3}+1$. The inequality $\gamma(Q_n) \geqslant \gamma_n$ follows from Theorem 5.3(*ii*), since Q_n is bipartite and has 2^n vertices and $n \cdot 2^{n-1}$ edges. The reverse inequality will be verified by an inductive proof of the following statement: there is an embedding of Q_n in the orientable surface of genus γ_n so that each quadrilateral in which adjacent vertices differ only in the first or last positions is a region. Clearly this is true for $n = 2$ and $n = 3$. Let $n \geqslant 3$, and assume that the statement holds. Take a copy of Q_n so embedded on a surface, and add the suffix 0 to the usual label of each vertex. Take also a "mirror-image" surface and copy of Q_n, and add a suffix 1 at each vertex. Now add a handle from each of the special regions on the first surface to its counterpart on the second surface; there are 2^{n-2} such regions. Since each of the original surfaces had genus γ_n, the new surface has genus $2\gamma_n + 2^{n-2} - 1 = \gamma_{n+1}$. The construction is completed by adding four edges across each handle, joining the vertices which differ in their last place. Since, in the new embedding, the quadrilaterals with adjacent vertices differing only in the first or last positions are regions, the result is proved. ||

We comment that the above inductive construction applies to several other classes of repeated Cartesian products of bipartite graphs (see White [**69**]).

There are several other classes of graphs whose genus has been found, notably certain classes of Cayley graphs. This leads naturally to the concept of the "genus of a group"; for a discussion of these matters, see White [**71**].

The Maximum Genus

A related natural question arises: in which orientable surfaces does a given graph have 2-cell embeddings? In order to investigate this, we define the **maximum genus** $\gamma_M(G)$ of a connected graph G to be the largest value of h for which G is 2-cell embeddable in S_h. It follows from a result of Duke [**17**] that G has a 2-cell embedding in S_h for all values of h between $\gamma(G)$ and $\gamma_M(G)$.

A frequently attained upper bound for $\gamma_M(G)$ can be given in terms of the first Betti number $\beta(G) = q-p+1$—namely, $\gamma_M(G) \leqslant [\frac{1}{2}\beta(G)]$. Equality holds here if and only if there is a 2-cell embedding with just 1 or 2 regions, depending on whether $\beta(G)$ is even or odd. Graphs for which equality holds have been investigated by a number of mathematicians, including Ringeisen [**49**], Jungerman [**41**], and Xuong [**74**]. Xuong has also given a formula for computing $\gamma_M(G)$ for an arbitrary connected graph G [**75**].

7. The Non-orientable Genus

Every graph is embeddable in a non-orientable surface, just as it is in an orientable one (since we may add a crosscap to an appropriate orientable surface). We define the **crosscap number**, or **non-orientable genus**, $\tilde{\gamma}(G)$ of a graph G to be the minimum number of crosscaps needed on a sphere to achieve embeddability. Even though the sphere itself is orientable, we shall allow 0 for its crosscap number, so that we do not need to state exceptions.

The theory for the non-orientable genus sometimes parallels that for the genus, and sometimes does not. The two parameters are related by the equation $\tilde{\gamma}(G) \leqslant 2\gamma(G)+1$. This result follows from the observation that the addition of one crosscap to the orientable surface S_h results in the non-orientable surface N_{2h+1}. One might expect to have some sort of bound in the opposite direction, but Auslander, Brown and Youngs [**3**] have shown that there are graphs of arbitrarily large genus which are embeddable in the projective plane.

One result for which the non-orientable analog does not hold without modification is Theorem 5.1—an embedding of a connected graph in the non-orientable surface $N_{\tilde{\gamma}(G)}$ need not be a 2-cell embedding. For example, K_7 is not embeddable in the Klein bottle N_2 (as shown by Franklin [**21**]), but since it is embeddable in the torus it has an embedding in N_3 which is not a 2-cell embedding. Thus we have no guarantee that Euler's formula holds for embeddings of a graph in surfaces of minimum non-orientable genus.

Youngs [76] surmounted this difficulty by defining an embedding of a connected graph G into a surface (orientable or non-orientable) to be **simplest** if there is no embedding of G into a surface of greater characteristic, and **maximal** if there is no embedding which gives more regions. He then established the following result:

Theorem 6.1. *An embedding of a connected graph is simplest if and only if it is both 2-cell and maximal.* ‖

A second theorem whose direct analog for non-orientable genus is false is Theorem 5.2—the crosscap number is not additive over blocks or components. For example, consider the graph $2K_7$. Since $\gamma(2K_7) = 2\gamma(K_7) = 2$, it follows that $\tilde{\gamma}(2K_7) \leqslant 5$; but, as we have seen, $\tilde{\gamma}(K_7) = 3$, so that $\tilde{\gamma}(2K_7) \neq 2\tilde{\gamma}(K_7)$. Nevertheless, there is a non-orientable analog of Theorem 5.2, due to Stahl and Beineke [63]. In order to state it, we define the **manifold number** $\mu(G) = \max\{2-2\gamma(G),\ 2-\tilde{\gamma}(G)\}$, and call a graph **orientably simple** if $\mu(G) \neq 2-\tilde{\gamma}(G)$; that is, if $\tilde{\gamma}(G) > 2\gamma(G)$.

Theorem 6.2. *Let G be a graph with blocks (or components) $G_1, G_2, \ldots, G_k$. If G is orientably simple, then*

$$\tilde{\gamma}(G) = 1-k+\sum_{i=1}^{k} \tilde{\gamma}(G_i);$$

if not, then

$$\tilde{\gamma}(G) = 2k-\sum_{i=1}^{k} \mu(G_i).\ \|$$

Lower bounds for the crosscap number follow from Corollary 3.4, just as those for the genus followed from Corollary 3.3:

Theorem 6.3. *If G is a connected graph with p ($\geqslant 3$) vertices and q edges, then*

(*i*) $\tilde{\gamma}(G) \geqslant \{\frac{1}{3}(q-3p+6)\}$;

(*ii*) $\tilde{\gamma}(G) \geqslant \{\frac{1}{2}(q-2p+4\}$, *if G has no triangles.* ‖

We mention just a few classes of graphs for which the non-orientable genus has been found. The first is due to Ringel [50] and will be discussed in Chapter 3, since it is crucial for the non-orientable version of the Heawood map-coloring theorem. The second is also due to Ringel [55], and the third to Jungerman [40]. For these three families of graphs, the results in the orientable and non-orientable cases are almost identical. Except for K_7, Q_4

and Q_5, the largest possible Euler characteristic gives the correct value, and $\gamma(G) = \{\frac{1}{2}\gamma(G)\}$.

Theorem 6.4. *The crosscap number of the complete graph* K_p $(p \geqslant 3)$ *is*

$$\tilde{\gamma}(K_p) = \{\tfrac{1}{6}(p-3)(p-4)\},$$

except that $\tilde{\gamma}(K_7) = 3.$ ‖

Theorem 6.5. *The crosscap number of the complete bipartite graph* $K_{r,s}(r, s \geqslant 2)$ is

$$\tilde{\gamma}(K_{r,s}) = \{\tfrac{1}{2}(r-2)(s-2)\}. \;\|$$

Theorem 6.6. *The crosscap number of the n-cube* Q_n $(n \geqslant 2)$ *is*

$$\tilde{\gamma}(Q_n) = (n-4)2^{n-2}+2,$$

except that $\tilde{\gamma}(Q_4) = 3$, *and* $\tilde{\gamma}(Q_5) = 11.$ ‖

Bouchet [**13**] has investigated the non-orientable, as well as the orientable, genus of the regular multipartite graphs $K_{n(m)}$. Again using "generative m-valuations", he considered the residue classes of m and n (modulo 6). The 36 cases split into 8 cases for which Euler's formula implies that there is no triangulation, 7 cases for which the theory does not apply, 3 cases for which it applies but the issue has not yet been resolved, and 18 cases for which triangulations have been found (12 of these may have one possible exceptional value of n). Thus it can be said that $\tilde{\gamma}(K_{n(m)})$ has been found for half of the cases.

The **maximum non-orientable genus** $\tilde{\gamma}_M(G)$ of a connected graph G is defined to be the maximum k for which G has a 2-cell embedding on N_k. In contrast to the orientable case, the value of $\tilde{\gamma}_M(G)$ can be given immediately for all connected graphs—it is simply the Betti number $\beta(G) = q-p+1$. This remarkable result was proved independently by Ringel [**57**] and Stahl [**62**], and it implies that there is always a 2-cell embedding with just one face. Stahl [**61**] also showed that there is a 2-cell embedding of G in N_k, for all k lying between $\tilde{\gamma}(G)$ and $\beta(G)$.

Before leaving embeddings to go on to an application, and then to other topics, we comment that there are many other interesting problems in this area which we have not discussed. One of these concerns the "n-irreducible graphs" (graphs of genus n all of whose proper subgraphs have smaller genus), and their non-orientable counterparts. For $n = 1$, the two 1-irreducible graphs K_5 and $K_{3,3}$ characterize those graphs of genus zero, by Kuratowski's theorem (see Chapter 1). However, for no value of $n > 1$ is it even known whether or not a finite class of this type exists. Results obtained to date indicate that, even for $n = 2$, such a class would be quite large. In the non-orientable case, Glover and Huneke [**24**] have claimed that the set of irreducible graphs for the projective plane is finite, and Glover, Huneke and Wang [**25**] have found 103 graphs in this class.

7. Block Designs

As in Chapter 1, a **(v, b, r, k, λ)-balanced incomplete block design** (*BIBD*) is an arrangement of v objects into b blocks so that

(*i*) each object appears in exactly r blocks;

(*ii*) each block contains exactly k $(< v)$ objects;

(*iii*) each pair of distinct objects appear together in exactly λ blocks.

Block designs are combinatorial structures that, apart from their intrinsic interest, have applications in scheduling problems and in experimental design.

In [**2**], Alpert established a one–one correspondence between $(v, b, r, 3, 2)$-*BIBD*s ("2-fold triple systems") and triangular imbeddings of complete graphs K_v. (Here the embeddings are into "generalized pseudosurfaces"—see [**72**].) The key observation is that any triangulation of K_v, with vertices as objects and regions as blocks, automatically has $k = 3$ and $\lambda = 2$, since each pair of distinct vertices constitutes an edge in K_v, and hence appears in the boundary of exactly two triangular regions; moreover, $b = \frac{1}{3}v(v-1)$, and $r = v-1$.

Simple counting arguments establish that for any *BIBD*, $vr = bk$, and $\lambda(v-1) = r(k-1)$. For 2-fold triple systems, the two necessary conditions coalesce into one: $v \equiv 0$ or 1 (modulo 3), where $v \geqslant 3$. It is well known that this necessary condition is also sufficient, and Alpert [**2**] has presented an elegant new proof of this fact: the non-orientable genus embeddings for K_n (where $n \equiv 0, 1, 3$ or 4 (modulo 6)) all give triangulation systems, except for $n = 3, 4$ and 7, where we use the orientable genus embeddings on the sphere and the torus. This exhausts all $v \equiv 0$ or 1 (modulo 3), where $v \geqslant 3$.

Related combinatorial designs are the "Steiner triple systems", or $(v, b, r, 3, 1)$-*BIBD*s. A triangulation system for K_v with bichromatic dual not only determines one 2-fold triple system, as seen above, but also splits naturally into two Steiner triple systems, each consisting of regions (blocks) of like color. It is well known that Steiner triple systems with v objects exist if and only if $v \equiv 1$ or 3 (modulo 6). Garman, Ringeisen and White [**23**] have observed that the orientable genus embeddings for K_n, where $n \equiv 3$ (modulo 12), all have bichromatic dual, so that Steiner triple systems are independently produced for these values of n.

We now illustrate these relationships between triangular embeddings of

K_n and *BIBD*s in the case $n = 7$. The voltage graph of Chapter 3, Fig. 19 has a bichromatic dual, and thus (it can be shown) so does every covering embedding of that voltage graph. For K_7 the 14 regions are:

0 – 1 – 3		0 – 3 – 2
1 – 2 – 4		1 – 4 – 3
2 – 3 – 5	and	2 – 5 – 4
3 – 4 – 6		3 – 6 – 5
4 – 5 – 0		4 – 0 – 6
5 – 6 – 1		5 – 1 – 0
6 – 0 – 2		6 – 2 – 1.

(These can be read off directly from the above mentioned voltage graph; see also Fig. 6 of this chapter.) Together these give a 2-fold triple system on 7 objects; when split in half, as indicated, the 14 regions yield two Steiner triple systems on 7 objects, each of which is isomorphic to that obtained from the Fano plane.

A triangular embedding for $K_{n(m)}$ $(m > 1)$ gives a $(v, b, r, 3; 0, 2)$-*PBIBD* (a partially-balanced incomplete block design with two association classes) determined by adjacency in the strongly regular graph $K_{n(m)}$ (see **[72]**, or Section 8 of Chapter 12). For example, the same voltage graph used above for K_7, when reinterpreted for $\Gamma = \mathbf{Z}_8$, gives a bichromatic dual embedding of $K_{4(2)}$ in S_1, yielding one $(8, 16, 6, 3; 0, 2)$-*PBIBD* and two $(8, 8, 3, 3; 0, 1)$-*PBIBD*s. Conversely (see **[72]**), each group-divisible *PBIBD* with object-set partitioned into n groups of m objects each (where $m > 1$ and $n > 2$), and with $k = 3$, $\lambda_1 = 0$, $\lambda_2 = 2$, determines a triangular embedding of $K_{n(m)}$ into some generalized pseudosurface. Such group-divisible designs have been shown by Hanani **[36]** to exist if and only if 3 divides $mn(n-1)$. This suffices to establish the existence of generalized pseudosurface triangular embeddings of $K_{n(m)}$ in *every* case which is consistent with Euler's formula.

8. Crossing Numbers

A more general concept than that of an embedding is that of a "drawing", where the restriction that the arcs be disjoint is removed. More formally, a **drawing** of a graph G in a surface S is a mapping of the vertices of G to distinct points of S, and the edges of G to (open) arcs in S, such that

(*i*) the image of no edge contains that of any vertex;

(*ii*) the image of an edge vw joins the points corresponding to v and w.

A drawing is called **good** if the edges (actually, their images) are such that

(*iii*) no two arcs with a common end point meet;

(*iv*) no two arcs meet in more than one point;

(*v*) no three arcs meet in a common point.

The configurations prohibited by these three conditions are indicated in Fig. 14. Since an arbitrary drawing can be converted to a good one, we shall consider only good drawings here.

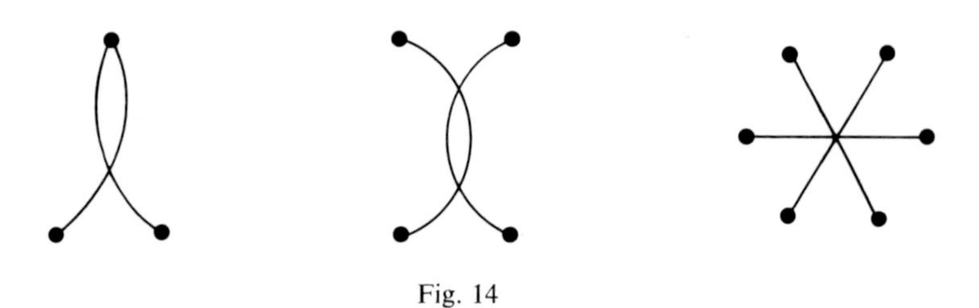

Fig. 14

A point of intersection of two edges in a drawing is called a **crossing**, and the **crossing number** $\nu(G)$ of a graph G is the minimum number of crossings in any good drawing of G in the plane. Surprisingly little is known about this parameter in general. (For a more extensive survey, see Erdős and Guy [**20**], and for the most exhaustive study of drawings, see Eggleton [**19**].)

For historical reasons, we begin with the problem of determining the crossing number of the complete bipartite graph $K_{r,s}$. This problem, which had its origin as "Turán's brick factory problem", is still open, but was believed to have been solved for a number of years (see Guy [**31**]). A simple construction gives the best-known upper bound. Let $r_1 = [\frac{1}{2}r]$, $r_2 = \{\frac{1}{2}r\}$, $s_1 = [\frac{1}{2}s]$ and $s_2 = \{\frac{1}{2}s\}$, so that $r_1+r_2 = r$, $s_1+s_2 = s$ and the numbers in each pair differ by at most 1. On the Cartesian coordinate system put r_1 vertices on the positive x-axis, r_2 vertices on the negative x-axis, s_1 vertices on the positive y-axis, and s_2 vertices on the negative y-axis. Now join vertices on different axes by a line segment to form $K_{r,s}$. (We show this construction for $K_{5,6}$ in Fig. 15). For each set of four vertices, with two of the same sign on the x-axis and two of the same sign on the y-axis, there is exactly one crossing, and hence the total number of crossings is

$$\binom{r_1}{2}\binom{s_1}{2}+\binom{r_1}{2}\binom{s_2}{2}+\binom{r_2}{2}\binom{s_1}{2}+\binom{r_2}{2}\binom{s_2}{2},$$

which equals the quantity given in the following theorem:

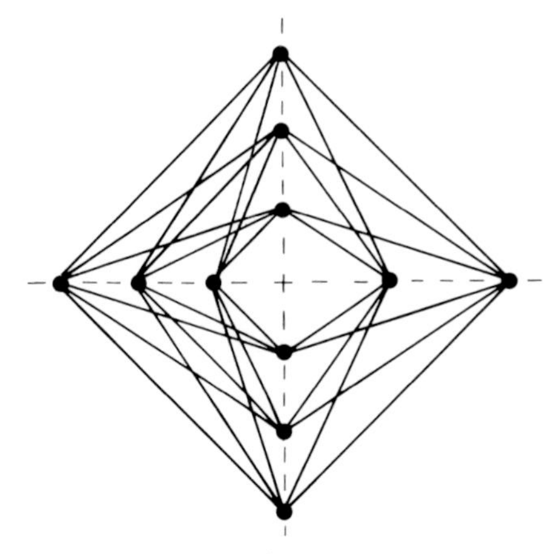

Fig. 15

Theorem 8.1. *The crossing number of the complete bipartite graph $K_{r,s}$ satisfies the inequality*

$$\nu(K_{r,s}) \leqslant [\tfrac{1}{2}r][\tfrac{1}{2}(r-1)][\tfrac{1}{2}s][\tfrac{1}{2}(s-1)]. \quad \|$$

One can show that if equality holds in Theorem 8.1 for any odd value of r (with s fixed), then it also holds for the next even value. For if $m \geqslant r$ and $n \geqslant s$, we have

$$\nu(K_{m,n}) \geqslant \binom{m}{r}\binom{n}{s}\ \nu(K_{r,s}) \Big/ \binom{m-2}{r-2}\binom{n-2}{s-2},$$

by the following simple counting argument: each drawing of $K_{m,n}$ includes $\binom{m}{r}\binom{n}{s}$ drawings of $K_{r,s}$, each obtained by selecting r and s vertices, respectively, from the two partite sets; but each crossing involves two vertices from each partite set and is thus counted $\binom{m-2}{r-2}\cdot\binom{n-2}{s-2}$ times. Hence if $\nu(K_{r,s})$ is equal to the upper bound in the theorem, then with $r = 2k-1$, $s = n$, and $m = 2k$, we have

$$\nu(K_{m,n}) \geqslant \frac{k}{k-1}\ \nu(K_{r,s}),$$

and this lower bound agrees with the upper bound for $K_{m,n}$.

Kleitman [**46**] has shown that equality holds in certain cases:

Theorem 8.2. *If* $\min\{r, s\} \leqslant 6$, *then*

$$\nu(K_{r,s}) = [\tfrac{1}{2}r][\tfrac{1}{2}(r-1)][\tfrac{1}{2}s][\tfrac{1}{2}(s-1)]. \quad \|$$

Using this work, Guy [**33**] has obtained the following lower bound:

Theorem 8.3. *If $5 \leqslant r \leqslant s$, then*

$$\nu(K_{r,s}) \geqslant \{\tfrac{1}{20}r(r-1)s(s-2)\}. \,\|$$

The smallest complete bipartite graph whose crossing number remains unknown is $K_{7,7}$, and it can be shown that it is equal to 77, 79 or 81.

Turning to complete graphs, we note again that it is not difficult to obtain an upper bound for the crossing number by means of a good drawing. Showing that this is the best-possible estimate (if in fact it is) has proved extremely difficult:

Theorem 8.4. *The crossing number of the complete graph K_p satisfies the inequality*

$$\nu(K_p) \leqslant \tfrac{1}{4}[\tfrac{1}{2}p][\tfrac{1}{2}(p-1)][\tfrac{1}{2}(p-2)][\tfrac{1}{2}(p-3)].$$

Proof. We first prove the result for p even ($=2n$, say). The surface used is a truncated right circular cylinder (clearly equivalent to a sphere). Around each end we put n vertices at equal distances. On each end we put the diagonals of that n-gon with a good drawing, and along shortest counter-clockwise helical curves we put the lateral edges. The number of crossings of diagonals of a convex n-gon is $\binom{n}{4}$, since any four vertices produce one crossing. An arbitrary base vertex can be shown to contribute to $\frac{1}{3}n(n-1)(n-2)$ intersections along the side, so that there are $\frac{1}{6}n^2(n-1)(n-2)$ lateral crossings. Hence the total number of crossings is $\frac{1}{4}n(n-1)^2(n-2)$, which agrees with the expression in the statement of the theorem.

For p odd ($=2n-1$), we use the construction for K_{2n} and delete one vertex. From the facts given in the even case, we note that each vertex is a part of $\binom{n-1}{3}$ end-crossings and $2\binom{n}{3}$ lateral crossings, so that the resulting drawing of K_{2n-1} has $\frac{1}{4}(n-1)^2(n-2)^2$ crossings altogether, which again agrees with the expression in the statement of the theorem. ‖

The construction used in this proof was given by Guy [**31**], and by Blažek and Koman [**12**]. There are also other constructions which give the same number of crossings.

As we mentioned earlier, the machinery for proving equality in general does not seem to be available, although it has been done step-by-step for $n \leqslant 10$ (see, for example, Guy [**33**]). However, one can get a lower bound for $\nu(K_p)$ in terms of $\nu(K_{r,p-r})$. To do this, we note that $K_{r,p-r}$ can be taken as a subgraph of K_p in $\binom{p}{r}$ ways. In counting crossings in all such subgraphs, we count a given crossing in $4\binom{p-4}{r-2}$ ways, since there are 4 ways of choosing two of the crossing contributors to be in the set of r vertices, and $\binom{p-4}{r-2}$ ways of completing that set. Thus, in particular with $r = 6$,

$$\nu(K_p) \geqslant [\tfrac{1}{2}(p-6)][\tfrac{1}{2}(p-7)] \binom{p}{6} \Big/ 4\binom{p-4}{4};$$

that is,

$$\nu(K_p) \geqslant \frac{p(p-1)(p-2)(p-3)}{20(p-6)(p-7)} [\tfrac{1}{2}(p-6)][\tfrac{1}{2}(p-7)].$$

Now it is not difficult to see that $\nu(K_p)/p^4$ is bounded and non-decreasing, and it follows that the limit

$$\lim_{p\to\infty} \nu(K_p)/p^4$$

exists and lies between $\frac{1}{80}$ and $\frac{1}{64}$.

Very few exact formulas for arbitrarily large crossing numbers are known. In addition to that of Theorem 8.2 we have the following results, due to Beineke and Ringeisen (see [**10**]):

Theorem 8.5. *The crossing numbers of the following Cartesian products are*

(*i*) $\nu(C_3 \times C_n) = n$, *for* $n \geqslant 3$;

(*ii*) $\nu(C_4 \times C_n) = 2n$, *for* $n \geqslant 4$;

(*iii*) $\nu(K_4 \times C_n) = 3n$, *for* $n \geqslant 3$. ||

There are also some results for crossing numbers on orientable surfaces of higher genus. The **crossing number** $\nu_n(G)$ is defined to be the minimum number of crossings in any good drawing of G in S_n; thus $\nu_0(G) = \nu(G)$.

Guy and Jenkins [**35**] have obtained a formula for the torus:

Theorem 8.6. *The "toroidal crossing number" of* $K_{3,s}$ *is*

$$\nu_1(K_{3,s}) = [\tfrac{1}{12}(s-3)^2]. \;\|$$

Gross [**27**] has studied the n-dimensional octahedral graph $K_{n(2)}$.

Theorem 8.7. *The crossing number of* $K_{n(2)}$ *in* S_h *is*

$$\nu_h(K_{n(2)}) = \tfrac{1}{2}n(n-1),$$

where $h = \frac{1}{4}(n-1)(n-4)$, *and n is a prime power congruent to* 1 (*modulo* 4). ||

Finally, Kainen and White [**44**] have found both orientable and non-orientable crossing numbers for the Cartesian products $Q_n \times K_{4,4}$. These graphs are Cayley graphs of minimum genus for all finite Hamiltonian p-groups (see, for example, [**73**]). In fact, $\gamma(Q_n \times K_{4,4}) = 1+n2^n$, and $\tilde{\gamma}(Q_n \times K_{4,4}) = 2+n2^{n+1}$. The crossing number of G on N_k is denoted by $\tilde{\nu}_k(G)$.)

Theorem 8.8. *Let* $h = \gamma(Q_n \times K_{4,4}) - m$, *and* $k = \tilde{\gamma}(Q_n \times K_{4,4}) - 2m$, *and let* $n \geqslant 0$; *then*

(*i*) $\nu_h(Q_n \times K_{4,4}) = 4m$, if $0 \leqslant m \leqslant 2^n$;

(*ii*) $\tilde{\nu}_k(Q_n \times K_{4,4}) = 4m$, if $0 \leqslant m < 2^n$. ‖

There is a lower bound for the crossing number of a graph G on a surface S, due to Kainen (see, for example [**44**]); not surprisingly, its derivation uses Euler's formula. This bound is attained in Theorems 8.7 and 8.8 above, but not for the other crossing number equalities given in this section. We define $\nu_S(G)$ to be $\nu_h(G)$ if $S = S_h$, and $\tilde{\nu}_k(G)$ if $G = N_k$, and obtain the following result:

Theorem 8.9. *If* G *is a graph with* $p(\geqslant 3)$ *vertices,* q *edges, and girth* $g(\geqslant 3)$, *then*

$$\nu_S(G) \geqslant q - \frac{g}{g-2}(p - \chi(S)). \; \|$$

We conclude this section by discussing the rectilinear crossing number $\bar{\nu}(G)$. It is well known that a simple planar graph can be embedded in the plane so that all edges are represented by line segments. In contrast, Guy (see [**33**]) confirmed a conjecture of F. Harary and A. Hill by showing that drawings of a given graph in the plane may of necessity have *more* crossings if line segments (rather than just arcs) are required for the edges. To illustrate this, we define the **rectilinear crossing number** $\bar{\nu}(G)$ of a graph G to be the fewest crossings in any drawing of G using only line segments. Guy showed that, although $\bar{\nu}(K_9) = \nu(K_9)$, we have $\bar{\nu}(K_8) = 1 + \nu(K_8)$. He also showed that for $p \geqslant 10$, $\bar{\nu}(K_p) > \nu(K_p)$. D. Singer (unpublished) has found a straight-line drawing of K_{10} with just 62 crossings, so that $\bar{\nu}(K_{10})$ is 61 or 62. For small values of p, the known crossing number results for K_p are given in Table 1.

Table 1. Crossing numbers.

p	4	5	6	7	8	9	10
$\nu(K_p)$	0	1	3	9	18	36	60
$\bar{\nu}(K_p)$	0	1	3	9	19	36	61 or 62

9. Thickness

Another measure of non-planarity which has proved interesting to many graph-theorists is the "thickness" of a graph. The concept originated in a question asked by J. Selfridge: can the complete graph K_9 be expressed as the union of two planar graphs? It is easy to find such a decomposition of K_8, and it is also easy to show that K_{11} cannot be so expressed (since, by Corollary 3.3(*i*), a planar graph with eleven vertices can have at most 27 edges, and K_{11} has 55 edges). That K_9 itself is not the union of two planar graphs was first shown by Battle, Harary and Kodama [**2**]. All known proofs of this result are quite lengthy, and the easiest to follow is probably that of Tutte [**65**].

For a graph G, we define the **thickness** $t(G)$ to be the minimum number of planar subgraphs of G whose union is G (so that, for example, $t(K_9) > 2$). Just as for the genus and crossing number, finding the thickness of an arbitrary graph is a seemingly intractable problem. Also as with the genus and crossing number, the thickness of a graph has applications in the theory of printed circuits (see [**71**, pp. 66–68]). Here we begin by deriving some bounds for the thickness of a graph, and then look at some specific classes of graphs.

Lower bounds for the thickness of a graph are easy consequences of Euler's formula; in particular, we have the following result:

Theorem 9.1. *If G is a graph with p (>2) vertices and q edges, then*

(*i*) $t(G) \geqslant \left\{\dfrac{q}{3p-6}\right\}$;

(*ii*) $t(G) \geqslant \left\{\dfrac{q}{2p-4}\right\}$, *if G has no triangles.* ||

If we apply part (*i*) of this theorem to the complete graph K_p, and then simplify the expression, we find that

$$t(K_p) \geqslant \left\{\frac{\frac{1}{2}p(p-1)}{3(p-2)}\right\} = \left[\frac{p(p-1)+6p-14}{6(p-2)}\right] = \left[\frac{p+7}{6}\right].$$

We have already seen that equality does not hold for $p = 9$ or 10, but that it does hold for $p \leqslant 8$ (since K_4 is planar, and K_5 is not). What is perhaps a bit surprising is that equality holds in all other cases.

Theorem 9.2. *The thickness of the complete graph K_p is $[\frac{1}{6}(p+7)]$, except that* $t(K_9) = t(K_{10}) = 3$. ||

This theorem has a rather lengthy history. The bulk of the result, for $p \not\equiv 4$ (modulo 6), was proved in 1965 by Beineke and Harary [**6**], who gave a decomposition of the octahedral graph $K_{3n(2)}$ into n triangulations of the plane, and modified this to show that $t(K_{6n+3}) \leqslant n+1$. This was then extended to the corresponding results for $p \equiv 0$, 1, 2 and 5 (modulo 6). Over a period of years, the cases $p = 16, 22, 28, 34, 40$ and 46 were handled on an individual basis. Finally, in 1976 Alekseev and Gonchakov [**1**] and Vasak [**66**] independently completed the proof for the remaining values of $p \equiv 4$ (modulo 6). Except for $p = 16$, the constructions are all modifications of the Beineke–Harary decomposition. (Because so many mathematicians were unsuccessful in their attempts to decompose K_{16} into three planar graphs, the consensus of opinion was that $t(K_{16}) = 4$, until Mayer [**47**] proved otherwise.)

As we implied above, the thickness of the octahedral graphs follows from the work on the complete graphs:

Theorem 9.3. *The thickness of the n-dimensional octahedron $K_{n(2)}$ is $\{\frac{1}{3}n\}$.* ‖

The thickness of the complete bipartite graph has also been investigated. From Theorem 9.1(*ii*) one gets a lower bound, and again it has been proved that equality usually holds (see Beineke, Harary and Moon [**9**]). Here there are no known exceptions.

Theorem 9.4. *The thickness of the complete bipartite graph $K_{r,s}$ is*

$$t(K_{r,s}) = \left\{\frac{rs}{2(r+s)-4}\right\},$$

except possibly when r and s are both odd, $r \leqslant s$, and there is an integer k satisfying $\frac{1}{4}(r+5) \leqslant k \leqslant \frac{1}{2}(r-3)$ for which

$$s = \left[\frac{2k(r-2)}{r-2k}\right]. \quad \|$$

The unknown cases are quite rare, and the smallest has 48 vertices. For $r < 30$ and $s \geqslant r$, there are just six unknown cases—namely, $K_{19,29}$, $K_{19,47}$, $K_{23,27}$, $K_{25,59}$, $K_{27,71}$ and $K_{29,129}$. While there are undoubtedly an infinite number of such cases, for given r there cannot be more than $\frac{1}{4}(r-7)$ of them. There does not seem to be a recognizable pattern to these unknown cases, so that any significant progress would probably require a technique different from the basic construction used to establish the theorem.

The only other major family of graphs for which the thickness is known consists of the cube graphs; the following result was established by Kleinert [**45**]:

Theorem 9.5. *The thickness of the n-cube Q_n is $[\frac{1}{4}n]+1$.* ‖

There has also been some investigation of "minimal" graphs of given thickness (that is, graphs all of whose proper subgraphs have smaller thickness), and in particular those which are complete bipartite graphs (see, for example, Bouwer and Broere [**14**]).

Thickness on Other Surfaces

The concept of thickness can readily be extended to surfaces other than the sphere. For a surface S, we define the ***S*-thickness** $t_S(G)$ of a graph G to be the minimum number of S-embeddable graphs whose union is G. Euler's formula can again be used to find lower bounds like those of Theorem 9.1:

Theorem 9.6. *For any surface S, and any graph G with p ($\geqslant 3$) vertices and q edges, we have*

(*i*) $t_S(G) \geqslant \dfrac{q}{3(p-\chi(S))}$;

(*ii*) $t_S(G) \geqslant \dfrac{q}{2(p-\chi(S))}$, *if G has no triangles.* ||

Ringel [**54**] has determined the "toroidal-thickness" of all complete graphs, and Beineke [**6**] independently found this as well as the solutions for the projective plane and double-torus. All three can be found by using the basic construction for the planar thickness of the octahedral graphs, and they are thus easier than the planar thickness of complete graphs.

Theorem 9.7. *The S-thickness of the complete graph K_p ($p > 2$) is*

$[\frac{1}{6}(p+5)]$, *for the projective plane*;

$[\frac{1}{6}(p+4)]$, *for the torus*;

$[\frac{1}{6}(p+3)]$, *for the double-torus.* ||

For the next two surfaces, the Klein bottle and triple-torus, one quickly encounters difficulties. We have already seen that Euler's bound for the genus does not give equality for K_7 and the Klein bottle. The next question is whether K_{13} is the union of two Klein bottle graphs; if so, both are triangulations. For the triple-torus, the question is whether or not K_{16} is the union of two graphs embeddable in that surface; again, both graphs would have to be triangulations. For both surfaces, it is not difficult to establish exact results for "five-sixths" of the complete graphs.

For complete bipartite graphs, the situation is similar to the planar case. For example, one can get a result for the toroidal-thickness, but the unknown cases occur rather more frequently. However, if one considers only the regular complete bipartite graphs $K_{r,r}$, exact results may be found for a number of surfaces (see [**6**]):

Theorem 9.8. *The S-thickness of the regular complete bipartite graph $K_{r,r}$ $(r > 1)$ is*

$[\frac{1}{4}(r+5)]$, *for the plane;*
$[\frac{1}{4}(r+4)]$, *for the projective plane;*
$[\frac{1}{4}(r+3)]$, *for the Klein bottle, torus, and double-torus;*
$[\frac{1}{4}(r+2)]$, *for the triple-torus.* ||

Restricted Planarity

Some investigations have also been made of problems involving decompositions of graphs into special types of planar graphs. The only one we consider in any detail could be called the "tree-thickness", but isn't!—the **arboricity** of a graph G is the minimum number of forests whose union is G. A remarkable theorem of Nash-Williams [**48**] gives the arboricity of all graphs. Clearly the arboricity of a graph with p vertices and q edges is at least $\{q/(p-1)\}$, and is also at least that of any subgraph. These facts give a lower bound for the arboricity of G, which turns out to be exact in all cases:

Theorem 9.9. *The arboricity of a graph G is*

$$\max_{2 \leqslant k \leqslant p} \left\{\frac{q_k}{k-1}\right\},$$

where q_k is the maximum number of edges in any k-vertex subgraph of G. ||

Other variations of thickness which have been studied include the "vertex-arboricity" (see, for example, Chartrand and Kronk [**16**]), the "bipartite-thickness" (Walther [**67**]), the "outerplanar-thickness" (Guy [**34**]), and the "book-thickness" (Bernhart and Kainen [**11**]). Space does not permit us to describe these or other variations of the topics in this chapter.

References

1. V. B. Alekseev and V. S. Gonchakov, Thickness of arbitrary complete graphs, *Mat. Sbornik* **101** (**143**) (1976), 212–230.
2. S. R. Alpert, Two-fold triple systems and graph imbeddings, *J. Combinatorial Theory* (*A*) **18** (1975), 101–107; *MR***51**#185.
3. L. Auslander, T. A. Brown and J. W. T. Youngs, The imbedding of graphs in manifolds, *J. Math. Mech.* **12** (1963), 629–634; *MR***27**#1940.
4. J. Battle, F. Harary and Y. Kodama, Every planar graph with nine points has a non-planar complement, *Bull. Amer. Math. Soc.* **68** (1962), 569–571; *MR***27**#5248.
5. J. Battle, F. Harary, Y. Kodama and J. W. T. Youngs, Additivity of the genus of a graph, *Bull. Amer. Math. Soc.* **68** (1962), 565–568; *MR***27**#5247.
6. L. W. Beineke, Minimal decompositions of complete graphs into subgraphs with embeddability properties, *Canad. J. Math.* **21** (1969), 992–1000; *MR***39**#6783.
7. L. W. Beineke and F. Harary, The genus of the *n*-cube, *Canad. J. Math.* **17** (1965), 494–496; *MR***31**#81.

8. L. W. Beineke and F. Harary, The thickness of the complete graph, *Canad. J. Math.* **21** (1969), 850–859; *MR***32**#4032.
9. L. W. Beineke, F. Harary and J. W. Moon, On the thickness of the complete bipartite graph, *Proc. Cambridge Phil. Soc.* **60** (1964), 1–6; *MR***28**#1611.
10. L. W. Beineke and R. D. Ringeisen, On crossing numbers of certain products of graphs (to appear).
11. F. Bernhart and P. Kainen, The book thickness of a graph, *J. Combinatorial Theory (B)* (to appear).
12. J. Blažek and N. Koman, A minimal problem concerning complete plane graphs, in *Theory of Graphs and its Applications* (ed. M. Fiedler), Czechoslovak Academy of Sciences, Prague, 1964, pp. 113–117; *MR***30**#4249.
13. A. Bouchet, Triangular imbeddings into surfaces of a join of equicardinal independent sets following an Eulerian graph, in *Theory and Applications of Graphs*, Lecture Notes in Mathematics **642** (ed. Y. Alavi and D. R. Lick), Springer-Verlag, Berlin, Heidelberg and New York, 1978, pp. 86–115.
14. I. Z. Bouwer and I. Broere, Note on t-minimal complete bipartite graphs, *Canad. Math. Bull.* **11** (1968), 729–732; *MR***39**#1347.
15. H. R. Brahana, Systems of circuits of two-dimensional manifolds, *Ann. of Math.* **30** (1923), 234–243.
16. G. Chartrand and H. V. Kronk, The point-arboricity of planar graphs, *J. London Math. Soc.* **44** (1969), 612–616; *MR***39**#1350.
17. R. A. Duke, The genus, regional number, and Betti number of a graph, *Canad. J. Math.* **18** (1966), 817–822; *MR***33**#4917.
18. J. R. Edmonds, A combinatorial representation for polyhedral surfaces, *Notices Amer. Math. Soc.* **7** (1960), 646.
19. R. Eggleton, Crossing numbers of graphs, Ph.D. thesis, University of Calgary, 1973.
20. P. Erdős and R. K. Guy, Crossing number problems, *Amer. Math. Monthly* **80** (1973), 52–58; *MR***52**#2894.
21. P. Franklin, A six-color problem, *J. Math. Phys.* **13** (1934), 363–369.
22. B. L. Garman, Cayley graph imbeddings and the associated block designs, Ph.D. thesis, Western Michigan University, 1976.
23. B. L. Garman, R. D. Ringeisen and A. T. White, On the genus of strong tensor products of graphs, *Canad. J. Math.* **28** (1976), 523–532; *MR***54**#123.
24. H. H. Glover and J. P. Huneke, There are finitely many Kuratowski graphs for the projective plane, in *Graph Theory and Related Topics* (ed. J. A. Bondy and U. S. R. Murty), Academic Press, New York, 1978 (to appear).
25. H. H. Glover, J. P. Huneke and C. S. Wang, 103 graphs which are irreducible for the projective plane, *J. Combinatorial Theory (B)* (to appear).
26. J. L. Gross, Voltage graphs, *Discrete Math.* **9** (1974), 239–246; *MR***50**#153.
27. J. L. Gross, An infinite family of octahedral crossing numbers, *J. Graph Theory* **2** (1978), 171–178.
28. J. L. Gross and S. R. Alpert, Branched coverings of graph imbeddings, *Bull. Amer. Math. Soc.* **79** (1973), 942–945; *MR***48**#10870.
29. J. L. Gross and S. R. Alpert, Components of branched coverings of current graphs. *J. Combinatorial Theory (B)* **20** (1976), 283–303; *MR***54**#7301.
30. W. Gustin, Orientable embeddings of Cayley graphs, *Bull. Amer. Math. Soc.* **69** (1963), 272–275; *MR***26**#3037.
31. R. K. Guy, The decline and fall of Zarankiewicz's theorem, in *Proof Techniques in Graph Theory* (ed. F. Harary), Academic Press, New York, 1969, pp. 63–69; *MR***40**#7144.
32. R. K. Guy, Sequences associated with a problem of Turán and other problems, in *Combinatorial Theory and its Applications, II* (ed. P. Erdős *et al.*), North-Holland, Amsterdam, 1970, pp. 553–569.
33. R. K. Guy, Crossing numbers of graphs, in *Graph Theory and Applications*, Lecture Notes in Mathematics **303** (ed. Y. Alavi *et al.*), Springer-Verlag, Berlin, Heidelberg and New York, 1972, pp. 111–124; *MR***49**#2442.
34. R. K. Guy, Outerthickness and outercoarseness of graphs, in *Combinatorics*, London

Mathematical Society Lecture Notes **13**, Cambridge University Press, London 1974, pp. 57–60; *MR*50#15.
35. R. K. Guy and T. A. Jenkyns, The toroidal crossing number of $K_{m,n}$, *J. Combinatorial Theory* **6** (1969), 235–250; *MR*38#5660.
36. H. Hanani, Balanced incomplete block designs and related designs, *Discrete Math.* **11** (1975), 255–369; *MR*52#2918.
37. L. Heffter, Über das Problem der Nachbargebiete, *Math. Ann.* **38** (1891), 477–508.
38. A. Jacques, Constellations et propriétés algèbriques des graphes topologiques, Ph.D. thesis, University of Paris, 1969.
39. M. Jungerman, The genus of the symmetric quadripartite graph *J. Combinatorial Theory* (*B*) **19** (1975), 181–187; *MR*52#1345.
40. M. Jungerman, The non-orientable genus of the n-cube, *Pacific J. Math.* (to appear).
41. M. Jungerman, A characterization of upper-embeddable graphs, *Trans. Amer. Math. Soc.* (to appear).
42. M. Jungerman and G. Ringel, The genus of the n-octahedron: regular cases, *J. Graph Theory* **2** (1978), 69–75.
43. I. N. Kagno, The triangulation of surfaces and the Heawood color formula, *J. Math. Phys.* **15** (1936), 179–186.
44. P. C. Kainen and A. T. White, On stable crossing numbers, *J. Graph Theory* **2** (1978), 181–187
45. M. Kleinert, Die Dicke des n-dimensionalen Würfel-Graphen, *J. Combinatorial Theory* **3** (1967), 10–15; *MR*35#2776.
46. D. J. Kleitman, The crossing number of $K_{5,n}$, *J. Combinatorial Theory* **9** (1970), 315–323; *MR*43#2776.
47. J. Mayer, Décomposition de K_{16} en trois graphes planaires, *J. Combinatorial Theory* (*B*) **13** (1972), 71; *MR*47#1672.
48. C. St. J. A. Nash-Williams, Edge-disjoint spanning trees of finite graphs, *J. London Math. Soc.* **36** (1961), 445–450; *MR*24#3087.
49. R. D. Ringeisen, Upper and lower embeddable graphs, in *Graph Theory and Applications*, Lecture Notes in Mathematics **303** (ed. Y. Alavi *et al.*), Springer-Verlag, Berlin, Heidelberg and New York 1972, pp. 261–268; *MR*48#10872.
50. G. Ringel, Bestimmung der Maximalzahl der Nachbargebiete auf nichtorientierbaren Flächen, *Math. Ann.* **127** (1954), 181–214; *MR*15–245.
51. G. Ringel, Über drei kombinatorische Probleme am n-dimensionalen Würfel und Würfelgitter, *Abh. Math. Sem. Univ. Hamburg* **20** (1955), 10–19; *MR*17#772.
52. G. Ringel, *Färbungsprobleme auf Flächen und Graphen*, VEB Deutscher Verlag der Wissenschaften, Berlin (1959); *MR*22#235.
53. G. Ringel, Das Geschlecht des vollständigen paaren Graphen, *Abh. Math. Sem. Univ. Hamburg* **38** (1965), 139–150; *MR*32#6439.
54. G. Ringel, Die toroidale Dicke der vollständigen Graphen, *Math. Z.* **87** (1965), 19–26; *MR*30#2489.
55. G. Ringel, Der vollständige paare Graph auf nichtorientierbaren Flächen, *J. Reine Angew. Math.* **220** (1965), 89–93; *MR*32#445.
56. G. Ringel, *Map Color Theorem*, Springer-Verlag, Berlin, 1974; *MR*50#1955.
57. G. Ringel, The combinatorial map color theorem, *J. Graph Theory* **1**, (1977), 141–155.
58. G. Ringel, Non-existence of graph embeddings, in *Theory and Applications of Graphs*, Lecture Notes in Mathematics **642** (ed. Y. Alavi and D. R. Lick), Springer-Verlag, Berlin, Heidelberg and New York, 1978, pp. 465–476.
59. G. Ringel and J. W. T. Youngs, Solution of the Heawood map-coloring problem, *Proc. Nat. Acad. Sci. U.S.A.* **60** (1968), 438–445; *MR*37#3959.
60. G. Ringel and J. W. T. Youngs, Das Geschlecht des vollständigen drei-farbaren Graphen, *Comment. Math. Helv.* **45** (1970), 152–158; *MR*41#6724.
61. S. Stahl, Self-dual embeddings of graphs, Ph.D. thesis, Western Michigan University, 1975.
62. S. Stahl, Generalized embedding schemes, *J. Graph Theory* **2** (1978), 41–52.

63. S. Stahl and L. W. Beineke, Blocks and the nonorientable genus of graphs, *J. Graph Theory* **1** (1977), 75–78.
64. S. Stahl and A. T. White, Genus embeddings for some complete tripartite graphs, *Discrete Math.* **14** (1976), 279–296; *MR***54**#10060.
65. W. T. Tutte, On the non-biplanar character of the complete 9-graph, *Canad. Math. Bull.* **6** (1963), 319–330; *MR***28**#253.
66. J. M. Vasak, The thickness of the complete graph, *Notices Amer. Math. Soc.* **23** (1976), A-479.
67. H. Walther, Über die Zerlegung des vollständigen Graphen in paare planare Graphen, in *Beitrage zur Graphentheorie* (ed. H. Sachs *et al.*), Teubner-Verlag, Leipzig, 1968, pp. 189–205; *MR***39**#6777.
68. A. T. White, The genus of the complete tripartite graph $K_{mn,n,n}$, *J. CombinatorialTheory* **7** (1969), 283–285; *MR***39**#6778.
69. A. T. White, The genus of repeated Cartesian products of bipartite graphs, *Trans. Amer. Math. Soc.* **151** (1970), 393–404; *MR***43**#7368.
70. A. T. White, On the genus of a group, *Trans. Amer. Math. Soc.* **173** (1972), 203–214; *MR***47**#6529.
71. A. T. White, *Graphs, Groups and Surfaces*, North-Holland, Amsterdam, 1973; *MR***49** #4783.
72. A. T. White, Block designs and graph imbeddings, *J. Combinatorial Theory* (*B*) (to appear).
73. A. T. White, Graphs of groups on surfaces, in *Combinatorial Surveys: Proceedings of the Sixth British Combinatorial Conference* (ed. P. J. Cameron), Academic Press, London, 1977, pp. 165–197.
74. N. H. Xuong, Sur quelques classes de graphes possédant des propriétés topologiques remarquables (English summary), *C.R. Acad. Sci. Paris* (*A–B*) **283** (1976), A813–A816, *MR***54**#12560.
75. N. H. Xuong, How to determine the maximum genus of a graph, *J. Combinatorial Theory* (*B*) (to appear).
76. J. W. T. Youngs, Minimal imbeddings and the genus of a graph, *J. Math. Mech.* **12** (1963), 303–315; *MR***26**#3043.
77. J. W. T. Youngs, The Heawood map-coloring conjecture, in *Graph Theory and Theoretical Physics* (ed. F. Harary), Academic Press, New York, 1967, pp. 313–354; *MR***38** #4357.

3
The Proof of the Heawood Conjecture

ARTHUR T. WHITE

1. Introduction

Map-coloring problems are notoriously easy to formulate, yet difficult to solve. Thus they have been intriguing, even for the novice, while frustrating, even for the expert. A map-coloring problem for a specific map generally takes the following form: the map is drawn on some surface, and it is desired to color the countries of the map so as to distinguish two countries whenever they have a stretch of their borders in common; the challenge is to minimize the number of colors required for this purpose.

Our problem in this chapter is to determine for each surface S_h (the sphere with h handles attached) the minimum number of colors required so that *every* map that might be drawn on the surface could be colored with that many colors; this number is called the **chromatic number** of S_h, and is written $\chi(S_h)$. Specifically, $\chi(S_h)$ is the *maximum* value of $\chi(M)$ for all maps M on S_h, where $\chi(M)$ is the *minimum* number of colors required for M.

It is a curiosity in mathematics that occasionally a problem apparently several orders of magnitude more difficult than a given problem nevertheless has a more ready solution. Such is the case here: one might expect that $\chi(S_0)$, the chromatic number of the sphere, would be relatively simple to determine, whereas each $\chi(S_h)$, for positive h, might be more difficult. On the contrary, the values of $\chi(S_h)$ for all positive h were determined by one theorem, whose proof was completed in 1968, whereas the single case $h = 0$ was not settled until 1976. The former result is known as the "Heawood map-coloring theorem", whereas the corresponding result for the sphere is the celebrated 'four-color theorem", discussed in the next chapter.

The **Heawood map-coloring theorem** states that

$$\chi(S_h) = [\tfrac{1}{2}(7+\sqrt{1+48h})], \qquad \text{for} \qquad h = 1, 2, \ldots, \tag{1}$$

where $[x]$ denotes the greatest integer not exceeding x. For notational convenience, we let $H(h)$ represent the right-hand side of (1). Since the four-color theorem tells us that $\chi(S_0) = 4$, and since $H(0) = 4$, we observe that (1) extends to all non-negative integers h; nevertheless, the Heawood map-coloring theorem is generally taken to exclude the case $h = 0$, and it is to the proof of this theorem that we devote this chapter. We first discuss the history of the problem, then its solution, and finally some related work stimulated by the original conjecture and its proof.

The author would like to thank B. L. Garman, J. L. Gross, M. Jungerman, H. V. Kronk, G. Ringel and S. Stahl for their helpful comments and suggestions regarding this chapter.

2. History

Interest in the four-color problem dates back to the mid-nineteenth century, and a "proof" by A. B. Kempe [**19**] in 1879 was accepted by the mathematical community until a fundamental error was pointed out by P. J. Heawood [**15**] in 1890 (see Chapter 4). Map-coloring results are established in two steps: in order to show that $\chi(S_h) = m$, say, the problem-solver must both

(*i*) show that $\chi(S_h) \leqslant m$, by some theoretical argument, and then

(*ii*) show that $\chi(S_h) \geqslant m$, by drawing an actual map M on S_h with chromatic number $\chi(M) = m$.

Kempe used an easy example (the tetrahedral map) to show that $\chi(S_0) \geqslant 4$, but his theoretical argument to show that $\chi(S_0) \leqslant 4$ was in error.

In his 1890 paper, Heawood correctly showed that $\chi(S_h) \leqslant H(h)$, for all positive h. (He was apparently motivated, in part, by the hope that his methods would carry over for $h = 0$ also; they do not, as we shall see in Section 4.) Heawood then produced a map on the torus showing that

$\chi(S_1) \geqslant H(1) = 7$ (see Fig. 1), and incorrectly claimed that constructions automatically existed to show that $\chi(S_h) \geqslant H(h)$ for all $h \geqslant 2$:

> . . . a single surface such as a sphere is the only one . . . where we consequently do not get an immediate definite solution of the problem,—apart from the verification figure, which we have indeed only given for the case of an anchor ring [torus], but for the more highly connected surfaces it will be observed that there are generally contacts enough and to spare for the above number of divisions each to touch each.

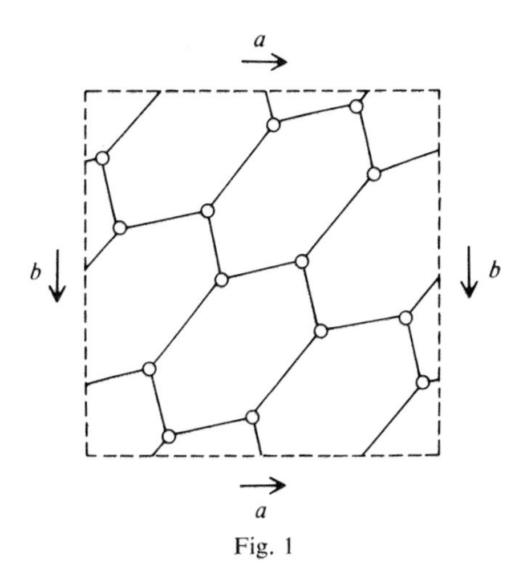

Fig. 1

Heawood apparently believed that any map compatible with Euler's formula would actually exist, in spite of his own statements such as "verification, by an actual figure, is requisite", and "duplicate contact between some pair of counties . . . would necessitate that some other pair of counties should fail to touch" (see [**15**, p. 335]). Franklin [**8**] dispelled this notion in 1934, by showing that the Klein bottle does not have chromatic number 7, even though this number is allowed by Euler's formula. (Thus Heawood erred in step (*ii*) described above, whereas Kempe erred in step (*i*), for their respective proofs.)

L. Heffter was aware of the shortcoming of Heawood's proof; in 1891 he commented [**16**]:

> . . . except in the case of genus one, the proof of Mr. Heawood is deficient, that on a surface of arbitrary genus the maximum number of neighboring regions actually occurs. For without this proof, it is not certain that both numbers agree; it could of course be that fewer neighboring regions occur on a surface than the maximum number of the upper bound. This deficit is however essential; the omitted proof is by no means obvious.

We comment that, at this stage in the history of the problem, the construction of a neighboring system of $H(h)$ regions (each region bordering on all the others) was clearly sufficient, but not known to be necessary, to complete the proof. In this connection, see Theorem 12.1.

Just as Heawood's disclaimer of Kempe's proof escaped notice in many quarters (see [**2**, p. 107]), so also did Heffter's denial of Heawood's proof for the lower bound. In 1941 the following statement appeared in Courant and Robbins [**5**, p. 248]: ". . . for surfaces more complicated than the plane or the sphere the corresponding theorems have actually been proved . . ."; this perhaps contributed to a false impression that Heawood's conjecture had already been established for all $h \geqslant 1$.

Many mathematicians assisted in completing the proof, with the bulk of the work being done by G. Ringel and J. W. T. Youngs; their announcement of the final solution came in 1968 [**29**], in the eighth decade following Heawood's paper. Other significant contributors include Heffter and Gustin, as we shall see.

During the period 1968 to 1976, the surface chromatic number $\chi(S_h)$ was known for all h except for $h = 0$. Finally, Appel and Haken [**1**] established the four-color theorem by showing that $\chi(S_0) \leqslant 4$—the (very large) step that Kempe had treated incorrectly. The work of Appel and Haken is treated in Chapter 4; here we shall discuss the Ringel–Youngs solution of the Heawood map-coloring theorem.

3. The Dual Formulation

One contribution of Heffter, incorporated by Ringel and Youngs into their final solution, was to focus attention not on the map drawn on a given surface, but on the *dual graph* of the map. For example, Fig. 2 again shows the

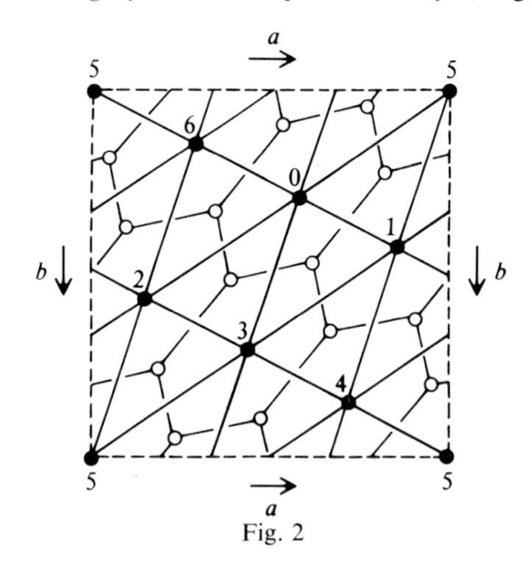

Fig. 2

Heawood map for the torus; a neighboring system of $H(1) = 7$ regions is shown with hollow vertices and dashed edges, and the dual graph K_7 is indicated by solid vertices and edges. (The vertex labels for K_7 will be used again in Section 9.)

Instead of seeking a neighboring system of $H(h)$ regions on S_h, Heffter sought an embedding of the complete graph $K_{H(h)}$ on S_h, and thus began the investigation of the genus parameter γ for K_p (see Chapter 2). One advantage of dualizing in this fashion is that Euler's formula can be conveniently applied, as we shall see in Sections 4 and 6. Heffter found the value of $\gamma(K_p)$ for $p \leqslant 12$ and for $p = 12s+7$, where $4s+3 = q$ is prime, and the order of 2 in the multiplicative group of $\mathbf{Z}_q$ is either $q-1$ or $\frac{1}{2}(q-1)$. (It is not known whether this is an infinite class of numbers.) He also gave a lower bound for $\gamma(K_p)$ in general, together with information about the regions that would result from a conjectured minimal embedding of K_p (see Section 6), and he initiated the use of rotation schemes for describing 2-cell embeddings of graphs algebraically (as further developed by Ringel in much of his work, and by Edmonds and Youngs; see Section 3 of Chapter 2).

The Heawood map-coloring theorem, in its dual formulation, is then:

$$\chi(S_h) = H(h), \qquad \text{for} \qquad h = 1, 2, \ldots, \tag{1*}$$

where $\chi(S_h)$ is the maximum chromatic number of any graph embeddable on the surface S_h. (The chromatic number of a map on S_h will equal that of its dual graph; conversely, the chromatic number of a graph G embedded on S_h will be that of the map formed on S_h by taking the dual of G.)

4. The Upper Bound

In this section we shall prove that $\chi(S_h) \leqslant H(h)$, if h is positive. We shall use Euler's formula $p-q+r = 2-2h$, for a 2-cell embedding of a graph G in the surface S_h with p vertices, q edges and r regions (see Section 3 of Chapter 2).

Lemma 4.1. *Let G be a graph of order p ($\geqslant 3$) which is 2-cell embedded in the surface S_h, and let a denote the average valency of the vertices of G. Then*

$$a \leqslant 6 - \frac{12-12h}{p}.$$

Proof. Since $3r \leqslant 2q$, and $p-q+r = 2-2h$, we have

$$q \leqslant 3q-3r = 3p+6h-6.$$

It follows that

$$a = \frac{2q}{p} \leqslant 6 - \frac{12-12h}{p}. \quad \|$$

Now let $f(h) = \frac{1}{2}(7+\sqrt{1+48h})$, so that $H(h) = [f(h)]$.

Theorem 4.2. *$\chi(S_h) \leqslant H(h)$, for $h > 0$.*

Proof. We use induction on p to show that $\chi(G) \leqslant H(h)$ for all graphs G of order p embedded in S_h. It is immediate that $\chi(G) \leqslant H(h)$, for $p \leqslant H(h)$. Thus we assume that $p > H(h)$, and that $\chi(G) \leqslant H(h)$ for all graphs G embedded in S_h with order less than p.

Consider now an embedding of a graph G of order p in S_h. From the definition of $f(h)$, it follows that $f^2(h) - 7f(h) - 12h + 12 = 0$, so that

$$6 - \frac{12-12h}{f(h)} = f(h) - 1.$$

If G is not connected, then each component has chromatic number at most $H(h)$, by the induction hypothesis, so that $\chi(G) \leqslant H(h)$ also. Thus we may assume G to be connected. If the embedding of G in S_h is a 2-cell embedding, then Lemma 4.1 applies, so that

$$\begin{aligned} a &\leqslant 6 - \frac{12-12h}{p} \\ &\leqslant 6 - \frac{12-12h}{f(h)} \\ &= f(h) - 1, \end{aligned}$$

where the second inequality depends upon h being *positive*. (This is precisely where Heawood's proof fails for the sphere.) If the embedding is not a 2-cell embedding, then it is not a minimal embedding, since G is connected (see Theorem 5.1 of Chapter 2), and we can find a 2-cell embedding of G in $S_{h'}$, for some $h' < h$. If $h' = 0$, then $a < 6 \leqslant f(h) - 1$, by Lemma 4.1 and the fact that $f(h) \geqslant 7$ for $h \geqslant 1$. If $h' > 0$, then we apply Lemma 4.1 as above, to find $a \leqslant f(h') - 1 \leqslant f(h) - 1$. Thus in every case $a \leqslant f(h) - 1$, so that we can find a vertex v of G whose valency is at most $H(h) - 1$. Then $\chi(G - v) \leqslant H(h)$ by the induction hypothesis, and we use at most $H(h) - 1$ of the $H(h)$ colors for vertices adjacent to v; we therefore have a color available for v, so that $\chi(G) \leqslant H(h)$. $\|$

This concludes the easy part of the proof of Heawood's theorem, the part provided by Heawood himself. We now turn our attention to the lower bound, still using the dual formulation.

5. The Lower Bound is Provided by the Complete Graph Theorem

In this short section we show that the complete graph theorem suffices to establish the lower bound $\chi(S_h) \geqslant H(h)$. The **complete graph theorem** is that the genus of the complete graph of order p is given by

$$\gamma(K_p) = \{\tfrac{1}{12}(p-3)(p-4)\}$$

for $p \geqslant 3$, where $\{x\}$ denotes the least integer not less than x. Thus we temporarily *assume* this genus result (in subsequent sections we shall discuss its proof), and we see that it is just what is needed to complete the proof of the Heawood map-coloring theorem.

Theorem 5.1. *$\chi(S_h) \geqslant H(h)$, for $h \geqslant 0$.*

Proof. Consider the surface S_h; let $p = H(h)$ and consider also $S_{\gamma(K_p)}$. Then, for $h \geqslant 0$, so that $p \geqslant 4$,

$$\begin{aligned} \gamma(K_p) &= \{\tfrac{1}{12}(p-3)(p-4)\} \\ &= \{\tfrac{1}{12}[\tfrac{1}{2}(\sqrt{1+48h}+1)][\tfrac{1}{2}(\sqrt{1+48h}-1)]\} \\ &\leqslant h, \end{aligned}$$

since $[x] \leqslant x$ for all x. Thus $\chi(S_{\gamma(K_p)}) \leqslant \chi(S_h)$, since $\chi(S_h)$ is a non-decreasing function. By definition, K_p can be embedded in $S_{\gamma(K_p)}$, and $\chi(K_p) = p$, so that

$$\chi(S_h) \geqslant \chi(S_{\gamma(K_p)}) \geqslant \chi(K_p) = p = H(h). \quad \|$$

This connection between the Heawood map-coloring theorem and the complete graph theorem, first observed by Heffter and heavily exploited by Ringel and Youngs, motivates the lavish attention paid to the latter theorem, as we now indicate.

6. The Complete Graph Theorem—Lower Bound

Let us review our position. We readily established the upper bound for the Heawood theorem, using Euler's formula for the dual formulation. The lower bound for the Heawood theorem is much more difficult to obtain, and its proof uses the complete graph theorem, still in the dual formulation. In contrast, the lower bound for the complete graph theorem is easy to come by, using Euler's formula; the upper bound will require a substantial effort.

We let r_i denote the number of regions in a 2-cell embedding of a graph G having as boundary a closed walk of length i in G, where $i \geqslant 3$ for 2-cell embeddings of graphs having order at least three.

Theorem 6.1. *Let K_p be minimally embedded in S_h, for $p \geqslant 3$; then*

$$\gamma(K_p) = h = \tfrac{1}{12}(p-3)(p-4)+\tfrac{1}{6}\sum_{i\geqslant 4}(i-3)r_i.$$

Proof. If K_p is minimally embedded in S_h, with $p \geqslant 3$, then $\gamma(K_p) = h$, and the embedding is a 2-cell embedding, by Theorem 5.1 of Chapter 2. Thus Euler's formula applies, and we have

$$p-q+r = 2-2h. \tag{i}$$

But $r = r_3+r_4+r_5+\ldots$, and $2q = 3r_3+4r_4+5r_5+\ldots$, so that

$$\sum_{i\geqslant 4}(i-3)r_i = 2q-3r. \tag{ii}$$

The result follows on multiplying both sides of (i) by 3 and using (ii), with $q = \frac{1}{2}p(p-1)$. ‖

Corollary 6.2. $\gamma(K_p) \geqslant \{\frac{1}{12}(p-3)(p-4)\}$ *for* $p \geqslant 3$; *moreover, equality holds if and only if an embedding can be found with* $\sum_{i\geqslant 4}(i-3)r_i \leqslant 5$. ‖

It is clear from Theorem 6.1 that triangular embeddings are possible only when $\frac{1}{12}(p-3)(p-4)$ is an integer—that is, when $p \equiv 0, 3, 4$ or 7 (modulo 12). For other values of p, the second inequality of Corollary 6.2 determines just how close to a triangulation the embedding must be; this was first observed by Heffter [**16**], and will be crucial in what follows.

7. The Complete Graph Theorem—Upper Bound, in Twelve Cases

The analysis in Section 6 suggests that a study of the upper bound might naturally be split into twelve cases, depending upon the residue class of p (modulo 12), and this is indeed the situation. We remark at the outset that, in addition to the original references given in our case-by-case analysis below, a thorough treatment of all twelve cases will be found in Ringel's fine book *Map Color Theorem* [**28**]. We refer to the case $p \equiv k$ (modulo 12) as *Case k*, $k = 0, 1, 2, \ldots, 11$.

In each case the analysis will begin with the **regular part** of the problem: *to construct a triangular embedding for K_p, if possible, or to construct a triangular embedding of a closely-related graph G_p otherwise.* In the previous section we saw that such triangular embeddings of K_p are possible only in Cases 0, 3,

4 and 7. These constructions use the "theory of current graphs", as applicable to minimal embeddings of K_p. We shall discuss this theory in Section 8, and apply it to Cases 7 and 10 in Section 9. (An alternate viewpoint, that of voltage graphs, will be provided in Section 10.) In the present section where, for each case, we give the solution of the regular part of the problem only in outline, it is enough to know that a current graph uses a group Γ of order n, together with a subgroup Ω of Γ, in order to embed a graph G of order n; the index of Ω in Γ is called the **index** of the embedding, or of the current graph, or of the solution for that case. Each right coset of Ω in Γ determines one row of a rotation scheme for G (see Section 3 of Chapter 2), and these $[\Gamma : \Omega]$ rows generate an entire rotation scheme and hence a 2-cell embedding of G in an orientable surface. (If the current graph is appropriately chosen, this embedding will be triangular.) Thus the higher the index is, the more complicated the solution is; fortunately, in several of the cases, an index 1 solution has been found, so that $\Omega = \Gamma$, and the top row of the rotation scheme determines the whole scheme.

In the cases $p \not\equiv 0$, 3, 4 or 7 (modulo 12), no triangular embedding is possible, and the analysis continues with the **additional adjacency part** of the problem: *to modify the triangular embedding of* G_p *so as to obtain a minimal embedding for* K_p. In practice the additional adjacency part is more naturally handled first, so that the problem-solver can ascertain what boundary conditions should be sought on the regular part of the problem. In this section we outline the additional adjacency constructions in all cases where such are necessary; in Section 9 we shall treat Case 10 thoroughly. We begin our analysis with Case 1, saving Case 0 (which is atypical) for the end.

Case 1

The group $\Gamma = \mathbf{Z}_{12s-2}$ is used to obtain an index 1 triangular embedding of $K_{12s+1} - K_3$, for $s \geqslant 2$. If s is even, the solution is uncomplicated, and was found in 1963; it appears, together with a separate treatment for s odd, in [**39**]. In 1964 Gustin found a unified approach for all $s \geqslant 2$, using the fact that $\mathbf{Z}_{12s-2} = \mathbf{Z}_2 \times \mathbf{Z}_{6s-1}$; this solution is found in [**28**]. In each case the three missing edges are added using one extra handle (after suitable modifications), giving a genus embedding of K_{12s+1} with $r_4 = r_5 = 1$. (By implication, $r_3 = r-2$.)

Case 2

If s is odd, the group $\Gamma = \mathbf{Z}_{12s}$ was shown by Ringel [**30**] to give an index 2 triangular embedding of $K_{12s+2}-K_2$. The missing edge is added using one extra handle to obtain a genus embedding of K_p with $r_8 = 1$. At first, no extension of this method was found for s even; instead, Ringel and Youngs, in 1967, decided to seek an index 1 solution, with a more difficult additional adjacency problem. The group $\Gamma = \mathbf{Z}_{12s-6}$, where $s \geqslant 1$, is employed to produce an index 1 triangular embedding of the graph L_{12s+2}, which has two vertices whose identification would give $K_{12s+2}-K_8$. This identification and the addition of the missing 28 edges are accommodated over six extra handles, to obtain a genus embedding of K_p. The region distribution that results from this process is not uniquely determined (see [**31**]). In 1974 Jungerman [**18**] found an index 2 triangular embedding for $K_{12s+2}-K_2$ for all $s \geqslant 1$, giving, upon the addition of one handle for the missing edge, a genus embedding of K_{12s+2} with $r_8 = 1$.

Case 3

Ringel found the first solution for this case in 1961 [**27**]. In 1963 Gustin [**14**] announced an index 3 solution, using the group $\Gamma = \mathbf{Z}_{12s+3}$ to produce a triangular embedding of K_{12s+3} directly; this solution appears in [**40**].

Case 4

In 1963 Gustin [**14**] announced a solution for this case; a simpler solution was found by Terry, Welch and Youngs the same year, but not published. The solution given in [**36**] uses the group $\Gamma = \mathbf{Z}_2 \times \mathbf{Z}_{6s+2}$ to find an index 1 triangular embedding of K_{12s+4}.

Case 5

Ringel solved Case 5 in 1954 [**25**]; this was the first case to be treated completely. Without explicitly using current graphs, and while seeking a *nonorientable* triangular embedding of $K_{12s+5}-K_2$, Ringel found an orientable triangular embedding of this graph as a bonus. The solution found in [**40**] is patterned after that of Case 3, using the group $\Gamma = \mathbf{Z}_{12s+3}$ to find an index 3 triangular embedding, but now of $K_{12s+5}-K_2$. The missing edge is added using one extra handle, giving a genus embedding of K_{12s+5} with $r_8 = 1$ (see Fig. 3; the eight-sided region has clockwise boundary $a, d, f, e, d, a, c, b, a$).

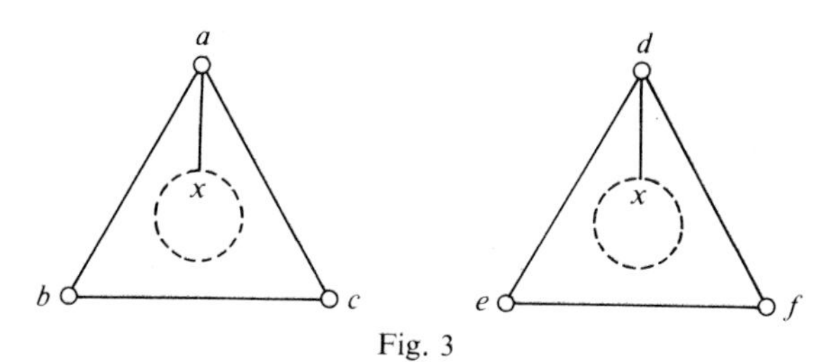

Fig. 3

Case 6

Youngs found the first solution to this case, for $s \geqslant 3$, in 1966; he used the group $\Gamma = \mathbf{Z}_{12s+3}$ to find an index 3 triangular embedding of $K_{12s+6} - K_3$. The analysis splits into four subcases, depending upon the residue class of s (modulo 4). (A unified solution appears in [**40**].) The three missing edges are then added using an extra handle (after suitable modifications), giving a genus embedding of K_{12s+6} with $r_4 = r_5 = 1$.

Case 7

As indicated in Section 3, Heffter found triangular embeddings of K_{12s+7} for certain values of s. Ringel [**27**] solved this case completely, in 1961. Gustin announced a current graph solution in 1963 [**14**], as detailed in [**39**]; the group $\Gamma = \mathbf{Z}_{12s+7}$ is used to obtain an index 1 triangular embedding of K_{12s+7}, $s \geqslant 0$. (This is carried out in full in Section 9.)

Case 8

Here the natural approach would be to start by obtaining a triangular embedding of $K_{12s+8} - K_5$; this has been done, but no solution has been found for the corresponding additional adjacency problem. Instead, the group $\Gamma = \mathbf{Z}_{12s+6}$ is used to find an index 1 triangular embedding of the graph M_{12s+8}, which has two vertices whose identification would give $K_{12s+8} - K_2$. This identification and the addition of the missing edge are accomplished with the addition of one handle; the result is a genus embedding of K_{12s+8} with either $r_4 = 2$ or $r_5 = 1$ (both can be constructed). The solution when s is even was found by Ringel and Youngs in 1967; later the same year a three-subcase solution was obtained for s odd. A unified solution was found during the next year [**32**] for $s \neq 0$, 1 or 3. The case $s = 3$ was treated separately, as it is in [**28**].

Case 9

The first complete solution for this case was found in 1965 by Gustin, but remains unpublished. In 1969 Ringel and Youngs [**30**] found an index 2 solution for s even. The index 3 solution given in [**40**] uses the group $\Gamma = \mathbf{Z}_{12s+6}$ ($s \geqslant 1$) to find a triangular embedding of $K_{12s+9} - K_3$. After suitable modifications, the three missing edges are added on one handle to give a genus embedding of K_{12s+9} with $r_4 = r_5 = 1$. The index 1 solution, due to Jungerman and presented in [**28**], uses the group $\Gamma = \mathbf{Z}_{12s+8}$ to find a triangular embedding of the graph N_{12s+8}, which has two vertices whose identification would produce K_{12s+9}. This identification is readily accomplished using one handle to give a genus embedding of K_{12s+9} with $r_6 = 1$ (see Fig. 4; the region on the handle has clockwise boundary a, b, c, d, e, f, a, where $d = a$).

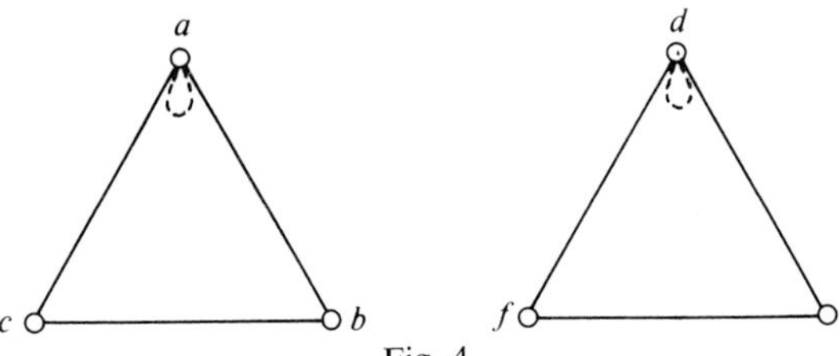

Fig. 4

Case 10

Ringel was the first to solve this case, in 1961 [**25**]. The index 1 current graph solution in [**39**] uses the group $\Gamma = \mathbf{Z}_{12s+7}$ ($s \geqslant 0$) and a simple modification of the solution for Case 7 to obtain a triangular embedding of $K_{12s+10} - K_3$. The three missing edges are added using one extra handle, after suitable modifications, to give a genus embedding of K_{12s+10} with $r_4 = r_5 = 1$. (This is carried out in full in Section 9.)

Case 11

The solution for this case uses the group $\Gamma = \mathbf{Z}_{12s+6}$ to find an index 1 triangular embedding of $K_{12s+11} - K_5$, for $s \geqslant 2$. The missing ten edges are then accommodated over two handles, after suitable modifications. This gives a genus embedding of K_{12s+11} with $r_4 = 2$. The first solution here was by Guy, Ringel and Youngs, in 1967; it treated $s \geqslant 5$ in six subcases. These subcases were later reduced to four, and in early 1968 a unified solution was found for $s \geqslant 2$; this appears in [**33**].

Case O

This is one of the "regular" cases, in that only the regular part of the problem need be treated, as no additional adjacency construction is required. As is customary, it is desirable to find an index 1 solution. However, in this case it can be shown that no such solution is possible using an Abelian group. Nevertheless, an index 1 solution has been obtained (by Terry, Welch and Youngs in 1963 [**35**]), using the following non-Abelian group Γ. For $p = 12s = 2^{m+2}3(2r+1)$, let $\Gamma = G(m) \times \mathbf{Z}_{2r+1}$ (direct product), where $G(m) = F^+(2^{m+2}) * A$ (semi-direct product); $F^+(2^{m+2})$ denotes the additive group of the field $GF(2^{m+2})$, and A is a cyclic group of order three generated by a particular automorphism of $F^+(2^{m+2})$. As an induction proof is employed in the verification of the current graph properties, this solution establishes the *existence* of triangular embeddings of K_{12s}, but does not provide an explicit *construction* for them, as was the situation with the other eleven cases. Recently, an index 4 constructive proof has been found by Pengelley and Jungerman [**23**].

The restrictions on s in the case-by-case analysis above leave the graphs K_p unaccounted for, when p = 8, 9, 11, 13, 18, 20, 23 and 30. (We accept as trivial the facts that $\gamma(K_p) = 0$ for $p \leqslant 4$, and that $\gamma(K_5) = \gamma(K_7) = 1$ forces $\gamma(K_6) = 1$.) But, as mentioned in Section 3, Heffter used rotation schemes constructed directly (without the use of current graphs) to find $\gamma(K_p)$ when $p \leqslant 12$, and Ringel [**24**] settled the case $p = 13$ in 1952. Finally, in 1968, Jean Mayer (Professor of French Literature at the University of Montpellier) treated all cases $p \leqslant 23$, and $p = 30$ [**22**].

8. Current Graphs

As discussed in the previous section, the theory of current graphs underlies the construction of triangular embeddings for complete, or nearly complete, graphs. This beautiful theory was initiated by Gustin in 1963; in [**14**] he outlined the theory and provided specific examples. The theory was developed more fully by Youngs (see, in particular, [**38**]), who also introduced the theory of vortices to handle the non-triangular cases $p \not\equiv 0$, 3, 4 and 7 (modulo 12). In 1969 Jacques [**17**] (see also [**37**]) unified the theory, putting the vortex theory into a more natural setting. In this section we shall incorporate features from each of these approaches to show how an appropriate current graph, in the simplest general case, determines a rotation scheme which in turn gives a triangular embedding of the graph in question.

But the development of the theory did not end here. In 1973, Gross and Alpert [**10**] (see also [**11**], [**12**]) interpreted the theory of current graphs in the topological context of (sometimes branched) covering spaces. Finally, in 1974, Gross [**9**] dualized to the even more natural setting of voltage graphs (see Section 4 of Chapter 2); also in 1974, Gross and Tucker [**13**] re-examined the Heawood map-coloring theorem and its proof from this vantage point. These recent developments will be discussed in Section 10.

The Gross–Alpert current graph theory applies to orientable 2-cell embeddings of arbitrary graphs; that of Jacques applies to orientable 2-cell embeddings of Cayley graphs (see [**37**]); and that of Gustin and Youngs applies to orientable triangular embeddings of complete graphs K_p or nearly complete graphs G_p, using a group Γ of order p and a subgroup Ω of Γ. The simplest general case of this latter theory occurs for $\Gamma = \Omega = \mathbf{Z}_p$, and it is just such special index 1 current graphs which we discuss here. For the unrestricted Gustin–Youngs theory, as well as the Jacques theory, see [**37**].

We start with a few definitions; examples illustrating these terms may be found in Section 9 (see, in particular, Figs 8–12).

A **current graph** for $\Gamma = \mathbf{Z}_n$ is a triple $(K, \mathbf{Z}_n, \lambda)$, where K is a general graph, and $\lambda\colon K^* \to \mathbf{Z}_n - \{0\}$ is an assignment of currents to the oriented edges in $K^* = \{(u, v) : uv \in E(K)\}$ satisfying $\lambda(v, u) = -\lambda(u, v)$ for each $(u, v) \in K^*$.

Now let the current graph $(K, \mathbf{Z}_n, \lambda)$ be 2-cell embedded in an orientable surface S_h. The **Kirchhoff current law** (KCL) is said to hold at a vertex $v \in V(K)$ if the sum $\sum_v$ (in $\mathbf{Z}_n$) of the currents directed away from v (taken in the order given by π_v, as in the rotation scheme for the embedding) is 0. (Of course, if Γ is Abelian, then the order of taking the sum is immaterial; however, in the general case, this is important.)

An **index 1 current graph** $(K, \mathbf{Z}_n, \lambda)$ has a 2-cell embedding in S_h such that

(i) there is exactly one 2-cell;

(ii) this 2-cell is an $(n-1)$-gon;

(iii) each element of $\mathbf{Z}_n - \{0\}$ appears exactly once as a current on some (clockwise, say) oriented edge of the $(n-1)$-gon;

(iv) each vertex of K is of valency 3;

(v) the KCL holds at each vertex.

An index 1 current graph is extremely useful in that it contains a rotation scheme for K_p, as follows. Label the vertices of K_p with the group elements from $\mathbf{Z}_p = \{0, 1, 2, \ldots, p-1\}$. By the above properties (i), (ii) and (iii) the clockwise-oriented boundary of the single 2-cell can serve as row 0 for the rotation scheme:

$$\pi_0: (a_1, a_2, \ldots, a_{p-1}),$$

say. We then obtain row i, $1 \leqslant i \leqslant p-1$, by adding i to each entry in row 0:

$$\pi_i: (a_1+i, a_2+i, \ldots, a_{p-1}+i).$$

Next, we use properties (*iv*) and (*v*) to show that the 2-cell embedding determined by this rotation scheme is necessarily triangular. (Property (*iv*) can be relaxed to allow vertices of valency 1, provided that the incident edges are assigned currents of order 3, but we treat only the trivalent case here.)

Theorem 8.1. *An index* 1 *current graph* $(K, \mathbf{Z}_p, \lambda)$ *determines an orientable triangular embedding for* K_p.

Proof. We have already remarked that an index 1 current graph determines a rotation scheme $\{\pi_i\}_{i \in \mathbf{Z}_p}$ for K_p, where

$$\pi_i: (a_1+i, a_2+i, \ldots, a_{p-1}+i), \qquad \text{for} \qquad 0 \leqslant i \leqslant p-1,$$

and thus determines a 2-cell embedding in an orientable surface; it remains to show that every region is triangular. To this end, consider two successive boundary edges (j, i) and (i, k) for an arbitrary region, as depicted in Fig. 5. We claim that the oriented edge (k, j) of K_p completes this region. Necessarily $\pi_i(j) = k$, so that $\pi_0(j-i) = k-i$. Thus in the current graph K we have a situation such as is depicted in Fig. 6, by property (*iv*). Then, by property (*v*), we must have $x = k-j$, so that $\pi_0(k-j) = i-j$, and therefore $\pi_j(k) = i$. Similarly, since $\pi_0(i-k) = j-k$, we find that $\pi_k(i) = j$, and our proof is complete. ||

We comment here that the verification above is equivalent to determining that Ringel's rule Δ^* ([**28**, p. 22]) is satisfied—namely,

if in row i one has i. $\quad \ldots, j, k, \ldots,$
then in row k one has k. $\quad \ldots, i, j, \ldots.$

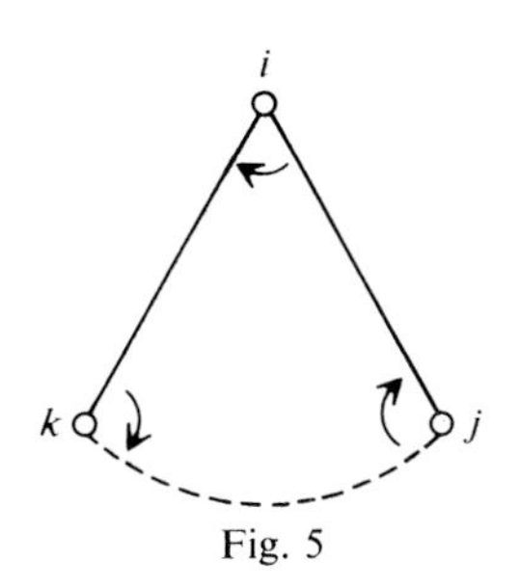

Fig. 5

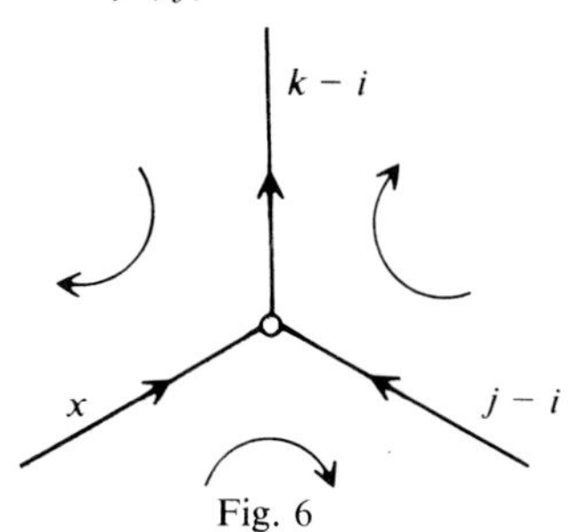

Fig. 6

For the non-triangular cases for K_p, some modification of the current graph theory is required. Youngs [38] introduced the "theory of vortices", used for finding triangular embeddings for graphs which are "nearly" K_p, as seen in Section 7. In the simplest application of the vortex theory (which we shall illustrate in the next section for Case 10), property (*iv*) above is modified to allow vertices of valency 1, with incident currents of order m in $\Gamma = \mathbf{Z}_m$; these vertices are called **vortices**. Each vortex is labelled with a letter (x, say); see Fig. 7, where δ_x is a generator of $\mathbf{Z}_m$. For $i \in \mathbf{Z}_m$, we write

$$\pi_i: (\ldots, a+i, -\delta_x+i, x, \delta_x+i, b+i, \ldots).$$

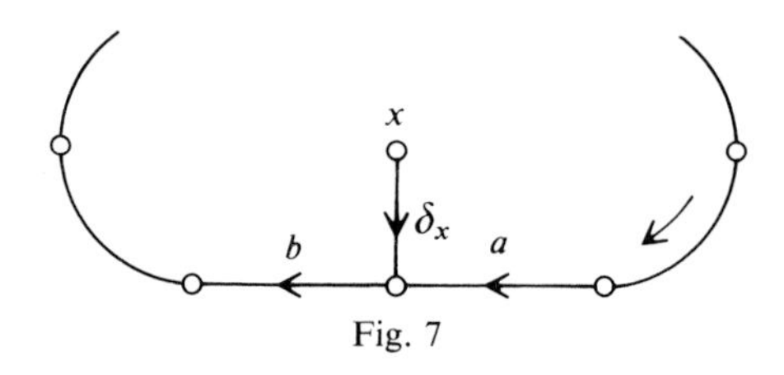

Fig. 7

In this fashion we generate $\{\pi_i\}_{i \in \mathbf{Z}_m}$. For each vortex x, corresponding to a generator δ_x of $\mathbf{Z}_m$, we append

$$\pi_x: ((m-1)\delta_x, \ldots, 2\delta_x, \delta_x, 0)$$

to the rotation scheme. If we have $p-m$ vortices $\{x_i\}$, $1 \leqslant i \leqslant p-m$, of this type, then the collection

$$\{\pi_i\}_{i \in \mathbf{Z}_m} \cup \{\pi_{x_i}\}_{i=1}^{p-m}$$

determines a 2-cell embedding of the join $K_m + \bar{K}_{p-m}$, which is "nearly" K_p, to the extent that m is "nearly" p. (Each vortex x_i contributes a vertex, also called x_i, of the embedded graph, adjacent to every vertex of K_m but to no other vortex vertices.) Moreover, the embedding is triangular; for example, $\pi_i(-\delta_x+i) = x$, $\pi_x(i) = -\delta_x+i$, and $\pi_{-\delta_x+i}(x) = i$, so that $-\delta_x+i$, i, x, $-\delta_x+i$, i constitutes an orbit under the permutation Π^* (see Section 3 of Chapter 2).

Jacques' contribution in this context was to leave the vortex (such as that labeled x above) unlabeled, and to consider the non-triangular 2-cell embedding of K_m determined by $\{\pi_i\}_{i \in \mathbf{Z}_m}$, with

$$\pi_i: (\ldots, a+i, -\delta_x+i, \delta_x+i, b+i, \ldots).$$

In this embedding, the vortex produces a region whose boundary is a Hamiltonian circuit in K_m; the vertices of this cycle are, in counter-clockwise order, $0, \delta, 2\delta, \ldots, (m-1)\delta$. Then a new vertex x is added in the interior of this region, and all of the edges $\{x, kd\}$, $0 \leqslant k \leqslant m-1$, are also added within this region. This process is repeated for each vortex. The result is again a triangular embedding for $K_m + \bar{K}_{p-m}$, where as before $p-m$ is the total number of vortices. This interpretation of Jacques will be examined more closely in Section 10.

9. Cases 7 and 10, in their Entirety

As further indications of the methods employed in solving the Heawood Conjecture, we now work Cases 7 and 10 completely. Admittedly these are two of the simplest cases, but they illustrate nicely the power of the current graph theory (both with and without vortices) and, in Case 10, the type of *ad hoc* argument used in the additional adjacency part of the solutions.

Case 7

Recall that Case 7 is one of the regular cases; that is, that a triangular embedding of K_{12s+7} is possible for $s = 0, 1, 2, \ldots$. By Theorem 8.1, it suffices to find an index 1 current graph $(K, \mathbf{Z}_p, \lambda)$ for $p = 12s+7$. It turns out that S_h (the surface into which the current graph is embedded) is actually S_{s+1}, which is inconvenient to picture directly on the printed page. We therefore use a device, which Youngs calls a *nomogram*, to simplify the representation. The efficacy of this device rests upon the trivalent nature of K, wherein any vertex rotation may be expressed as being either in the clockwise (●) or counter-clockwise (○) sense. Thus each nomogram gives a rotation scheme, and hence describes a 2-cell embedding of K into S_{s+1}.

If $s = 0$, then $S_h = S_1$, and the actual embedding of K is shown in Fig. 8; Fig. 9 gives the corresponding planar representation (the nomogram). In either case, the current graph K is trivalent, and the current graph properties

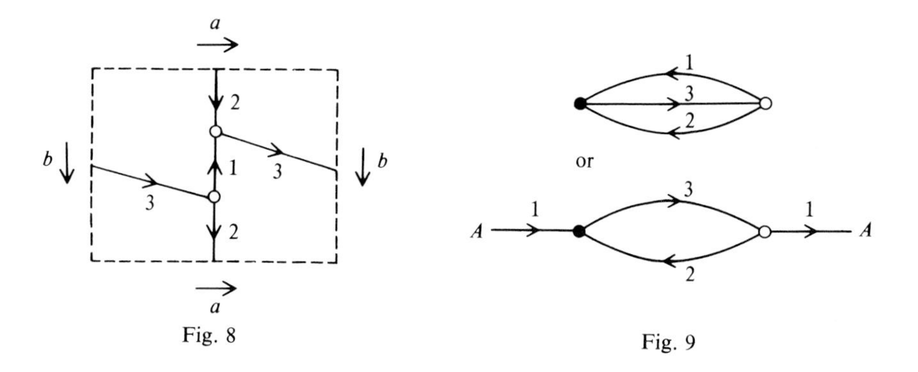

Fig. 8

Fig. 9

(i)–(v) of the previous section are completely satisfied; Theorem 8.1 thus establishes the existence of a triangular embedding for K_7. Moreover, this embedding is constructed as follows: from the single (clockwise-oriented) region boundary, we read:

$$\pi_0: (1, 3, 2, 6, 4, 5).$$

Since this is an index 1 embedding (that is, $[\mathbf{Z}_7:\mathbf{Z}_7] = 1$, or $r = r_6 = 1$ for K in $S_h = S_1$), this rotation determines all others, as:

$$\begin{aligned} &\pi_0: (1, 3, 2, 6, 4, 5),\\ &\pi_1: (2, 4, 3, 0, 5, 6),\\ &\pi_2: (3, 5, 4, 1, 6, 0),\\ &\pi_3: (4, 6, 5, 2, 0, 1),\\ &\pi_4: (5, 0, 6, 3, 1, 2),\\ &\pi_5: (6, 1, 0, 4, 2, 3),\\ &\pi_6: (0, 2, 1, 5, 3, 4). \end{aligned}$$

Although it is not nececessary to do so, it can be checked from this algebraic description using the Π^* algorithm (see Section 3 of Chapter 2) that $r = r_3 = 14$ for this embedding of K_7; that is, 0–1–5–0–1 gives (0, 1, 5) as one triangular region, and so on. Note that this is exactly the embedding of Fig. 2 in Section 3.

There are two ways of viewing this construction.

1. The seven permutation cycles of length six given above determine the boundaries of seven regions for a map of seven countries on the torus; this map is constructed by identifying labeled directed region boundary edges in the standard manner of surface topology. (Since each region is adjacent to all the other six, seven colors are necessary for the map.) Then the dual graph of this map is K_7 embedded in S_1; since the map is trivalent, the K_7 embedding is triangular.
2. The seven permutation cycles of length six are the vertex rotations for K_7, and determine a triangular embedding of K_7 in S_1 directly.

In Figs 10, 11 and 12 we give the nomograms for $s = 1$, $s = 2$, and general s ($s \geqslant 1$) respectively. In each case, the properties (ii)–(v) of the index 1 current graph are readily verified. Property (i) will be seen to hold by a careful tracing out of the single region boundary for the current graph embedding, using the solid and the hollow vertices to determine the vertex rotations. (In Fig. 12, the vertical edges are directed alternately and carry the currents $1, 2, \ldots, 2s$ consecutively; all other currents not shown in the middle portion of the figure are determined by the KCL requirement of property (v).) Thus Theorem 8.1 is applicable, and the proof is complete, for Case 7.

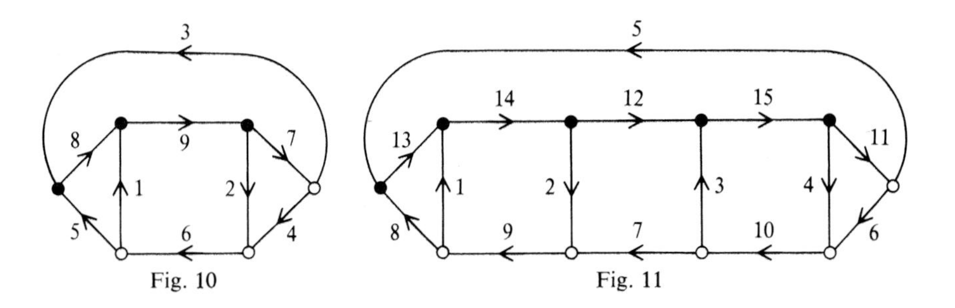

Fig. 10 Fig. 11

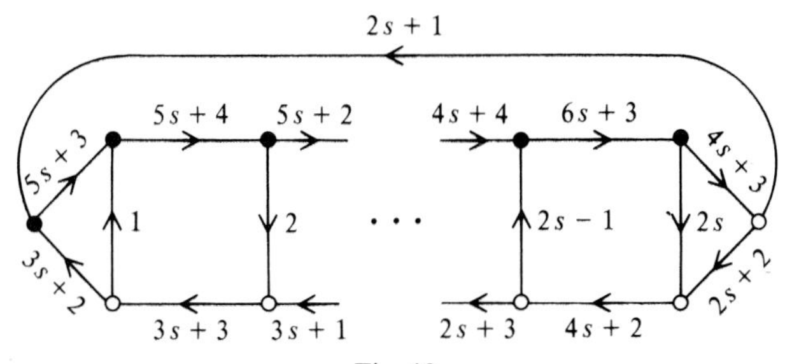

Fig. 12

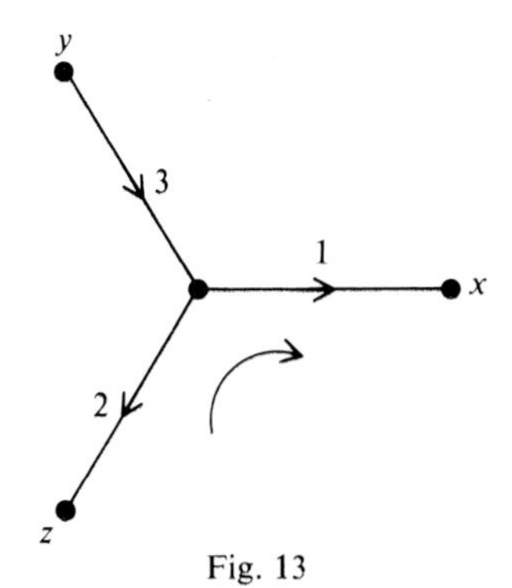

Fig. 13

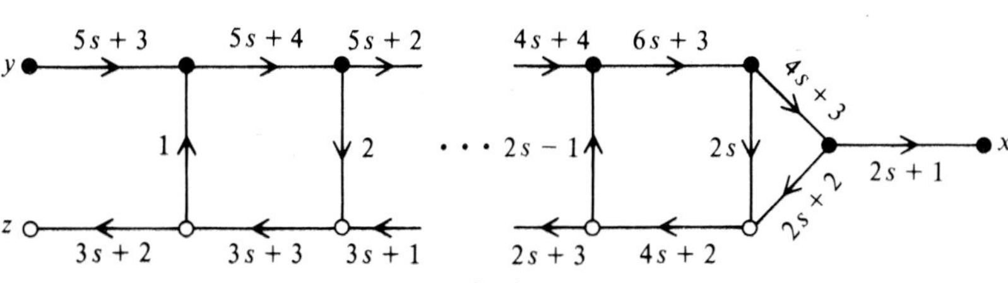

Fig. 14

Case 10

The regular part of the "irregular" Case 10 follows from a ready modification of the Case 7 current graph. Figure 9 is modified as in Fig. 13 (still for $\Gamma = \mathbf{Z}_7$); Fig. 12 is modified as in Fig. 14 (still for $\Gamma = \mathbf{Z}_{12s+7}$, $s \geqslant 1$). For each value of s, there are exactly three vortices, labeled x, y and z; the current at each vortex is a generator of the group Γ. (In Fig. 14,

$$\gcd(2s+1,\ 12s+7) = 1,$$

since
$$-6(2s+1)+1(12s+7) = 1;$$

$$\gcd(5s+3,\ 12s+7) = 1,$$

since
$$12(5s+3)-5(12s+7) = 1;$$

and
$$\gcd(3s+2,\ 12s+7) = 1,$$

since
$$4(3s+2)-1(12s+7) = 1.)$$

Then the theory of vortices, as explained in Section 8, establishes that a triangular embedding of the graph $K_{12s+7}+\overline{K}_3$ has been constructed, in $S_{12s^2+13s+3}$. This completes the regular part of this case.

(We illustrate, for $s = 0$, the embedding of $K_7+\overline{K}_3$, as described by:

$$\begin{aligned}
\pi_0&: (1, x, 6, 4, y, 3, 2, z, 5),\\
\pi_1&: (2, x, 0, 5, y, 4, 3, z, 6),\\
\pi_2&: (3, x, 1, 6, y, 5, 4, z, 0),\\
\pi_3&: (4, x, 2, 0, y, 6, 5, z, 1),\\
\pi_4&: (5, x, 3, 1, y, 0, 6, z, 2),\\
\pi_5&: (6, x, 4, 2, y, 1, 0, z, 3),\\
\pi_6&: (0, x, 5, 3, y, 2, 1, z, 4),\\
\pi_x&: (0, 1, 2, 3, 4, 5, 6),\\
\pi_y&: (0, 4, 1, 5, 2, 6, 3),\\
\pi_z&: (0, 2, 4, 6, 1, 3, 5).
\end{aligned}$$

Figure 15 depicts the region of the *map* on S_3 whose *dual* is $K_7+\overline{K}_3$ corresponding to the group element (0). Note the role of the vortices x, y and z.)

To repeat, we have a triangular embedding of $K_{12s+7}+\overline{K}_3$ in $S_{12s^2+13s+3}$, whereas we are trying to embed K_{s12+10} in $S_{12s^2+13s+4}$. We thus seek to add the three missing edges xy, xz and yz with the addition of only one handle; this is the additional adjacency part of the problem.

We refer to Fig. 16. The essential information at region $0 = (0)$ (in the dual of the $K_7+\overline{K}_3$ embedding) is depicted in Fig. 16(a). Now consider the supplementary toroidal map given in Fig. 16(b). Excise region 0 from S_3 in Fig. 16(a), and region C from S_1 in Fig. 16(b), and identify the boundaries of the two resulting surfaces as indicated by the arrows. This produces a new region

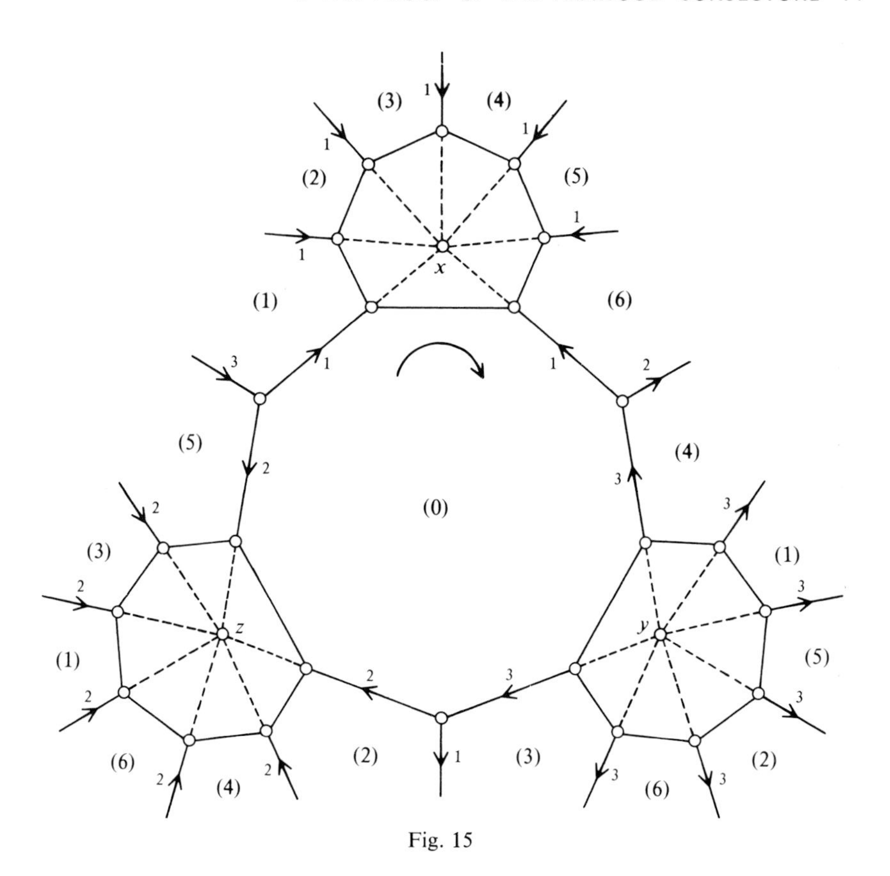

Fig. 15

(a)

(b)

Fig. 16

0 all of whose adjacencies are possessed by the excised region 0. Moreover, regions x, y and z are now mutally adjacent. Thus the *dual graph* (after eliminating three redundant edges) of this map is K_{10}, embedded in S_4.

The same method works directly for K_{12s+10}, $s \geqslant 1$, and in fact is also applicable to the additional adjacency parts of the solutions for Cases 1, 6 and 9.

10. A Second Viewpoint

We now re-examine the heart of the proof of the Heawood map-coloring theorem—the construction of genus embeddings of complete graphs by the use of current graphs—in the context of the covering spaces of algebraic topology (see Section 4 of Chapter 2). This re-examination pertains to the regular part of each case, but never to the additional adjacency parts, which continue to be handled as before.

Let G_p denote either K_p, or the related graph for which an embedding was constructed using current graphs (before extension to a triangulation using vortex vertices) and then modified to obtain a genus embedding for K_p, in each case as described in Section 7. Let m denote the genus of the surface in which G_p was embedded. (Often $m > \gamma(G_p)$, since the embedding of G_p may be far removed from a triangulation.) Our remark for the example $p = 7$ applies equally to G_p in general; that is, the constructed embeddings of G_p can be obtained in two ways: (*i*) as the dual of a map M whose regions are obtained from the regions of the embedding for the current graph K, or (*ii*) as a graph whose vertex rotations are determined from those of the dual of K in its embedding surface S_h (see Fig. 17).

Gross and Alpert [**10**], [**11**], [**12**] established that the embedding of G_p in S_m can be described using a (possibly branched) covering projection from

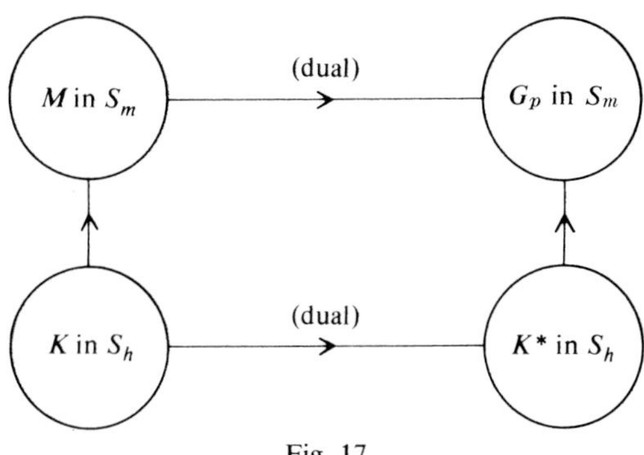

Fig. 17

S_m to S_h, with G_p projecting to the dual K^* of the current graph K. With this interpretation, vertices, edges and regions of G_p in S_m cover vertices, edges and regions (respectively) of K^* in S_h, but in turn are "determined" by regions, dual edges and vertices (respectively) of K in S_h. Thus it is more natural to focus on K^*, rather than on the current graph $K = (K, \Gamma, \lambda)$. This is what Gross did in [**9**], where in fact the technique extends to embeddings of arbitrary graphs, as indicated in Section 4 of Chapter 2; here we follow this approach, and call K^* a **voltage graph**. (In obtaining K^* from K, we orient and label edges as indicated in Fig. 18.)

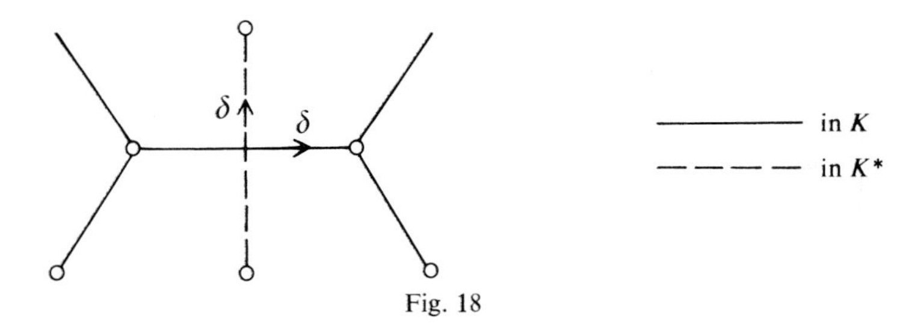

Fig. 18

The entire Gross theory of voltage graphs is dual to the theory of current graphs (as being branch-covered, in the dual, by graph embeddings), as developed by Gross and Alpert. For example, if the KCL holds at a vertex v of K, then the KVL (**Kirchhoff voltage law**) holds around the region v^* of K^* in S_h dual to v: the sum of the voltages around the region boundary is 0 in $\mathbf{Z}_p$. (If Γ is a non-Abelian group, then the sum (group product) must be taken in the order given by the orientation on the surface.)

Let us now consider the current graph solutions of Section 9 for Cases 7 and 10 in their dual form as voltage graphs. The dual of Fig. 8 appears in Fig. 19; the single vertex has been shifted to the corners of the rectangle to simplify the viewing of the regions. The single region in Fig. 8 has clockwise boundary (1, 3, 2, 6, 4, 5); this becomes the clockwise rotation at the single vertex in Fig. 19. The counter-clockwise vertex rotations in Fig. 8 are (6, 3, 5) and (1, 4, 2); these describe the counter-clockwise region boundaries in Fig. 19. (Recall that a clockwise vertex orientation induces a counter-clockwise region orientation, using the permutation Π^* of Section 3 of Chapter 2.) The theory of voltage graphs, when applied to Fig. 19 with $\Gamma = \mathbf{Z}_7$, gives rise to a seven-fold covering; since the KVL holds on each region, there is no branching. This guarantees that the covering graph has $p = 7 \times 1 = 7$, $q = 7 \times 3 = 21$, and $r = r_3 = 7 \times 2 = 14$. Moreover, since the six voltages emanating from the vertex in Fig. 19 are exactly the six non-identity elements

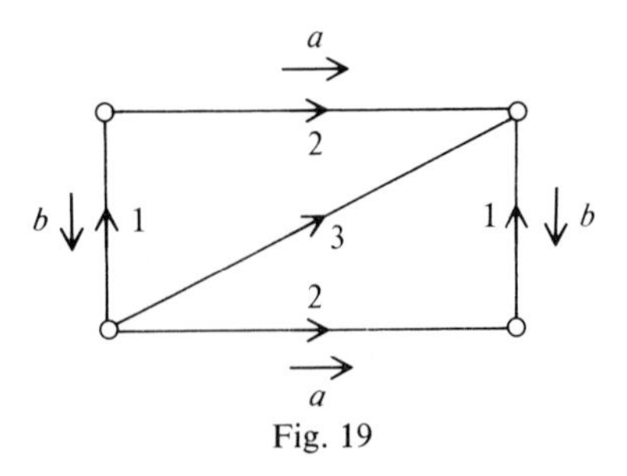

Fig. 19

of $\mathbf{Z}_7$, the covering graph is K_7, and a triangular embedding of K_7 is determined in S_1 (by a result of Section 4 of Chapter 2).

In general, the current graph $K(s)$ for K_{12s+7} (as given in Section 9) has as its dual a voltage graph for $\Gamma = \mathbf{Z}_{12s+7}$, with $p = 1$, $q = 6s+3$, and $r = r_3 = 4s+2$; the voltage graph is embedded in S_{s+1}, as was the current graph. The covering is unbranched and $(12s+7)$-fold, giving an embedding of K_{12s+7} in S_{12s^2+7s+1} with, of course, $p = 12s+7$, $q = (12s+7)(6s+3)$, and $r = r_3 = (12s+7)(4s+2)$.

The current graph for K_{10} (Fig. 13) has as its dual the voltage graph of Fig. 20. The single region in Fig. 13 has clockwise boundary (1, 6, 4, 3, 2, 5); this is also the clockwise vertex rotation for Fig. 20. The trivalent vertex in Fig. 13 has counter-clockwise rotation (1, 4, 2), and this describes the boundary of the triangular (outer) region of Fig. 20. The KVL holds for this region ($\Gamma = \mathbf{Z}_7$), so that the region lifts to seven triangles in the covering embedding of K_7. However, each of the three monogons (regions with only one boundary edge) has a voltage sum of order 7, and by Theorem 4.1 of Chapter 2, each of these regions is covered by one seven-sided region (a heptagon). Moreover, since each voltage in question generates $\mathbf{Z}_7$, each heptagonal region boundary is a Hamiltonian circuit for K_7. (There is a branch point in the interior of each monogon; in each case, the covering heptagon is "wrapped around" the

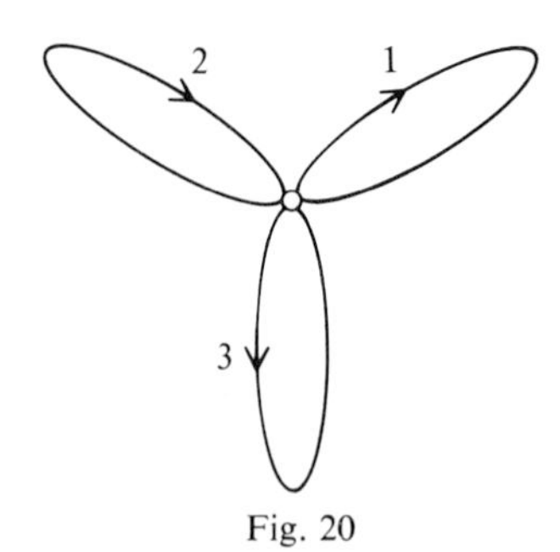

Fig. 20

monogon in seven-fold fashion, as if by the complex map $z \to z^7$. Thus this is a branched covering projection, and we have K_7 embedded in S_3 as a branched covering space over S_0.) Now one vertex is added in the interior of each heptagon, and joined to each vertex in the boundary. (Recall that Fig. 15 shows the dual of this construction.) The result is a triangular embedding of $K_7+\bar{K}_3$, still in S_3, and the additional adjacency construction is now implemented as before. The three vortices for each current graph in Case 10 become monogons in the voltage graph embedding, with the single boundary edge always carrying as voltage a generator of $\mathbf{Z}_{12s+7}$. Thus each monogon is branch-covered by a single $(12s+7)$-gon, bounded by a Hamiltonian circuit in K_{12s+7}; the construction proceeds as above.

We observe that S_h is S_{s+1} for Case 7, but S_s for Case 10; the covering spaces are respectively S_{12s^2+7s+1} and $S_{12s^2+13s+3}$, and the genus surfaces are S_{12s^2+7s+1} and $S_{12s^2+13s+4}$.

Gross and Tucker [**13**] have discussed all twelve cases in this voltage graph setting.

11. The Non-orientable Heawood Theorem

Nearly every embedding or map-coloring problem which can be asked for the closed orientable 2-manifolds can also be posed for the closed non-orientable 2-manifolds. In Sections 2 and 3 of Chapter 2 we discussed non-orientable surfaces, and the embedding theory for these surfaces is very similar to that for the orientable case. Here we briefly outline the solution of the non-orientable version of Heawood's theorem, which is likewise analogous to that of the orientable situation.

Let $\chi(N_k)$ denote the chromatic number of the sphere with k crosscaps, where $k \geqslant 1$, and let $R(k) = [\frac{1}{2}(7+\sqrt{1+24k})]$. The upper bound is established by an argument similar to that of Theorem 4.2, except that N_1 (the projective plane) must be treated separately, as it is the only non-orientable surface with positive characteristic. (Recall that for the only orientable surface with positive characteristic, the inequality $\chi(S_0) \leqslant H(0) = 4$ was by no means trivial; nevertheless, the inequality $\chi(N_1) \leqslant R(1) = 6$ follows by a simple induction argument.)

Theorem 11.1. $\chi(N_k) \leqslant R(k)$, *for* $k \geqslant 1$. ‖

The non-orientable complete graph theorem is also established in two stages; the lower bound is an easy consequence of Euler's formula, and embeddings attaining this bound are constructed in twelve cases using non-orientable current graphs (see, for example, [**28**]). Note, however, that K_7 does not embed on the Klein bottle N_2, as we indicated in Section 2.

Theorem 11.2. $\tilde{\gamma}(K_p) = \{\frac{1}{6}(p-3)(p-4)\}$, *for* $p \geqslant 5$, *except that* $\tilde{\gamma}(K_7) = 3$. ||

Then the lower bound for $\chi(N_k)$, $k \neq 2$, follows from the non-orientable complete graph theorem, in a manner directly analogous to that of Theorem 5.1:

Theorem 11.3. $\chi(N_k) \geqslant R(k)$, *for* $k \geqslant 1$, *except that* $\chi(N_2) = 6$. ||

Theorems 11.1 and 11.3 now combine to give:

Theorem 11.4. $\chi(N_k) = R(k)$, *for* $k \geqslant 1$, *except that* $\chi(N_2) = 6$. ||

Finally, we note that both Heawood theorems (the orientable version, which includes the four-color theorem, and the non-orientable version) can be combined into one, as follows. Let M_n denote a closed 2-manifold of characteristic n; thus $n = 2-2h$ if $M_n = S_h$, and $n = 2-k$ if $M_n = N_k$. Then, combining the formulas for $\chi(S_h)$, $h \geqslant 0$, and $\chi(N_k)$, $k \geqslant 1$, we find:

Theorem 11.5. $\chi(M_n) = [\frac{1}{2}(7+\sqrt{49-24n})]$, *for* $M_n \neq N_2$; $\chi(N_2) = 6$. ||

12. Other Work Stimulated by the Heawood Conjecture

Much of the work in graph theory (and in particular in topological graph theory) in the twentieth century has been stimulated by the Heawood conjecture and by its close relative, the four-color problem. Two results by Dirac [7] deserve mention in this connection. We state these in their dual formulation, letting $g(n) = [\frac{1}{2}(7+\sqrt{49-24n})]$.

Theorem 12.1. *If G is a graph embedded in a closed 2-manifold M_n of characteristic n ($n \neq 2$, 1, or -1) with $\chi(G) = g(n)$, then G contains a complete subgraph of order $g(n)$. If $n = 1$, G contains either a subgraph K_6, or a subgraph contractible to K_6; similarly, if $n = -1$, G contains either a subgraph K_7, or a subgraph contractible to K_7.* ||

When Dirac published this result in 1952 the Heawood theorem was not yet established. We now know that such graphs G do exist, unless $M_n = N_2$. The importance of this theorem was that, in order to attain the Heawood upper bound, one really could not avoid the study of complete graph embeddings. In addition, Dirac proved that if $n \leqslant 1$, then a graph G embedded on M_n must have minimum valency $\rho_{\min}(G) < g(n)$.

White [37] strengthened this result as follows:

Theorem 12.2. *If $n \leqslant 1$, then a graph G embedded on M_n must have*

$\rho_{\min}(G) \leqslant g(n)-1$; *moreover, there exists a graph* G, *embeddable on* M_n, *such that* $\rho_{\min}(G) = g(n)-1$. ‖

Complete graphs attain this maximum value of $\rho_{\min}$ in all cases except when $M_n = N_2$, where nevertheless a graph was constructed for this purpose. Here we have another curiosity: whereas the Klein bottle is the only exception to the general formula for $\chi(M_n)$ (since K_7 *is not* embeddable on N_2), the sphere is the only exception to the formula of Theorem 12.2 (since the octahedral graph—or worse yet, the icosahedral graph—*is* embeddable on S_0).

In this section we also explore two classes of problems directly related to the Heawood conjecture: (a) determining vertex-partition numbers, and (b) coloring maps on pseudosurfaces. For a third class—determining $\gamma(G)$ or $\tilde{\gamma}(G)$, when $G \neq K_p$—see Sections 5 and 6 of Chapter 2.

(a) Vertex-partition Numbers

The Heawood theorem addresses itself to the chromatic number of a graph G, which can be regarded as the minimum number of subsets into which the vertex-set $V(G)$ can be partitioned so that each subset induces a null graph. A related concept is the **vertex-arboricity** $a(G)$, defined as the minimum number of subsets into which $V(G)$ can be partitioned so that each subset induces a forest. In 1969, Kronk [20] defined the vertex-arboricity $a(S_h)$ of a closed orientable 2-manifold S_h as the maximum $a(G)$ such that G is embeddable in S_h, and showed that

$$a(S_h) = [\tfrac{1}{4}(9+\sqrt{1+48h})], \qquad \text{if } h \geqslant 1.$$

(Note the striking similarity to the Heawood formula.) Also in 1969, Chartrand and Kronk [4] showed that $a(S_0) = 3$, in disagreement with the formula for $h \geqslant 1$. In 1972 Lick and White [21] extended these results in a natural way, first defining a graph G to be ***k*-degenerate** if every subgraph of G has minimum valency at most k, and then defining the **vertex-partition number** $\rho_k(G)$ to be the minimum number of subsets into which $V(G)$ can be partitioned so that each subset induces a k-degenerate graph. Thus $\rho_0(G) = \chi(G)$, and $\rho_1(G) = a(G)$. Finally, $\rho_k(M_n)$, the vertex-partition number of the closed 2-manifold M_n of characteristic n, is the maximum $\rho_k(G)$ among all graphs G embeddable in M_n. The following theorem thus generalizes the two theorems of Heawood, as well as that of Kronk:

Theorem 12.3. *The vertex-partition numbers for the closed 2-manifolds* M_n *of* *characteristic n are given by*

$$\rho_k(M_n) = \left[\frac{2k+7+\sqrt{49-24n}}{2k+2}\right],$$

where $k = 0, 1, 2, \ldots,$ *and* $n = 2, 1, 0, -1, -2, \ldots,$ *except that*

(*i*) *in the orientable case,* $\rho_1(S_0) = 3$, $\rho_3(S_0) = \rho_4(S_0) = 2$;

(*ii*) *in the non-orientable case* $\rho_0(N_2) = 6$, $\rho_1(N_2) = 3$ *or* 4, *and* $\rho_2(N_2) = 2$ *or* 3. ‖

It is worth commenting that the general proof of the upper bound in the above theorem fails only for $M_n = S_0$, and the general proof of the lower bound fails only for $M_n = N_2$; these respective cases are treated by *ad hoc* arguments. Note also that, with at most six, and possibly only four, exceptions, both the orientable and the non-orientable cases are combined into a single formula which holds for all non-negative values of k.

It only remains to settle whether $\rho_1(N_2) = 3$ or 4, and whether $\rho_2(N_2) = 2$ or 3. In the first case one must either find a graph G embeddable in the Klein bottle with $\rho_1(G) = 4$ (so that no partition of $V(G)$ into three subsets, each inducing an acyclic subgraph of G, can be found), or prove that no such graph exists. In the second case, one seeks a Klein bottle graph G with $\rho_2(G) = 3$. Note that $\rho_1(K_7) = 4$, and $\rho_2(K_7) = 3$, and that K_7 should embed on the Klein bottle (since it embeds in the torus, which has the same characteristic)—but it doesn't! Does the general formula of Theorem 12.3 fail for $k = 1$ and 2 and $M_n = N_2$, just as it does for $k = 0$? (Recall that this established failure is due to the fact that K_7 is not a Klein bottle graph.) Is the determination between $\rho_1(N_2) = 3$ or 4, or between $\rho_2(N_2) = 2$ or 3, a problem of comparable magnitude to that of settling that $\rho_0(S_0)$ is 4 and not 5? The answers to these questions remain unknown.

(b) Coloring Maps on Pseudosurfaces

For all maps considered thus far in this chapter, each country has been connected. Now suppose that disconnected countries are also allowed, but that no country can have more than two components. We desire to color both components of a given disconnected country alike, but we must still color two different countries differently whenever they share a one-dimensional portion of their borders. In the simplest case here, we assume that all maps are drawn on the surface of the sphere (or, equivalently, in the plane). Passing to the dual formulation, we obtain a graph G with certain pairs of distinguished vertices, each pair representing one country. It is thus natural to identify each such pair, and we thus obtain an embedding of a graph G' on a *pseudosurface*—in this case, a sphere with n pairs of points identified, if the map has n disconnected countries of two components each. This topological object is a 2-manifold at every point except the n points, called *singular points*, of identification; we denote it by $S(0; n(2))$. Then the map-coloring number in question here is given by the largest chromatic number $\chi(G')$ among all graphs G' embeddable in $S(0; n(2))$, subject to the restriction that each singular point is occupied by a vertex.

We can also ask: "what is the largest chromatic number among *all* spherical maps, in which no country has more than two components?"—that is, what is $\chi_2(S_0)$, the maximum value of $\chi(S(0; n(2)))$, if this number indeed exists? In **[15]** Heawood produced a map on the sphere with twelve countries, each with exactly two components, and with each country bordering on all the others. Thus K_{12} is embeddable in $S(0; 12(2))$, with each vertex occupying a singular point, so that $\chi_2(S_0) \geqslant \chi(S(0; 12(2))) \geqslant 12$. In fact, Heawood showed also that $\chi_2(S_0) \leqslant 12$, so that $\chi_2(S_0) = 12$. (This result may also be found in Ringel **[26]**, and is another instance of an early solution to a problem which is at face value more difficult than the four-color problem.)

In 1972 Dewdney **[6]** showed that $\chi(S(0; n(2))) \leqslant n+4$, for $n > 0$, and that equality holds for $n = 1$ and 2. In 1973, M. O'Bryan and J. Williamson (unpublished) showed that equality also holds for $n = 3$ and 4, and of course we now know that both the bound and equality hold also for $n = 0$. In fact, this particular pseudosurface map-coloring problem has been completely solved recently by Borodin and Mel'nikov **[3]** (see also Sulanke **[34]**, where only the case $\chi(S(0; 7(2))) = 9$ or 10 is left in doubt); note the familiar bracketed expression in the statement of the theorem:

Theorem 12.4.
$$\chi(S(0; n(2))) = \begin{cases} n+4, & \textit{if } 0 \leqslant n \leqslant 4, \\ 8, & \textit{if } n = 5, \\ [\frac{1}{2}(7+\sqrt{1+24n})], & \textit{if } 3 \leqslant n \leqslant 12,\ n \neq 5, \\ 12, & \textit{if } n \geqslant 12. \end{cases}$$ ‖

Borodin and Mel'nikov also found the value of $\chi(S(0; n(2)))$ in the case where an edge of the embedded graph can pass through a singular point. Such embeddings might have the following interpretation as the dual of a map to be colored: an edge passing through a singular point joins two vertices in the dual which represent two regions of the map which *must* be colored differently (for some political reason, say), even though they may not share a one-dimensional border segment.

The map-coloring problem has been settled for spherical maps whose countries have at most two components. What about maps on S_h $(h > 0)$

having countries with at most two components? How about maps on S_0 having countries with at most three components? Or at most c ($\geqslant 4$) components? The non-orientable analogs could also be considered. There seems to have been practically no investigation whatsoever into these questions to date. Perhaps this is because Heawood [**15**] claimed that the expression

$$[\tfrac{1}{2}(6c+1+\sqrt{(6c+1)^2-24n})]$$

gives the exact map-coloring number for a surface (orientable or non-orientable) of characteristic n, where each country could have at most c components—except where $n = 2$ and $c = 1$ (the case of the four-color problem). Heawood "proved" this generalization of his map-coloring theorems (the cases $c = 1$), just as he did the map-coloring theorems themselves, by establishing this number as the upper bound (this is correct) and then "assuming the verification figure" ([**15**, p. 336]). It would appear that all verification figures for $c > 1$, except for $c = 2$ on S_0, remain to be constructed.

In conclusion, we note that we now have two substantial generalizations of the Heawood formula for the chromatic number of closed 2-manifolds: that of Theorem 12.3 for arbitrary vertex-partition numbers ρ_k (where $c = 1$), and the formula above (which is still a conjecture) for ρ_0 and arbitrary c; there may well be *one* even more general formula containing both of these as special cases.

References

1. K. Appel and W. Haken, Every planar map is four colorable, *Bull. Amer. Math. Soc.* **82** (1976), 711–712; *MR***54**#12561.
2. N. L. Biggs, E. K. Lloyd, and R. J. Wilson, *Graph Theory 1736–1936*, Clarendon Press, Oxford, 1976.
3. O. V. Borodin and L. S. Mel'nikov, The chromatic number of a pseudosurface, *Diskret. Analiz* **24** (1974), 8–20; *MR***51**#5361.
4. G. Chartrand and H. V. Kronk, The point-arboricity of planar graphs, *J. London Math. Soc.* **44** (1969), 612–616; *MR***39**#1350.
5. R. Courant and H. Robbins, *What is Mathematics?*, Oxford University Press, New York, 1941; *MR***3**–144.
6. A. K. Dewdney, The chromatic number of a class of pseudo-2-manifolds, *Manuscripta Math.* **6** (1972), 311–319; *MR***45**#5022.
7. G. A. Dirac, Map-colour theorems, *Canad. J. Math.* **4** (1952), 480–490; *MR***14**–394.
8. P. Franklin, A six-color problem, *J. Math. Phys.* **13** (1934), 363–369.
9. J. L. Gross, Voltage graphs, *Discrete Math.* **9** (1974), 239–246; *MR***50**#153.
10. J. L. Gross and S. R. Alpert, Branched coverings of graph imbeddings, *Bull. Amer. Math. Soc.* **79** (1973), 942–945; *MR***48**#10870.
11. J. L. Gross and S. R. Alpert, The topological theory of current graphs, *J. Combinatorial Theory* (*B*) **17** (1974), 218–233; *MR***51**#226.
12. J. L. Gross and S. R. Alpert, Components of branched coverings of current graphs, *J. Combinatorial Theory* (*B*) **20** (1976), 283–303.

13. J. L. Gross and T. W. Tucker, Quotients of complete graphs, revisiting the Heawood map-coloring theorem, *Pacific J. Math.* **55** (1974), 391–402; *MR***52**#10466.
14. W. Gustin, Orientable embedding of Cayley graphs, *Bull. Amer. Math. Soc.* **69** (1963), 272–275; *MR***26**#3037.
15. P. J. Heawood, Map-colour theorem, *Quart. J. Pure Appl. Math.* **24** (1890), 332–338.
16. L. Heffter, Über das Problem der Nachbargebiete, *Math. Ann.* **38** (1891), 477–508.
17. A. Jacques, Constellations et propriétés algèbriques des graphes topologiques, Ph.D. thesis, University of Paris, 1969.
18. M. Jungerman, Orientable cascades and related embedding techniques, Ph.D. thesis, University of California, Santa Cruz, 1974.
19. A. B. Kempe, On the geographical problem of the four colours, *Amer. J. Math.* **2** (1879), 193–200.
20. H. V. Kronk, An analogue to the Heawood map-colouring problem, *J. London Math. Soc.* (*2*) **1** (1969), 750–752; *MR***40**#4167.
21. D. R. Lick and A. T. White, The point partition numbers of closed 2-manifolds, *J. London Math. Soc.* (*2*) **4** (1972), 577–583; *MR***45**#5021.
22. J. Mayer, Le problème des régions voisines sur les surfaces closes orientables, *J. Combinatorial Theory* **6** (1969), 177–195; *MR***38**#3177.
23. D. Pengelley and M. Jungerman, Index four orientable embedding and case zero of the Heawood conjecture, *J. Combinatorial Theory* (*B*) (to appear).
24. G. Ringel, Farbensatz für orientierbaren Flächen vom Geschlecht $p > 0$, *J. Reine Angew. Math.* **193** (1954), 11–38; *MR***16**#387.
25. G. Ringel, Bestimmung der Maximalzahl der Nachbargebiete auf nichtorientierbaren Flächen, *Math. Ann.* **127** (1954), 181–214; *MR***16**-58.
26. G. Ringel, *Färbungsprobleme auf Flächen und Graphen*, VEB Deutscher Verlag der Wissenschaften, Berlin, 1959; *MR***22**#235.
27. G. Ringel, Über das Problem der Nachbargebiete auf orientierbaren Flächen, *Abh. Math. Sem. Univ. Hamburg* **25** (1961), 105–127; *MR***23**#A2876.
28. G. Ringel, *Map Color Theorem*, Springer-Verlag, Berlin, 1974; *MR***50**#1955.
29. G. Ringel and J. W. T. Youngs, Solution of the Heawood map-coloring problem, *Proc. Nat. Acad. Sci. U.S.A.* **60** (1968), 438–445; *MR***37**#3959.
30. G. Ringel and J. W. T. Youngs, Lösung des Problems der Nachbargebiete auf orientierbaren Flächen, *Arch. Math.* (*Basel*) **20** (1969), 190–201; *MR***39**#4049.
31. G. Ringel and J. W. T. Youngs, Solution of the Heawood map-coloring problem—case 2, *J. Combinatorial Theory* **7** (1969), 342–352; *MR***42**#128.
32. G. Ringel and J. W. T. Youngs, Solution of the Heawood map-coloring problem—case 8, *J. Combinatorial Theory* **7** (1969), 353–363; *MR***41**#6723.
33. G. Ringel and J. W. T. Youngs, Solution of the Heawood map-coloring problem—case 11, *J. Combinatorial Theory* **7** (1969), 71–93; *MR***39**#1360.
34. T. Sulanke, Coloring pseudo-surfaces and maps with colonies (preprint).
35. C. M. Terry, L. R. Welch and J. W. T. Youngs, The genus of K_{12s}, *J. Combinatorial Theory* **2** (1967), 43–60; *MR***34**#6755.
36. C. M. Terry, L. R. Welch and J. W. T. Youngs, Solution of the Heawood map-coloring problem—case 4, *J. Combinatorial Theory* **8** (1970), 170–174; *MR***41**#3321.
37. A. T. White, *Graphs, Groups and Surfaces*, North-Holland, Amsterdam, 1973; *MR***49** #4783.
38. J. W. T. Youngs, The Heawood map-coloring conjecture, in *Graph Theory and Theoretical Physics* (ed. F. Harary), Academic Press, London and New York, 1967, pp. 313–354; *MR***38**#4357.
39. J. W. T. Youngs, The Heawood map-coloring problem—cases 1, 7, and 10, *J. Combinatorial Theory* **8** (1970), 220–231; *MR***41**#3323.
40. J. W. T. Youngs, Solution of the Heawood map-coloring problem—cases 3, 5, 6, and 9, *J. Combinatorial Theory* **8** (1970), 175–219; *MR***41**#3322.

4
The Appel–Haken Proof of the Four-Color Theorem

D. R. WOODALL AND ROBIN J. WILSON

1. Introduction

In July 1976, Kenneth Appel and Wolfgang Haken, two mathematicians at the University of Illinois, announced the solution [3] of what was probably the best-known unsolved problem in the whole of mathematics—the four-color map problem:

The Four-Color Map Problem. *Can the regions of every map in the plane be colored with four colors in such a way that no two neighboring regions have the same color?*

Here, "neighboring regions" are regions that share a length of common border. We do not require that two regions which meet only at a finite number of points (such as regions D and F in Fig. 1) should necessarily have different colors.

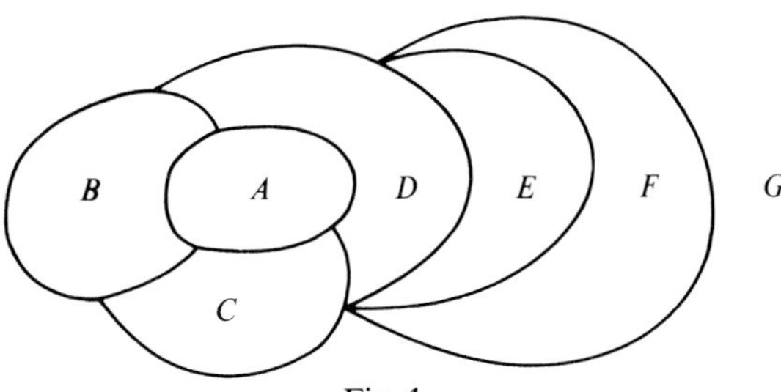

Fig. 1

This problem was first proposed in 1852 by a London student, Francis Guthrie, who thought of it while coloring a map of the counties of England. He noticed that four colors are sometimes needed (for example, for regions *A*, *B*, *C* and *D* of Fig. 1), and conjectured that four colors are always sufficient, but he was unable to prove this. His brother Frederick brought the problem to the attention of Augustus De Morgan, Professor of Mathematics at University College, London, who mentioned it on a number of occasions, giving credit to Francis Guthrie for proposing it. However, it was not until 1878 (after De Morgan's death) that the problem became widely known. At a meeting of the London Mathematical Society in that year, Arthur Cayley asked whether the problem had been solved, and shortly afterwards wrote a note **[9]** in which he attempted to explain where the difficulty lies.

The first serious attempt at a proof of the "four-color conjecture" seems to have been made by A. B. Kempe **[13]**, a barrister and keen amateur mathematician who was Treasurer, and later President, of the London Mathematical Society. In 1879, he published a "proof" in the newly-founded *American Journal of Mathematics*, and for more than ten years this proof was generally accepted as correct. But in 1890, P. J. Heawood wrote an important paper **[11]** pointing out that Kempe's proof contained a flaw, although he was able to salvage enough from Kempe's argument to prove that every map can be colored with five colors. For some years afterwards, the flaw seems not to have been recognized as serious, but as the years went by and nobody found a satisfactory way around the difficulty, it gradually became realized that the problem was much harder than originally supposed. Since then, many mathematicians have tried to prove the conjecture, and so Appel and Haken's achievement in doing so is a very fine one.

The authors would like to thank Kenneth Appel for his helpful comments regarding this chapter.

2. Kempe's "Proof"

Kempe's argument, although incorrect, contained several important ideas that were to become the foundation for almost all subsequent attempts on the problem, including Appel and Haken's successful attempt. It will therefore be helpful to begin by studying Kempe's argument. However, we shall translate it into the language that Appel and Haken used in their proof.

Suppose we are given a map that we want to color. As in Chapter 1, we can form its dual graph by placing a vertex inside each region of the map and joining two of these vertices by an edge whenever the corresponding two regions are neighbors. If we can four-color the vertices of this dual graph in such a way that adjacent vertices have different colors, then (as Kempe himself pointed out) we can certainly four-color the regions of the original map. The converse implication also holds, and so the main theorem can be restated as follows:

The Four-Color Theorem. *Every planar graph is four-colorable.*

Throughout this chapter we shall be discussing the theorem in this dual form, coloring the vertices rather than the regions.

Furthermore, it suffices to consider plane triangulations—that is, connected plane graphs whose edges divide the plane into regions bordered by exactly three edges. As an illustration, Fig. 2(a) shows the graph corresponding to the map of Fig. 1, and Fig. 2(b) shows the same graph made into a triangulation.

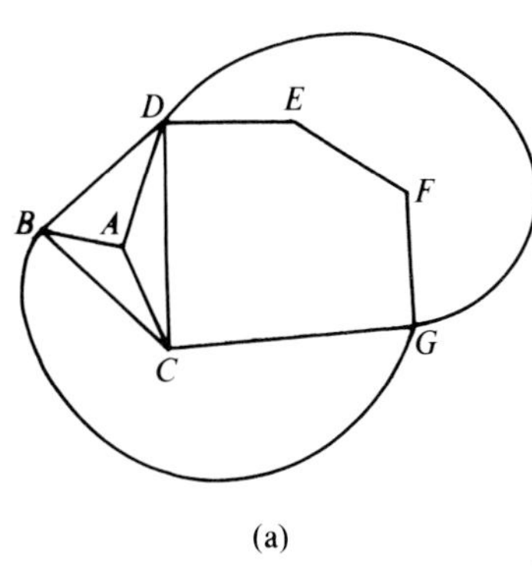

(a)

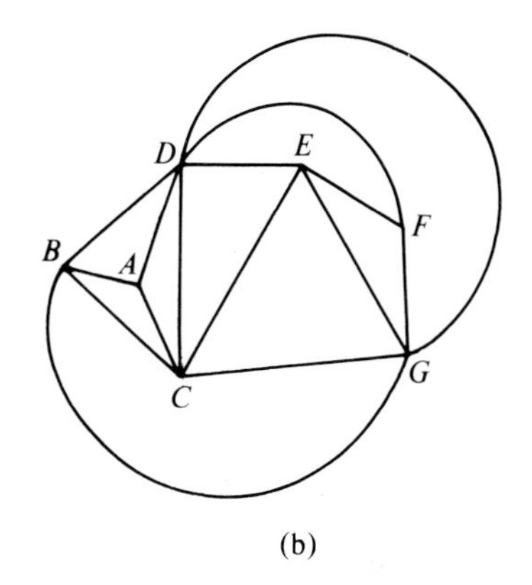

(b)

Fig. 2

It is clear that every plane graph can be triangulated in this way, and that if we can four-color the vertices of the triangulation, then the same coloring will do for the original plane graph. So from now on we shall confine our attention to plane triangulations.

So suppose that T is a plane triangulation with p vertices, q edges and r regions, and suppose further that there are p_k vertices of valency k, for each k. (Note that $p_0 = p_1 = 0$.) By Euler's polyhedral formula (see Chapter 1), $p-q+r = 2$. Clearly the total number of vertices of T is given by

$$p = \sum_k p_k,$$

and, since each edge has two ends,

$$2q = \sum_k kp_k.$$

Since every edge borders two regions, and every region has three edges,

$$2q = 3r.$$

Combining these four equations, we obtain the following result:

Theorem 2.1. *If, for each k, p_k is the number of vertices of valency k in a plane triangulation T, then*

$$\sum_k (6-k)p_k = 12;$$

that is,

$$4p_2+3p_3+2p_4+p_5 \qquad -p_7-2p_8-3p_9- \ldots = 12. \;\|$$

Corollary 2.2. *T contains a vertex of valency at most* 5. ‖

From Corollary 2.2, we deduce that T must contain at least one of the four configurations shown in Fig. 3.

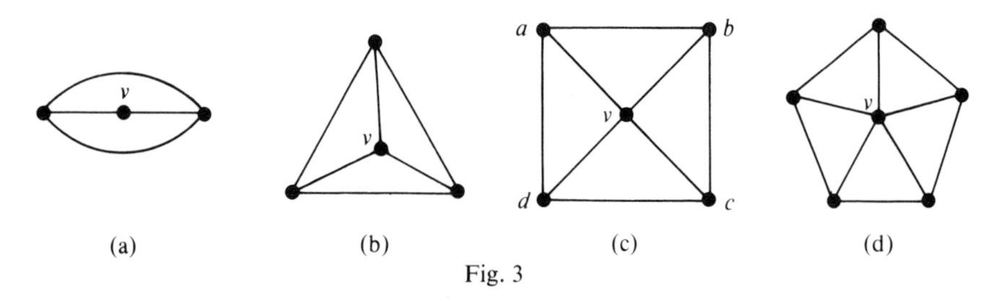

Fig. 3

Let us now suppose that there exists a counter-example to the four-color conjecture; we shall try to prove that this is impossible by obtaining a contradiction. We may choose a counter-example with the minimum number of vertices. If this is not already a triangulation, we can certainly add extra edges to turn it into a triangulation T, without increasing the number of vertices. So every planar graph with fewer vertices than T is four-colorable, but T itself is not.

If T contains either of the configurations in Fig. 3(a) or (b), we need only remove the vertex v from T, color what is left with four colors, and reinstate v. Since v is adjacent to at most three other vertices, there is always a spare color that can be used to color v. So we have a four-coloring for T, contradicting our assumption that T is not four-colorable. This contradiction shows that T cannot in fact contain either of the configurations in Fig. 3(a) or (b).

In the case of the configuration of Fig. 3(c) we can try the same thing, but this time we are in trouble if a, b, c and d are all differently colored; in this case we cannot immediately color v. In order to get round this difficulty,

Kempe used an ingenious argument (of the type that we now call a **Kempe-chain argument**) to show that in this case we can modify the coloring scheme of $T-v$ so that either a and c, or b and d, end up with the same color; there is then a spare color for v, so that T can be four-colored, and we get a contradiction as before. In order to carry this out in practice, we suppose that a, b, c and d are colored red, green, blue and yellow, respectively. Now the graph $T-v$ may contain a path from a to c consisting only of vertices colored red or blue (a "Kempe-chain"), and it may contain a path from b to d consisting only of vertices colored green or yellow, but it cannot contain both, since these two paths would then have to cross at a vertex colored red or blue *and* green or yellow (see Fig. 4). So we may suppose without loss of generality that there is no red-blue path from a to c. In that case it is easy to see that we can alter the coloring of $T-v$ so that a becomes blue without changing the colors of b, c and d, and then we can color v red. This gives the desired contradiction.

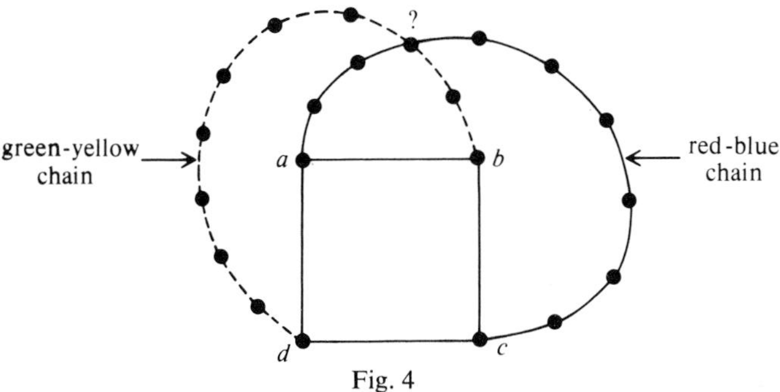

Fig. 4

Thus Kempe showed that T cannot contain any of the configurations in Fig. 3(a), (b) or (c). If he had also been able to show that T cannot contain the configuration in Fig. 3(d), then he would have completed the proof, since we know that if T exists, it must contain at least one of the configurations in Fig. 3(a), (b), (c) or (d), and we have just seen that this is impossible. Unfortunately, Kempe's discussion of the configuration in Fig. 3(d) was incorrect, since he tried to do two color-interchanges simultaneously. That this is not permissible was shown by Heawood, since it can result in two adjacent vertices elsewhere in the graph being re-colored with the same color. However, Heawood was able to use a Kempe-chain argument (similar to that used for the configuration in Fig. 3(c)) to prove the weaker result that every plane graph (or map) can be colored with just five colors—the so-called *five-color theorem*.

Although Kempe's proof was fallacious (and hence technically worthless), he nevertheless made a very fine contribution towards the solution of the problem, and this has often been underestimated by later writers. The arguments he introduced have formed the foundation for most of the subsequent work on the problem, and have led to the two main ideas involved in the eventual proof of the four-color theorem, as we now see.

3. The Two Main Ideas

In order to describe Kempe's ideas in modern terminology, we need a few definitions. In the context of this chapter, a **configuration** C in a plane triangulation T consists of the part of T that lies within some circuit, called the **ring bounding the configuration**; the number of vertices in this ring is then called the **ring-size** of C. Note that the diagrams in Fig. 3 represent configurations consisting of a single vertex v, and the ring-sizes are respectively 2, 3, 4 and 5; Fig. 5 shows a configuration with seven vertices and ring-size 12. (For clarity, we shall sometimes represent the edges of T joining the configuration to the circuit by dashed edges.)

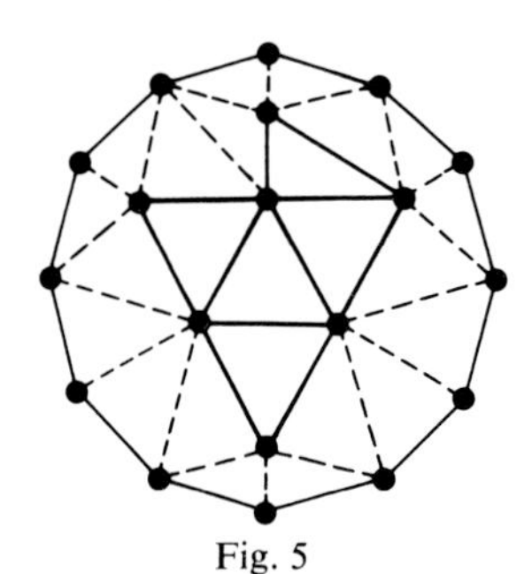

Fig. 5

A set of configurations is said to be **unavoidable** if every plane triangulation contains at least one of them; for example, the four configurations in Fig. 3 form an unavoidable set, since every plane triangulation contains a vertex of valency two, three, four or five. A more complicated example is the unavoidable set shown in Fig. 6; in the next section we shall describe a method to prove that sets such as these are unavoidable.

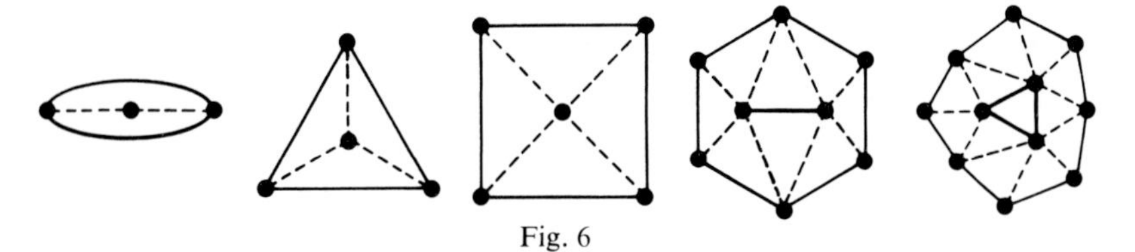

Fig. 6

We also define a configuration C to be **reducible** if it cannot be contained in any minimum counter-example to the conjecture. It follows that if there were a counter-example containing C, then we could find a smaller counter-example. (The idea behind the name is that a counter-example containing C can be "reduced" to a smaller counter-example.) We showed in the previous section that the configurations of Fig. 3(a), (b) and (c) are all reducible, but were unable to show that the configuration of Fig. 3(d) was reducible. Another configuration that can be proved reducible by standard arguments, and which we shall discuss at length in Section 5, is the so-called *Birkhoff diamond*, illustrated in Fig. 7.

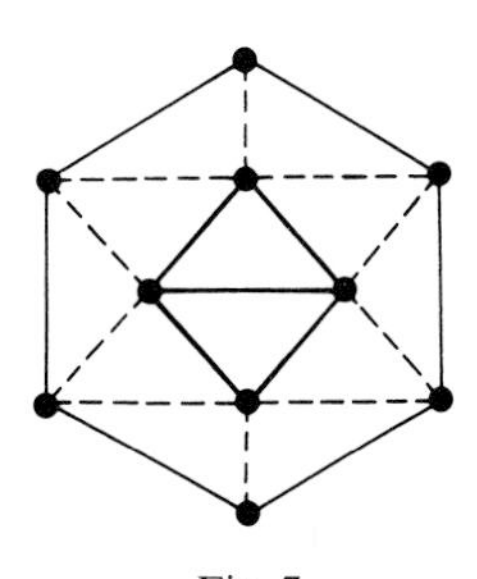

Fig. 7

In order to prove the four-color theorem, it suffices to find an unavoidable set U of reducible configurations. Since U is an unavoidable set, every plane triangulation must contain at least one of these configurations; but since all of the configurations are reducible, none of them can occur in a minimum counter-example. It follows that no counter-example can exist, and hence that the theorem is proved. If Kempe had been able to prove that the configuration of Fig. 3(d) was reducible, he would have obtained an unavoidable set containing just four reducible configurations, and the four-color theorem would have been proved. By way of contrast, Appel and Haken proved the theorem by constructing an unavoidable set of almost 2000 reducible configurations. (They have since brought this number down to about 1500.) These configurations are exhibited in **[6]**, and include the configurations in Figs 3(a), (b), (c), 5 and 7. Some of them are quite large, going up to ring-size 14. To prove that configurations of this size are reducible is very complicated, involving massive reliance on the computer. In fact it can be shown (see Section 6) that Appel and Haken certainly needed to go up to ring-size 12 in order to have any chance of success by their methods, and if they had had to consider many configurations with ring-size greater than 14, then the resulting calculations would probably have been too lengthy for present-day computers.

So Appel and Haken's proof involves two main steps:

(*i*) the construction of an unavoidable set of configurations;

(*ii*) the proof that all of these configurations are reducible.

Each of these steps is comparatively straightforward on its own; it is the interplay between them that is sophisticated, and in which Appel and Haken's work goes qualitatively, and not just quantitatively, way beyond anything that had been done before.

4. Unavoidable Sets

In the previous section we gave an example of an unavoidable set—namely, the four configurations of Fig. 3. Since the last of these configurations has not been shown to be reducible, it is natural to ask whether it can be replaced by any other configurations to form another unavoidable set. This was accomplished in 1904 by Wernicke [**17**], who produced the unavoidable set shown in Fig. 8. Note that the configuration of Fig. 3(d) has been replaced by two new configurations, one consisting of two adjacent vertices of valency 5, and the other consisting of a vertex of valency 5 adjacent to a vertex of valency 6.

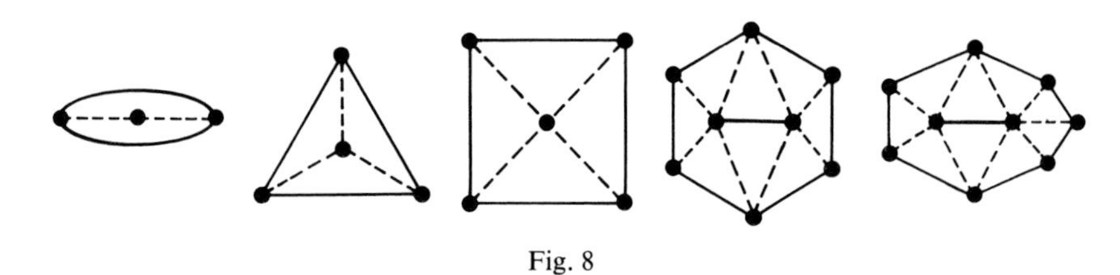

Fig. 8

A slightly more complicated example was shown in Fig. 6, where the last configuration of Wernicke's example is replaced by a triangle consisting of one vertex of valency 5 and two vertices of valency 6. We shall prove the unavoidability of these two sets later in this section.

It is clear that these diagrams are going to become exceedingly cumbersome as the configurations increase in size and complexity. To alleviate this, we shall use the following simplifications:

(*i*) we shall omit the configurations of Fig. 3(a), (b) and (c), since they will always be assumed to occur;

(*ii*) we shall omit the bounding ring and the edges linking it to the configuration;

(*iii*) we shall represent each vertex of valency 5 by a solid circle (●), each vertex of valency 6 by a hollow circle (○), and each vertex of valency $k > 6$ by a circle containing the appropriate valency (ⓚ).

Using these simplifications, we see that the unavoidable sets of Figs 8 and 6 become simply { , } and { , }. Another unavoidable set is { , , }, given by Franklin **[10]** in 1922, and several further examples were given by Lebesgue **[14]** in 1940.

In order to prove that a given set U of configurations is unavoidable, Appel and Haken developed the technique of "discharging", first introduced by Heesch **[12]** in 1969. The idea of this method is as follows. Suppose that there exists a triangulation T that contains none of the configurations in U; we want to obtain a contradiction. Let us assign to each vertex of T a number, called its **charge** (which can be thought of as akin to an electrical charge). More precisely, to each vertex of valency k we assign charge $6-k$, so that vertices of valency 5 receive unit charge, vertices of valency 6 receive zero charge, and vertices of valency 7 or more receive negative charge. Note that, by Theorem 2.1, the total charge assigned to T is positive (12 units). We then try to redistribute the charge among the vertices in such a way that no charge is created or destroyed, and that each vertex ends up with negative or zero charge; this is called "discharging" T. Since no charge is created or destroyed, T must still have total positive charge, and so we have obtained a contradiction; it follows that every triangulation must contain one of the configurations in U, and hence that U is an unavoidable set.

The main difficulty in carrying out this procedure is in deciding how to redistribute the charges. To show how this is done in practice, we consider three simple examples:

Example 1 (Wernicke). { , } *is an unavoidable set.*

Proof. Let T be a triangulation containing no vertices of valency 2, 3 or 4, and neither of the configurations shown, and let each vertex of valency k receive a charge of $6-k$. In order to carry out the discharging process, we transfer *one fifth* of a unit of charge from each positively-charged vertex to each of its negatively-charged neighbors. Since, by assumption, no vertex of valency 5 is adjacent to any vertex of valency 6 or less, every vertex of valency 5 must have five negatively-charged neighbors, and must therefore end up with zero charge. Vertices of valency 6 are unaffected by the discharging process, and end up with the zero charge they started with. A vertex of valency 7 can end up with positive charge only if it has at least six neighbors of valency 5, which cannot happen since at least two of them would be adjacent. In fact, any vertex of valency $k \geqslant 7$ can have at most $\frac{1}{2}k$ neighbors of valency 5, and so can receive at most $\frac{1}{10}k$ units of charge, which are insufficient to overcome the initial negative charge of $6-k$. So the total charge on T is non-positive, contradicting the result of Theorem 2.1. ‖

Example 2 (Appel and Haken). { , } *is an unavoidable set.*

Proof. Let T be a triangulation containing no vertices of valency 2, 3 or 4, and neither of the configurations shown, and allocate charge as before. This time we carry out the discharging process by transfering *one third* of a unit of charge from each vertex of valency 5 to each of its neighbors of valency 7 or more. Now each vertex of valency 5 has at least three neighbors of valency 7 or more, since otherwise it would have either a neighbor of valency 5, or two adjacent neighbors of valency 6, contrary to our assumption; it follows that every vertex of valency 5 ends up with zero or negative charge. As in Example 1, vertices of valency 6 are unaffected, ending up with the zero charge they started with, and vertices of valency 7 or more can never receive enough contributions of $\frac{1}{3}$ to overcome their initial negative charge. So the total charge on T is non-positive, and the result follows from Theorem 2.1, as before. ‖

Example 3 (Appel and Haken). {○, ⑦, ⑧, , , , } *is an unavoidable set.*

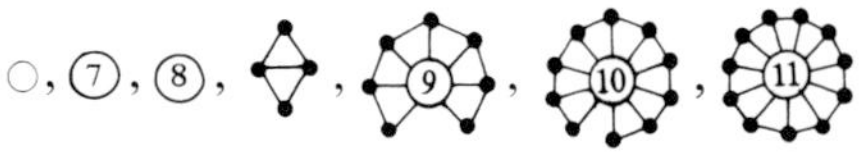

Proof. Let T be a triangulation containing no vertices of valency 2, 3, 4, 6, 7 or 8, and not containing the Birkhoff diamond  or any of the last three configurations shown, and allocate charge as before. Note that every vertex of valency five must be adjacent to at least two vertices of higher valency, since otherwise T would contain a Birkhoff diamond, contrary to our assumption. This time we carry out the discharging process by distributing the charge on each vertex of valency five equally to its neighbors of valency 9 or more. It is clear that each vertex of valency 9 or more can receive no more than half a unit of charge from each of its neighbors of valency 5, and one can use this fact to show, as in Examples 1 and 2 above, that such vertices can never receive enough contributions from their neighbors of valency 5 to overcome their initial negative charge. So the total charge on T is non-positive, and the result follows from Theorem 2.1, as before. ‖

One difficulty we skated over in these examples is that our proofs did not show that one of the given configurations must necessarily occur in T *with all of its vertices distinct*; it is conceivable that some vertices may coincide. It is easy to get round this difficulty when the configurations are small, but for large configurations this is a serious technical problem. Appel and Haken termed this the "immersion problem", and an account of how they tackled it may be found in [**6**].

We conclude this section with a historical note. As a result of using a discharging process similar to that in Example 3 above, in which the charge on each vertex of valency 5 is equally distributed among its neighbors of valency 7 or more, Heesch in 1970 believed that the four-color problem could be reduced to the consideration of only finitely many (about 8900) configurations, which he exhibited explicitly. This would have required the consideration of reducible configurations of ring-size 18, which were too large to be handled explicitly, and it encouraged Appel and Haken to look for discharging algorithms that avoided certain "obstacles" which seemed to prevent the configurations containing them from being reducible. After much trial and error, and a good deal of insight, they obtained a discharging algorithm that worked. This algorithm is far more complicated than any of the ones we have described, although the underlying ideas are basically the same. Its construction, together with a description of the "obstacles" that had to be avoided, will be discussed in Section 6.

5. Reducible Configurations

The study of reducibility can be traced back at least as far as 1913, when Birkhoff wrote an important paper [**8**] on the subject. In this paper he proved, among other things, that the configuration of Fig. 7 (the Birkhoff diamond) is reducible. Since then, many mathematicians have joined the search for reducible configurations, and thousands of such configurations are now known. Our aim in this section is to describe Birkhoff's method, as refined by Heesch, and to outline a proof of the reducibility of the Birkhoff diamond.

So suppose that T is a minimum counter-example to the four-color conjecture, and suppose that it contains the configuration C of Fig. 9(a)—the Birkhoff diamond enclosed in a circuit of length 6. Let T' be the graph obtained from T by removing the Birkhoff diamond—that is, replacing C by the empty hexagon in Fig. 9(b). Since T was a minimum counter-example, T' must be four-colorable. In order to obtain a contradiction, we shall show that every four-coloring of T' can be "extended" to a four-coloring of T.

In order to do this, let us consider a particular four-coloring of T', and let us list the possible color schemes for the vertices of the hexagon in this four-coloring. There are 31 of them, as follows:

1 2 1 2 1 2	1 2 1 3 2 4 √	1 2 3 1 4 3	1 2 3 4 1 2
1 2 1 2 1 3 √	1 2 1 3 4 2 √	1 2 3 2 1 2 √	1 2 3 4 1 3
1 2 1 2 3 2	1 2 1 3 4 3 √	1 2 3 2 1 3 √	1 2 3 4 1 4 √
1 2 1 2 3 4 √	1 2 3 1 2 3	1 2 3 2 1 4 √	1 2 3 4 2 3
1 2 1 3 1 2 √	1 2 3 1 2 4	1 2 3 2 3 2 √	1 2 3 4 2 4 √
1 2 1 3 1 3	1 2 3 1 3 2 √	1 2 3 2 3 4	1 2 3 4 3 2 √
1 2 1 3 1 4	1 2 3 1 3 4	1 2 3 2 4 2	1 2 3 4 3 4 √
1 2 1 3 2 3 √	1 2 3 1 4 2	1 2 3 2 4 3	

Here, the six numbers in each scheme refer to the colors of the vertices a to f in Fig. 9(b). Note that schemes such as 1 2 1 2 1 1 and 1 2 1 2 3 1 are not listed since they give adjacent vertices the same color, and 1 2 1 2 1 4 is not listed since it is of the same type as 1 2 1 2 1 3 (that is, it can be made the same by relabeling the colors). Of course, it may be that not all of these 31 color schemes will actually be possible for T', but since we do not know what T' looks like outside the hexagon, we have got to consider them all.

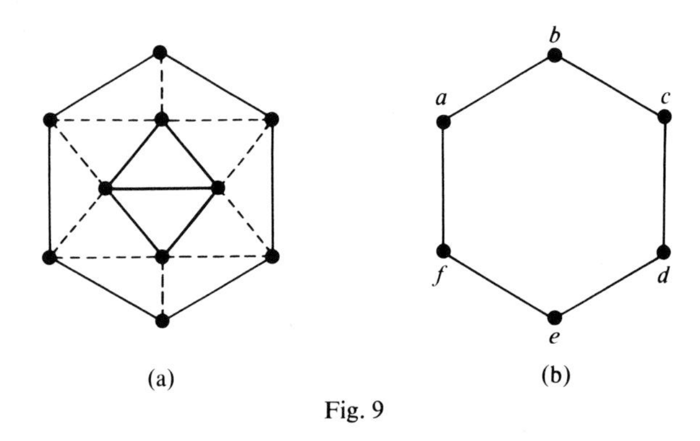

(a) (b)

Fig. 9

Now, some of these color schemes can immediately be extended to colorings of C, and so would give rise to four-colorings of T. These color schemes are called *good* colorings, and are denoted by ticks (√) in the above list. For example, Fig. 10 shows how the color scheme 1 2 1 2 1 3 can be extended to C, and so this color scheme is good. If every one of the 31 possible color schemes were good, then every possible coloring of T' could be extended to give a four-coloring of T; this contradiction would prove that C is reducible. However, this never happens in practice. The next step is to try to use Kempe-chain arguments to convert the bad color schemes into good ones. For example, it is not difficult to check that the color scheme 1 2 1 3 1 3 (which is bad) can always be converted into one of the color schemes 1 2 1 2 1 3,

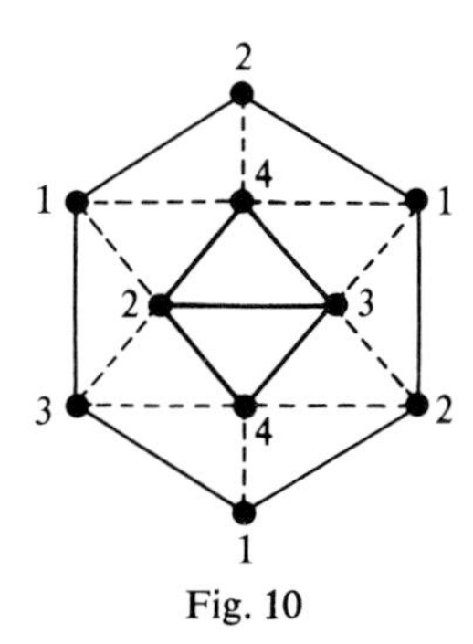

Fig. 10

1 2 1 3 1 2 and 1 2 1 3 4 3 (all of which are good) by interchanging the colors 1 and 4, or 2 and 3, on appropriate Kempe chains; it follows that the scheme 1 2 1 3 1 3 can be made good. Similarly, the scheme 1 2 1 3 1 4 (which is also bad) can always be converted, by interchanges of colors 1 and 2, or 3 and 4, into either 1 2 1 3 2 4 (which is good) or 1 2 1 3 1 3 (which can be made good, as we have just seen); it follows that 1 2 1 3 1 4 can also be made good. If it can be proved that every bad color scheme can be converted into a good one in this way, then, just as before, T can be four-colored and the configuration is reducible; in this case the configuration is called ***D*-reducible**.

The first thing that the computer does when checking for reducibility is to use the above method to see whether or not the configuration is D-reducible. If it is not, then the next step is to note that it is unnecessary to consider all 31 color schemes in the list. Since we are supposing that any graph with fewer vertices than T is four-colorable, we can replace C by any other configuration with fewer vertices, such as the configuration of Fig. 11 (still leaving the rest of the graph unchanged); this gives us another four-colorable graph T''.

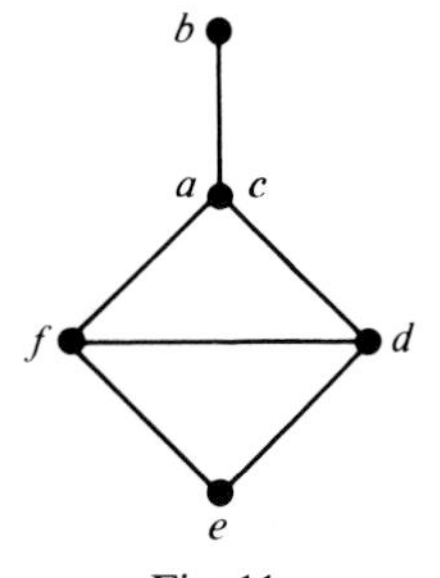

Fig. 11

The effect of this particular substitution is that we need only consider color schemes in which a and c have the same color, and d and f have different colors; this rules out all but six of the color schemes on the list—namely, 1 2 1 2 1 3, 1 2 1 2 3 4, 1 2 1 3 1 2, 1 2 1 3 1 4, 1 2 1 3 2 4 and 1 2 1 3 4 2. Moreover, five of these remaining color schemes are good, and we have just seen that the sixth (1 2 1 3 1 4) can be made good. If this happens (that is, if all the remaining color schemes are good, or are among those that can be made good by Kempe-chain arguments), then again T is four-colorable, and the configuration is reducible. There are clearly many different possible configurations that can be substituted for C; anything that leaves the hexagon intact (with nothing outside it) and contains fewer vertices will do. If any one of these possible substitute configurations achieves the desired result, then the original configuration is called ***C*-reducible**.

The computer programs used by Appel and Haken (written largely by a postgraduate student, John Koch, using algorithms of Heesch) first checked all the possible color schemes to see whether or not the configuration was D-reducible, and then, if it was not, tried a few ways of proving that it was C-reducible. If it was not shown to be C-reducible fairly quickly, then it was abandoned and the unavoidable set was modified appropriately. This may seem a very cumbersome approach, especially since they had to cope with configurations bordered by rings of up to 14 vertices, and not just 6. In fact, it might seem that it would be quicker to test for C-reducibility first. But it turns out in practice that this often involves a lot of duplication of effort if the first substitute configuration does not work, and it is usually found quicker to start by listing all the color schemes, and then seeing which of them can be made good, in the manner described above.

6. Combining the Two Steps

The main point that we have not yet explained is the method by which the discharging procedure and the unavoidable set of configurations were modified each time a configuration in the set turned out to be irreducible (or not quickly proved reducible). It is clear that these progressive modifications relied on a large number of empirical rules, which enabled an unwanted configuration to be excluded from the avoidable set at the expense of possibly introducing one or more further configurations. Appel and Haken carried out about 500 such modifications in all, for a variety of reasons that we now discuss. Further discussion of these matters may be found in Section 2 of Chapter 15.

Reduction Obstacles

While Heesch was testing various configurations for reducibility, he observed that there are certain features that appear to prevent a configuration from being reducible. Although it has never been proved that configurations possessing these features cannot be reducible, it nevertheless seemed sensible to avoid them. The three obstacles to be avoided were as follows:

(*i*) a "4-legger vertex" (such as vertex v in Fig. 12(a)), which is a vertex inside the bounding circuit that is joined to four consecutive vertices of the circuit;

(*ii*) a "3-legger articulation vertex" (such as vertex v in Fig. 12(b)), which is a vertex inside the bounding circuit that is joined to three vertices of the circuit that are not all consecutive;

(*iii*) a "hanging 5–5 pair" (such as vertices v and w in Fig. 12(c)), which is a pair of adjacent vertices of valency 5 inside the bounding circuit, each of which has only one (and the same) neighbor u inside the bounding circuit.

A configuration is said to be **geographically good** if it does not contain either of the obstacles (*i*) and (*ii*).

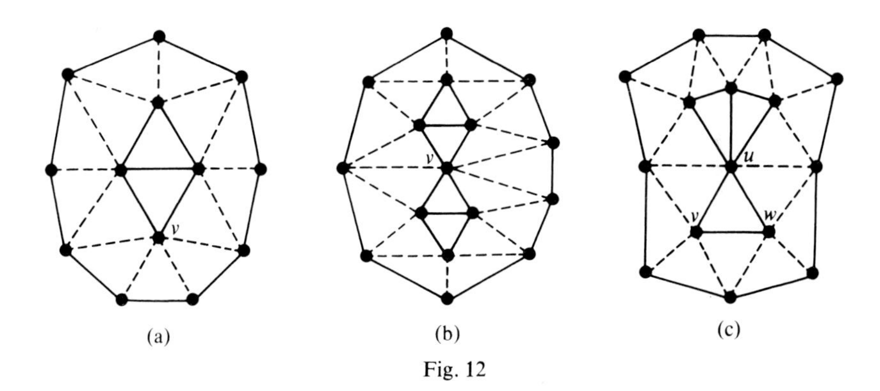

(a) (b) (c)

Fig. 12

Appel and Haken felt that if there was a manageable set of configurations which were all obstacle-free, then there was a very good chance of finding such a set consisting entirely of reducible configurations. In 1972, they wrote a computer program using the discharging process that seemed most natural to them—namely, distributing the positive charges on vertices of valency 5 equally to all negatively-charged neighbors. The result of this was that geographically good configurations of reasonable size were found to occur near to most vertices which ended up with positive charge. Moreover, the same configurations occurred often enough for them to hope that the eventual list of configurations might be of manageable size.

Using the results of their first computer run, Appel and Haken were able to make several adjustments to their discharging process to overcome some of the problems that had been indicated. A man–machine dialogue ensued, which continued for several months, with the discharging process modified at each stage. Eventually Appel and Haken obtained a discharging process that yielded an unavoidable set of geographically good configurations. Replacing this set by an unavoidable set of obstacle-free configurations (avoiding obstacles of type (*iii*)) only doubled the size of the unavoidable set, and the stage was ready for obtaining an unavoidable set of reducible configurations.

Ring-size

The trouble with some of the configurations obtained by the process just described was that they were too large. If there was to be any hope of using the method of Section 5 to prove these configurations reducible, it was essential to keep the ring-size as small as possible. Although it is a reasonably simple matter to check configurations with ring-size 11 on the computer, the computer time required increases by a factor of about four whenever the ring-size increases by one. In fact, although the average time taken for a configuration of ring-size 14 was less than 30 minutes, one particularly troublesome configuration of this size took no less than 26 hours when it was first run, and even then the results were inconclusive. If Appel and Haken had needed to go up to ring-size 18, as they had originally expected, the computer time required would have been in excess of what was available—possibly even a matter of years!

On the other hand, it was known that it would be necessary to go up to ring-size 12, since some years earlier, E. H. Moore had constructed a map in which the ring-size of the smallest *C*-reducible configuration is 12. (The illustration of such a map which appears in [**5**] is incorrect; however, Moore has produced a new map with the required properties.) Using a probabilistic argument, Appel and Haken showed that it was "very likely" that they could restrict their attention to an unavoidable set of configurations with ring-size 14 or less, and this they did. Whether or not there is an unavoidable set of *C*-reducible configurations with ring-size 13 or less remains an open question.

Computer Time

Appel and Haken also used probabilistic ideas to decide whether or not a given configuration was likely to be proved reducible. Using the computer, they listed the possible color-schemes of the bounding circuit, as we did in Section 5, and calculated the proportion of these color-schemes that were good. They found, as a useful rule-of-thumb, that if this proportion is less than 10%, then the configuration has practically no chance of being *D*-reducible; if it is about 20%, then there is a very good chance; and if it is over 30%, then *D*-reducibility is almost certain. (Further details of this type of argument may be found in [**4**].)

As we mentioned in Section 5, the computer first tested for *D*-reducibility. If this failed, then a few ways of proving *C*-reducibility were tried. Any configuration which was not fairly quickly proved reducible in this way (within 90 minutes on an IBM 370-158, or 30 minutes on an IBM 370-168) was abandoned, and replaced by one or more other configurations. It is probable that many configurations that are actually *C*-reducible were excluded from the unavoidable set, but it proved more efficient to exclude any configuration which did not yield quickly, and to replace it with other configurations, than to carry the analysis of any one particular configuration to its limit.

As remarked above, the initial unavoidable set of geographically good configurations was obtained as the result of a lengthy process of trial and error with the aid of a computer, lasting over a year. By the end of this time, however, Appel and Haken had developed such a feeling for what was likely to work (even though they could not always explain why) that they were eventually able to carry out the discharging process by hand, and so construct the final unavoidable set without using the computer at all. This is the crux of their achievement. Unavoidable sets had been constructed before, and configurations had been proved reducible before, but no-one before had been able to complete the monumental task of constructing an unavoidable set of reducible configurations.

7. Conclusions

The length of Appel and Haken's proof is unfortunate, for two reasons. The first is that it makes it difficult to verify. Clearly a proof has no value unless it can be checked by other mathematicians, and ideally one would like a proof to be verifiable by as many people, and as quickly, as possible. A very long proof is likely to take a long time to check, and may be intellectually accessible to only comparatively few people. (Of course, a short proof may have these defects as well.) This is particularly true of a proof that uses the computer. Before the introduction of computers into pure mathematics, every proof could be checked by anyone possessing the necessary mental apparatus. Now a further requirement is added—namely, access to a high-speed computer. Appel estimates that to check all the details of their proof would take about 300 hours on a big machine. There must be few mathematicians who have access to that sort of computer time.

The other big disadvantage of a long proof is that it tends not to give very much understanding of why the result is true. This is particularly true of a proof that involves looking at a large number of separate cases, whether or not it uses the computer. Many mathematicians would consider that proving theorems is only a means to the true end of pure mathematics, which is to understand what is going on. Sometimes a proof is so illuminating that one feels immediately that it explains the real reason for the result being true. It may be unreasonable to expect every theorem to have a proof of this sort, but it seems nonetheless to be a goal worth aiming for. So undoubtedly much work will be done in the next few years in an attempt to shorten Appel and Haken's proof, and possibly to find a more illuminating method of proof altogether, since it seems unlikely that their method of proof can be shortened sufficiently to avoid massive reliance on the computer.

In fact, there remain a number of conjectures that, if they could be proved, would imply the truth of the four-color theorem, but do not follow from it. One of these in particular (Hadwiger's conjecture, that every k-chromatic graph is contractible to a graph containing K_k) is unlikely to be provable by the sort of technique that Appel and Haken have used, and it is possible that a shorter proof of the four-color theorem will result from attempts to prove the more general conjecture. On the other hand, it is conceivable that there are limits to what can be achieved by theoretical methods alone. It may be that there are theorems of great mathematical interest that can be proved only by computer methods, and that the four-color theorem is one of these. Whatever the case, there is no doubt that Appel and Haken's proof is a magnificent achievement which will cause many mathematicians to think afresh (or possibly for the first time) about the role of the computer in mathematics.

References

For an account of the early history of the problem, see [**7**]; for a general survey of the solution, see [**1**] or [**5**]; for background reading, see [**15**]; and for further work by Appel and Appel and Haken, see [**2**].

1. K. Appel, The proof of the four-colour theorem, *New Scientist* **72** (21 October 1976), (1976), 711–712; *MR***54**#12561.
2. K. Appel and W. Haken, The existence of unavoidable sets of geographically good configurations, *Illinois J. Math.* **20** (1976), 218–297.

3. K. Appel and W. Haken, Every planar map is four colorable, *Bull. Amer. Math. Soc.* **82** (1976), 711–712; *MR***54**#12561.
4. K. Appel and W. Haken, Every planar map is four colorable: Part 1, Discharging, *Illinois J. Math.* **21** (1977), 429–490.
5. K. Appel and W. Haken, The solution of the four-color-map problem, *Scientific American* **237**, No. 4 (October 1977), 108–121.
6. K. Appel, W. Haken and J. Koch, Every planar map is four colorable: Part 2, Reducibility, *Illinois J. Math.* **21** (1977), 491–567.
7. N. L. Biggs, E. K. Lloyd and R. J. Wilson, *Graph Theory 1736–1936*, Clarendon Press, Oxford, 1976.
8. G. D. Birkhoff, The reducibility of maps, *Amer. J. Math.* **35** (1913), 115–128 = *Collected Mathematical Papers*, Vol. 3, American Mathematical Society, New York, 1950, pp. 6–19.
9. A. Cayley, On the colouring of maps, *Proc. Roy. Geog. Soc.* (*New Ser.*) **1** (1879), 259–261 = *Math. Papers*, Vol. 11, Cambridge University Press, Cambridge, 1896, pp. 7–8.
10. P. Franklin, The four color problem, *Amer. J. Math.* **44** (1922), 225–236.
11. P. J. Heawood, Map-colour theorem, *Quart. J. Pure Appl. Math.* **24** (1890), 332–338.
12. H. Heesch, Untersuchungen zum Vierfarbenproblem, B. I. Hochschulscripten, 810/810a/810b, Bibliographisches Institut, Mannheim-Vienna-Zürich, 1969; *MR***40**#1303.
13. A. B. Kempe, On the geographical problem of the four colours, *Amer. J. Math.* **2** (1879), 193–200.
14. H. Lebesgue, Quelques conséquences simples de la formule d'Euler, *J. de Math. Pure Appl.* **19** (1940), 27–43; *MR***1**–316.
15. O. Ore, *The Four-Color Problem*, Academic Press, New York, 1967; *MR***36**#74.
16. T. L. Saaty and P. C. Kainen, *The Four-Color Problem*, Mc-Graw Hill, New York, 1977.
17. P. Wernicke, Über den kartographischen Vierfarbensatz, *Math. Ann.* **58** (1904), 413–426.

5
Edge-Colorings of Graphs

STANLEY FIORINI AND ROBIN J. WILSON

1. Introduction

In the past hundred years, problems involving the coloring of the vertices or the regions of graphs and maps have received considerable attention—mainly because of one problem, the four-color problem. In view of this, it is somewhat surprising that various closely-related problems involving the coloring of the *edges* of a graph have, until recently, received comparatively little attention. In this chapter we shall describe some of the progress that has been made in connection with edge-coloring problems.

Although the origins of chromatic graph theory may be traced back to 1852, with the birth of the four-color problem, the first papers to be written on edge-colorings did not appear until 1880, when P. G. Tait published two brief abstracts in the *Proceedings of the Royal Society of Edinburgh* [**36**]. In these papers, Tait showed that if the four-color conjecture is true, then the edges of every trivalent planar graph can be properly colored using only three colors; we shall prove this result and its converse in Section 2. This approach was developed by P. J. Heawood in his "Heawood congruences" paper [**19**] in 1898, but after this, little was done until D. König proved in 1916 [**25**] that if G is a bipartite graph whose maximum valency is ρ, then its edges can be properly colored using at most ρ colors. Somewhat later, in 1949, C. E.

Shannon [**33**] proved that if G is an arbitrary graph or multigraph with maximum valency ρ, then its edges can be properly colored with $[\frac{3}{2}\rho]$ colors. The great breakthrough occurred in 1964, however, when V. G. Vizing [**39**] proved that if G is a graph (without multiple edges), then the number of colors needed to color the edges of G is always either ρ or $\rho+1$—a surprisingly strong result. Much of this chapter will be concerned with the problem of finding which graphs can be colored with ρ colors, and which need $\rho+1$ colors.

Just as in vertex-coloring problems, it turns out that arbitrary graphs are often too clumsy to deal with in general, and we find it convenient to deal with graphs which are "critical" in some sense. We shall discuss the most important properties of critical graphs in Sections 5 and 6. This study gives rise to various interesting conjectures, and we shall describe the progress made on these conjectures. Graphs with particular topological and chromatic properties are considered in Sections 7 and 8, and Section 9 is devoted to the "multichromatic index" (a parameter related to the chromatic index), the study of which provides further evidence for some of the above-mentioned conjectures.

In a short account such as this, it is impossible to do justice to the wide variety of results relating to edge-colorings of graphs. For example, we shall not be discussing applications of the theory to such areas as electrical network theory, scheduling problems, matrix algebra, and the design of experiments. For details of these and other topics we refer the interested reader to our recent book [**15**].

2. The Chromatic Index

If G is a graph or multigraph, we define its **chromatic index** (or **edge-chromatic number**) to be the least number of colors needed to color the edges of G in such a way no two adjacent edges are assigned the same color; we denote the chromatic index of G by $\chi'(G)$. It is clear that if G is a graph with at least one edge, then the chromatic *index* of G is equal to the chromatic *number* $\chi(L(G))$ of $L(G)$, the line graph of G. Note also that if G contains a vertex of valency k, then $\chi'(G) \geqslant k$; it follows that if ρ is the maximum valency of G, then ρ is a lower bound for the chromatic index of G. Examples of the chromatic index, for certain well-known types of graph, are stated without proof in the following theorem:

Theorem 2.1. (*i*) *The chromatic index of the circuit graph C_p is given by*

$$\chi'(C_p) = \begin{cases} 2, & \textit{if } p \textit{ is even}, \\ 3, & \textit{if } p \textit{ is odd}; \end{cases}$$

(*ii*) *the chromatic index of the complete graph* K_p *is given by*

$$\chi'(K_p) = \begin{cases} p-1, & \text{if } p \text{ is even}, \\ p, & \text{if } p \text{ is odd}; \end{cases}$$

(*iii*) *the chromatic index of the complete bipartite graph* $K_{r,s}$ *is* $\max\{r, s\}$. ||

Remark. Part (*i*) is trivial, and a proof of part (*ii*) may be found in (for example) [**5**] or [**15**]. Part (*iii*) follows as an immediate corollary of König's theorem [**25**], mentioned in the Introduction, and proved below (Theorem 2.2). König's proof uses induction on the number of edges of the graph, and employs a "Kempe-chain argument"; this type of argument was discussed in Chapter 4 in connection with vertex-colorings, and involves looking at a two-colored subgraph $H(\alpha, \beta)$ and interchanging the colors. Note that in the case of edge-colorings, any such two-colored subgraph has a particularly simple form—namely, a union of open paths and even circuits.

Theorem 2.2 (König's Theorem). *If* G *is a bipartite graph with maximum valency* ρ, *then* $\chi'(G) = \rho$.

Proof. We use induction on the number of edges of G. It is clearly sufficient to prove that if all but one of the edges of G have been colored with at most ρ colors, then there is a ρ-coloring of the edges of G.

So suppose that each edge of G has been colored, except for the edge vw. Then there is at least one color missing at the vertex v, and at least one color missing at the vertex w. If there is some color missing from *both* v and w, then the result follows by coloring the edge vw with this color. If this is not the case, then let α be a color missing at v, and β be a color missing at w, and let $H(\alpha, \beta)$ be the connected subgraph of G consisting of the vertex w and all those edges and vertices of G which can be reached from w by a path consisting entirely of edges colored α or β. Since G is bipartite, the subgraph $H(\alpha, \beta)$ cannot contain the vertex v, and so we can interchange the colors α and β in this subgraph without affecting v or the rest of the coloring. The edge vw can then be colored α, thereby completing the coloring of the edges of G. ||

Several results concerning vertex-colorings of graphs have natural analogs for edge-colorings. As an example, we present the following analog of the theorem of Nordhaus and Gaddum (see Chapter 1) on the chromatic numbers of a graph and its complement; this result was proved by Vizing [**40**], and (somewhat later) by Alavi and Behzad [**1**]:

Theorem 2.3. *Let G be a graph of order p, and let $\bar{G}$ be its complement. If p is even, then*

$$p-1 \leqslant \chi'(G)+\chi'(\bar{G}) \leqslant 2(p-1), \quad \text{and} \quad 0 \leqslant \chi'(G)\cdot\chi'(\bar{G}) \leqslant (p-1)^2;$$

if p is odd, then

$$p \leqslant \chi'(G)+\chi'(\bar{G}) \leqslant 2p-3, \quad \text{and} \quad 0 \leqslant \chi'(G)\cdot\chi'(\bar{G}) \leqslant (p-1)(p-2). \;\|$$

We conclude this section by discussing those graphs which initiated the whole subject—namely, trivalent maps. If G is a trivalent map, then its line graph $L(G)$ is a regular 4-valent graph which is not isomorphic to K_5. It follows from Brooks' theorem (see Chapter 1) that $\chi(L(G)) \leqslant 4$, and hence that $\chi'(G) \leqslant 4$. The fact that $\chi'(G) = 3$ for all trivalent maps G follows as a direct consequence of Tait's classic result of 1880 **[36]**, relating the chromatic index of trivalent maps to the four-color theorem; it is because of this theorem that a 3-coloring of the edges of a trivalent map is sometimes referred to as a *Tait coloring*:

Theorem 2.4. *The four-color theorem is equivalent to the statement that every trivalent map has chromatic index* 3.

Proof. Suppose first that G is a trivalent map whose regions are colored with the four colors A, B, C and D. We can obtain a 3-coloring of the edges of G by coloring with color 1 those edges which separate regions colored A and B, or C and D; by coloring with color 2 those edges which separate regions colored A and C, or B and D; and by coloring with color 3 those edges which separate regions colored A and D, or B and C (see Fig. 1). It is easy to see that this coloring assigns the colors, 1, 2 and 3 to the edges of G in such a way that the edges meeting at each vertex are differently colored.

To prove the converse result, it is sufficient to prove that if G is any trivalent map whose edges have been colored with the colors 1, 2 and 3, then the regions of G can be colored with four colors. But the subgraph of G determined by those edges colored 1 or 2 is a regular 2-valent graph whose regions can be colored with two colors α and β. In a similar way, the regions of the subgraph determined by those edges colored 1 or 3 can be colored with the

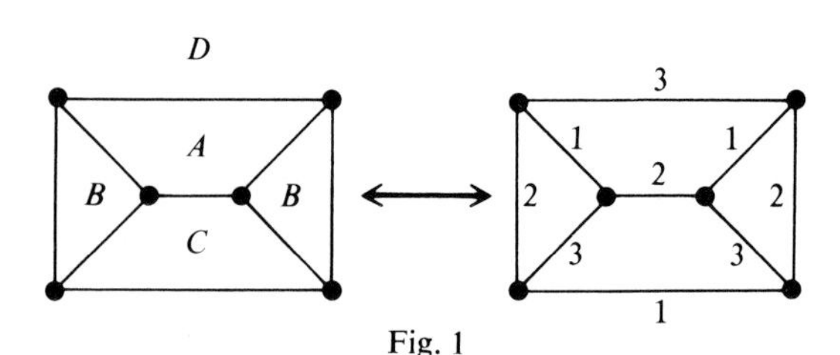

Fig. 1

colors γ and δ. It follows that we can assign to each region of G two coordinates (x, y), where x is either α or β, and y is either γ or δ. Since the coordinates assigned to two adjacent regions of G must differ in at least one place, it follows that these coordinates $A = (\alpha, \gamma)$, $B = (\beta, \delta)$, $C = (\beta, \gamma)$ and $D = (\alpha, \delta)$, give the required 4-coloring of the regions of G. ‖

Corollary 2.5. *Every trivalent map has chromatic index* 3. ‖

3. Vizing's Theorem

Until 1964, the most that one could say in general about the chromatic index of an arbitrary graph G with maximum valency ρ was that $\rho \leqslant \chi'(G) \leqslant [\frac{3}{2}\rho]$. The left-hand inequality is trivial, and the right-hand inequality was proved in 1949 by Shannon [**33**] for an arbitrary multigraph. In 1964, Vizing [**39**] showed that if we restrict our attention to graphs, rather than multigraphs, then the right-hand inequality can be considerably strengthened to $\rho+1$. In this section we shall prove the results of both Vizing and Shannon, starting with Vizing's Theorem.

Theorem 3.1 (Vizing's Theorem). *If G is a graph with maximum valency ρ, then $\rho \leqslant \chi'(G) \leqslant \rho+1$.*

Proof. It is clear that $\chi'(G) \geqslant \rho$. In order to prove that $\chi'(G) \leqslant \rho+1$, we use induction on the number of edges of G; more precisely, we prove that if all but one of the edges of G have been colored with at most $\rho+1$ colors, then there is a $(\rho+1)$-coloring of all of the edges of G. So suppose that each edge of G has been colored with one of the $\rho+1$ given colors, with the single exception of the edge $e_1 = vw_1$. Then there must be at least one color missing at the vertex v, and at least one color missing at the vertex w_1. If some color is missing at both v and w_1, then this color can be used to color the edge e_1, and the proof is complete. If this is not the case, let α be any color missing at v, and let β_1 $(\neq\alpha)$ be any color missing at w_1. The proof now proceeds in three steps, as follows:

Step 1. Let $e_2 = vw_2$ be the edge incident with v which has been assigned the color β_1; such an edge must exist, since otherwise the color β_1 would be missing from both v and w_1, contrary to our assumption. We now un-color the edge e_2, and assign the color β_1 to the edge e_1 (see Fig. 2). We may assume that the vertices v, w_1 and w_2 all belong to the same component of $H(\alpha, \beta_1)$, since otherwise we could interchange the colors of the edges in the component containing w_2 without altering the color of e_1; this would mean that the edge e_2 could then be colored α, thereby completing the coloring of the edges of G. So the situation is now as depicted in Fig. 3.

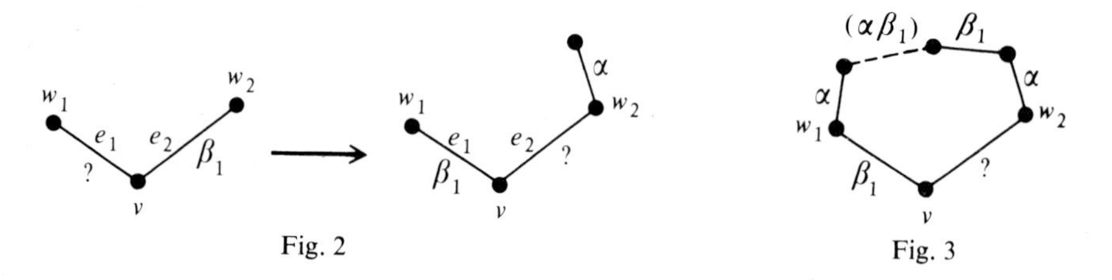

Fig. 2

Fig. 3

Step 2. Now let β_2 ($\neq \beta_1$) be a color missing at w_2. We may assume that β_2 occurs at v, since otherwise we can complete the proof by using β_2 to color the edge e_2. So let $e_3 = vw_3$ be the edge incident to v colored β_2. Then we can un-color the edge e_3, and assign the color β_2 to the edge e_2. By the same argument as we used in Step 1, we may assume that the vertices v, w_2 and w_3 all belong to the same component of $H(\alpha, \beta_2)$. The situation is now as depicted in Fig. 4.

Step 3. If we repeat the above procedure, we shall eventually reach a vertex w_k adjacent to v such that the edge vw_k is un-colored, and some color β_i $(i < k-1)$ is missing from the vertex w_k. As before, we may assume that the vertices v, w_i and w_{i+1} all belong to the same component Γ of $H(\alpha, \beta_i)$. Since α does not appear at v, nor β_i at w_{i+1}, Γ must be a path from v to w_{i+1}, passing through w_i and consisting entirely of edges colored alternately β_i and α (see Fig. 5). Clearly this path does not contain w_k, since β_i does not

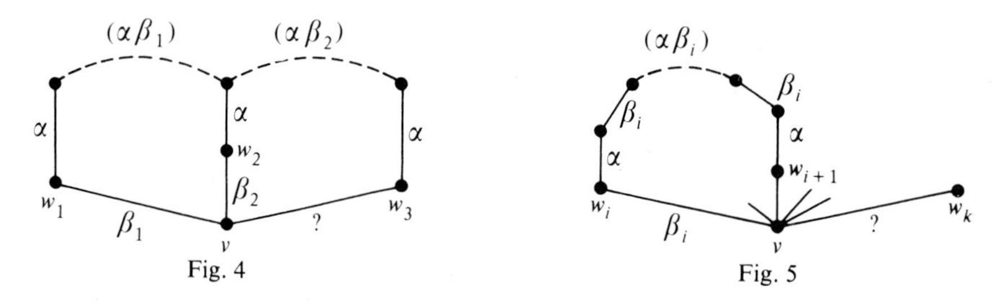

Fig. 4

Fig. 5

appear at w_k. It follows that if $\tilde{\Gamma}$ is the component of $H(\alpha, \beta_i)$ containing the vertex w_k, then Γ and $\tilde{\Gamma}$ must be disjoint. It is therefore possible to interchange the colors of the edges in $\tilde{\Gamma}$, thereby enabling us to color the edge vw_k with color α. This completes the proof. ||

Vizing also obtained a corresponding bound for multigraphs, which is sometimes better than that given by Shannon. It involves the **maximum multiplicity** μ of a multigraph M, defined to be the maximum number of edges joining any pair of vertices in M. This result, obtained independently by Gupta [**18**], is proved in [**39**] (see also [**5**], [**20**], [**30**]), and may be stated as follows:

Theorem 3.2. *If M is a multigraph with maximum valency ρ and maximum multiplicity μ, then $\rho \leqslant \chi'(M) \leqslant \rho+\mu$.* ||

We conclude this section by showing how Vizing's theorem for multigraphs can be used to prove Shannon's result:

Theorem 3.3 (Shannon's Theorem). *If M is a multigraph with maximum valency ρ, then $\chi'(M) \leqslant [\frac{3}{2}\rho]$.*

Proof. Let M be a multigraph for which $\chi'(M) = k$, where $k > [\frac{3}{2}\rho]$. By removing sufficiently many edges from M (if necessary), we may assume that $\chi'(M-e) = k-1$, for each edge e of M. It follows from Theorem 3.2 that $k \leqslant \rho+\mu$, where μ is the maximum multiplicity of M, and so there must exist vertices v and w which are joined by at least $k-\rho$ edges.

We now color all of the edges of M except one of the edges joining v and w; since $\chi'(M-e) = k-1$, this coloring can be done with $k-1$ colors. Now the number of colors missing from v or w (or both) cannot exceed $(k-1)-(\mu-1)$, which in turn cannot exceed ρ, since $k \leqslant \rho+\mu$. But the number of colors missing from v is at least $(k-1)-(\rho-1) = k-\rho$, and similarly the number of colors missing from w is at least $k-\rho$. It follows that the number of colors missing from *both* v and w is at least $(2k-\rho)-\rho$, which is positive since $k > [\frac{3}{2}\rho]$. By assigning one of these missing colors to the un-colored edge joining v and w, we have colored all of the edges of M using only $k-1$ colors, thereby contradicting the fact that $\chi'(M) = k$. This contradiction establishes the theorem. ||

4. The Classification Problem

Vizing's theorem (Theorem 3.1) gives us a simple way of classifying graphs into two classes. A graph G is said to be of **class 1** if $\chi'(G) = \rho$, and of **class 2** if $\chi'(G) = \rho+1$. We have already seen that even complete graphs K_{2k} and bipartite graphs are of class 1, and that odd circuit graphs C_{2k+1} and odd complete graphs K_{2k+1} are of class 2, but the general problem of deciding which graphs belong to which class (the so-called **Classification Problem**) is unsolved. The importance and difficulty of this problem become apparent when we realize that its solution would imply the four-color theorem, in view of Tait's theorem (Theorem 2.4). Although the classification problem is in general unsolved, it seems that graphs of class 2 are relatively scarce. For example, of the 143 connected graphs with at most six vertices, only eight are of class 2 (see Fig. 6). A more general result of this kind is due to Erdős and Wilson [**11**], who have proved that almost all graphs are of class 1, in the

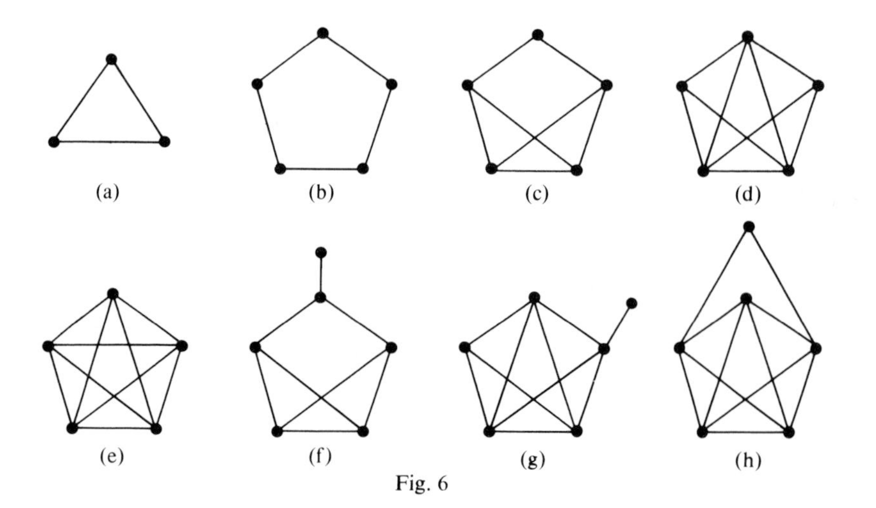

Fig. 6

sense that if $P(p)$ is the probability that a random graph of order p is of class 1, then $P(p) \to 1$ as $p \to \infty$. However, no progress has been made on the more difficult problem of deciding which class contains almost all graphs with a given maximum valency ρ; even for $\rho = 3$ this is unknown.

It seems natural to expect that the more edges a graph has, the more likely it is to be of class 2. This idea is made precise in the following result, which gives a sufficient condition for a graph to be of class 2. This elementary result was proved by Beineke and Wilson [**4**], but is implicit in the work of Vizing [**40**]; note that it applies only when p is odd.

Theorem 4.1. *Let G be a graph with p vertices, q edges, and with maximum valency ρ; then G is of class 2 if $q > \rho[\frac{1}{2}p]$.*

Proof. If G is of class 1, then any ρ-coloring of the edges of G partitions the set of edges into ρ independent subsets. But the number of edges in each independent subset cannot exceed $[\frac{1}{2}p]$, since otherwise two of these edges would be adjacent. It follows that $q \leqslant \rho[\frac{1}{2}p]$, giving the required contradiction. ‖

Using Theorem 4.1, we can immediately deduce the following corollary:

Corollary 4.2. (*i*) *Every regular graph of odd order is of class* 2;

(*ii*) *if H is a regular ρ-valent graph of odd order, and if G is any graph obtained from H by deleting at most $\frac{1}{2}\rho - 1$ edges, then G is of class* 2;

(*iii*) *if H is a regular graph of even order, and if G is any graph obtained from H by inserting a new vertex into any edge of H, then G is of class* 2;

(*iv*) *if G is any graph obtained by taking an odd circuit* C_{2k+1} *and adding to it not more than* $2k-2$ *independent sets of k edges, then G is of class* 2. ‖

We can also deduce the following result of Vizing [**40**]:

Corollary 4.3. *If G is a regular graph containing a cut-vertex, then G is of class* 2.

Proof. If G is of odd order, then the result follows from Corollary 4.2(*i*). If G is of even order, let $G = H \cup K$, where $H \cap K = \{v\}$. We may assume that H has odd order (k, say), and that every vertex of H has valency ρ, except for v whose valency in H is less than ρ. It follows that the number of edges of H is $\frac{1}{2}\{(k-1)\rho+\rho_H(v)\} > \rho[\frac{1}{2}k]$, and the result follows from Theorem 4.1. ‖

Corollaries 4.2 and 4.3 provide us with several examples of graphs of class 2. Further examples are given by the following theorem, part (*i*) of which is due to Laskar and Hare [**27**], and part (*ii*) to Parker [**31**]; the graphs $K_{3(2)}$ and $C_{3(2)}$ are both illustrated in Fig. 7.

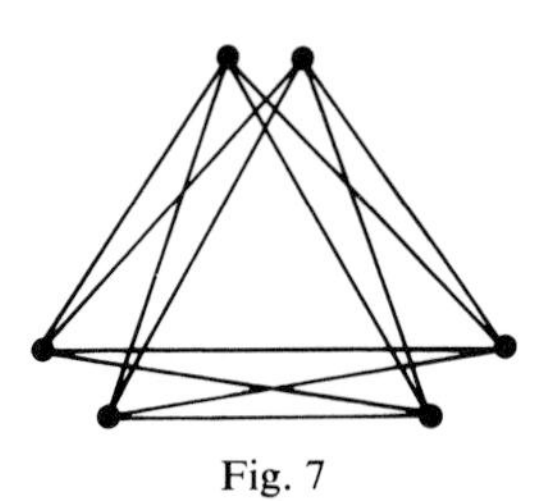

Fig. 7

Theorem 4.4. (*i*) *Let* $K_{r(k)}$ *be the complete r-partite graph each of whose parts has exactly k vertices; then* $K_{r(k)}$ *is of class* 2 *if and only if k and r are both odd.*

(*ii*) *Let* $C_{r(k)}$ *be the generalized circuit graph obtained by arranging r copies of the null graph with k vertices into a cycle, and joining two vertices if and only if they lie in neighboring members of the cycle; then* $C_{r(k)}$ *is of class* 2 *if and only if k and r are both odd.* ‖

In studying the classification problem, it is natural to consider the problem of classifying regular graphs, and trivalent graphs in particular. By Corollary 4.2(*i*), we may restrict our attention to trivalent graphs of even order. Much of the interest in these graphs arises from the following conjecture of Tutte [**38**], showing that the Petersen graph has an important role to play in such investigations:

Conjecture. *Every bridgeless trivalent graph of class* 2 *contains a subgraph contractible* (*or homeomorphic*) *to the Petersen graph.*

There has been much interest in the edge-chromatic properties of graphs related to the Petersen graph. Although the Petersen graph itself is of class 2 with chromatic index 4, its generalizations can lie in either class. We conclude this section by looking at two classes of graphs—the generalized Petersen graphs, introduced by Watkins [**43**], and the so-called "snarks", which have been hunted by several people over the years. Further generalizations may be found in the papers of Meredith [**28**], Meredith and Lloyd [**29**], and Biggs [**7**], or in [**15**, Chapter 7].

The **generalized Petersen graph** $P(n, k)$ consists of an outer n-circuit, n spokes incident to the vertices of this n-circuit, and an inner n-circuit attached by joining its vertices to every kth spoke. The generalized Petersen graph $P(9, 2)$ is shown in Fig. 8; we shall be returning to it in Section 8. Note that $P(n, k)$ is isomorphic to $P(n, n-k)$.

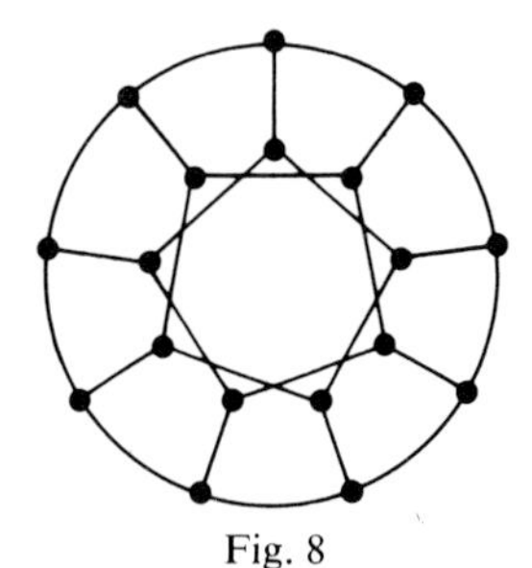

Fig. 8

In 1969, Watkins conjectured that, apart from the Petersen graph itself, all of these graphs are of class 1. Although he managed to settle several cases of this conjecture himself, he was unable to solve it in general, and it was not until four years later that the conjecture was finally proved by Castagna and Prins [**8**]. We state their result as a theorem:

Theorem 4.5. *All of the generalized Petersen graphs are of class* 1, *except the Petersen graph itself.* ‖

There has also been a dramatic interest in snark-hunting in recent years. A **snark** is a bridgeless trivalent graph of class 2, usually taken (to avoid trivial cases) to contain no separating sets with at most three edges, and to have girth at least 5. (Because such graphs are so difficult to find, Martin Gardner, in a delightful popular article on the subject [**16**], christened them "snarks" after

Lewis Carroll's *The Hunting of the Snark*.) Part of the interest in snark-hunting arose from trying to prove or disprove the existence of a planar graph of this type, since this would settle the four-color problem (Theorem 2.4). Now that the four-color theorem has been proved, and planar snarks are known not to exist, the search is directed towards non-planar snarks.

The smallest of all snarks is the Petersen graph, which dates from 1898. About fifty years later, two further snarks were uncovered—the Blanuša snark of order 18, discovered in 1946, and the Descartes snark of order 210, discovered in 1948. The fourth snark to be found was the Szekeres snark of order 50, discovered in 1973. Up until 1973, these were the only snarks known.

The art of snark-hunting underwent a dramatic change in 1975, when Isaacs published an important article [**21**] in which he described two infinite families of snarks, one of which included all of the known snarks, and the other of which was completely new. He also discovered one extra snark—the "double-star snark" of order 30—which does not belong to either family (see Fig. 9). Since then, some more infinite families of snarks have been found, notably by Loupekhine (see [**22**]). Further details may be found in [**21**], [**22**] and [**15**].

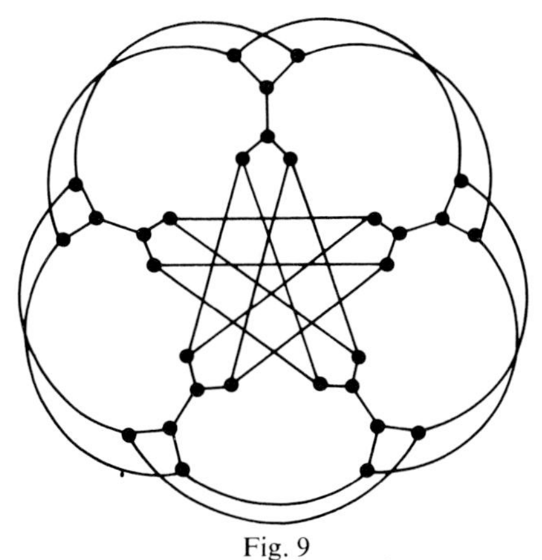

Fig. 9

5. Critical Graphs

In the study of vertex-colorings of graphs, those graphs which are critical in some sense have played an important role. The reason for this is not hard to find—since every graph contains a critical graph, and since critical graphs generally contain more structure than arbitrary graphs, we certainly lose nothing, and often gain quite a lot, by restricting our attention to critical graphs. We now show that a corresponding situation holds for edge-colorings.

We define a graph G to be **critical** if G is connected and of class 2, and if the removal of any edge of G lowers its chromatic index; if G has maximum valency ρ, we say that G is **ρ-critical**. For example, every odd circuit graph is 2-critical, and the graph obtained by removing any edge from the complete graph K_5 is 4-critical. Some elementary properties of critical graphs are summarized in the following theorem, whose proof can be found in [**15**] or [**23**]:

Theorem 5.1. *If G is a ρ-critical graph, then*

(*i*) *G has no cut-vertices;*

(*ii*) *if I is any independent set of edges in G, then $\chi'(G-I) = \chi'(G)-1$;*

(*iii*) *G contains a k-critical subgraph, for each k satisfying $2 \leqslant k \leqslant \rho$;*

(*iv*) *if v and w are adjacent vertices in G, then $\rho(v)+\rho(w) \geqslant \rho+2$.* ‖

It is clear from Corollary 4.2(*ii*) that if G is a critical graph of odd order (other than an odd circuit), then G cannot be regular. We now prove that a corresponding result holds for critical graphs in general:

Theorem 5.2. *There are no regular ρ-critical graphs with $\rho \geqslant 3$.*

Proof. By our remarks above, it is sufficient to prove this result for graphs of even order. If G is a regular ρ-critical graph of even order, and if H is the graph obtained from G by inserting a new vertex z into the edge $e = vw$ of G, then $\chi'(H) = \rho+1$, by Corollary 4.2(*iii*). Now $\chi'(G-e) = \rho$, since G is critical, and in any coloring of the edges of $G-e$ there is at least one color α missing at v and at least one color β missing at w; we may assume that $\alpha \neq \beta$, since otherwise we could immediately obtain a ρ-coloring of the edges of G. It follows that we can obtain a ρ-coloring of the edges of H by coloring the edge vz with color α and the edge wz with color β. This contradiction establishes the result. ‖

Now let vw be an edge of a ρ-critical graph G, and suppose that all of the edges of G except vw have been colored with at most ρ colors. If v has valency k, then every color missing from v must appear at w, since otherwise we should be able to color all of the edges of G with ρ colors. So the valency of w must be at least $1+\rho-(k-1) = \rho-k+2$. This simple result has been generalized by Vizing [**41**] in what we call **Vizing's adjacency lemma**. This lemma is of central importance, in that most of the structural properties of critical graphs can be derived from it. Before stating it, and presenting a sketch of its proof, we need a definition.

Let w be a vertex of a graph H whose edges have been colored in such a way that adjacent edges are assigned different colors. A **fan-sequence** F at w with initial edge wa_1 is a sequence of distinct edges of the form $wa_1, wa_2, wa_3, \ldots$, such that, for each $i \geqslant 1$, the edge wa_{i+1} is colored with a color missing from the vertex a_i.

Theorem 5.3 (Vizing's Adjacency Lemma). *Let G be a ρ-critical graph, and let v and w be adjacent vertices of G with $\rho(v) = k$. Then*

(*i*) *if $k < \rho$, then w is adjacent to at least $\rho - k + 1$ vertices of valency ρ;*

(*ii*) *if $k = \rho$, then w is adjacent to at least two vertices of valency ρ.*

Sketch of Proof. It follows from our earlier remarks that $\rho(w) \geqslant \rho - k + 2$. We now make two assertions without proof:

(*1*) if F and F' are two distinct fan-sequences at w in $G - vw$, and if the initial edges of F and F' are distinct and colored with colors which appear at w but not at v, then F and F' have no edges in common.

(*2*) If $F = wa_1, \ldots, wa_s$ is a fan-sequence of maximum length at w, starting with an edge wa_1 whose color does not appear at v, then the vertex a_s has valency ρ.

Assuming these results, we can complete the proof as follows. For each of the $\rho - k + 1$ colors appearing at w but not at v, there is a fan-sequence of maximum length at w, starting with an edge of that color. By (*1*), these fan-sequences are all disjoint, and by (*2*), each one ends with an edge incident to a vertex of valency ρ. The result now follows immediately. ||

A full proof of Vizing's adjacency lemma may be found in **[15]**. Using this result we can immediately deduce the following corollary:

Corollary 5.4. *If G is a ρ-critical graph with minimum valency k, then*

(*i*) *G has at least $\rho - k + 2$ vertices of valency ρ;*

(*ii*) *each vertex of G is adjacent to at least two vertices of valency ρ.* ||

We conclude this section by presenting a result of Vizing on circuit length in critical graphs, which can also be deduced from Vizing's adjacency lemma; proofs of this result can be found in **[15]** or **[41]**:

Theorem 5.5. *If G is a ρ-critical graph, then G has a circuit whose length is at least $\rho + 1$.* ||

6. The Critical Graph Conjecture

In this section, we look at critical graphs whose maximum valency is small.

This will lead us to make a conjecture about critical graphs in general, and we shall then describe what progress has been made on this conjecture.

We start by observing that there are clearly no critical graphs with maximum valency 1, and that the only ones with maximum valency 2 are the odd circuits C_{2k+1} ($k \geqslant 1$). The problem of determining all critical graphs with maximum valency 3 is unsolved, although substantial progress has been made on this

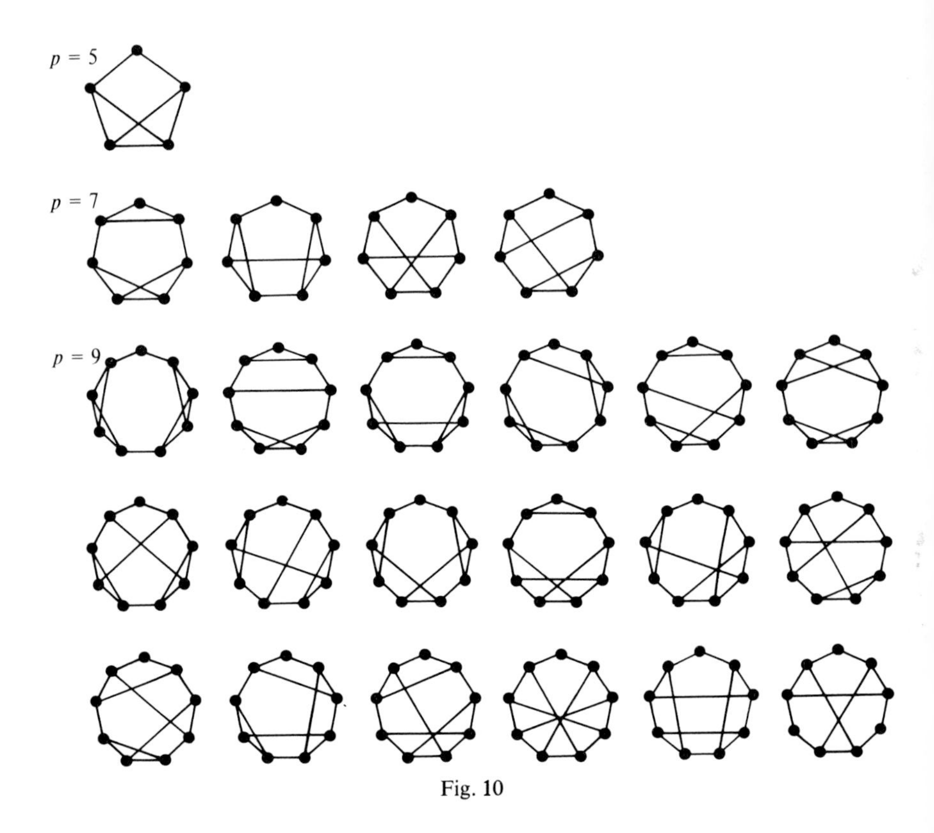

Fig. 10

problem. If G is a critical graph with maximum valency 3, then the valency of each vertex must be either 2 or 3, by Theorem 5.1(*i*); moreover, by Corollary 5.4(*ii*), if v is a vertex of valency 2, then v must be adjacent to two vertices of valency 3, whereas if v is a vertex of valency 3, then v can be adjacent to only one vertex of valency 2. It follows that the distance between any two vertices of valency 2 must be at least three, and hence that the number of vertices of valency 3 is at least twice the number of valency 2. From this, Jakobsen deduced the following theorem [**24**]:

Theorem 6.1. *Let G be a 3-critical graph of order p. Then*

(*i*) *the number of vertices of valency* 2 *is at most* $\frac{1}{3}p$;

(*ii*) *G cannot have exactly two vertices of valency* 2;

(*iii*) *the number of edges of G lies between* $\frac{4}{3}p$ *and* $\frac{1}{2}(3p-1)$. ‖

Using these results, Jakobsen was able to construct all of the 3-critical graphs on at most ten vertices; these graphs are depicted in Fig. 10.

The most interesting thing about these graphs is that they all have odd order. This leads to the following conjecture, formulated independently by Jakobsen [**24**], and Beineke and Wilson [**4**].

Critical Graph Conjecture. *There are no critical graphs of even order.*

A study of critical graphs of small order by Beineke and Fiorini [**3**] shows that the conjecture is true for all ρ-critical graphs of order $p \leqslant 10$, and if we restrict our attention to graphs with $\rho = 3$, then a recent result of Andersen and Fiorini [**2**] states that there are no 3-critical graphs of even order $p \leqslant 16$. The arguments used in obtaining these results make use of the existence of 1-factors in critical graphs of even order. In particular, it is shown in [**3**] and [**12**] that every 3-critical graph of even order $p \leqslant 26$, and every ρ-critical graph of even order $p \leqslant 10$, contains a 1-factor. We make the following related conjectures the second of which is due to Vizing [**40**]:

Factorization Conjectures. (*i*) *Every critical graph of even order contains a* 1-*factor*;

(*ii*) *every critical graph contains a* 2-*factor.*

Further results relating to the critical graph conjecture will be given in Section 9.

7. Planar Graphs

Although the classification problem is far from solved in general, appreciable progress has been achieved in the particular case of planar graphs.

If *G* is a planar graph whose maximum valency ρ is at most 5, then *G* can lie in either class 1 or class 2. Examples of planar graphs of class 1 are the even circuits ($\rho = 2$), and the graphs of the tetrahedron ($\rho = 3$), the octahedron ($\rho = 4$), and the icosahedron ($\rho = 5$), whereas examples of planar graphs of class 2 are the odd circuits ($\rho = 2$), and the graphs obtained by inserting a vertex into any edge of the graphs of the tetrahedron ($\rho = 3$), the octahedron ($\rho = 4$), and the icosahedron ($\rho = 5$). On the other hand, Vizing [**41**] has proved the surprising result that *every planar graph with* $\rho \geqslant 8$ *is necessarily of class* 1. We shall prove a similar, but weaker, result which may be found in his earlier paper [**40**]:

Theorem 7.1. *If G is a planar graph whose maximum valency is at least* 10, *then G is of class* 1.

Proof. Suppose that G is a planar graph of class 2 with $\rho \geqslant 10$, and suppose, without loss of generality, that G is ρ-critical. Since G is planar, there must be at least one vertex in G whose valency is at most 5. If S is the set of all such vertices, then the subgraph of G induced by those vertices which are not in S is also planar, and must therefore contain a vertex (w, say) which is adjacent in G to at most five vertices which are not in S. But if v is a vertex in S to which w is adjacent in G, then the valency of v is at most 5. It follows from Theorem 5.2 that w must be adjacent to at least $\rho - 4 > 5$ vertices of valency ρ, giving the required contradiction. ||

As we mentioned above, this result can be improved to show that every planar graph with $\rho \geqslant 8$ is of class 1. However, the problem of determining what happens when the maximum valency is either 6 or 7 remains open. In this connection, the following conjecture was formulated by Vizing [**42**]:

Planar Graph Conjecture. *Every planar graph with maximum valency* 6 *or* 7 *is of class* 1.

If we impose restrictions on the girth of a planar graph, then sharper results can be obtained. The following theorem was stated by Kronk, Radlowski and Franen [**26**], and its proof can be found in [**12**]:

Theorem 7.2. *Let G be a planar graph with girth g and maximum valency* ρ; *then G is of class* 1 *if any of the following conditions hold*:

(*i*) $\rho \geqslant 3$ *and* $g \geqslant 8$; (*ii*) $\rho \geqslant 8$ *and* $g \geqslant 3$;
(*iii*) $\rho \geqslant 4$ *and* $g \geqslant 5$; (*iv*) $\rho \geqslant 5$ *and* $g \geqslant 4$. ||

If we restrict ourselves to outerplanar graphs, then the classification problem is completely solved:

Theorem 7.3. *An outerplanar graph G is of class* 2 *if and only if it is a circuit of odd length.*

Proof. If G is a circuit of odd length, then G is clearly an outerplanar graph of class 2.

Conversely, let G be an outerplanar graph of class 2. We may assume that its maximum valency ρ is at least 3, and that the order and maximum valency of G are chosen to be as small as possible, subject to these restrictions. Since the order of G is minimal, G must be 2-connected and hence have a Hamiltonian circuit.

If G is of even order, then G has a 1-factor F consisting of alternate edges of the Hamiltonian circuit, so that $G-F$ has maximum valency $\rho-1$. Thus $G-F$ is of class 1, by our assumption on the minimality of ρ. It follows that G is of class 1.

If, on the other hand, G is of odd order $2k+1$, then G must have a vertex v of valency 2, by outerplanarity. It follows that G has an independent set M of alternate edges of the Hamiltonian circuit, covering all vertices except v. Clearly, $G-M$ is outerplanar and has maximum valency $\rho-1$, so that $G-M$ is of class 1, by our assumption on the minimality of ρ. It follows that G is also of class 1, thereby completing the proof. ‖

A partial solution to the planar graph conjecture can be obtained as a simple consequence of the following conjecture on the number of edges of critical graphs. This conjecture is due to Vizing ([**40**], [**42**]), and its truth implies that every planar graph with maximum valency 7 is of class 1.

Vizing's Conjecture. *If G is a ρ-critical graph with p vertices and q edges, then* $q \geqslant \frac{1}{2}p(\rho-1)+3$.

Using his adjacency lemma (Theorem 5.3), Vizing obtained the following lower bound on the number of edges of a critical graph:

Theorem 7.4. *If G is a ρ-critical graph with q edges, then* $q \geqslant \frac{1}{8}(3\rho^2+6\rho-1)$.

Proof. If k is the minimum valency in G, then G has at least $\rho-k+2$ vertices of valency ρ, by Corollary 5.3(i). Since G has at least $\rho+1$ vertices, the number of edges of G must satisfy

$$\begin{aligned} 2q &\geqslant \rho(\rho-k+2)+k\{(\rho+1)-(\rho-k+2)\} \\ &= k^2-(\rho+1)k+\rho(\rho+2). \end{aligned}$$

This last expression is smallest when $k = [\frac{1}{2}(\rho+1)]$, as may easily be seen by taking derivatives. It follows that $2q \geqslant \frac{1}{4}(3\rho^2+6\rho-1)$, as required. ‖

This bound is a good one when p, the order of G, is relatively small in comparison with ρ, but if this is not the case, then a bound involving p is desirable. Such a bound is provided by the following theorem of Fiorini [**14**]:

Theorem 7.5. *If G is a ρ-critical graph with p vertices and q edges, then*

$$q \geqslant \tfrac{1}{4}p(\rho+1), \quad \text{if } \rho \text{ is odd,}$$

and

$$q \geqslant \tfrac{1}{4}p(\rho+2), \quad \text{if } \rho \text{ is even.} \ ‖$$

We conclude this section by presenting without proof an upper bound on the number of edges of a critical graph; note that it can be used to give an alternative proof of Theorem 5.2:

Theorem 7.6. *If G is a ρ-critical graph with p vertices, q edges and minimum valency k, then*

$$q \leqslant \tfrac{1}{2}(p-1)\rho+1, \text{ if } p \text{ is odd,}$$

and

$$q \leqslant \tfrac{1}{2}(p-2)\rho+k-1, \text{ if } p \text{ is even.} \quad \|$$

8. Uniquely Colorable Graphs

If G is a graph with chromatic index k, then G is said to be **uniquely k-colorable** if every k-coloring of the edges of G induces the same partition of the edge-set —in other words, there is only one way of coloring the edges, apart from permutations of the colors. For example, all even circuits C_{2s} are uniquely 2-colorable, the graphs K_3 and K_4 are uniquely 3-colorable, and for each t the star graph $K_{1,t}$ is uniquely t-colorable. Some simple properties of uniquely colorable graphs are summarized in the following theorem:

Theorem 8.1. *If G is a uniquely k-colorable graph with p vertices and q edges, then*

(*i*) *each edge of G is adjacent to edges of every other color;*

(*ii*) *$H(\alpha, \beta)$, the subgraph induced by edges colored α or β, is either an open path or an even circuit;*

(*iii*) *q satisfies the inequalities*

$$\tfrac{1}{2}pk - \binom{k}{2} \leqslant q \leqslant \tfrac{1}{2}pk,$$

and both bounds can be attained;

(*iv*) *if G is a regular k-valent graph ($k \geqslant 3$), and if H is a graph obtained from G by inserting a vertex into any edge of G, then H is k-critical.* $\|$

It was conjectured in [**13**] that if $k \geqslant 4$, then the only uniquely k-colorable graphs are the star graphs $K_{1,k}$. This conjecture has recently been proved by Thomason [**37**], and we shall present a sketch of his proof. Note that it is sufficient to prove the result only in the case $k = 4$, since if G is any uniquely k-colorable graph with $k \geqslant 5$ (other than $K_{1,k}$), then the subgraph of G induced by those edges colored with any four of the colors must be uniquely 4-colorable.

We start by stating without proof a result of Thomason on the decomposition of a regular 4-valent graph or multigraph into two edge-disjoint Hamiltonian circuits; such a pair of circuits is called a **Hamiltonian pair**:

Theorem 8.2. *Let G be a regular 4-valent graph or multigraph with at least three vertices, and suppose that G has a Hamiltonian pair; then G has at least four such pairs.* ||

Using the above results, Thomason showed that if there exists a uniquely 4-colorable graph G other than $K_{1,4}$, then there must exist a uniquely 4-colorable multigraph H which is either 4-valent, or has two vertices of valency 2 and the rest of valency 4. In this latter case, the vertices of valency 2 must be incident to edges colored with the same pair of colors. Using these results, we can now outline the main part of Thomason's proof:

Theorem 8.3. *The only uniquely 4-colorable graph is the graph $K_{1,4}$.*

Sketch of Proof. Let G be a uniquely 4-colorable graph other than $K_{1,4}$. By virtue of the above remarks, we can assume that G is either a 4-valent graph, or that G has exactly two vertices of valency 2 and the rest of valency 4.

In the first case, the union of any two of the four color-classes determines a Hamiltonian pair in G, so that G must have exactly three Hamiltonian pairs determined by these color-classes. But by Theorem 8.2, G has at least four such pairs, and so cannot be uniquely colorable.

In the second case, G has two vertices v and w of valency 2 with the same pair of colors appearing at each. By deleting v and w, and joining their neighbors appropriately, we can obtain from G another graph G' which also contradicts Theorem 8.2. Thus in either case the proof is complete. ||

It follows that uniquely k-colorable graphs are completely characterized when $k \geqslant 4$. Since the only uniquely 2-colorable graphs are the open paths and the even circuits, there remains only the case $k = 3$ to consider. We can construct an infinite family of uniquely 3-colorable graphs by successively taking a uniquely 3-colorable trivalent graph and replacing any vertex v by a triangle T. For example, if we start with K_4, then we get the uniquely 3-colorable graph shown in Fig. 11; the process can then be applied to this

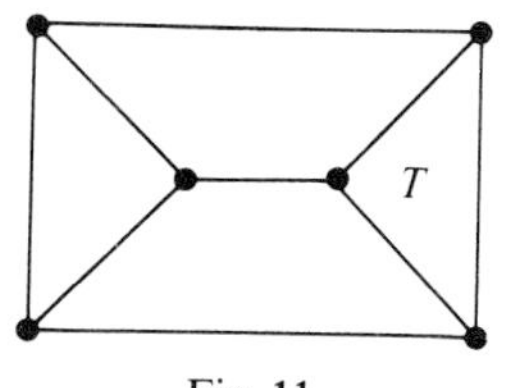

Fig. 11

graph, and so on. All of the graphs obtained in this way are planar, and this leads to the following conjecture:

Conjecture A. *Every planar uniquely* 3-*colorable graph other than* $K_{1,3}$ *contains a triangle.*

If this conjecture were true, it would follow immediately that *every* planar trivalent uniquely 3-colorable graph can be obtained from K_4 by successively applying the above construction. Note that Conjecture A is false if the word "planar" is omitted; a counter-example is the generalized Petersen graph $P(9, 2)$ (see Fig. 8). This leads to a second conjecture:

Conjecture B. *The only non-planar trivalent uniquely* 3-*colorable graph is the graph* $P(9, 2)$.

A third conjecture involving uniquely 3-colorable graphs is due to Greenwell and Kronk [**17**]:

Conjecture C. *Every trivalent graph with exactly three Hamiltonian circuits is uniquely* 3-*colorable.*

It is clear that every uniquely 3-colorable graph has exactly three Hamiltonian circuits. Since every bipartite trivalent graph can be shown to have an even number of Hamiltonian circuits, it follows that every uniquely 3-colorable graph must contain an odd circuit—a result of some relevance to Conjecture A.

We conclude this section by noting that, with only one exception, uniquely k-colorable graphs are necessarily of class one. This was first proved by Greenwell and Kronk [**17**], and was later obtained independently by Fiorini [**13**]. In view of the above results of Thomason, we may restrict our attention to the case $k = 3$.

Theorem 8.4. *If* G *is a uniquely* 3-*colorable graph other than* K_3, *then* G *is of class* 1. ‖

9. The Multichromatic Index

In this final section, we consider a parameter related to the chromatic index. Although this parameter is more difficult to define than the chromatic index, it turns out that the analogs of some difficult or unsolved questions for the chromatic index are completely solved in this case. In particular, we shall prove the analogs of the planar graph conjecture (Section 7) and the critical graph conjecture (Section 6), thereby strengthening our belief in the truth of these conjectures.

If G is a graph, we denote by G^k the multigraph obtained from G by replacing each edge by a set of k multiple edges. The **multichromatic index** $\chi^*(G)$ of G is then defined by

$$\chi^*(G) = \inf \frac{1}{k} \chi'(G^k),$$

where the infimum is taken over all positive integers k. It can be shown that this infimum is actually attained—that is, that there is some integer k for which $\chi^*(G) = (1/k)\chi'(G^k)$.

The multichromatic index has been studied by several authors, notably Berge **[6]**, Chen and Hilton **[9]**, Seymour **[32]**, and Stahl **[34]**, **[35]**. In order to discuss its properties, we shall need the following parameters. If H is a subgraph of G with p vertices and q edges, then we define

$$\tau(H) = \frac{2q}{p-1}, \qquad \text{and} \qquad \Phi(G) = \max\{\tau(H)\},$$

where the maximum is taken over all subgraphs H of G with odd order, and where $\{x\}$ denotes the least integer not less than x. (The parameter $\Phi(G)$ is very closely related to the arboricity of G.)

The first fundamental result on the multichromatic index is a consequence of a difficult theorem on matchings due to Edmonds **[10]**; proofs of this theorem may be found in **[32]** or **[34]**:

Theorem 9.1. *If G is a graph with maximum valency ρ, then*

$$\chi^*(G) = \max\{\rho, \Phi(G)\}. \;\|$$

We remark that Theorem 9.1 implies the following result of Stahl, which is of relevance to the classification problem, and which generalizes Theorem 4.1:

Corollary 9.2. *If G is a graph with maximum valency ρ, which contains an odd subgraph H satisfying $\tau(H) > \rho$, then G is of class* 2. ‖

We next show how the result of Theorem 9.1 can be used to give a simple proof of the multichromatic analogue of the planar graph conjecture. The following result is due to Stahl **[34]**:

Theorem 9.3. *If G is a planar graph whose maximum valency ρ is at least* 6, *then $\chi^*(G) = \rho$.*

Proof. It follows from Euler's polyhedral formula that G has at most $3p-6$ edges, where p is the order of G. It follows that $\tau(G) \leqslant 6(p-2)/(p-1)$.

But if H is any subgraph of G with p' vertices and q' edges, then H is a planar graph with $p' \leqslant p$. It follows that

$$\tau(H) = \frac{2q'}{p'-1} \leqslant \frac{6(p'-2)}{p'-1} \leqslant \frac{6(p-2)}{p-1},$$

and hence that

$$\Phi(G) \leqslant \left\{\frac{6(p-2)}{p-1}\right\} = 6.$$

The required result now follows from Theorem 9.1. ||

We now show how the result of Theorem 9.1 can be used to give a simple proof of the multichromatic analog of the planar graph conjecture. In this context, we define a connected graph G to be **critical*** if $\chi^*(G) > \rho$ and, for each edge e of G, $\chi^*(G-e) < \chi^*(G)$. The following result is also due to Stahl:

Theorem 9.4. *There are no critical* graphs of even order.*

Proof. Since $\chi^*(G) > \rho$, it follows from Theorem 9.1 that $\chi^*(G) = \Phi(G)$. Hence there exists an induced subgraph H of odd order for which $\chi^*(G) = \{\tau(H)\}$. If $G \neq H$, then there exists an edge e which lies in G but not in H, and for any such edge,

$$\chi^*(G-e) \geqslant \{\tau(H)\} = \chi^*(G),$$

contradicting the fact that G is critical*. It follows that $G = H$, and hence that G has odd order. ||

We remark finally that Theorem 9.1 can be used to prove a multichromatic extension of the four-color theorem (in the form given by Corollary 2.5). The following result has been obtained independently by Berge [**6**], Chen and Hilton [**9**], and Stahl [**34**]:

Theorem 9.5. *If G is a trivalent bridgeless graph, then $\chi^*(G) = 3$.* ||

References

1. Y. Alavi and M. Behzad, Complementary graphs and edge chromatic numbers, *Siam J. Appl. Math.* **20** (1971), 161–163; *MR***44**#3923.
2. L. D. Andersen and S. Fiorini, Note on the non-existence of certain non-3-edge-colorable graphs (to appear).
3. L. W. Beineke and S. Fiorini, On small graphs critical with respect to edge-colourings, *Discrete Math.* **16** (1976), 109–121; *MR***55**#2631.
4. L. W. Beineke and R. J. Wilson, On the edge-chromatic number of a graph, *Discrete Math.* **5** (1973), 15–20; *MR***47**#4836.
5. C. Berge, *Graphs and Hypergraphs*, North-Holland, Amsterdam, 1973; *MR***50**#9640.

6. C. Berge, The multicolorings of graphs and hypergraphs, *Theory and Applications of Graphs*, Lecture Notes in Mathematics **642** (ed. Y. Alavi and D. R. Lick), Springer-Verlag, Berlin, Heidelberg and New York, 1978, pp. 23–36.
7. N. L. Biggs, Three remarkable graphs, *Canad. J. Math.* **25** (1973), 397–411; *MR***48**#156.
8. F. Castagna and G. Prins, Every generalized Petersen graph has a Tait coloring, *Pacific J. Math.* **40** (1972), 53–58; *MR***46**#3358.
9. C. C. Chen and A. J. W. Hilton, A 4-color conjecture for planar graphs, in *Combinatorics* (ed. A. Hajnal and V. T. Sós), North-Holland, Amsterdam, 1978, pp. 205–212.
10. J. Edmonds, Maximum matching and a polyhedron with (0, 1) vertices, *J. Res. Nat. Bur. Standards* **69B** (1965), 125–130; *MR***32**#1012.
11. P. Erdős and R. J. Wilson, On the chromatic index of almost all graphs, *J. Combinatorial Theory* (*B*) **26** (1977), 255–257.
12. S. Fiorini, The chromatic index of simple graphs, Ph.D. thesis, The Open University, 1974.
13. S. Fiorini, On the chromatic index of a graph, III: Uniquely edge-colourable graphs, *Quart. J. Math.* (*Oxford*) (*3*) **26** (1975), 129–140; *MR***51**#7925.
14. S. Fiorini, Some remarks on a paper by Vizing on critical graphs, *Math. Proc. Cambridge Phil. Soc.* **77** (1975), 475–483; *MR***51**#10146.
15. S. Fiorini and R. J. Wilson, *Edge-Colourings of Graphs*, Research Notes in Mathematics **16**, Pitman Publishing, London, 1977.
16. M. Gardner, Mathematical games, *Scientific American* **234**, No. 4 (April 1976), 126–130, and **234**, No. 9 (September 1976), 210–211.
17. D. Greenwell and H. Kronk, Uniquely line-colorable graphs, *Canad. Math. Bull.* **16** (1973), 525–529; *MR***50**#158.
18. R. P. Gupta, Studies in the theory of graphs, Ph.D. thesis, Tata Institute for Fundamental Research, Bombay, 1967.
19. P. J. Heawood, On the four-colour map theorem, *Quart. J. Pure Appl. Math.* **29** (1898), 270–285.
20. A. J. W. Hilton, On Vizing's upper bound for the chromatic index of a graph, *Cahiers Centre Études Recherches Opér.* **17** (1975), 225–233.
21. R. Isaacs, Infinite families of non-trivial trivalent graphs which are not Tait colorable, *Amer. Math. Monthly* **82** (1975), 221–239; *MR***52**#2940.
22. R. Isaacs, Loupekhine's snarks: a bifamily of non-Tait-colorable graphs (to appear).
23. I. T. Jakobsen, Some remarks on the chromatic index of a graph, *Arch. Math.* (*Basel*) **24** (1973), 440–448; *MR***48**#10874.
24. I. T. Jakobsen, On critical graphs with chromatic index 4, *Discrete Math.* **9** (1974), 265–276; *MR***50**#161.
25. D. König, Über Graphen und ihre Anwendung auf Determinantentheorie und Mengenlehre, *Math. Ann.* **77** (1916), 453–465.
26. H. Kronk, M. Radlowski and B. Franen, On the line chromatic number of triangle-free graphs, Abstract in *Graph Theory Newsletter* **3**, No. 3 (1974), 3.
27. R. Laskar and W. Hare, Chromatic numbers of certain graphs, *J. London Math. Soc.* (*2*) **4** (1971), 489–492; *MR***45**#6680.
28. G. H. J. Meredith, Regular *n*-valent, *n*-connected, non-Hamiltonian, non-*n*-edge-colourable graphs, *J. Combinatorial Theory* (*B*) **14** (1973), 55–60; *MR***47**#65.
29. G. H. J. Meredith and E. K. Lloyd, The footballers of Croam, *J. Combinatorial Theory* (*B*) **15** (1973), 161–166; *MR***48**#149.
30. O. Ore, *The Four Color Problem*, Academic Press, New York, 1967; *MR***36**#74.
31. E. T. Parker, Edge-coloring numbers of some regular graphs, *Proc. Amer. Math. Soc.* **37** (1973), 423–424; *MR***47**#1658.
32. P. D. Seymour, On multi-colourings of cubic graphs and conjectures of Fulkerson and Tutte (to appear).
33. C. E. Shannon, A theorem on coloring the lines of a network, *J. Math. Phys.* **28** (1949), 148–151; *MR***10**–728.
34. S. Stahl, Edge multicolorings of graphs (preprint).

35. S. Stahl, Fractional edge colorings (to appear).
36. P. G. Tait, [Remarks on the colouring of maps], *Proc. Royal Soc. Edinburgh* **10** (1880), 501–503, 729.
37. A. G. Thomason, Hamiltonian cycles and uniquely edge colourable graphs, in *Advances in Graph Theory* (ed. B. Bollobás), North-Holland, Amsterdam, 1978, pp. 259–268.
38. W. T. Tutte, A geometrical version of the four color problem, in *Combinatorial Mathematics and its Applications* (ed. R. C. Bose and T. A. Dowling), University of North Carolina Press, Chapel Hill (1969), pp. 553–560; *MR*41#8298.
39. V. G. Vizing, On an estimate of the chromatic class of a p-graph (Russian), *Diskret. Analiz* **3** (1964), 25–30; *MR*31#856.
40. V. G. Vizing, The chromatic class of a multigraph, *Cybernetics* **3** (1965), 32–41; *MR*32 #7333.
41. V. G. Vizing, Critical graphs with a given chromatic class (Russian), *Diskret. Analiz* **5** (1965), 9–17; *MR*34#17.
42. V. G. Vizing, Some unsolved problems in graph theory, *Uspekhi Mat. Nauk* **23** (1968), 117–134 = *Russian Math. Surveys* **23** (1968), 125–142; *MR*39#1354.
43. M. E. Watkins, A theorem on Tait colorings with an application to the generalized Petersen graphs, *J. Combinatorial Theory* (*B*) **6** (1969), 152–164; *MR*38#4360.

6
Hamiltonian Graphs

J.-C. BERMOND

1. Introduction

If G is a graph, a **Hamiltonian circuit** in G is a circuit which contains every vertex of G, and a **Hamiltonian path** in G is a path which contains every vertex of G. So if G has order p, then a Hamiltonian circuit has length p and a Hamiltonian path has length $p-1$. (Throughout this chapter we shall assume, usually without saying so, that $p \geqslant 3$.) A graph which contains a Hamiltonian circuit is called a **Hamiltonian graph**, and a graph which contains a Hamiltonian path is called a **traceable graph**.

Hamiltonian graphs are named after William Rowan Hamilton, although they were studied earlier by Kirkman. In 1856, Hamilton invented a mathematical game, the "icosian game", consisting of a dodecahedron each of whose twenty vertices was labeled with the name of a city. The object of the game was to travel along the edges of the dodecahedron, visiting each city exactly once and returning to the initial point. In graph-theoretical terms, the aim is to find a Hamiltonian circuit in the dodecahedral graph (see Fig. 1, where such a circuit is indicated by heavy lines). For a picture of the icosian game, and more details concerning the origin of this problem, see [**10**, Chapter 2].

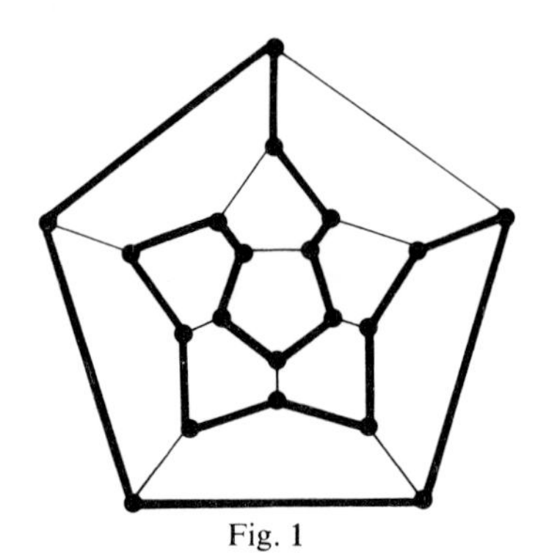

Fig. 1

Another example concerns a "squirrel cage", which consists of a $3 \times 3 \times 3$ cube made up from 27 small cubes. Is it possible for a child to start at one corner and visit each cube in turn, ending at the central cube? If we consider the graph whose vertices represent the small cubes, and whose edges join those pairs of vertices which correspond to adjacent cubes, then the problem asks whether this graph has a Hamiltonian path with given initial and terminal vertices—the answer will be given in Section 3. Many other problems can also be expressed in terms of Hamiltonian paths and circuits, such as the knight's-tour problem on a chessboard, and the "Tower of Hanoi"; the reader is referred to **[6]**, **[10]**, or **[19]** for further details.

It is clear that the complete graph K_p is Hamiltonian, since we can start at any vertex and go successively to any other vertex not yet visited. However, if we "weight" the edges of K_p, then the problem of finding a minimum weight Hamiltonian circuit is a difficult one. It is usually called the "traveling salesman problem", and represents the problem of finding how a traveling salesman can visit each of p towns exactly once in the shortest possible time. More precisely, the aim is to find a "good" or "efficient" algorithm which produces the required Hamiltonian circuit, but although there is an extensive literature on this problem, no efficient algorithm is known. However, this is primarily an integer programming problem, rather than a problem on Hamiltonian graphs, and so we shall not consider it here (see, for example, **[4]**, or **[20**, Chapter 10]).

Another example we shall be considering is the Petersen graph (see Fig. 2). It is not difficult to check that it is a non-Hamiltonian graph which is traceable. (In fact, it is "homogeneously traceable", in the sense that it contains a Hamiltonian path starting at any given vertex.) The Petersen graph forms the basis of most of the counter-examples to conjectures on Hamiltonian graphs. We shall discuss further non-Hamiltonian graphs in Sections 3 and 6.

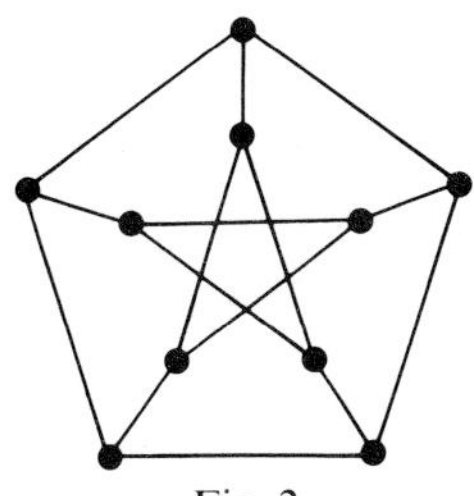

Fig. 2

Unlike the situation for Eulerian graphs, there is no known non-trivial characterization of Hamiltonian graphs, although several necessary conditions (see Section 3) and several sufficient conditions (see Sections 4 and 5) have been discovered. In fact, the problem of determining which graphs are Hamiltonian is one of the major unsolved problems of graph theory. As a consequence, a very large literature exists on the subject, and there are several interesting expository accounts. The interested reader is referred in particular to the surveys of Berge [**6**, Chapter 10], Bondy and Murty [**17**, Chapters 4 and 9], Nash-Williams [**49**], [**50**], [**52**], Chartrand, Kapoor and Kronk [**19**], Chvátal [**21**] and Lesniak [**43**]. In this chapter, we shall often find it convenient to refer to these articles, rather than to the original papers.

Finally, we should like to thank all those who helped in the preparation of this chapter—in particular, J. A. Bondy, M. Chein and C. Thomassen.

2. A Simple Theorem

In this section we shall show that the problems of determining which graphs, digraphs and bipartite graphs are Hamiltonian or traceable are essentially the same. In this context, a digraph D is a **Hamiltonian digraph** if it contains a directed circuit which passes through every vertex of D, and is a **traceable digraph** if it contains a directed path which passes through every vertex of D. For example, consider the digraph whose vertices represent jobs to be done, and where there is an arc from vertex i to vertex j if the work for job i can be done before that for job j; if this digraph is traceable, then the jobs can be done in order, and the determination of such an ordering optimal under certain conditions is an important problem in operations research.

We can now state and prove the main result of this section, first proved by Nash-Williams [**49**]:

Theorem 2.1. *The following problems are equivalent:*

(*i*) *the determination of all Hamiltonian graphs;*

(*ii*) *the determination of all traceable graphs*;
(*iii*) *the determination of all Hamiltonian digraphs*;
(*iv*) *the determination of all traceable digraphs*;
(*v*) *the determination of all Hamiltonian bipartite graphs.*

Proof. (*i*) ⇔ (*ii*). Let G be a graph, and let H be the graph obtained by taking a new vertex and joining it to every vertex of G; then G is traceable if and only if H is Hamiltonian, and so if we know which graphs are Hamiltonian, we can determine which graphs are traceable. (Note that most of the known conditions for a graph to be traceable are obtained in this way from conditions for a graph to be Hamiltonian.) Conversely, let H be a graph, let v be a vertex of H, and let G be the graph obtained by taking three new vertices x, y and z, joining z to all the neighbors of v, and adding the edges vx and yz; then H is Hamiltonian if and only if G is traceable, and so if we know which graphs are traceable, we can determine which graphs are Hamiltonian. So (*i*) ⇔ (*ii*).

(*iii*) ⇔ (*iv*). This is very similar to the above proof.

(*i*) ⇔ (*iii*). Let G be a graph, and let D be the digraph obtained by replacing each edge of G by two opposite arcs; then G is a Hamiltonian graph if and only if D is a Hamiltonian digraph, and so if we know which symmetric digraphs are Hamiltonian, we can determine which graphs are Hamiltonian. Conversely, let D be a digraph, and let G be a graph constructed from D in the following manner: to each vertex v of D associate a path of length three with initial and terminal vertices a_v and b_v, and choose these paths to be vertex-disjoint; let G consist of these paths, together with the edges $b_v a_w$ for every arc vw of D. Then it is easily shown that D is a Hamiltonian digraph if and only if G is a Hamiltonian graph, and so if we know which graphs are Hamiltonian, we can determine which digraphs are Hamiltonian. So (*i*) ⇔ (*iii*).

(*i*) ⇔ (*v*). The graph G in the preceding part is bipartite, so that (*iii*) ⇒ (*v*). But problem (*v*) is a special case of problem (*i*), and so (*i*) ⇔ (*v*). ‖

Note that the problem of recognizing Hamiltonian graphs is NP-complete. This has been established by Karp, Lawler and Tarjan (see [**1**, Chapter 10]).

3. Some Necessary Conditions

We now look at some necessary conditions for Hamiltonian graphs. Further details can be found in Chvátal's survey [**21**], or in his articles [**22**], [**23**]. We start with the following simple, but important, theorem:

Theorem 3.1. *Let G be a Hamiltonian graph, let S be a non-empty proper subset of the vertex-set $V(G)$, and let $c(G-S)$ be the number of components of the graph $G-S$. Then*

$$c(G-S) \leqslant |S|.$$

Proof. Let C be a Hamiltonian circuit of G. Then for every non-empty proper subset S of $V(G)$ we have $c(C-S) \leqslant |S|$. But $C-S$ is a spanning subgraph of $G-S$, and so $c(G-S) \leqslant c(C-S)$. The result follows. ‖

If we consider only sets S of cardinality 1, we obtain the following corollary:

Corollary 3.2. *Every Hamiltonian graph is 2-connected.* ‖

Chvátal has defined a graph to be **1-tough** if $c(G-S) \leqslant |S|$ for every non-empty proper subset S of $V(G)$. It follows from Theorem 3.1 that every Hamiltonian graph is 1-tough.

The converse of Corollary 3.2 is not true. For example, if $G = K_{r,s}$, where $r < s$, then G is r-connected but not 1-tough, and therefore not Hamiltonian. Using this we can deduce that the squirrel-cage problem of Section 1 is impossible. Since the "squirrel-cage graph" is easily seen to be bipartite, every Hamiltonian path starting at a corner and ending at the center induces a Hamiltonian circuit in $K_{13,14}$ (on adding one extra edge joining the starting cube and the center cube), giving the required contradiction.

The converse of Theorem 3.1 is also false. For example, the Petersen graph is a 1-tough graph which is not Hamiltonian. The smallest such graph is shown in Fig. 3.

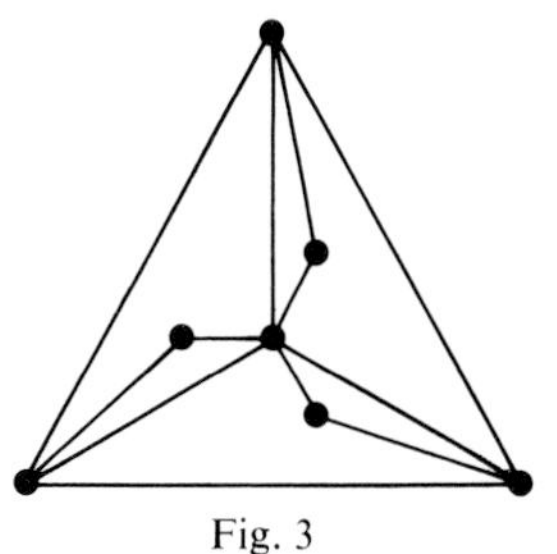

Fig. 3

In [22], Chvátal extended these ideas by defining the **toughness** $t(G)$ of a graph G (other than a complete graph) by

$$t(G) = \min \frac{|S|}{c(G-S)},$$

where the minimum extends over all separating sets S of G—that is, all subsets S of $V(G)$ for which $c(G-S) > 1$. So G is 1-tough if $t(G) \geqslant 1$. If we know the connectivity κ and the independence number α of G, we can easily obtain a lower bound for $t(G)$; note that this bound is exact for the complete bipartite graphs $K_{r,s}$:

Theorem 3.3. *Let G be a graph with connectivity κ and independence number α. Then $t(G) \geqslant \kappa/\alpha$.*

Proof. If S is a separating set of vertices of G, then $|S| \geqslant \kappa$. Also it is clear that $c(G-S) \leqslant \alpha$. The result follows. ‖

In [**22**] Chvátal obtained various other results relating the toughness of G to other parameters. He also conjectured the existence of a number t_0 such that every graph with toughness $t \geqslant t_0$ is necessarily Hamiltonian. If this conjecture is true for $t_0 = 2$, then we can deduce Fleischner's theorem on the squares of blocks (see Theorem 5.6), since $t(G^2) \geqslant \kappa$, by a result of [**22**]. Chvátal also conjectured that every graph with toughness $t > \frac{3}{2}$ is Hamiltonian, but this was disproved by Thomassen, who gave the following counter-example: take a trivalent 3-connected graph with no Hamiltonian path (such graphs exist—see Theorem 6.4(*iii*)), and replace every vertex by a triangle; then the resulting graph H can be shown to have toughness $\frac{3}{2}$ (see [**22**] for a proof). If we now let G be the graph obtained by adding a new vertex and joining it to every vertex of H, then $t(G) > \frac{3}{2}$ and G is non-Hamiltonian.

Another conjecture of Chvátal was proved by Jung [**40**]:

Theorem 3.4. *If G is 1-tough, then either G is Hamiltonian, or its complement $\bar{G}$ contains the graph of Fig. 4 as a subgraph.* ‖

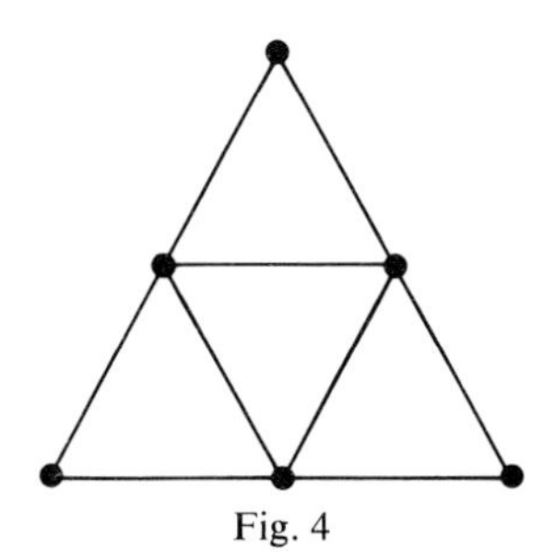

Fig. 4

We now turn our attention to "weakly Hamiltonian graphs" introduced by Chvátal [**23**]. If G is any graph with vertex-set V and edge-set E, and if C is any Hamiltonian circuit in G, then the **characteristic function** of C is the function $f: E \to \{0, 1\}$, defined by $f(e) = 1$ if $e \in C$, and $f(e) = 0$ otherwise. For any $v \in V$, we define $f(v)$ to be the sum of the values of $f(e)$, taken over all edges incident to v, and for any $T \subseteq V$, we define $f[T]$ to be the sum of the values of $f(e)$, taken over all edges e with both incident vertices in T. We can now prove the following simple results:

$$f(v) \leqslant 2, \quad \text{for all } v \in V;$$
$$f[T] \leqslant |T|-1, \quad \text{for all non-empty proper subsets } T \text{ of } V. \tag{1}$$

Conversely, let $f : E \to [0, \infty)$ be an integer-valued function satisfying both (1) and

$$f[V] = p, \tag{2}$$

where p is the order of G. Then f is the characteristic function of a Hamiltonian circuit of G.

In **[23]**, Chvátal proved that if f is an integer-valued function satisfying (1), then f must satisfy what he called a "comb inequality". Let $W_0, \ldots, W_n$ be non-empty proper subsets of V satisfying $|W_i \cap W_0| = 1$ for $i = 1, \ldots, n$, and let $K = W_0 \cup W_1 \cup \ldots \cup W_n$. Then

$$f_K \leqslant |W_0| + \sum_{i=1}^{n} |W_i - 1| - [\tfrac{1}{2}(n+1)], \tag{3}$$

where f_K is the sum of the values of $f(e)$, taken over all edges e with both incident vertices in the same W_i.

Chvátal defined a graph to be **weakly Hamiltonian** if there exists a function $f : E \to [0, \infty)$ (not necessarily integer-valued) satisfying (1), (2) and (3). It follows that every Hamiltonian graph is weakly Hamiltonian. But the duality theorem of linear programming can be used to give a characterization of weakly Hamiltonian graphs, which in turn gives a necessary condition for a graph to be Hamiltonian. A weaker, but more easily stated, version of this condition is the following:

Theorem 3.5. *If $G = (V, E)$ is a weakly Hamiltonian graph, then there is no partition $V = R \cup S \cup T$ into pairwise disjoint (possibly empty) sets with $T \neq V$, and*

$$|S| + \sum [\tfrac{1}{2}m(C, T)] < c(T),$$

where the summation extends over all components C of the subgraph induced by R, $m(C, T)$ is the number of edges joining C to T, and $c(T)$ is the number of components of the subgraph induced by T. ‖

The smallest weakly Hamiltonian graph which is not Hamiltonian is the Petersen graph. To see that it is weakly Hamiltonian, let $f(e) = \frac{2}{3}$ for each edge e. Some idea of how closely weakly Hamiltonian graphs approximate Hamiltonian graphs is given by the following theorem, also due to Chvátal **[23]**:

Theorem 3.6. *Every weakly Hamiltonian graph is 1-tough, has a 2-factor, and contains a circuit passing through any three given vertices.* ‖

4. Some Sufficient Conditions on Valencies

Most of the known sufficient conditions for a graph G to be Hamiltonian assert that if G is "large enough", or if G "has enough edges", then G is Hamiltonian. We shall essentially follow the method of Bondy and Chvátal [**16**], which unifies most of the known conditions and can also be applied to other problems in graph theory. We begin with a lemma whose proof contains the basic idea of the method:

Lemma 4.1. *Let G be a graph of order p, and let v and w be two non-adjacent vertices whose valencies satisfy*

$$\rho(v)+\rho(w) \geqslant p.$$

Then G is Hamiltonian if and only if $G+vw$ is Hamiltonian.

Proof. If G is Hamiltonian, then $G+vw$ is clearly Hamiltonian also. Conversely, suppose that $G+vw$ is Hamiltonian, but that G is not. Then G contains a Hamiltonian path from v to w—say, $v_1(=v), v_2, v_3, \ldots, v_{p-1}, v_p(=w)$. Let $S = \{v_k\colon vv_{k+1} \in E\}$ and $T = \{v_k\colon v_k w \in E\}$; then, by hypothesis,

$$|S|+|T| = \rho(v)+\rho(w) \geqslant p.$$

But $|S \cup T| < p$ since $w \notin S \cup T$, and so there exists k such that v is adjacent to v_{k+1} and w is adjacent to v_k. So G contains the Hamiltonian circuit

$$v, v_{k+1}, v_{k+2}, \ldots, v_{p-1}, w, v_k, v_{k-1}, \ldots, v_2, v$$

(see Fig. 5), giving the required contradiction. ‖

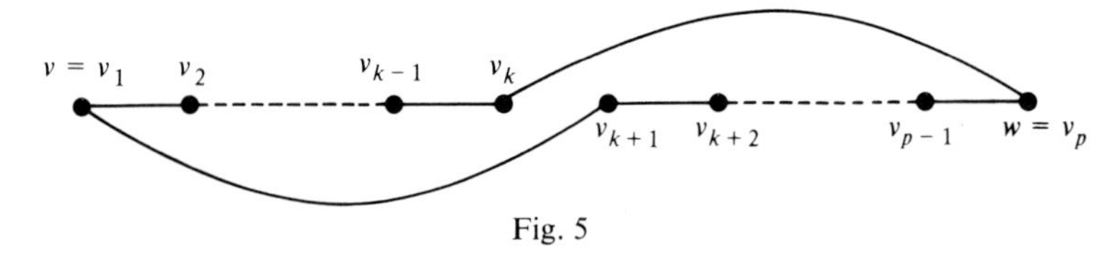

Fig. 5

As corollaries we obtain the following theorems of Ore (1960) and Dirac (1952):

Corollary 4.2 (Ore's Theorem). *Let G be a graph of order p ($\geqslant 3$). If $\rho(v)+\rho(w) \geqslant p$ for each pair of non-adjacent vertices v and w in G, then G is Hamiltonian.* ‖

Corollary 4.3 (Dirac's Theorem). *Let G be a graph of order p ($\geqslant 3$). If $\rho(v) \geqslant \frac{1}{2}p$ for each vertex v in G, then G is Hamiltonian.* ‖

(In fact, Nash-Williams (1969) has proved that under the conditions of Dirac's theorem, G contains at least $[\frac{5}{224}(n+10)]$ edge-disjoint Hamiltonian circuits. Jung [**41**] has also proved that if G is 1-tough, and if for each pair of non-adjacent vertices v and w in G, $\rho(v)+\rho(w) \geqslant p-4$, where $p \geqslant 11$, then G is Hamiltonian.)

Motivated by the result of Lemma 4.1, Bondy and Chvátal defined the **closure** cl(G) of a graph G to be the smallest graph H such that (i) G is a spanning subgraph of H, and (ii) $\rho_H(v)+\rho_H(w) < p$ for every pair of non-adjacent vertices v and w in H. (In [16], cl(G) is called the "p-closure".) The closure cl(G) can be obtained from G by the recursive procedure of joining two vertices whenever the sum of their valencies is at least p—this recursive procedure can be effected in $O(p^4)$ steps. For example, the closure of the graph G_0 in Fig. 6(a) is K_6, as can be seen by successively adding edges between non-adjacent vertices with valency-sum 6 or more. The various steps are shown in Fig. 6(b), (c) and (d).

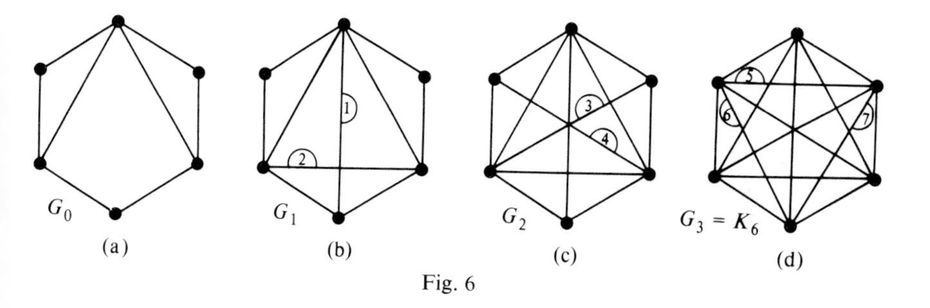

Fig. 6

The importance of the closure operation lies in the following theorem and corollary, due to Bondy and Chvátal [**16**]:

Theorem 4.4. *G is Hamiltonian if and only if* cl(G) *is Hamiltonian.*

Proof. The result follows by applying Lemma 4.1 each time an edge is added to form the closure. ‖

Corollary 4.5. *If* cl(G) *is a complete graph, then G is Hamiltonian.* ‖

This corollary incorporates most of the known conditions on the valencies of Hamiltonian graphs. Before stating these conditions we outline an algorithm in $O(p^3)$ steps for finding a Hamiltonian circuit in G from a Hamiltonian circuit in cl(G). We first assign to each added edge vw of cl(G) a label $\alpha(vw)$

corresponding to the order in which it was added in the above recursive procedure (see Fig. 6). Now let C be a Hamiltonian circuit in cl(G). If every edge of C lies in G, then we have a Hamiltonian circuit in G. If not, we choose the edge vw of C with the highest label. By the proof of Lemma 4.1, there exist two consecutive vertices v_k and v_{k+1} of C such that v is adjacent to v_{k+1} and v_k is adjacent to w, and such that $\alpha(vv_{k+1}) < \alpha(vw)$ and $\alpha(wv_k) < \alpha(vw)$—such vertices can be found in $O(p)$ steps. By deleting from C the edges v_kv_{k+1}, and adding the edges vv_{k+1} and wv_k, we obtain a Hamiltonian circuit in cl(G) all of those labeled edges have label less than $\alpha(vw)$. By repeating this procedure at most $O(p^2)$ times, we obtain a Hamiltonian circuit in G.

We can now use Corollary 4.5 on the closure of G to prove a result of Las Vergnas (1971):

Theorem 4.6. *Let G be a graph with vertex-set $\{v_1, \ldots, v_p\}$, and suppose that there do not exist integers i, j satisfying*

$$\begin{cases} i < j,\ i+j \geqslant p,\ v_iv_j \notin E,\ \rho(v_i) \leqslant i,\ \rho(v_j) \leqslant j-1, \\ \text{and } \rho(v_i)+\rho(v_j) \leqslant p-1. \end{cases} \tag{4}$$

Then G is Hamiltonian.

Proof. By Corollary 4.5 we need only prove that cl(G) is a complete graph. So suppose that $H = \text{cl}(G)$ is not a complete graph, and choose in H two non-adjacent vertices v_i and v_j such that

(*i*) j is as large as possible, and

(*ii*) i is as large as possible subject to condition (*i*).

We shall obtain a contradiction by showing that i and j satisfy all of the properties (4) in the statement of the theorem.

First of all, $i < j$, by (*i*). Next, since H is the closure, we have

$$\rho_H(v_i)+\rho_H(v_j) \leqslant p-1 \tag{5}$$

(where ρ_H denotes the valency in H), and so $\rho(v_i)+\rho(v_j) \leqslant p-1$. By (*i*), v_i must be adjacent in H to all those v_k with $k > j$, and so

$$\rho_H(v_i) \geqslant p-j. \tag{6}$$

By (*ii*), v_j must be adjacent in H to all those v_k with $k > i$, $k \neq j$, and so

$$\rho_H(v_j) \geqslant p-i-1. \tag{7}$$

Now (5) and (6) imply that

$$\rho(v_j) \leqslant \rho_H(v_j) \leqslant p-1-(p-j) = j-1,$$

and (5) and (7) imply that

$$\rho(v_i) \leqslant \rho_H(v_i) \leqslant p-1-(p-i-1) = i.$$

Finally, (6) and (7) imply that

$$\begin{aligned} i+j &\geqslant 2p-1-\rho_H(v_i)-\rho_H(v_j), \\ &\geqslant p, \text{ by (5).} \; \| \end{aligned}$$

The graph G of Fig. 6(a) shows that Theorem 4.4 is stronger than Theorem 4.6, but Theorem 4.6 is stronger than the following result of Chvátal (1972) which involves only the valencies. The deduction of Chvátal's theorem from Las Vergnas' theorem is not immediate (see, for example, [**16**, Appendix 2]); alternatively, Chvátal's result can be proved by imitating the proof of Las Vergnas' theorem.

Theorem 4.7 (Chvátal's Theorem). *Let G be a graph with non-decreasing valency-sequence $\rho_1, \rho_2, \ldots, \rho_p$. If $\rho_k \leqslant k < \frac{1}{2}p \Rightarrow \rho_{p-k} \geqslant p-k$, for each k, then G is Hamiltonian.* ||

Note that Chvátal's theorem (and hence Las Vergnas' theorem) generalizes the earlier results of Ore and Dirac (Corollaries 4.2 and 4.3), as well as other results of Pósa, Bondy and Nash-Williams. However, if we consider only conditions on the valencies, then Chvátal's theorem is in some sense the best possible result. To see this, we consider the graphs $G(r, p)$ defined as follows: if $1 \leqslant r < \frac{1}{2}p$, let $G(r, p)$ be the graph of order p with vertex-set $S \cup T \cup U$, where $|S| = |T| = r$ and $|U| = p-2r$, and where two vertices are joined if either of them belongs to S, or if both of them belong to U (see Fig. 7). Note that the subgraphs induced by S, T and U are K_r, $\bar{K}_r$ and K_{p-2r}, respectively.

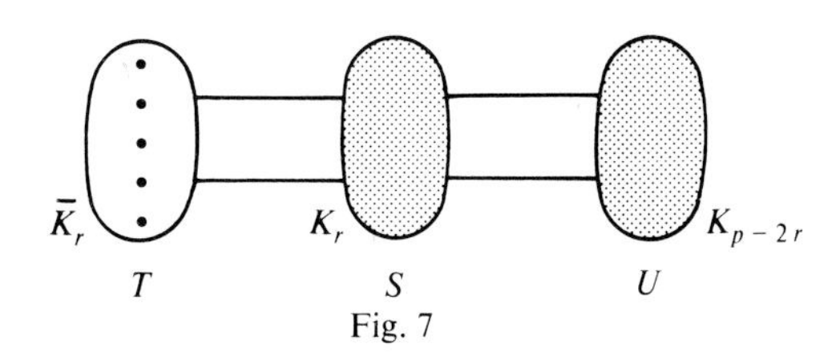

Fig. 7

Also, if $v \in S$ then $\rho(v) = p-1$, if $v \in T$ then $\rho(v) = r$, and if $v \in U$ then $\rho(v) = p-r-1$; it follows that the (non-decreasing) valency-sequence of $G(r, p)$ is

$$\underbrace{r, r, \ldots, r,}_{r \text{ terms}} \quad \underbrace{p-r-1, p-r-1, \ldots, p-r-1,}_{p-2r \text{ terms}} \quad \underbrace{p-1, p-1, \ldots, p-1.}_{r \text{ terms}}$$

It can be shown that $G(r, p)$ is the only graph with this valency-sequence. Moreover, $G(r, p)$ is not Hamiltonian since $c(G-S) = p+1 > |S|$, and hence $G(r, p)$ is not 1-tough.

A sequence $\rho_1, \rho_2, \ldots, \rho_p$ is said to be **majorized** by a sequence $\rho'_1, \rho'_2, \ldots, \rho'_p$ if $\rho_i \leqslant \rho'_i$ for $i = 1, 2, \ldots, p$. A graph G is **valency-majorized** by a graph H if the non-decreasing valency-sequence of G is majorized by the non-decreasing valency-sequence of H. (Note that G and H must have the same order.) We can now prove the following result of Chvátal (1972):

Theorem 4.8. *If G is a non-Hamiltonian graph of order p, then G is valency-majorized by some $G(r, p)$.*

Proof. Let $\rho_1 \leqslant \rho_2 \leqslant \ldots \leqslant \rho_p$ be the non-decreasing valency sequence of G. By Chvátal's theorem (Theorem 4.7), there is an integer $k < \frac{1}{2}p$ such that $\rho_k \leqslant k$ and $\rho_{p-k} < p-k$, and so G is valency-majorized by $G(k, p)$. ‖

There are also various theorems which relate to the number of edges in the graph. One of these is the following result, due to Ore and Bondy:

Theorem 4.9. *If G is a graph with p vertices and more than $\binom{p-1}{2}+1$ edges, then G is Hamiltonian. Furthermore, the only non-Hamiltonian graphs with p vertices and exactly $\binom{p-1}{2}+1$ edges are the graphs $G(1, p)$ and (for $p = 5$) $G(2, 5)$ (see Fig. 8).*

Proof. By Theorem 4.8, every non-Hamiltonian graph of order p can be valency-majorized by $G(r, p)$, for some r, and so the number of edges of G is at most $|E(G(r, p))|$, which is $\binom{p-r}{2}+r^2$. But an easy calculation shows that

$$\binom{p-r}{2}+r^2 \leqslant \binom{p-1}{2}+1,$$

with equality if $r = 1$, or if $r = 2$ and $p = 5$. (Recall that $G(r, p)$ is the only graph with its valency-sequence.) The result follows. ‖

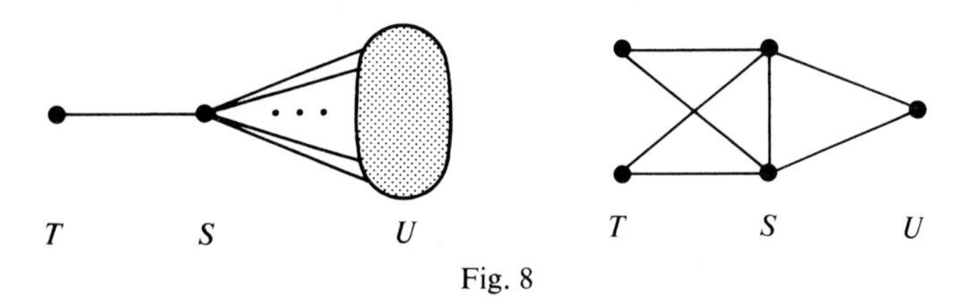

Fig. 8

Another result on the number of edges of a Hamiltonian graph is of a probabilistic nature, and is due to Komlós and Szemerédi (1978)—private communication of Chvátal—and improved an earlier result of Pósa (1976).

Theorem 4.10. *Suppose that we are given p vertices, and that we place* $[\frac{1}{2}p \log p + \frac{1}{2}p \log \log p] + O(p)$ *edges between them at random. If* $P(p, c)$ *is the probability that the resulting graph is Hamiltonian, then for all sufficiently large c,*

$$\lim_{p \to \infty} P(p, c) = 1. \; \|$$

There are some valency sequences which do not satisfy the hypothesis of Chvátal's theorem (Theorem 4.7), but which are necessarily the valency-sequences of Hamiltonian graphs. A valency-sequence is called **forcibly Hamiltonian** if every graph with this valency-sequence is Hamiltonian. Some of these sequences have been characterized by Nash-Williams [**51**], who proved in particular that every regular k-valent graph of order $2k+1$ is Hamiltonian. Further results on regular graphs have been obtained by Bollobás, Erdős, and Hobbs, and recently by Jackson [**37**], who obtained the following result:

Theorem 4.11. *Let G be a regular 2-connected graph of order p and valency* ρ, *where* $\rho \geqslant \frac{1}{3}p$. *Then G is Hamiltonian.* ‖

This result is best possible when $\rho = 3$, because of the Petersen graph, and is almost best possible when $\rho \geqslant 4$, since there exist infinite families of regular ρ-valent 2-connected non-Hamiltonian graphs of order $3\rho+4$ (if ρ is even) or $3\rho+5$ (if ρ is odd). Häggkvist has made an analogous conjecture for bipartite graphs—namely, that if G is a regular 2-connected bipartite graph of order p and valency ρ, where $\rho \geqslant \frac{1}{6}p$, then G is Hamiltonian.

In the case of regular graphs, Jackson [**38**] has improved Nash-Williams' result on the number of edge-disjoint Hamiltonian circuits (see after Corollary 4.3), by proving that if G is a regular graph of order p ($\geqslant 14$) and valency ρ, where $\rho \geqslant \frac{1}{2}(p-1)$, then G contains at least $[\frac{1}{6}(3\rho - p + 1)]$ edge-disjoint Hamiltonian circuits (see also Section 11).

Our next result is due to Woodall [**68**]; he has also obtained many interesting results concerning the "binding number" of a graph.

Theorem 4.12. *If, for every non-empty subset S of V, the number of vertices adjacent to some vertex in S is at least* $\frac{1}{3}(|S|+p+3)$, *then G is Hamiltonian.* ‖

We conclude this section by mentioning that several of the above results can be modified to yield sufficient conditions for a graph to be traceable, by means of the technique described in the proof of Theorem 2.1. As an example, we present the analog of Chvátal's theorem (Theorem 4.7); this analog can be used to prove a result of Clapham (1974) and Camion (1975) that every self-complementary graph is traceable:

Theorem 4.13. *Let G be a graph with non-decreasing valency-sequence* $\rho_1, \rho_2, \ldots, \rho_p$. *If* $\rho_k \leqslant k-1 < \frac{1}{2}(p-1) \Rightarrow \rho_{p-k+1} \geqslant p-k$, *for each k, then G is traceable.* ‖

5. Other Sufficient Conditions

In this section we consider three topics—the independence number of a graph, powers of graphs, and line graphs. References not given explicitly in this section may be found in [**43**]. We start with a theorem of Chvátal and Erdős (1972):

Theorem 5.1. *Let G be a graph with connectivity* κ *and independence number* α. *If* $\alpha \leqslant \kappa$, *then G is Hamiltonian.*

Proof. Suppose that G is a non-Hamiltonian graph with connectivity κ and independence number $\alpha \leqslant \kappa$; we shall derive a contradiction. Since G is κ-connected, G must contain a circuit of length at least κ. Let C be a circuit of maximum length in G, and let v be a vertex not belonging to C. Since G is κ-connected, there exist κ paths from v to C which are vertex-disjoint apart from v, and intersect C only at the terminal vertices $v_1, \ldots, v_\kappa$. (It suffices to add a new vertex w joined to all the vertices of C, and to use Menger's theorem to deduce the existence of κ paths from v to w.) If for each i, w_i is the successor of v_i in some fixed cyclic ordering of C, then no w_i is joined to v, since otherwise, we can obtain a circuit of greater length than C by replacing the edge v_iw_i by the path from v_i to v together with the edge vw_i. Since $\alpha \leqslant \kappa$, the set $\{v, w_1, \ldots, w_\kappa\}$ is not an independent set, and so there exists an edge w_sw_t. By deleting the edges v_sw_s and v_tw_t, and adding the edge w_sw_t together with the paths joining v to v_s and v to v_t (see Fig. 9), we obtain a circuit of greater length than C, thereby giving the required contradiction. ‖

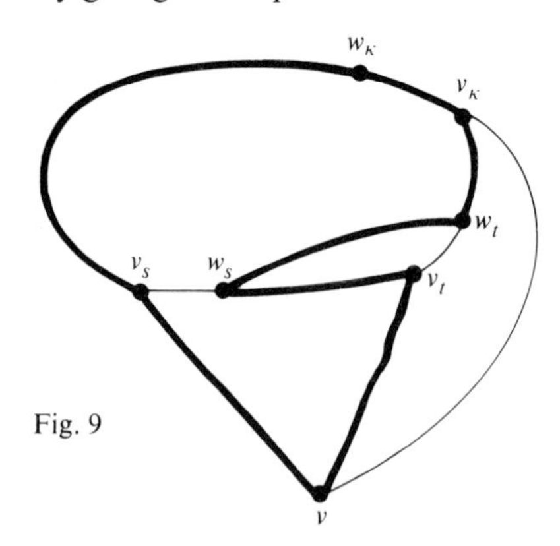

Fig. 9

Theorem 5.1 is best possible, in the sense that there exist non-Hamiltonian graphs for which $\alpha = \kappa+1$. Examples of such graphs are the complete bipartite graphs $K_{\kappa,\kappa+1}$, and the Petersen graph (with $\alpha = 4$, $\kappa = 3$). However, many of these non-Hamiltonian graphs G satisfy $t(G) < 1$. In this connection, the following results of Bilgáke are interesting (see [**9**]):

Theorem 5.2. *Let G be a κ-connected 1-tough graph of order p with independence number $\alpha \leqslant \kappa+1$.* Then

(*i*) *if $\kappa = 3$ and $p \geqslant 11$, then G is Hamiltonian*;

(*ii*) *if $\kappa = 4$, then G is Hamiltonian*;

(*iii*) *for each κ there exists a number $p_0(\kappa)$ such that if $p \geqslant p_0(\kappa)$, then G is Hamiltonian.* ||

From Theorems 3.1 and 3.3 we see that if $\kappa/\alpha \leqslant t(G) < 1$, then G is non-Hamiltonian, and by comparing this with the result of Theorem 5.1 we see that the only graphs which we have not classified as being Hamiltonian or not are those satisfying $\kappa/\alpha < 1 \leqslant t(G)$.

It is worth noting that Theorem 5.1 can be viewed as a generalization of Ore's theorem, as was observed by Bondy [**15**], who proved the following result:

Theorem 5.3. *If $\rho(v)+\rho(w) \geqslant p$ for each pair of non-adjacent vertices v and w in a graph, then $\alpha \leqslant \kappa$.* ||

Bondy and Nash-Williams (1971) have also found a result which is stronger than Theorem 5.1 when the minimal valency $\rho_{\min}$ of G is not too small.

Theorem 5.4. *If G is a 2-connected graph with $\rho_{\min} \geqslant \max\{\alpha, \frac{1}{3}(p+2)\}$, then G is Hamiltonian.* ||

Powers of Graphs

The kth power of a graph G, denoted by G^k, is the graph whose vertices correspond to those of G, and where two distinct vertices are joined whenever the distance between them in G is at most k. There are several results relating Hamiltonian graphs to powers of graphs, including the following theorem of Sekanina (1960):

Theorem 5.5. *The cube of every connected graph is Hamiltonian.*

Remark. Using induction on p (the order of the graph G), one can prove the stronger result that G^3 is "Hamiltonian-connected"—that is, that between any two vertices of G^3 there is a Hamiltonian path; such a proof may be found in [**6**, Chapter 10]. The proof we shall outline here is an algorithmic proof due essentially to Rosenstiehl (1971); the construction given here yields a Hamiltonian circuit of G^3 in $O(p)$ steps.

Proof. We note first that the theorem can be rephrased as: the vertices of any connected graph G can be cyclically ordered so that the distance between any two consecutive vertices is at most 3. Note also that it is sufficient to restrict our attention to the case where G is a tree—the general result follows by regarding this tree as a spanning tree of G.

So let T be a tree of order p, and let T^* be the symmetric digraph obtained by replacing each edge of T by two opposite arcs. Let $\vec{P}_{2p-2}$ be an Eulerian directed path in T^*; such a path can be constructed by means of the following algorithm (an example is given at the end of the proof):

Step 1. Choose an arbitrary arc a_1, and let $\vec{P}_1 = a_1$.

Step 2. Suppose that $\vec{P}_k = a_1, a_2, \ldots, a_k$, and that there exists an arc a_{k+1} whose initial vertex is the end-vertex of a_k, and such that neither a_{k+1} nor its opposite occurs in $\vec{P}_k$; then we define $\vec{P}_{k+1} = a_1, a_2, \ldots, a_k, a_{k+1}$, and repeat Step 2 with $\vec{P}_{k+1}$ instead of $\vec{P}_k$. If no such arc a_{k+1} exists, go to Step 3.

Step 3. Suppose that $\vec{P}_k = a_1, a_2, \ldots, a_k$, and that there exists an arc a_{k+1} not belonging to $\vec{P}_k$ whose initial vertex is the end-vertex of a_k; then we define $\vec{P}_{k+1} = a_1, a_2, \ldots, a_k, a_{k+1}$, and repeat Step 2 with $\vec{P}_{k+1}$ instead of $\vec{P}_k$. If no such arc a_{k+1} exists, go to Step 4.

Step 4. STOP.

It is easy to show that when Step 4 is applied, the directed path obtained is Eulerian. Also, when Step 3 is applied, the resulting arc a_{k+1} is uniquely determined, since T is a tree. Moreover, if an arc appears in an odd position in the path, then its opposite must appear in an even place—in fact, if the part of the path between an arc and its opposite contains an arc, then it necessarily contains its opposite. Thus the sequence of $p-1$ "odd arcs" $a_1, a_3, a_5, \ldots, a_{2p-3}$ gives an ordering of the edges of T. (Note that this ordering has the property that two edges with consecutive labels are either adjacent, or incident with the end vertices of a third one; this proves essentially that the square of the line graph of T is Hamiltonian—a special case of Theorem 5.10(*ii*).)

From the sequence of arcs $a_1, a_3, \ldots, a_{2p-3}$ we can deduce the required ordering of the vertices, as follows: label the initial vertex of the arc a_1 with label 0, and the terminal vertex of a_1 with label 1; then assign each label k to the initial or terminal vertex of the arc a_{2k-1} according as this arc comes after, or before, its opposite in $\vec{P}_{2p-2}$. This labeling can be shown to have the required properties, and the proof is complete. ‖

As an example of the above procedure, let T be the tree shown in Fig. 10; the direction of the arcs a, b, c, d, e, f in T^* is taken to be from left to right. Then the algorithm above gives us, for example,

$$\vec{P}_{2p-2} = \vec{a}, \vec{b}, \vec{c}, \vec{d}, \overleftarrow{d}, \overleftarrow{c}, \vec{e}, \vec{f}, \overleftarrow{f}, \overleftarrow{e}, \overleftarrow{b}, \overleftarrow{a}.$$

The sequence $a_1, a_3, \ldots, a_{2p-3}$ is $\vec{a}, \vec{c}, \overleftarrow{d}, \vec{e}, \overleftarrow{f}, \overleftarrow{b}$, and we get the ordering of the vertices indicated in Fig. 10. (Note that, for example, 3 appears as the initial vertex of $\overleftarrow{d}$, since $\overleftarrow{d}$ appears after its opposite $\vec{d}$ in $\vec{P}_{2p-2}$.)

It follows from the example of Fig. 10 that the square of a connected graph graph is not necessarily Hamiltonian. In fact, if T is a tree, then Neumann

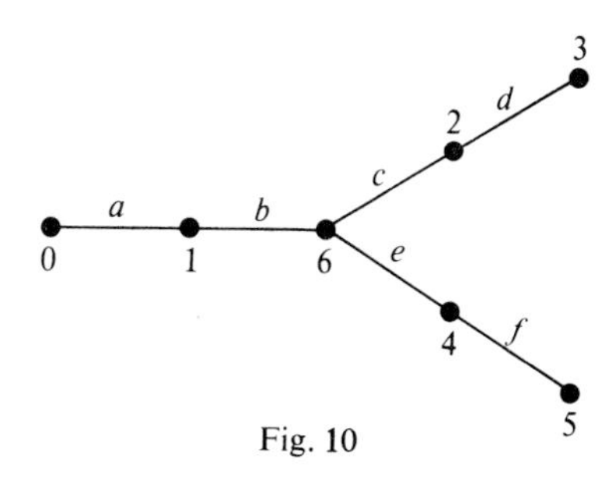

Fig. 10

(1964), and Harary and Schwenk (1971), proved that T^2 is Hamiltonian if and only if T does not contain the tree of Fig. 10 as a subgraph. The characterization of those graphs whose square is Hamiltonian is an unsolved problem*, but Fleischner (1974) has obtained the following deep result, thereby answering a conjecture of Plummer and Nash-Williams:

Theorem 5.6 (Fleischner's Theorem). *The square of every* 2*-connected graph is Hamiltonian.* ||

The most difficult and important part of the proof of Fleischner's theorem is the proof of the fact that every connected bridgeless graph contains an "*EPS*-subgraph"—that is, a connected spanning subgraph S which is the edge-disjoint union of a (not necessarily connected) graph E, all of whose vertices have even valency, with a (possibly empty) forest P each of whose components is a path. The proof of this result, and its use in the proof of Fleischner's theorem, are too complicated to be given here. As we remarked in Section 3, Fleischner's theorem would also follow from Chvátal's conjecture that every graph G with toughness $t(G) \geqslant 2$ is Hamiltonian.

Nebeský has observed that Fleischner's theorem implies the following result:

* See P. Underground, *Discrete Math.* **21** (1978), 323.

Theorem 5.7. *If G is any graph, then either G^2 or $(\bar{G})^2$ is Hamiltonian.*

Proof. If G is 2-connected, then G^2 is Hamiltonian, by Fleischner's theorem. If G is connected, but not 2-connected, then G contains a cut-vertex v. If $\rho(v) = p-1$, where p is the order of G, then G^2 is a complete graph, and the result is clear. If $\rho(v) < p-1$, then $(\bar{G}-v)^2$ is a complete graph, and so $(\bar{G})^2$ is Hamiltonian. ‖

Fleischner and Hobbs [**27**] have shown that if G^2 is Hamiltonian, then G must have a connected spanning subgraph which resembles an *EPS*-subgraph, except that the components of P are those which do not contain the tree of Fig. 10 as a subgraph. They also obtained new classes of graphs whose square is Hamiltonian (see, for example, [**36**]).

They have also reformulated some of their results in terms of total graphs (see Chapter 1), whose relevance here derives from the fact that if $T(G)$ is the total graph of a graph G, then $T(G)$ is the square of the subdivision graph $S(G)$ obtained by inserting a vertex into each edge of G—for example, the tree in Fig. 10 is $S(K_{1,3})$. Note that Fleischner's theorem implies that if G is 2-connected, then $T(G)$ is Hamiltonian. It can also be shown that if G is 2-edge-connected or planar, then $T(G)$ is Hamiltonian. These results follow from the following characterization of Hamiltonian total graphs obtained by Fleischner and Hobbs [**28**].

Theorem 5.8. *$T(G)$ is Hamiltonian if and only if G contains an EPS-subgraph.* ‖

Hamiltonian Line Graphs

Although line graphs are studied in detail in Chapter 10, we include here a few results which pertain to Hamiltonian graphs (for full references, see Chapter 10, or [**43**]). One of these is the following theorem of Harary and Nash-Williams (1965):

Theorem 5.9. *The line graph $L(G)$ is Hamiltonian if and only if either G is isomorphic to $K_{1,s}$ for some $s \geqslant 3$, or G contains a closed path C with the property that every edge of G is incident to at least one vertex of C.* ‖

Using Theorem 5.9, we can easily deduce that if G is either Hamiltonian or Eulerian, then $L(G)$ is Hamiltonian. One can also prove the following results, due to Nebeský (1973) and Bermond and Rosenstiehl (1973):

Theorem 5.10. *Let G be a connected graph of order p. Then*

(*i*) *if $p \geqslant 5$, then at least one of the graphs $L(G)$ and $L(\bar{G})$ is Hamiltonian;*

(*ii*) *if* $p \geqslant 4$, *then* $L^2(G)$ *is Hamiltonian*;

(*iii*) *if* $p \geqslant 3$, *then* $L(G^2)$ *is Hamiltonian.* ||

The proof of part (*ii*) is obtained by a slight refinement of the algorithm used in the proof of Theorem 5.5.

We conclude this section with a result of Chartrand (1968) on the "iterated line graph", defined for each $k > 1$ by $L^k(G) = L(L^{k-1}(G))$:

Theorem 5.11. *If G is a connected graph of order p (other than a path), then* $L^k(G)$ *is Hamiltonian for all* $k \geqslant p-3$. ||

6. Hamiltonian Planar Graphs

The interest in Hamiltonian planar graphs arises partly from a result of Whitney (1931), that in order to prove the four-color theorem it is sufficient to consider only Hamiltonian planar graphs, and partly from a related conjecture of Tait (1880), that every trivalent 3-connected planar graph is Hamiltonian. (Fuller references for the results in this section may be found in Grünbaum [**34**].) If Tait's conjecture had been proved true, we could have obtained a very simple proof of the four-color theorem, but a counter-example of order 46 was found by Tutte (1946) (see Fig. 11). The importance of 3-con-

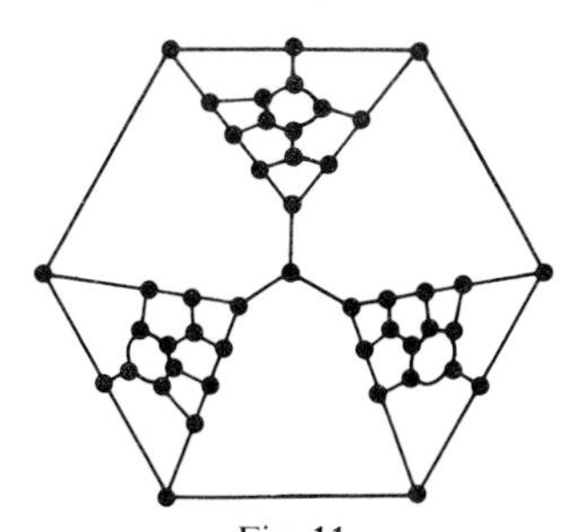

Fig. 11

nected planar graphs is due primarily to their relationship with polyhedra. A graph G is a ***d*-polytopal graph** if there exists a d-dimensional convex polytope P whose vertices and edges are in one-to-one correspondence with the vertices and edges of G. A 3-polytopal graph is called a **polyhedral graph**, and such graphs have been characterized by Steinitz (1922):

Theorem 6.1. *A graph G is polyhedral if and only if it is planar and 3-connected.* ||

It follows from Tutte's counter-example (Fig. 11) that there exist trivalent polyhedral graphs which are not Hamiltonian; the smallest example of a polyhedral non-Hamiltonian graph is the "Herschel graph" shown in Fig. 12. Note also that the Petersen graph is a trivalent 3-connected non-Hamiltonian graph. A method for generating polyhedral non-Hamiltonian graphs has

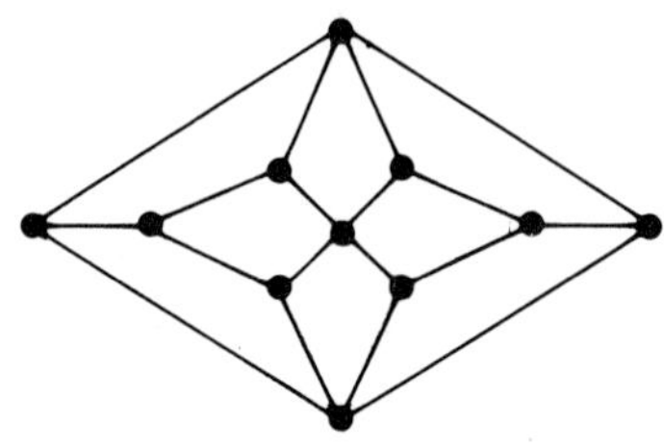

Fig. 12

been described by Grinberg (1968), who has also given the following necessary condition for a planar graph to be Hamiltonian:

Theorem 6.2. *Let G be a Hamiltonian planar graph, and let C be a Hamiltonian circuit in G. Let f_k denote the number of regions bounded by k edges in the interior of C, and let g_k denote the number of regions bounded by k edges in the exterior of C. Then*

$$\sum_k (k-2)(f_k - g_k) = 0.$$

Proof. If m denotes the number of edges of G which lie in the interior of C, then the total number of regions in the interior of C (that is, $\sum_k f_k$) is $m+1$, since C is a Hamiltonian circuit. But each such edge lies on the boundary of two regions in the interior of C, and each edge of C lies on the boundary of one region in the interior of C, so that $\sum_k kf_k = 2m+p$, where p is the order of G. It follows that

$$\sum_k (k-2) f_k = p-2,$$

and similarly that

$$\sum_k (k-2) g_k = p-2.$$

The result follows immediately. ||

As an example of the use of this theorem, we shall prove that the planar graph of Fig. 13 is non-Hamiltonian. In this graph, every region is either a pentagon or an octagon, except for one which is a square, and so, if the graph were Hamiltonian, we should have the equation

$$3(f_5-g_5)+6(f_8-g_8)+2(f_4-g_4) = 0.$$

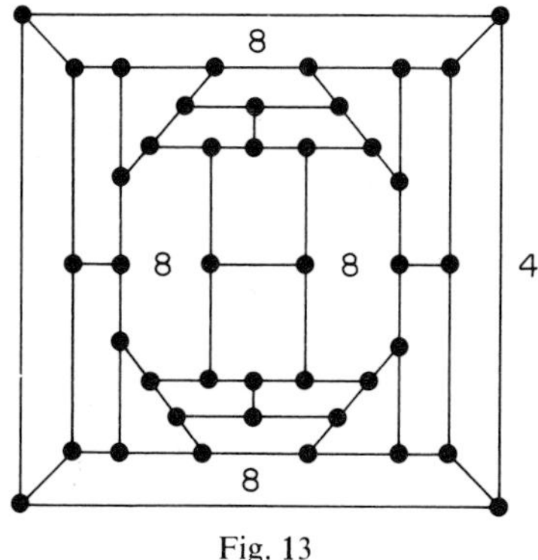

Fig. 13

But this is impossible, as can be seen by reducing modulo 3; the graph of Fig. 13 is therefore non-Hamiltonian.

Related to the above ideas is that of a cyclically edge-connected graph. A graph G is called **cyclically k-edge connected** if by deleting fewer than k edges, we cannot disconnect G into components each of which contains a circuit; for example, the graphs in Figs 11 and 13 are cyclically 3-edge connected, and cyclically 4-edge connected, respectively. The interest in such graphs arises from the fact that if one could prove that every trivalent 3-connected planar graph which is cyclically 5-connected is also Hamiltonian, then one could deduce the four-color theorem. However, this is not the case—an example of a trivalent 3-connected planar graph which is both cyclically 5-connected and non-Hamiltonian is given in Fig. 14; it is due to Grinberg.

Whitney (1931) proved that every plane triangulation without separating

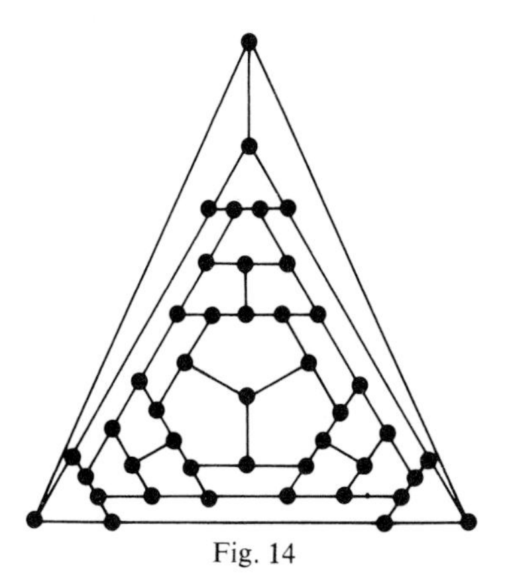

Fig. 14

triangles (triangles which do not bound a region) is Hamiltonian. Tutte (1956) generalized this to give the following important theorem, whose proof is much too complicated to be given here; a comprehensive proof is given in [**64**], and an algorithm in $O(p^3)$ steps to find a Hamiltonian circuit may be found in [**32**]:

Theorem 6.3. *Every* 4-*connected planar graph is Hamiltonian.* ‖

The condition of planarity cannot be omitted here, as shown by the graph $K_{4,5}$. Meredith (see [**17**, p. 239]) has exhibited a regular 4-valent, 4-connected non-Hamiltonian graph, thereby disproving a conjecture of Nash-Williams. The depth of Tutte's theorem is shown by the fact that Malkevitch (1971) and others have constructed 4-connected planar graphs which are not pancyclic (see Section 8), and which contain no pair of disjoint Hamiltonian circuits (see Grünbaum and Malkevitch [**35**], Martin [**46**] and Owens [**54**]).

Various other results of a similar nature have been obtained. In the following theorem we summarize some of these results, referring the reader to the surveys of Grünbaum [**33**], [**34**] and Owens [**54**] for references and further details:

Theorem 6.4. (*i*) *There exist d-polytopal non-Hamiltonian graphs, for every* $d \geqslant 3$;

(*ii*) *there exist regular* 4-*valent and* 5-*valent graphs which are* 3-*connected, cyclically* 6-*edge connected, planar, and non-Hamiltonian*;

(*iii*) *there exist regular* 3-*valent,* 4-*valent and* 5-*valent graphs which are* 3-*connected, planar, and non-traceable*;

(*iv*) *there exist constants* $\alpha < 1$ *and* $c > 0$ *with the property that there is a* 3-*connected planar graph of order p the length of whose longest path is less than* cp^{α}. ‖

For bipartite planar graphs, Barnette [**3**] has formulated the following conjecture:

Conjecture. *Every regular* 3-*valent* 3-*connected bipartite planar graph is Hamiltonian.*

Partial results have been obtained by Goodey [**31**] and Peterson [**55**]. Horton has shown that this conjecture becomes false if the condition of planarity is omitted, thereby settling a conjecture of Tutte [see **17**, p. 240].

We conclude this section by mentioning briefly some results involving Hamiltonian graphs on surfaces of positive genus g. The first of these is due to Duke (1972):

Theorem 6.5. *For each $g \geqslant 1$, there exists an integer $c(g)$ such that every $c(g)$-connected graph of genus g is Hamiltonian. Furthermore, $c(g)$ satisfies the inequalities*

$$[\tfrac{1}{2}(5+\sqrt{1+16g})] \leqslant c(g) \leqslant \{3+\sqrt{3+6g}\}. \;\|$$

A special case of this result (for $g = 1$) had previously been proved by Altshuler (1972), and Bloom and Schmeichel **[11]** have also shown that Altshuler's result follows from Chvátal's conjecture in Section 3 that every graph of toughness $t \geqslant 2$ is Hamiltonian. In fact, Bloom and Schmeichel's observation is the special case $g = 1$, $k = 6$ of the following more general theorem **[11]**:

Theorem 6.6. *If G is a k-connected graph of genus g, then*

$$c(G-S) \leqslant \frac{2}{k-2}(|S|-2+2g),$$

for all subsets $S \subseteq V(G)$ such that $|S| > k$. ‖

7. Hamiltonian Digraphs

In view of Theorem 2.1, one might expect that most of the results in this chapter extend easily to digraphs. In fact, this is not the case—some results have no analogs for digraphs, while others can be extended but the proofs are more complicated and the methods are different. In this section we shall denote by $\rho_{\text{out}}(v)$ the out-valency of a vertex v in a digraph D (the number of arcs of D of the form vw), and by $\rho_{\text{in}}(v)$ the in-valency of v (the number of arcs of D of the form wv). The valency $\rho(v)$ of v is defined to be the sum of $\rho_{\text{out}}(v)$ and $\rho_{\text{in}}(v)$.

We first give the extension to digraphs of Ore's theorem (Corollary 4.2), which was obtained by Meyniel **[47]**; simpler proofs have been given by Overbeck-Larisch **[53]**, and Bondy and Thomassen **[18]**. The proof presented here is a constructive version of Bondy and Thomassen's proof, following an idea of Minoux **[48]**; it shows that a Hamiltonian circuit can be constructed in $O(p^4)$ steps.

Theorem 7.1 (Meyniel's Theorem). *Let D be a strongly-connected digraph of order p. If $\rho(v)+\rho(w) \geqslant 2p-1$ for each pair of non-adjacent vertices v and w in D, then D is Hamiltonian.*

Proof. We shall need the following notation: if S is any subset of V, the vertex-set of D, then $\rho_S(v)$ will denote the number of arcs joining v with vertices in S (in either direction); an *S-path* is a directed path of length two or more whose ends belong to S, but whose vertices are otherwise disjoint from S. A directed path $v_0 \to v_1 \to \ldots \to v_k$ is called *quasi-maximum* if there exists no other vertex v such that $v_i v$ and vv_{i+1} are both arcs of D, for some i; similarly, a directed circuit $v_0 \to v_1 \to \ldots \to v_k \to v_0$ is called quasi-maximum if there exists no other vertex v such that $v_i v$ and vv_{i+1} are both arcs of D, for some i, or such that $v_k v$ and vv_0 are both arcs of D.

(1) We first prove that if $P = v_0 \to v_1 \to \ldots \to v_k$ is a quasi-maximum directed path in D, then for every vertex v not included in P,

$$\rho_{V(P)}(v) \leqslant |V(P)|+1.$$

In order to see this, we let $A = \{v_i : 1 \leqslant i \leqslant k-1$, and $v_i v$ is an arc of $D\}$ and $B = \{v_i : 1 \leqslant i \leqslant k-1$, and vv_{i+1} is an arc of $D\}$. Then $A \cap B = \varnothing$, by hypothesis, and $v_k \notin A \cup B$, so that $|A|+|B| \leqslant |V(P)|-1$. It follows that $\rho_{V(P)}(v) \leqslant |A|+|B|+2 \leqslant |V(P)|+1$, as required.

(2) In order to construct a Hamiltonian directed circuit in D, we start by constructing a quasi-maximum directed circuit C. Note that the construction of such a directed circuit involves $O(p^3)$ steps, since in $O(p^2)$ steps one can check whether a circuit is quasi-maximum, or construct a directed circuit containing one more vertex, and there are p vertices to consider. If C is Hamiltonian, the construction is complete. If not, we can construct a longer circuit C' in $O(p^3)$ steps, and then a quasi-maximum circuit C'' longer than C, also in $O(p^3)$ steps; the details of these constructions are given in (3)–(6) below. Repeating the construction with C'', we obtain a Hamiltonian directed circuit in $O(p^4)$ steps.

(3) We now show that if C is a quasi-maximum directed circuit which is not Hamiltonian, then D contains an S-path, where S is the set of vertices of C. If no such path exists, then since D is strongly connected, D must contain a directed circuit C' having exactly one vertex z in common with C. Let S' be the set of vertices of C'. Since D has no S-path, no vertex in $S'-\{z\}$ can be joined to a vertex in $S-\{z\}$. Thus, for any $v \in S'-\{z\}$ and $w \in S-\{z\}$, we have $\rho_S(v) \leqslant 2$ and $\rho_S(w) \leqslant 2|S|-2$. Also, since D has no S-path, there cannot exist paths of the form $w \to u \to v$ or $v \to w$, for any $u \in V-S-\{v\}$, and so $\rho_{V-S}(v)+\rho_{V-S}(w) \leqslant 2(p-|S|-1)$. It follows that $\rho(v)+\rho(w) \leqslant 2p-2$, contradicting the hypothesis.

Note that the family of S-paths starting at a vertex v of C can be constructed in $O(p^2)$ steps.

(4) By (3), there exists an S-path in D. We shall take this path to be $P = x \to \ldots \to y$, where x and y lie in C, and we shall choose P so that the directed path $C(x, y)$ included in C with initial and terminal vertices x and y is as short as possible. (This choice can be made in $O(p^2)$ steps.) Let $S_1 = V(C(x, y))-\{x\}-\{y\}$, and $S_2 = S-S_1$, and let Q be a quasi-maximum

directed path joining y to x included in the subgraph generated by S and containing all of the vertices of S_2 (by (2), such a directed path can be chosen in $O(p^3)$ steps). We shall show that Q contains every vertex of S, so that $C'' = Q \cup P$ is a directed circuit which is longer than C. To do this, let $S_3 = V(Q)$, so that $S_2 \subseteq S_3 \subseteq S$; we shall assume that $S_3 \neq S$, and derive a contradiction.

(5) If $S_3 \neq S$, we have $S_2 \neq S$, and hence $S_1 \neq \varnothing$. By our choice of P, no vertex in $V(P)-\{x\}-\{y\}$ can be adjacent to any vertex in S_1, and so if $v \in V(P)-\{x\}-\{y\}$, we have $\rho_{S_1}(v) = 0$. By applying the result of (1) to the directed path included in C with initial and terminal vertices y and x, we get $\rho_{S_2}(v) \leqslant |S_2|+1$. Similarly, by applying the result of (1) to Q and the subgraph induced by S, we get $\rho_{S_3}(w) \leqslant |S_3|+1$, for any $w \in S-S_3$, and hence

$$\rho_S(w) \leqslant |S_3|+1+2(|S|-|S_3|-1) = 2|S|-|S_3|-1.$$

Finally, by our choice of P, there cannot exist paths $v \to u \to w$ or $w \to u \to v$ with $u \in V-S-\{v\}$, and so

$$\rho_{V-S}(v)+\rho_{V-S}(w) \leqslant 2(p-|S|-1).$$

Combining all these inequalities gives

$$\rho(v)+\rho(w) \leqslant |S_2|+1+2|S|-|S_3|-1+2(p-|S|-1) \leqslant 2p-2,$$

which contradicts the hypothesis. This contradiction proves the theorem. ||

Using Theorem 7.1 we can easily deduce a theorem of Ghouila-Houri (1960), which is the digraph analog of Dirac's theorem (Corollary 4.3); for other corollaries, see **[6]**, **[47]**, or Woodall (1972):

Corollary 7.2. *Let D be a strongly-connected digraph of order p. If $\rho(v) \geqslant p$ for each vertex in D, then D is Hamiltonian.* ||

The next corollary is due to Camion (1959); for a stronger result, with a direct proof, see Theorem 3.3 of Chapter 7:

Corollary 7.3. *Every strongly connected tournament is Hamiltonian.*

Proof. Since every two vertices are adjacent in a tournament, there is no pair of non-adjacent vertices, and so the result follows trivially from Theorem 7.1. ||

The requirement in Theorem 7.1 and Corollaries 7.2 and 7.3 that D be strongly connected is necessary, as can be seen by considering a digraph consisting of two complete symmetric digraphs joined by arcs all in the same direction. If we do remove this requirement, then we get the following weaker result:

Corollary 7.4. *Let D be a digraph of order p. If $\rho(v)+\rho(w) \geqslant 2p-3$ for each pair of non-adjacent vertices v and w in D, then D is traceable.*

Proof. We use the trick indicated in the proof of Theorem 2.1. Let $\tilde{D}$ be the digraph obtained from D by adding a new vertex and joining it to every vertex of D by two opposite arcs. Then $\tilde{D}$ is a strongly connected digraph which satisfies the conditions of Theorem 7.1 (with p replaced by $p+1$). It follows that $\tilde{D}$ is Hamiltonian, and hence that D is traceable. ||

As a corollary of this last result, we can also obtain Corollary 3.2 of Chapter 7, that every tournament is traceable.

One can also try to obtain digraph analogs of various other theorems for Hamiltonian graphs—for example, Chvátal's theorem (Theorem 4.7). In particular, we can ask whether every strongly connected digraph whose non-decreasing valency-sequence $\rho_1, \ldots, \rho_p$ satisfies the following conditions is Hamiltonian:

(*i*) $\rho_k \leqslant 2k < p \Rightarrow \rho_{p-k} \geqslant 2(p-k)$, for each k.

Similarly, one can ask whether every strongly connected digraph whose non-decreasing out-valency and in-valency sequences $\rho_1^+, \ldots, \rho_p^+$ and $\rho_1^-, \ldots, \rho_p^-$ satisfy the following conditions is Hamiltonian:

(*ii*) $\rho_k^+ \leqslant k < \frac{1}{2}p \Rightarrow \rho_{p-k}^+ \geqslant p-k$ and $\rho_k^- \leqslant k < \frac{1}{2}p \Rightarrow \rho_{p-k}^- \geqslant p-k$.

In fact, both of these are false in general, as is shown by the following simple example: let D be a complete symmetric digraph of order $p-2$ containing a vertex u, and let $\tilde{D}$ be the digraph obtained by adding two new vertices v and w dominated by u, and dominating every other vertex of D. Then $\tilde{D}$ is a strongly connected digraph whose valency-sequences satisfy (*i*) and (*ii*), but which is not Hamiltonian.

A third possible analogue of Chvátal's theorem, which survives the above example, is the following conjecture due to Nash-Williams [**52**]:

Conjecture. *If D is a strongly connected digraph whose non-decreasing out-valency and in-valency sequences $\rho_1^+, \ldots, \rho_p^+$ and $\rho_1^-, \ldots, \rho_p^-$ satisfy the conditions*

$$\rho_k^+ \leqslant k < \tfrac{1}{2}p \Rightarrow \rho_{p-k}^- \geqslant p-k \qquad \textit{and} \qquad \rho_k^- \leqslant k < \tfrac{1}{2}p \Rightarrow \rho_{p-k}^- \geqslant p-k$$

for each k, then D is Hamiltonian.

By using Corollary 7.2, Lewin [**44**] has shown that every strongly connected digraph of order p with at least $(p-1)(p-2)+3$ arcs is Hamiltonian, thereby providing a partial analog of Theorem 4.9. Furthermore, it can be shown that the only non-Hamiltonian digraphs of order p with $(p-1)(p-2)+2$ arcs are the symmetric digraphs corresponding to the extremal graphs $G(1, p)$ and $G(2, 5)$ of Theorem 4.9, and the above example. On the other hand, the example of Fig. 15 (due to Las Vergnas), representing a strongly 2-connected digraph whose independence number is 2, shows the difficulty of extending the theorem of Chvátal and Erdős (Theorem 5.1). Similarly, Fouquet **[29]** has exhibited, for every k, a strongly k-connected digraph D with the property that D^k is non-Hamiltonian, thereby showing that there is no analog of Theorem 5.5.

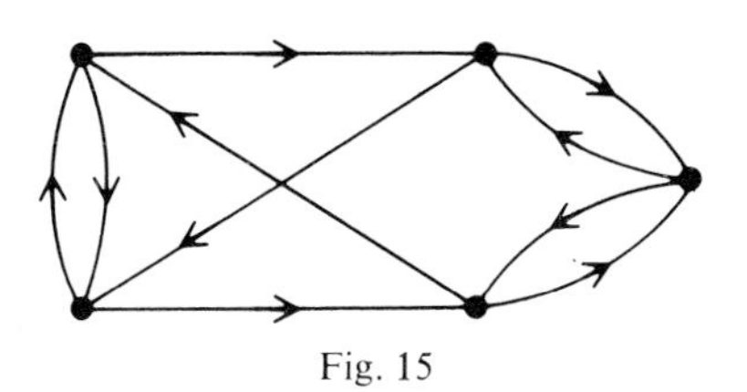

Fig. 15

It seems that the study of Hamiltonian digraphs is generally much more difficult than the corresponding study for graphs, but we conclude this section with a result of Kasteleyn (1963) on line digraphs, where the digraph version is the simpler one (see also Chapter 10, Section 10):

Theorem 7.5. *$L(D)$ is Hamiltonian if and only if D is Eulerian.* ||

8. Pancyclic and Panconnected Graphs

A graph G of order p is said to be **pancyclic** if G contains a circuit of length l, for each l satisfying $3 \leqslant l \leqslant p$; in particular, every pancyclic graph is Hamiltonian. The concept of a pancyclic graph was introduced by Bondy, who has written two expository papers on the subject **[13]**, **[14]** (see also **[43]** for references), and who has made the following "meta-conjecture": every condition which implies that a graph is Hamiltonian also implies that it is also pancyclic, with the possible exception of a simple family of exceptional graphs. Although this meta-conjecture is sometimes false, it turns out to be accurate in a surprisingly large number of cases.

We start with a result of Hakimi and Schmeichel (1974) which extends Chvátal's theorem (Theorem 4.7); the argument for the case when p is odd is due to Bondy (see **[21**, p. 92]):

Theorem 8.1. *Let G be a graph with non-decreasing valency-sequence $\rho_1, \rho_2, \ldots, \rho_p$. If $\rho_k \leqslant k < \frac{1}{2}p \Rightarrow \rho_{p-k} \geqslant p-k$, for each k, then G is either pancyclic or bipartite.*

Sketch of Proof. Suppose first that p is odd. Then $\rho_{\frac{1}{2}(p+1)} > \frac{1}{2}p$, since either $\rho_{\frac{1}{2}(p-1)} > \frac{1}{2}p$, in which case the result is obvious, or $\rho_{\frac{1}{2}(p-1)} \leqslant \frac{1}{2}(p-1)$, in which case (by the hypothesis) $\rho_{\frac{1}{2}(p+1)} \geqslant \frac{1}{2}(p+1) > \frac{1}{2}p$.

Now, by Chvátal's theorem, G is Hamiltonian. Let $v_1, v_2, \ldots, v_p$ be a Hamiltonian circuit, and suppose that G is not pancyclic and thus does not contain a circuit of length l, for some l satisfying $3 \leqslant l < p$. It follows that we cannot have simultaneously in G the pairs of edges

$$v_jv_k \quad \text{and} \quad v_{j+1}v_{k-l+3}, \quad \text{for} \quad j+l-1 \leqslant k \leqslant j-1,$$

and

$$v_jv_k \quad \text{and} \quad v_{j+1}v_{k-l+1}, \quad \text{for} \quad j+2 \leqslant k \leqslant j+l-2$$

(where all suffixes are taken modulo p—see Fig. 16), and so we have $\rho(v_j)+\rho(v_{j+1}) \leqslant p$.

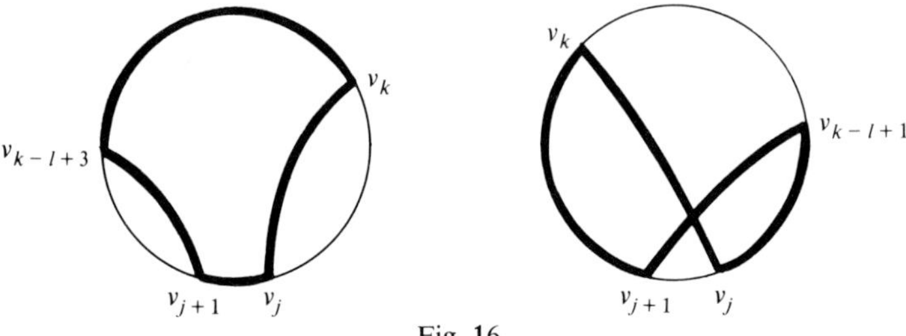

Fig. 16

But since $\rho_{\frac{1}{2}(p+1)} > \frac{1}{2}p$, more than half of the vertices of G have valency greater than $\frac{1}{2}p$, and so two such vertices (v_j and v_{j+1}, say) must be adjacent in the Hamiltonian circuit and $\rho(v_j)+\rho(v_{j+1}) > p$, which is the required contradiction.

If p is even and $\rho_{\frac{1}{2}p} \neq \frac{1}{2}p$, then the proof is the same as the one just given, but if $\rho_{\frac{1}{2}p} = \frac{1}{2}p$, then the proof is much more complicated, and will be omitted. ||

It is interesting to note that although Chvátal's theorem generalizes, the result of Corollary 4.5 on the closure of a graph does not. In particular, Thomassen has exhibited graphs G whose closure cl(G) is complete, but which are neither pancyclic nor bipartite. However, Ore's theorem (Corollary 4.2) does generalize; the following result was proved by Bondy (1971) and can be deduced from Theorem 8.1:

Corollary 8.2. *Let G be a graph of order p ($\geqslant 3$). If $\rho(v)+\rho(w) \geqslant p$ for each pair of non-adjacent vertices v and w in G, then G is either pancyclic or the graph $K_{\frac{1}{2}p,\frac{1}{2}p}$.* ||

There is also a generalization of the theorem of Chvátal and Erdős (Theorem 5.1), proved by Erdős in 1974:

Theorem 8.3. *Let G be a graph with connectivity κ and independence number α, and suppose that $\alpha \leqslant \kappa$. Then there exists a number c such that if G has at least $c\kappa^4$ vertices, then G is pancyclic.* ‖

In the case $\kappa = 2$, Bondy has shown that if $\alpha \leqslant 2$, then either G is pancyclic, or $G \cong C_4$ or C_5. More generally, he has shown that there exist two families of non-pancyclic graphs with $\alpha = \kappa$—namely, $K_{\kappa,\kappa}$, and the graph G_κ consisting of a circuit of length $2\kappa+2$ together with the edges $v_i v_{i+2j}$ for $1 \leqslant i \leqslant 2\kappa+2$, $2 \leqslant j \leqslant \kappa-1$.

Several of the results on powers of graphs (Section 5) also have pancyclic analogs. We summarize these in the following theorem (see **[13]** and **[26]**); part (*iii*) follows from part (*ii*), with the aid of Theorem 5.6:

Theorem 8.4. *Let G be a connected graph. Then*

(*i*) *G^3 is pancyclic;*

(*ii*) *if G^2 is Hamiltonian, then G^2 is pancyclic;*

(*iii*) *if G is 2-connected, then G^2 is pancyclic.* ‖

A corresponding concept also exists for digraphs. A digraph D of order p is said to be **pancyclic** if D contains a directed circuit of length l, for each l satisfying $3 \leqslant l \leqslant p$; in particular, every pancyclic digraph is Hamiltonian. Moon has shown that every strongly connected tournament is pancyclic, and this result is proved in Chapter 7 (Theorem 3.3). The following theorem was obtained by Thomassen **[62]**:

Theorem 8.5. *Let D be a strongly connected digraph of order p ($\geqslant$3). If $\rho(v)+\rho(w) \geqslant 2p$ for each pair of non-adjacent vertices v and w in D, then D is either pancyclic or the symmetric digraph obtained by replacing each edge of $K_{\frac{1}{2}p,\frac{1}{2}p}$ by two opposite arcs.* ‖

This theorem does not directly generalize Meyniel's theorem (Theorem 7.1) since there exist non-pancyclic digraphs satisfying the hypotheses of Theorem 7.1. However, it does contain as special cases Moon's tournament results, and earlier results of Overbeck-Larish, Häggkvist and Thomassen.

We now turn our attention to panconnected graphs. Following Ore, we define a graph G to be **Hamiltonian-connected** if each pair of vertices of G can be connected by a Hamiltonian path. A graph G of order p is **panconnected** if each pair v, w of vertices of G can be connected by a path of length l, for each l satisfying $d(v, w) \leqslant l \leqslant p-1$ (where $d(v, w)$ is the distance between v

and w). Note that every panconnected graph is Hamiltonian-connected and pancyclic. Faudree and Schelp (1976) have conjectured that if G is a Hamiltonian-connected graph of order p, then each pair of vertices of G can be connected by a path of length l, for each l satisfying $l \geqslant \frac{1}{2}p$*. Some properties of these graphs are summarized in the following theorem (see [**26**]); part (*i*) is due to Alavi and Williamson (1975) and part (*iii*) is due to Chartrand, Hobbs, Jung, Kapoor and Nash-Williams (1974):

Theorem 8.6. *Let G be a connected graph. Then*

(*i*) *G^3 is panconnected;*

(*ii*) *if G^2 is Hamiltonian-connected, then G^2 is panconnected;*

(*iii*) *if G is 2-connected, then G^2 is panconnected.* ||

In order to obtain analogs for Hamiltonian-connected graphs of our earlier theorems on Hamiltonian graphs, we usually need to take a slightly stronger hypothesis. In many cases the proofs are very similar to the original ones, or the new results can easily be deduced from the original results. As an example, we give the following analog of Chvátal's theorem (Theorem 4.7):

Theorem 8.7. *Let G be a graph with non-decreasing valency sequence $\rho_1, \rho_2, \ldots, \rho_p$. If $\rho_k \leqslant k+1 < \frac{1}{2}(p+1) \Rightarrow \rho_{p-k-1} \geqslant p-k$, for each k, then G is Hamiltonian-connected.*

Proof. Let v and w be any pair of vertices of G, and let G' be the graph obtained from G by adding a new vertex z, joining it to v and w, and deleting the edge vw (if such an edge exists). Then there is a Hamiltonian path connecting v and w in G if and only if there is a Hamiltonian circuit in G'. Since G' satisfies the hypotheses of Chvátal's theorem (with p replaced by $p+1$), G' is Hamiltonian, and hence G is Hamiltonian-connected. ||

9. Strongly Hamiltonian Graphs

A great number of generalizations of the ideas of the previous section have been investigated. Most of these are included in the following definition, proposed by Skupien and Wojda in 1974 (see the survey of Wojda [**65**]). A graph G is said to be **strongly t-edge Hamiltonian** if, for every system S of vertex-disjoint paths of total length t of the complete graph with vertex-set $V(G)$, there exists in the graph $G' = (V(G), E(G) \cup S)$ a Hamiltonian circuit containing S. In particular, a strongly 1-edge Hamiltonian graph is simply a Hamiltonian-connected graph. If all the edges of S belong to $E(G)$, then G is said to be t-edge Hamiltonian, a concept introduced by Pósa in 1963.

* This conjecture has been disproved by Thomassen.

Furthermore, G is said to be **strongly (s, t)-Hamiltonian** if the deletion of any s' vertices of G (where $0 \leqslant s' \leqslant s$) results in a strongly t-edge Hamiltonian graph; a strongly $(s, 0)$-Hamiltonian graph is called an **s-Hamiltonian graph**.

Note that these definitions are not equivalent. Although every Hamiltonian-connected graph is 1-edge Hamiltonian, the converse is not true, as may be seen by considering the circuit graph C_p. There also exist Hamiltonian-connected graphs which are not 1-Hamiltonian, and 1-Hamiltonian graphs which are not 1-edge Hamiltonian (see Theorem 10.3). Most of the results in Section 4 can be extended to give sufficient conditions for a graph to be strongly (s, t)-Hamiltonian. For example, in order to generalize Theorem 4.4, it is sufficient to consider the "$p+s+t$-closure" of G, defined to be the smallest graph H such that (*i*) G is a spanning subgraph of H, and (*ii*) $\rho_H(v)+\rho_H(w) < p+s+t$ for every pair of non-adjacent vertices v and w in H (see [**16**]). In a similar way, Chvátal's theorem (Theorem 4.7) becomes the following:

Theorem 9.1. *Let G be a graph with non-decreasing valency-sequence $\rho_1 \leqslant \rho_2 \leqslant \ldots \leqslant \rho_p$. If $\rho_k \leqslant k+s+t \leqslant \frac{1}{2}(p+s+t) \Rightarrow \rho_{p-s-t-k} \geqslant p-k$, for each k, then G is strongly (s, t)-Hamiltonian.* ||

Many other results have been obtained. For example, Benhocine [**5**] has obtained the following theorem on powers of graphs:

Theorem 9.2. *If G is a connected graph, and k is a positive integer, then G^k is strongly (s, t)-Hamiltonian for all s and t satisfying $t > 0$ and $s+t \leqslant k-2$.* ||

However, some of these generalizations remain to be proved. For example, the following conjecture seems reasonable, and reduces to Chvátal and Erdős's theorem (Theorem 5.1) when $s = t = 0$:

Conjecture A. *Let s, t and κ be positive integers satisfying $s+t \leqslant \kappa-1$. If G is a κ-connected graph, and if the independence number α of G satisfies $\alpha \leqslant \kappa-s-t$, then G is strongly (s, t)-Hamiltonian.*

In fact, it can easily be shown that it suffices to prove the conjecture for $s = 0$, and that a proof similar to that of Theorem 5.1 works if one can first prove the following conjecture:

Conjecture B. *Let $t \leqslant \kappa-1$. If G is a κ-connected graph, and if $\alpha \leqslant \kappa-t$, then every set of t disjoint edges is contained in a circuit of G.*

This second conjecture is a special case of the following conjecture of

Lovász [**45**], on which some progress has been made by Thomassen [**61**] and Woodall [**67**]:

Conjecture C. *If G is a κ-connected graph which is $\kappa+1$-edge-connected, then every set of κ vertex-disjoint edges is contained in a circuit of G.*

Sometimes the natural generalizations turn out to be false. For example, one can define a "Hamiltonian-connected digraph" to be a digraph in which there exists a Hamiltonian directed path from any vertex to any other. Ghouila-Houri (1960) has shown that there exist strongly 2-connected digraphs of order p satisfying $\rho(v) \geqslant p+1$ for every vertex v, but with the property that there is no directed path joining some pairs of vertices. However, generalizations with stronger hypotheses have been obtained by Overbeck-Larish [**53**].

We conclude this section by noting that $(p-1)$-edge-Hamiltonian graphs have been characterized. These graphs are called **randomly-Hamiltonian graphs**, and can be described as those graphs in which every path can be completed to a Hamiltonian circuit. Their characterization was carried out by Chartrand and Kronk (1969), and is as follows:

Theorem 9.3. *A graph G is randomly-Hamiltonian if and only if G is either a complete graph, a circuit graph, or a regular complete bipartite graph.* ||

Similar results have been obtained by Thomassen [**59**], who has characterized those graphs in which every path can be extended to a Hamiltonian path, and by Chartrand, Kronk and Lick (1969), who have characterized randomly-Hamiltonian digraphs.

10. Hypohamiltonian Graphs

A graph G is **hypohamiltonian** if G is not Hamiltonian, but the vertex-deleted subgraph $G-v$ is Hamiltonian for every vertex v. For example, the Petersen graph is hypohamiltonian, and is in fact the smallest such graph. The problem of determining those values of p for which a hypohamiltonian graph of order p exists is almost solved, and we summarize the results known in the following theorem (see [**24**], [**58**] and [**60**]):

Theorem 10.1. (*i*) *There exist hypohamiltonian graphs of order p if and only if $p = 10$ or 13, or $p \geqslant 15$ (with the possible exception of $p = 17$);*

(*ii*) *there exist trivalent hypohamiltonian graphs of order p if and only if p is even and $p = 10$ or $p \geqslant 18$ (with the possible exception of $p = 24$ and $p = 32$);*

(*iii*) *there exist planar hypohamiltonian graphs, and hypohamiltonian graphs containing a triangle.* ||

Many constructions for hypohamiltonian graphs have been described, and we now present a construction due to Thomassen [**58**] which creates new ones from old:

Let G_1 and G_2 be disjoint hypohamiltonian graphs, and suppose that G_1 has a vertex z_1 of valency 3, and that G_2 has a vertex z_2 of valency 3. Moreover, for $i = 1, 2$, let u_i, v_i and w_i be the vertices adjacent to z_i in G_i. We can suppose that G_i does not contain the edges u_iv_i, u_iw_i and v_iw_i, since if G_1 contains the edge u_1v_1 (for example), as $G_1 - u_1$ is Hamiltonian, therefore there exists a Hamiltonian path in $G_1 - u_1 - z_1$ joining v_1 to w_1; by taking this Hamiltonian path, together with the edges w_1z_1, z_1u_1 and u_1v_1, we get a Hamiltonian circuit for G_1, contradicting the fact that G_1 is not Hamiltonian.

Now let G be the graph obtained from G_1 and G_2 by deleting the vertices z_1 and z_2, and identifying the pairs of vertices u_1 and u_2, v_1 and v_2, and w_1 and w_2 (see Fig. 17); we shall call the identified vertices u, v and w, respectively.

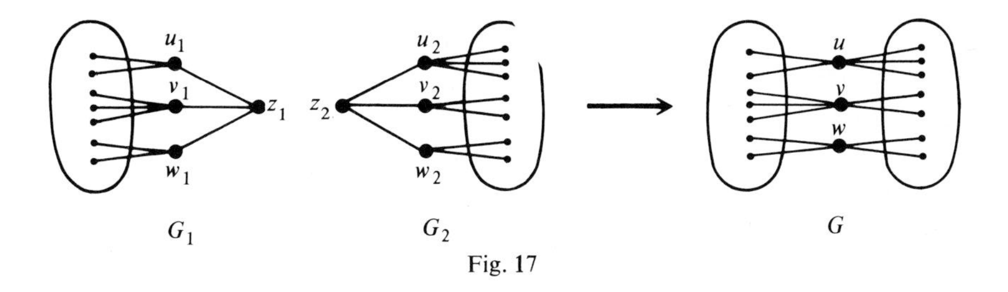

Fig. 17

Theorem 10.2. *The graph G just constructed is hypohamiltonian.*

Proof. We first prove that G is not Hamiltonian. If this is not the case, then G contains a Hamiltonian circuit, and this circuit must pass through u, v and w. The circuit must therefore be the union of a uv-path, a vw-path, and a wu-path, and two of these paths must be contained in $G_1 - z_1$ or $G_2 - z_2$. Without loss of generality, we may assume that the uv-path and the vw-path are contained in $G_1 - z_1$. The union of these two paths is a Hamiltonian path in $G_1 - z_1$ from u_1 to w_1, and by adding z_1 and the edges z_1u_1 and z_1w_1 we obtain a Hamiltonian circuit in G_1, which is a contradiction. So G is not Hamiltonian.

We now prove that if y is any vertex of G, then $G - y$ is Hamiltonian. We may assume without loss of generality that y belongs to $G_1 - z_1$. Since $G_1 - y$ is Hamiltonian, $G_1 - z_1 - y$ has a Hamiltonian path P_1 joining two of the vertices u_1, v_1 and w_1 (u_1 and v_1, say). But $G_2 - w_2$ is also Hamiltonian, and so $G_2 - z_2 - w_2$ has a Hamiltonian path P_2 joining u_2 and v_2. Then $P_1 \cup P_2$ is the required Hamiltonian circuit of $G - y$. This completes the proof. $\|$

Now if G_1 contains a vertex y_1 of valency 3 which is not adjacent to z_1, and if G_2 contains a vertex y_2 of valency 3 which is not adjacent to z_2, then G contains two non-adjacent vertices of valency 3. So if there exist two hypohamiltonian graphs (of orders p_1 and p_2, say) with two non-adjacent vertices of valency 3, then there exists a hypohamiltonian graph of order p_1+p_2-5 with two non-adjacent vertices of valency 3. For example, by starting with the Petersen graph, we can construct hypohamiltonian graphs of order $5k$, for every $k \geqslant 2$. Similarly, the existence of hypohamiltonian graphs of orders 10, 13, 16 and 22 with the above property implies the existence of hypohamiltonian graphs of all orders $p \geqslant 13$, except possibly 14, 17 and 19.

A connection between hypohamiltonian graphs and the graphs in the previous section is summarized in the following theorem:

Theorem 10.3. *If there exists a hypohamiltonian graph of order p, then there exists a Hamiltonian-connected graph of order $p+1$ which is not 1-Hamiltonian. Furthermore, if the hypohamiltonian graph contains a vertex of valency 3, then there exists a 1-Hamiltonian graph of order $p-1$ which is not 1-edge-Hamiltonian.*

Proof. Let G be a hypohamiltonian graph, and let G' be the graph obtained from G by adding a new vertex u joined to all the vertices of G. Then G' is Hamiltonian-connected, but not 1-Hamiltonian. To see this, let v and w be vertices of G', and suppose that both of them belong to G (the case in which one of them is equal to u is similar); then in $G-v$ there is a Hamiltonian path from w to w' (say), which together with the edges $w'u$ and uv gives a Hamiltonian path from w to v.

Now let z be a vertex of valency 3, and let $\bar{u}$, $\bar{v}$ and $\bar{w}$ be the vertices adjacent to z. Let G' be the graph obtained from G by deleting the vertex z and adding the edges $\bar{u}\bar{v}$, $\bar{u}\bar{w}$ and $\bar{v}\bar{w}$ (which are not in G, as we proved just before Theorem 10.2). Now G' is 1-Hamiltonian, since if y is a vertex of G', then $G-y$ has a Hamiltonian circuit containing two of the three edges zu, zv and zw, and so $G'-y$ is Hamiltonian. However, G' is not 1-edge-Hamiltonian, since there is no Hamiltonian path from u to v. This completes the proof. ||

One can similarly define a graph G to be **hypotraceable** if G is not traceable, but the vertex-deleted subgraph $G-v$ is traceable for every vertex v. The best results in this direction have been obtained by Thomassen [**58**], [**60**], who has proved the existence of hypotraceable graphs of order p, for $p = 34$, $p = 37$ and $p \geqslant 39$. However, several problems remain open; for example, Grünbaum has asked whether there exists a non-Hamiltonian graph G with the property that every vertex-deleted subgraph $G-v$ is non-Hamiltonian, but every "two-vertex-deleted subgraph" $G-v-w$ is Hamiltonian.

Finally, we remark that Fouquet and Jolivet [**30**] have proved that if $p \geqslant 6$, then there exists a hypohamiltonian digraph (defined in the obvious way) of order p. However, if we impose the extra restriction that pairs of opposite arcs are not allowed, then the set of numbers p for which such a digraph exists is unknown, although Thomassen has obtained partial results.

11. Some Miscellaneous Results

There are a great number of articles in which Hamiltonian paths and circuits play an important role, but whose results do not fit easily into any of the preceding sections. In this section we shall mention some of these results.

Hamiltonian Decompositions

We shall say that a graph G can be decomposed into Hamiltonian circuits if the edge-set of G can be partitioned into Hamiltonian circuits; similarly, we say that G can be decomposed into Hamiltonian paths if the edge-set of G can be partitioned into Hamiltonian paths (see [**8**] for a survey on this topic). The best-known result of this type is the following:

Theorem 11.1. (*i*) *if p is even, K_p can be decomposed into $\frac{1}{2}p$ Hamiltonian paths*;

(*ii*) *if p is odd, K_p can be decomposed into $\frac{1}{2}(p-1)$ Hamiltonian circuits.*

Proof. (*i*) Let the vertices of K_p be $v_1, v_2, \ldots, v_p$. Then the $\frac{1}{2}p$ Hamiltonian paths required are

$$v_{1+i}, v_{p+i}, v_{2+i}, v_{p+i-1}, \ldots, v_{j+i+1}, v_{p+i-j}, \ldots, v_{\frac{1}{2}p+i}, v_{\frac{1}{2}p+i+1},$$

for $i = 1, 2, \ldots, \frac{1}{2}p$, where the subscripts are all taken modulo p.

(*ii*) The Hamiltonian circuits required are obtained from the paths of part (*i*) by joining an extra vertex to each end of these paths. ||

One can also use a similar method to decompose K_p into $\frac{1}{2}(p-2)$ Hamiltonian circuits and a 1-factor, if p is even.

The corresponding results for digraphs are more difficult to prove, and have been obtained only recently by Tillson [**63**]:

Theorem 11.2. *If $p \neq 4$ or 6, then the complete symmetric digraph of order p can be decomposed into $p-1$ Hamiltonian directed circuits.* ||

However, many similar problems in this area are unsolved, including the following conjectures of Jackson [**38**], and Kelly:

Conjecture A. *If G is a regular ρ-valent graph of order p, where $\rho \geqslant \frac{1}{2}(p-1)$, then G has $[\frac{1}{2}p]$ edge-disjoint Hamiltonian circuits.*

Conjecture B. *Every regular tournament can be decomposed into Hamiltonian directed circuits.*

Enumeration Problems

How many Hamiltonian circuits does a given graph possess? This question and related problems seem to be very difficult, and results are few and far between. For example, the number $h(n)$ of Hamiltonian circuits in the n-cube is not known, except for the values $h(2) = 1$, $h(3) = 6$, $h(4) = 1344$, and $h(5) = 906\,545\,760$ (probably); for references and bounds on $h(n)$, see **[25]**.

Other results are concerned with the parity of the number of Hamiltonian circuits in a graph. Results of this kind include the following, due respectively to Smith (1946), and Kotzig (1966):

Theorem 11.3. *Let G be a trivalent graph. Then*

(*i*) *the number of Hamiltonian circuits of G containing a given edge is even*;

(*ii*) *if G is bipartite, then G contains an even number of Hamiltonian circuits.* ‖

For a proof of these results, see (for example) Thomason **[57]** (not to be confused with Thomassen!) who has also proved the following result (see Section 8 of Chapter 5):

Theorem 11.4. *If G can be expressed as the union of two edge-disjoint Hamiltonian circuits, then G can also be expressed as the union of two other edge-disjoint Hamiltonian circuits.* ‖

Results on the number of Hamiltonian circuits in a planar graph have also been obtained by various authors.

Vertex-transitive Graphs

A graph is **vertex-transitive** if, given any two vertices v and w of G there is an automorphism of G which maps v to w. Lovász has formulated the following conjecture:

Conjecture. *Every connected vertex-transitive graph is traceable.*

A related conjecture was made by Thomassen. While studying connected vertex-transitive graphs, he found only four which fail to be Hamiltonian—

namely, the Petersen graph, the Coxeter graph (see, for example, [**17**, p. 241]), and the graphs obtained by replacing each vertex of these graphs by a triangle, and on the strength of this, he formulated the following conjecture:

Conjecture. *There are only a finite number of connected vertex-transitive graphs which are non-Hamiltonian.*

Comparability Graphs

If D is any digraph, then the comparability graph of D is the (undirected) graph G whose vertex-set is the same as that of D, and whose edges join those pairs of vertices which are connected by a directed path in D. If D is a tree in which every arc is directed towards some root-vertex, then it can be shown that a graph G is the comparability graph of D only if for every path u, v, w, x, y in G, either uw or vx is an edge. Using the concepts of a "good valuation" and a "pseudo-valuation", Arditti and Cori have completely characterized those trees whose comparability graph is Hamiltonian; details of their results may be found in [**2**].

12. Generalizations

When a graph is not Hamiltonian, it is natural to ask "how far" it is from being Hamiltonian. This has given rise to many possible generalizations, several of which are extensions of the results of Section 4. We shall mention some of these generalizations here.

The Circumference of a Graph

Here one asks for the length of the longest circuit in the graph. When this circuit includes every vertex of the graph, we have a Hamiltonian circuit. For results on the circumference of a graph, see [7], [**12**] and [**66**].

Pseudo-Hamiltonian Circuits

If G is a graph of order p, a "pseudo-Hamiltonian circuit", or "Hamiltonian walk", is a closed walk (not necessarily a circuit) of minimum length which includes every vertex of G. Clearly, the length of such a walk is at least p, and is equal to p if and only if G is Hamiltonian. It is easy to see that if G contains a circuit of length c, then there is a pseudo-Hamiltonian circuit whose length is at most $2p-c$; for further details, see [7] and [**39**].

Vertex-partitions

Here the problem is to determine the minimum number of vertex-disjoint circuits (or paths) which cover all of the vertices of the graph G, and therefore partition its vertex-set. When this minimum number is 1, the graph is Hamiltonian. If G is not Hamiltonian, then the minimum number of paths which partition the vertex-set is equal to the minimum number of edges that must be added to G to make it Hamiltonian; for further details, see [**39**] and [**56**].

Spanning Trees

We can also investigate spanning trees of a graph, and ask for the spanning trees with the minimum number of vertices of valency 1. If this minimum number is 2, the graph is traceable; for further details, see [**42**].

Several theorems on Hamiltonian graphs have been extended to these concepts, although many problems remain open. We conclude with a theorem generalizing Ore's theorem (Corollary 4.2) which includes all of the above generalizations, and some related conjectures:

Theorem 12.1. *Let G be a connected graph of order p, and let c be a number satisfying $c \leqslant p$. If $\rho(v)+\rho(w) \geqslant c$ for each pair of non-adjacent vertices v and w in G, then*

(*i*) *the circumference of G is at least $\frac{1}{2}c$;*

(*ii*) *if G is 2-connected, the circumference of G is at least c;*

(*iii*) *G contains a pseudo-Hamiltonian circuit whose length is at most $2p-c$;*

(*iv*) *the vertices of G can be partitioned into at most $p-c+1$ circuits, or (if $c < p$) at most $p-c$ paths;*

(*v*) *if $c < p$, there exists a spanning tree with at most $p-c+1$ vertices of valency 1;*

(*vi*) *if G has connectivity κ and independence number α, where $\kappa \geqslant \alpha - c$, then G has a spanning tree with at most $c+1$ vertices of valency 1.* ||

The following conjectures are due, respectively, to Woodall and Jolivet:

Conjecture A. *If G is a 2-connected graph of order p, and if G has at least $k+\frac{1}{2}p$ vertices whose valency is at least k, then the circumference of G is at least $2k$.*

Conjecture B. *If G has connectivity κ and independence number α, then the circumference of G is at least* $\{\kappa(\alpha+p-\kappa)/\alpha\}$,

Conjecture C. *If* $\kappa \geqslant \alpha - c$, *then G has a spanning tree with at most* $c+1$ *vertices of valency* 1.

Recently S. Win announced a proof of Conjecture C (*Graph Theory Newsletter* **7** (1978)).

References

For general surveys of the topics covered in this chapter, the reader is referred in particular to [**6**, Chapter 10], [**17**, Chapters 4 and 9], [**19**], [**43**], [**49**], [**50**] and [**51**]. Necessary conditions for Hamiltonian graphs are discussed in [**21**], [**22**] and [**23**], sufficient conditions on valencies in [**16**] and [**43**], planar graphs in [**34**], pancyclic graphs in [**13**] and [**14**], and strongly Hamiltonian graphs in [**65**]. Recently A. Bondy has written an interesting survey: "Hamilton cycles in graphs and digraphs" presented at the Ninth Southeastern Conference on Combinatorics, 1978.

1. A. V. Aho, J. E. Hopcroft and J. D. Ullman, *The Design and Analysis of Computer Algorithms*, Addison-Wesley, Reading, Mass., 1974.
2. J.-C. Arditti and R. Cori, Hamilton circuits in the comparability graph of a tree, in *Combinatorial Theory and its Applications*, Vol. I (ed. P. Erdős *et al.*), North-Holland, Amsterdam, 1970, pp. 41–53; *MR*46#3361.
3. D. Barnette, Conjecture 5, in *Recent Progress in Combinatorics* (ed. W. T. Tutte), Academic Press, New York, 1969, p. 343.
4. M. Bellmore and G. L. Nemhauser, The traveling salesman problem: A survey, *Operations Res.* **16** (1968), 538–558; *MR*38#3027.
5. A. Benhocine, Hamiltonisme dans les puissances de graphes et les graphes adjoints, *C.R. Acad. Sci. Paris* (*A*) **285** (1977), 219–220.
6. C. Berge, *Graphs and Hypergraphs*, North-Holland, Amsterdam, 1973; *MR*50#9640.
7. J.-C. Bermond, On hamiltonian walks, in *Proceedings of the Fifth British Combinatorial Conference* (ed. C. St. J. A. Nash-Williams and J. Sheehan), Congressus Numerantium **XV**, Utilitas Mathematica, Winnipeg, 1976, pp. 41–51; *MR*53#2742.
8. J.-C. Bermond, Hamiltonian decompositions of graphs and hypergraphs, in *Advances in Graph Theory* (ed. B. Bollobás), North-Holland, Amsterdam, 1978, pp. 21–28.
9. A. Bilgáke, Kriterien von Ramsey-typ für Hamiltonkreise in Graphen, Ph.D. thesis, Berlin, 1977.
10. N. L. Biggs, E. K. Lloyd and R. J. Wilson, *Graph Theory 1736–1936*, Clarendon Press, Oxford, 1976.
11. G. S. Bloom and E. F. Schmeichel, Component inequalities and a characterization of connectivity in planar and other graphs, *J. Combinatorial Theory* (*B*) (to appear).
12. J. A. Bondy, Large cycles in graphs, *Discrete Math.* **1** (1971), 121–132; *MR*44#3903.
13. J. A. Bondy, Pancyclic graphs, in *Proceedings of the Second Louisiana Conference on Combinatorics, Graph Theory and Computing* (ed. R. C. Mullin *et al.*), Congressus Numerantium **III**, Utilitas Mathematica, Winnipeg, 1971, pp. 80–84. *MR*48#3805.
14. J. A. Bondy, Pancyclic graphs: recent results, in *Infinite and Finite Sets*, Vol. 1 (ed. A. Hajnal *et al.*), North-Holland, Amsterdam, 1975, pp. 181–187; *MR*51#10157.
15. J. A. Bondy, A remark on two sufficient conditions for Hamilton cycles, *Discrete Math.* **22** (1978), 191–193.
16. J. A. Bondy and V. Chvátal, A method in graph theory, *Discrete Math.* **15** (1976), 111–135; *MR*54#2531.
17. J. A. Bondy and U. S. R. Murty, *Graph Theory with Applications*, American Elsevier, New York, and MacMillan, London, 1976; *MR*54#117.

18. J. A. Bondy and C. Thomassen, A short proof of Meyniel's theorem, *Discrete Math.* **19** (1977), 195–197.
19. G. Chartrand, S. F. Kapoor and H. V. Kronk, The Hamiltonian hierarchy, Western Michigan University Mathematics Report **32**, Kalamazoo, 1973; part I also appeared as: The many facets of hamiltonian graphs, *Math. Student* **41** (1973), 327–336; *MR***53** #5367.
20. N. Christofides, *Graph Theory. An Algorithmic Approach*, Academic Press, New York, 1975; *MR***55**#2623.
21. V. Chvátal, New directions in Hamiltonian graph theory, in *New Directions in the Theory of Graphs* (ed. F. Harary), Academic Press, New York 1973, pp. 65–95; *MR***50** #9689.
22. V. Chvátal, Tough graphs and Hamiltonian circuits, *Discrete Math.* **5** (1973), 215–228; *MR***47**#4849.
23. V. Chvátal, Edmonds' polytopes and weakly Hamiltonian graphs. *Math. Programming* **5** (1973), 29–40; *MR***48**#5909.
24. J. B. Collier and E. F. Schmeichel, Systematic searches for hypohamiltonian graphs, *Networks* (to appear).
25. P. Defert, R. Devillers and J. Doyen, Non-equivalent hamiltonian circuits on the n-dimensional cube (to appear).
26. H. Fleischner, In the square of graphs, Hamiltonicity and pancyclicity, Hamiltonian connectedness and panconnectedness are equivalent concepts, *Monatsh Math.* **82** (1976), 125–149; *MR***55**#171.
27. H. Fleischner and A. M. Hobbs, A necessary condition for the square of a graph to be hamiltonian, *J. Combinatorial Theory* (*B*) **19** (1975), 97–118; *MR***54** #2535.
28. H. Fleischner and A. M. Hobbs, Hamiltonian total graphs, *Math. Nachr.* **68** (1975), 59–82; *MR***52**#5475.
29. J.-L. Fouquet, Problèmes hamiltoniens dans les puissances de graphes orientés, in *Problèmes Combinatoires de Théorie des Graphes* (ed. J.-C. Bermond *et al.*), C.N.R.S., Paris, 1978, 145–147.
30. J.-L. Fouquet and J.-L. Jolivet, Graphes hypohamiltoniens orientés, in *Problèmes Combinatoires et Théorie des Graphes* (ed. J.-C. Bermond *et al.*), C.N.R.S., Paris, 1978.
31. P. R. Goodey, Hamiltonian circuits in polytopes with even sided faces, *Israel J. Math.* **22** (1975), 52–56.
32. D. Gouyou Beauchamps, Un algorithme de recherche de circuits hamiltoniens dans les graphes 4-connexes planaires, in *Problèmes Combinatoires et Théorie des Graphes* (ed. J.-C. Bermond *et al.*), C.N.R.S., Paris, 1978, 185–187.
33. B. Grünbaum, Polytopes, graphs and complexes, *Bull. Amer. Math. Soc.* **76** (1970), 1131–1201; *MR* **42**#959.
34. B. Grünbaum, Polytopal graphs, in *Studies in Graph Theory, Part II*, Studies in Mathematics **12**, Mathematical Association of America, Washington D.C., 1975, pp. 201–224; *MR***53**#10654.
35. B. Grünbaum and J. Malkevitch, Pairs of edge disjoint hamiltonian circuits, *Aequationes Math.* **14** (1976), 191–196; *MR***54**#2544b.
36. A. M. Hobbs, Hamiltonian squares of cacti, *J. Combinatorial Theory* (*B*) (to appear).
37. W. Jackson, Hamilton cycles in regular 2-connected graphs (to appear).
38. W. Jackson, Edge-disjoint Hamilton cycles in regular graphs of large degree (to appear).
39. J.-L. Jolivet, Indice de partition en chaînes d'un graphe simple et pseudo-cycles hamiltoniens, *C.R. Acad. Sci. Paris* (*A*) **279** (1974), 479–481; *MR***53**#10647.
40. H. A. Jung, Note on Hamiltonian graphs, in *Recent Advances in Graph Theory* (ed. M. Fiedler), Academia, Prague, 1975, pp. 315–321; *MR***52**#13509.
41. H. A. Jung, On maximal circuits in finite graphs, in *Advances in Graph Theory* (ed. B. Bollobás), North-Holland, Amsterdam, 1978, 129–144.
42. M. Las Vergnas, Sur les arborescences dans un graphe orienté, *Discrete Math.* **15** (1976), 27–39; *MR* **54**#7300.
43. L. Lesniak-Foster, Some recent results in Hamiltonian graphs, *J. Graph Theory* **1** (1977), 27–36.

44. M. Lewin, On maximal circuits in directed graphs, *J. Combinatorial Theory* (*B*) **18** (1975), 175–179; *MR***51**#5375.
45. L. Lovász, Problem 5, *Periodica Math. Hungar.* **4** (1973), 82.
46. P. Martin, Cycles hamiltoniens dans les graphes 4-réguliers 4-connexes, *Aequationes Math.* **14** (1976), 37–40; *MR***54**#2544a.
47. H. Meyniel, Une condition suffisante d'existence d'un circuit Hamiltonien dans un graphe orienté, *J. Combinatorial Theory* (*B*) **14** (1973), 137–147; *MR***47**#6546.
48. M. Minoux, Démonstration constructive du théorème de Meyniel: un algorithme en $O(n^4)$ pour la détermination d'un circuit hamiltonien (to appear).
49. C. St. J. A. Nash-Williams, Hamiltonian circuits in graphs and digraphs, in *The Many Facets of Graph Theory*, Lecture Notes in Mathematics **110** (ed. G. Chartrand and S. F. Kapoor), Springer-Verlag, Berlin, Heidelberg and New York, 1969, pp. 237–243; *MR***40**#5484.
50. C. St. J. A. Nash-Williams, Hamiltonian arcs and circuits, in *Recent Trends in Graph Theory*, Lecture Notes in Mathematics **186** (ed. M. Capobianco *et al.*), Springer-Verlag, Berlin, Heidelberg and New York, 1971, pp. 197–210; *MR***43**#3150.
51. C. St. J. A. Nash-Williams, Valency sequences which force a graph to have Hamiltonian circuits, Report of the University of Waterloo, 1972.
52. C. St. J. A. Nash-Williams, Hamiltonian circuits, *Studies in Graph Theory, Part II*, Studies in Mathematics **12**, Mathematical Association of America, Washington D.C., 1975, pp. 301–360; *MR***53**#10649.
53. M. Overbeck-Larisch, Hamilton paths in oriented graphs, *J. Combinatorial Theory* (*B*) **21** (1976), 76–80.
54. P. J. Owens, On regular graphs and Hamiltonian circuits, including answers to some questions of Joseph Zaks (to appear).
55. D. Peterson, Hamiltonian cycles in bipartite plane cubic maps, Ph.D. thesis, Texas A. & M. University, 1977.
56. Z. Skupien, Hamiltonian shortage, path partitions of vertices and matchings in graphs, *Colloq. Math.* **36** (1976), 305–318.
57. A. G. Thomason, Hamiltonian cycles and uniquely edge colourable graphs, in *Advances in Graph Theory* (ed. B. Bollobás), North-Holland, Amsterdam, 1978, pp. 259–268.
58. C. Thomassen, Hypohamiltonian and hypotraceable graphs, *Discrete Math.* **9** (1974), 91–96; *MR***50**#184.
59. C. Thomassen, Graphs in which every path is contained in a Hamilton path, *J. Reine Angew. Math.* **268**/**269** (1974), 271–282; *MR***50**#1977.
60. C. Thomassen, Planar and infinite hypohamiltonian and hypotraceable graphs, *Discrete Math.* **14** (1976), 377–389; *MR***54**#10078.
61. C. Thomassen, Note on circuits containing specified edges, *J. Combinatorial Theory* (*B*) **22** (1977), 279–280.
62. C. Thomassen, An Ore-type condition implying a digraph to be pancyclic, *Discrete Math.* **19** (1977), 85–92.
63. T. Tillson, A hamiltonian decomposition of K_{2m}^*, $2m \geqslant 8$ *J. Combinatorial Theory* (to appear).
64. W. T. Tutte, Bridges and hamiltonian circuits in planar graphs, *Aequationes Math.* **15** (1977), 1–33.
65. A. P. Wojda, On strongly (p, q) hamiltonian graphs, *Graph Theory Newsletter* **6**, No. 5 (1977).
66. D. R. Woodall, Maximal circuits of graphs II, *Studia Sci. Math. Hungar.* **10** (1975), 103–109; *MR***55**#10325.
67. D. R. Woodall, Circuits containing specified edges, *J. Combinatorial Theory* (*B*) **22** (1977), 274–278; *MR***55**#12572.
68. D. R. Woodall, A sufficient condition for Hamiltonian circuits, *J. Combinatorial Theory* (*B*) (to appear).

7
Tournaments

K. B. REID AND LOWELL W. BEINEKE

1. Introduction

Tournaments form perhaps the most interesting class of directed graphs and have a very rich theory, a theory which has no analog in the theory of undirected graphs. Tournament scheduling problems first appeared over a hundred years ago, but the study of the combinatorial structure of tournaments has primarily taken place over the last few decades. Initially, most of the research was statistical in nature, and results were widely scattered. In 1965 Harary, Norman and Cartwright [**49**] presented an introductory survey of the graphical aspects of tournaments in their book on directed graphs, and in 1966 a similar survey by Harary and Moser [**48**] appeared. In 1968 Moon's book *Topics in Tournaments* [**67**] appeared, and it brought together virtually all the results on tournaments known up to that time. This important book stimulated further work on tournaments, and one of the purposes of this chapter is to cover these developments.

Our treatment will be less involved than Moon's, and interested readers are strongly encouraged to consult his book. Although our emphasis will be on recent work, we shall include sufficient background to make the survey self-contained. Because of their length or reliance on other material, we omit most proofs; however, we have tried to include enough to give an idea of the nature of the subject.

In Section 2, we present many of the necessary definitions and give some examples. The next two sections give basic results on paths, circuits, and scores (out-valencies), after which there is a section on enumeration. Extremal problems constitute an interesting area of the theory of tournaments, and in Section 6 we consider problems involving extreme numbers of paths, circuits, and transitive and strong subtournaments within tournaments. The following two sections cover some of the more algebraic aspects of tournaments—automorphisms and regularity. The chapter concludes with an extended section (Section 9) covering a variety of topics, including circuit-free subgraphs, path problems, and isomorphism questions.

2. Definitions and Examples

A **tournament** is a directed graph in which every pair of vertices is joined by exactly one arc. Thus, assigning an orientation to each arc of a complete graph results in a tournament. In Fig. 1 we show the non-isomorphic tournaments with at most five vertices.

Because of the unique structure of tournaments as complete irreflexive asymmetric relations, terminology has evolved which is in many cases peculiar to them.

If the arc joining vertices v and w is directed *from v to w*, then v is said to **dominate w**, as well as to be **adjacent to w**. The set of vertices dominated by v is denoted by $N(v)$, and the set of vertices which dominate v is denoted by $N'(v)$. The **score** (or **out-valency**) $s(v)$ of the vertex v is the number of vertices dominated by v; $s'(v)$ is the **co-score** (or **in-valency**) of v, and equals the number of vertices dominating v. The **score-list** of a tournament is the list of the scores of the vertices, usually arranged in non-decreasing order. For example, the four tournaments of order 4 in Fig. 1 have successive score lists (0, 1, 2, 3), (1, 1, 1, 3), (0, 2, 2, 2), (1, 1, 2, 2).

A tournament T of order n is frequently called an ***n*-tournament** (the letter "n" is usually used instead of "p"), and the subtournament of T with S as its set of vertices will be denoted by $\langle S \rangle$. The **converse** T' of a tournament T has the same vertex-set as T, but every arc is reversed—that is, if vw is an arc in T, then wv is an arc in T'. Among the four 4-tournaments in Fig. 1, the first and last are isomorphic to their converses, whereas the middle two are converses of each other. The concept of converse tournaments provides a duality in the theory of directed graphs, and of tournaments in particular. Each concept which involves directedness has a dual, and therefore each

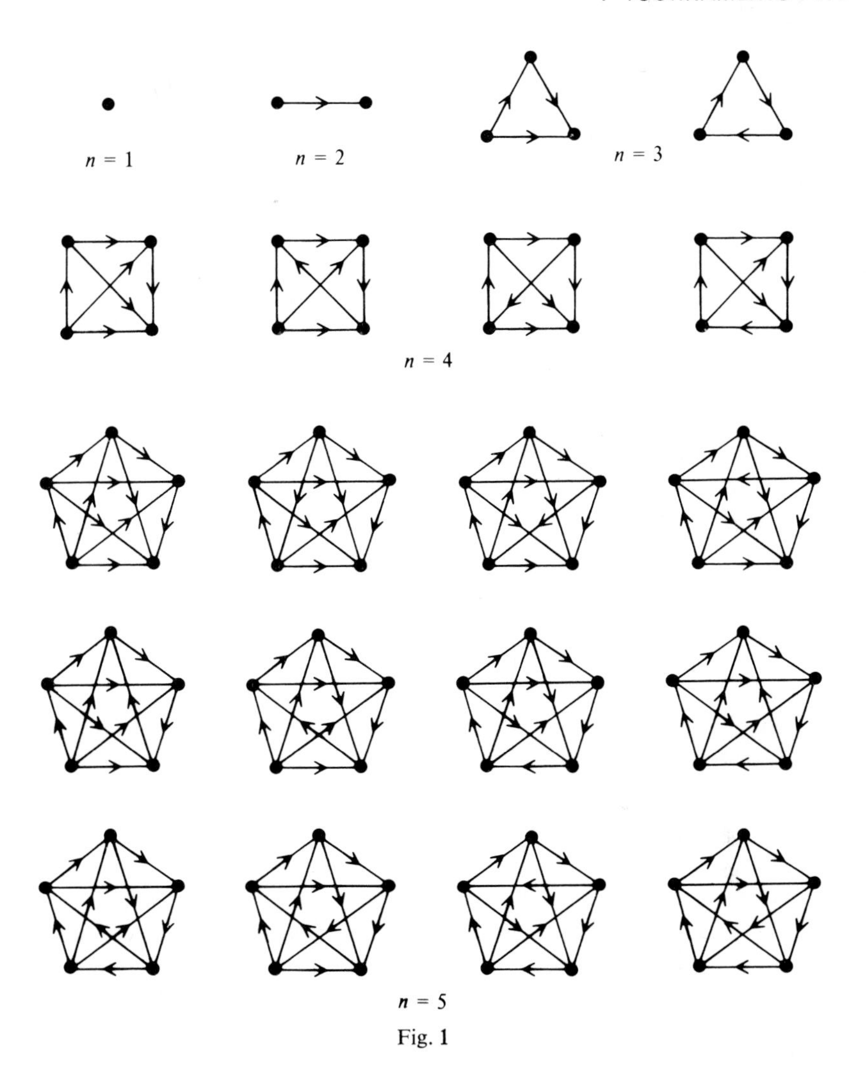

Fig. 1

theorem which depends on directedness yields a dual theorem requiring no further proof.

Special Types of Tournaments

A number of special kinds of tournaments will be of interest; we define some of these now, for future reference.

(a) *Transitive tournaments*: A tournament is called **transitive** if, whenever vertex u dominates v, and v dominates w, then u dominates w. Alternatively, an n-tournament is transitive if its vertices can be labeled $1, 2, \ldots, n$ in such a way that vertex i dominates vertex j if and only if $i < j$. Clearly, there is only one transitive n-tournament (up to isomorphism), and it has no directed circuits; we denote it by TT_n.

(b) *Reducible tournaments*: A tournament is **reducible** if its vertex-set can be partitioned into non-empty sets V_0 and V_1 in such a way that every vertex in V_0 dominates all vertices in V_1.

(c) *Strong tournaments*: A tournament is **strong** (or **strongly-connected**) if from each vertex there are directed paths to all other vertices.

(d) *Regular and near-regular tournaments*: A tournament is **regular** if all vertices have equal scores, and **near-regular** if the maximum difference between its scores is 1. We observe that all regular tournaments must have odd order, and all near-regular ones must have even order.

(e) *Rotational tournaments*: A regular tournament is called **rotational** if its vertices can be labeled $0, 1, 2, \ldots, r$ in such a way that, for some subset S of $\{1, 2, \ldots, r\}$, vertex i dominates $i+j$ (modulo $r+1$) if and only if $j \in S$. A set S of m positive integers gives rise to such a tournament if and only if, for all j, k in S, $j, k \leqslant 2m$ and $j+k \neq 2m+1$. For such a set S, the resulting tournament is denoted by $R(S)$, and S is called the "*symbol*" of the tournament.

(f) *Quadratic residue tournaments*: Let k be an odd positive integer and p a prime of the form $4r+3$, and let S be the set of non-zero squares, called quadratic residues, in the finite field $\mathrm{GF}(p^k)$. Then S is a set of $\frac{1}{2}(p^k-1)$ elements, which never contains both an element and its additive inverse, using addition in $\mathrm{GF}(p^k)$. The **quadratic residue tournament** QT_{p^k} is the tournament whose vertices can be labeled with the elements in $\mathrm{GF}(p^k)$ in such a way that i dominates $i+s$ (addition in $\mathrm{GF}(p^k)$) if and only if $s \in S$.

Our first theorem shows that every tournament is either strong or reducible:

Theorem 2.1. *A tournament is irreducible if and only if it is strong.*

Proof. If T is reducible it is clearly not strong, since if $\{V_0, V_1\}$ is a proper partition of the vertex-set with every vertex in V_0 dominating every vertex in V_1, then no vertex in V_1 can reach any vertex in V_0. Now assume that T is not strong, and let v and w be vertices such that w cannot reach v. Let W_1 be the set of vertices which are reachable from w, and let W_0 be those which are not reachable from w. Then no vertex of W_1 can dominate any vertex in W_0. It follows that T is reducible. ||

There are some other relationships among the various types of tournaments. Clearly, for $n > 1$, every transitive tournament is reducible. Furthermore, every rotational tournament is regular, and, for $n \neq 2$, every regular and near-regular tournament is strong (see Section 4).

In Fig. 1 we have examples of some of these tournaments. The first tournament of each order is transitive. The other tournament of order 3 is sometimes called the **cyclic triple**, and is the quadratic residue tournament QT_3 (and as such is rotational, regular, and strong). The only strong 4-tournament is the last one in Fig. 1, and it is near-regular. Of the 5-tournaments, the first six are reducible, and the last one is rotational. There are three regular 7-tournaments: the quadratic residue tournament $QT_7 = R(1, 2, 4)$, the rotational tournament $R(1, 2, 3)$, and a third tournament which is not rotational (see Kotzig [**58**]).

3. Spanning Paths and Circuits

Two classical results on tournaments are

(*i*) *every tournament has a spanning path* (that is, it is "traceable", in the terminology of Chapter 6);

(*ii*) *every strong tournament has a spanning circuit* (that is, it is "Hamiltonian" in the terminology of Chapter 6).

The first of these can be proved by using a simple induction proof (see, for example Harary and Moser [**48**]), and the second, due to Camion [**21**], is also not difficult. Both have generalizations (of very different natures) which we present here. The first result is due to Rédei [**76**], and as its proof is quite difficult we refer the reader to [**67**, p. 22] for the details:

Theorem 3.1. *Every tournament has an odd number of spanning paths.* ||

Corollary 3.2. *Every tournament has a spanning path.* ||

It is somewhat surprising that twenty-five years passed between the publication of this result and that of Camion. We give here a generalization of Camion's theorem, due to Moon [**67**, p. 6].

Theorem 3.3. *Each vertex of a strong n-tournament is contained in a circuit of length k, for $k = 3, 4, \ldots, n$.*

Proof. The proof is by induction on k. Let T be an n-tournament, and let v_0 be a vertex of T. First, v_0 must be on a 3-circuit, since there must be a vertex u dominated by v_0, and a vertex w dominating v_0 and dominated by u, because T is irreducible.

Now assume that v_0 lies on a k-circuit C: $v_0, v_1, \ldots, v_{k-1}, v_0$, with $k < n$. Let N and N' denote the sets of vertices dominated by, and dominating, all vertices of C, and let W be the set of vertices not in C, N or N'. If $W = \varnothing$, then since T is irreducible, N and N' cannot be empty, and some vertex u in N must dominate a vertex u' in N'; in this case v_0 lies on a $(k+1)$-circuit $v_0, u, u', v_2, \ldots, v_{k-1}, v_0$. On the other hand, if there exists a vertex $w \in W$, then there must be vertices v_i and v_{i+1} in C such that v_i dominates w and w dominates v_{i+1}; in this case v_0 again lies on a $(k+1)$-circuit. The result now follows by induction. ‖

Corollary 3.4. *Every strong n-tournament has circuits of lengths* 3, 4, . . ., *n.* ‖

A slight modification of the above proof yields another generalization of the corollary; this one is due to Douglas [**28**]:

Theorem 3.5. *If C is a circuit in a strong tournament T, then T has a spanning circuit in which the cyclic order of the vertices of C is preserved.* ‖

Yet another extension of Camion's theorem is due to Goldberg and Moon [**42**]. They defined a tournament to be ***r*-strong** if every non-empty proper subset of the vertex-set has at least r arcs to other vertices. One can show that being "strong" and "1-strong" are equivalent properties.

Theorem 3.6. *In any r-strong n-tournament, each vertex lies on r circuits of length k, for k* = 3, 4, . . ., *n.* ‖

In a later section, we shall consider some extremal problems involving paths and circuits.

A conjecture attributed to P. Kelly (see Moon [**67**, p. 7]) is that the arcs of every regular tournament can be partitioned into spanning circuits. Alspach (unpublished) has confirmed this for tournaments of order 9 or less. Furthermore, if T is a rotational tournament of prime order p and with symbol S, then the circuits of the form $0, s, 2s, \ldots, (p-1)s, 0$, for s in S, form such a partition. Kotzig [**60**] has obtained the following related result:

Theorem 3.7. *The arcs of every regular n-tournament can be partitioned into* $\frac{1}{2}(n-1)$ *sets of size n, each of which is an arc-disjoint union of circuits.* ‖

Since every tournament has a spanning path, it seems likely that if the arcs are partitioned into a relatively small number of sets, then one of these sets should contain a fairly long path. This is the idea of a theorem on digraphs in general due to Chvátal [**24**], which he proved using a result of Gallai [**37**]. The tournament version which we state here was subsequently proved by Gyárfas and Lehel [**45**] using different methods:

Theorem 3.8. *Let $A_1, A_2, \ldots, A_m$ be a partition of the arcs of an n-tournament T, and let $k_1, k_2, \ldots, k_m$ be positive integers such that $n = k_1 k_2 \ldots k_m + 1$. Then for some j satisfying $1 \leqslant j \leqslant m$, the set A_j contains the arcs of a path of length k_j.* ‖

A sequence of distinct vertices $v_0, v_1, \ldots, v_k$ induces an **antidirected path** in a tournament if, for $i = 1, 2, \ldots, k-1$, the vertex v_i dominates v_{i+1} if and only if it dominates v_{i-1}. In other words, in an antidirected path the orientation of consecutive arcs alternates. An **antidirected circuit** is defined similarly, and must obviously have even length. Grünbaum [**44**] first established which tournaments have antidirected paths, and conjectured that all n-tournaments of even order ($n \geqslant 10$) have an antidirected circuit. Thomassen [**92**] established this for $n \geqslant 50$, and this bound was later improved to 28 by Rosenfeld [**83**]; the other cases remain unknown. Rosenfeld [**82**] and Forcade [**34**] have also investigated the existence of "paths" and "circuits" with other specified orientations.

Theorem 3.9. *Every tournament has a spanning antidirected path, except the three rotational tournaments $R(1)$, $R(1, 2)$ and $R(1, 2, 4)$.* ‖

Theorem 3.10. *Every tournament of even order $n \geqslant 28$ has a spanning antidirected circuit.* ‖

4. Scores

Our chief interest in scores lies in determining which lists of integers can be the score-list of a tournament. The following result provides a useful reduction step:

Theorem 4.1. *Let L be a list of n integers between 0 and $n-1$ inclusive, and let L′ be obtained from L by deleting one entry r and reducing each of the largest $n-r-1$ remaining entries by 1. Then L is a score-list if and only if L′ is.*

Proof. Suppose first that, L' is the score-list of some tournament T. Then a tournament with list L can be obtained by adding a new vertex which dominates just those r vertices of T whose scores were not reduced in going from L to L'.

For the converse, we show that for any score r in a score-list L, there is a tournament with list L in which a vertex of score r is dominated by the (other) vertices with the largest scores. To prove this we suppose that this is not the case for a vertex v of score r. Then there exist vertices u dominating v, and w dominated by v, whose scores satisfy $s(u) < s(w)$. It follows that there is a vertex x which dominates u and not w. Reversing the arcs of the 4-circuit uv, vw, wx, xu results in a tournament with the same score list, but in which the sum of the scores of the vertices dominating v is greater than before. Iteration of this procedure gives the desired tournament. ‖

This theorem provides an algorithm for determining whether or not a list is a score-list, and for finding a corresponding tournament. This procedure is illustrated in Fig. 2, where a tournament with score-list (1, 2, 2, 3, 3, 4) is built up from tournaments with reduced score-lists (1, 1, 1), (1, 1, 2, 2), (1, 2, 2, 2, 3).

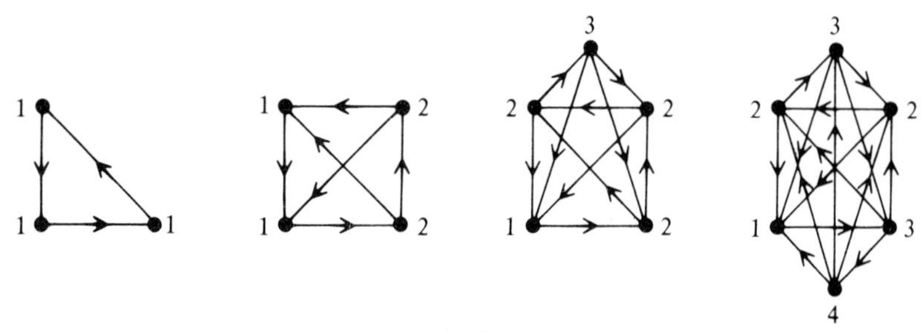

Fig. 2

The number $L(n)$ of different score-lists of length n can be obtained by a recursive technique of Bent and Narayana [**15**] (see also Moon [**67**, p. 67]), but an explicit formula is unknown. Kleitman [**56**] has shown (improving results of Erdős and Moon) that there exist positive constants c_1 and c_2 such that

$$\frac{c_1 4^n}{n^{\frac{5}{2}}} < L(n) < \frac{c_2 4^n}{n^{\frac{3}{2}}}.$$

The following simple non-constructive criterion for determining whether or not a list belongs to a tournament was found by Landau [**63**]. Since the subtournament induced by any set of k vertices must have $\binom{k}{2}$ arcs, the necessity of the inequalities is obvious; we omit the proof of the sufficiency (see [**67**, p. 61]):

Theorem 4.2. *A non-decreasing list of non-negative integers* $(s_1, s_2, \ldots, s_n)$ *is the score-list of a tournament if and only if, for* $k = 1, 2, \ldots, n$,

$$\sum_{i=1}^{k} s_i \geqslant \binom{k}{2},$$

with equality for $k = n$. ‖

As a corollary of Theorem 4.2, we obtain a criterion for a list to be the score-list of a strong tournament. It follows from an earlier observation that a tournament is strong if and only if every non-empty proper subset of the vertex-set has at least one out-going arc.

Corollary 4.3. *A non-decreasing list of non-negative integers* $(s_1, s_2, \ldots, s_n)$ *is the score-list of a strong tournament if and only if, for* $k = 1, 2, \ldots, n-1$,

$$\sum_{i=1}^{k} s_i > \binom{k}{2} \quad \textit{and} \quad \sum_{i=1}^{n} s_i = \binom{n}{2}. \quad \|$$

It has been observed by Beineke and Eggleton (unpublished) that in applying this theorem and corollary, one need not check all values of $k < n$, but only those for which $s_k < s_{k+1}$.

Some additional observations relating scores and reducibility were made by Moser and Beineke (see [**48**]):

Theorem 4.4. *Let* T *be a tournament with scores* $s_1 \leqslant s_2 \leqslant \ldots \leqslant s_n$ *and let* v_i *be a vertex with score* s_i. *Then*

(*i*) *if* $s_n - s_1 < \frac{1}{2}n$, *then* T *is strong*;

(*ii*) *every regular or near-regular n-tournament* ($n \geqslant 3$) *is strong*;

(*iii*) *if* $0 < s_j - s_i < \frac{1}{2}(j-i+1)$, *then* v_i *and* v_j *are in the same strong component*;

(*iv*) *if* $s_{i+1} = s_i + 1$, *then* v_i *and* v_{i+1} *are in the same strong component, or each constitutes a trivial strong component.* ‖

Our next result, which will be used later, follows from the fact that a vertex of score s is the dominating vertex in $\binom{s}{2}$ transitive triples:

Theorem 4.5. *A tournament* T *with scores* $s_1, s_2, \ldots, s_n$ *has* $\sum_i \binom{s_i}{2}$ *transitive triples and* $\binom{n}{3} - \sum_i \binom{s_i}{2}$ *cyclic triples.* ‖

Our theorems on scores have some additional interesting consequences. Let T and T' be converse tournaments in which, for $i = 1, 2, \ldots, n$, the vertex v_i has scores s_i and s_i', respectively. It follows from the preceding theorem that since T and T' have the same number of transitive triples, $\sum_i \binom{s_i}{2} = \sum_i \binom{s_i'}{2}$, and therefore that $\sum_i s_i^2 = \sum_i s_i'^2$.

It was observed in the proof of Landau's theorem that the sum of k scores in a tournament must be at least $\binom{k}{2}$. If we apply this to both T and T', and use the fact that $s_i + s_i' = n-1$, we obtain for any set K of k vertices,

$$\binom{k}{2} \leqslant \sum_{i \in K} s_i \leqslant nk - \binom{k+1}{2}.$$

Our next result provides an example of the use of scores in proving a statement which does not, on the surface at least, involve them:

Theorem 4.6. *If no vertex in a tournament T dominates every other vertex, then there are at least three vertices which reach all others in one or two steps.*

Proof. Let T be a tournament in which no vertex dominates all others, and let u be a vertex of maximum score in T. Among those vertices dominating u, let v be one of maximum score; and among those vertices dominating v, let w be one of maximum score. Now suppose that for one of these three vertices (call it x) some vertex y is not reachable within two steps. Then y must dominate x and all vertices dominated by x (including u if $x = v$, and v if $x = w$), so that $s(y) > s(x)$. In any of the three cases this is a contradiction. ‖

We can also consider not the entire list of scores of a tournament but only the set D of distinct scores, which we call the **score-set**. Clearly any single positive integer constitutes a score-set (for a regular tournament). It is also true that any set of non-negative integers which has 2 or 3 elements, or which forms a finite arithmetic or geometric progress is a score-set (see Reid [**80**]).

The **frequency of a score** s_i in a list is the number of occurrences of that score, and the **frequency-set** of a tournament is the set of frequencies of the scores in its list. For example, the frequency set of a tournament with score-list $(1, 2, 2, 3, 3, 4)$ is $\{1, 2\}$. Alspach and Reid [**8**] have determined the minimum order of a tournament with a given frequency set:

Theorem 4.7. *Let $F = \{f_1, f_2, \ldots, f_m\}$ be a set of positive integers, and let $N(F)$ denote the minimum order of a tournament with frequency-set F. Then*

(*i*) *for $m = 2$, $N(F) = 2f_1 + f_2$ with $f_1 < f_2$, except that $N(\{1, 2\}) = 3$;*

(*ii*) *for $m > 2$, let 2^p be the largest power of 2 dividing all of the f_i, and let $k_i = f_i/2^p$; then*

$$N(F) = \begin{cases} \sum f_i, & \text{if } p = 0 \text{ or a positive even number of the } k_i \text{ are odd,} \\ f_j + \sum f_i, & \text{otherwise, where } f_j \text{ is the least integer in } F \text{ for which } k_j \text{ is odd.} \end{cases}$$ ‖

5. Enumeration of Tournaments

A question that may have occurred to the reader is—how many tournaments of order n are there? The answer depends on how one distinguishes between tournaments. If one considers all tournaments on a fixed set of n vertices, then there are two choices for the arc between any pair of vertices, so that there are $2^{\binom{n}{2}}$ possible tournaments on these vertices. However, many pairs of these tournaments are isomorphic, and it is much more difficult to determine the number of isomorphism classes. The tournaments of the type considered first are called **labeled**, while the latter ones are called **unlabeled** (see Chapter 14). We observe that there is a crude inequality relating the two types (beyond the fact that unlabeled n-tournaments constitute a subset of the labeled ones) in that there are at least $2^{\binom{n}{2}}/n!$ unlabeled tournaments, since a given unlabeled tournament can be labeled in at most $n!$ ways. (In fact, it can be labeled in precisely $n!/a$ ways, where a is the number of automorphisms of the tournament—see Theorem 2.1 of Chapter 14; automorphisms will be discussed later.)

The enumeration of unlabeled tournaments is due to Davis [26], and we omit the proof; a similar proof appears in Section 6 of Chapter 14:

Theorem 5.1. *The number $T(n)$ of unlabeled tournaments of order n is*

$$T(n) = \frac{1}{n!}\sum_{\mathbf{j}} h(\mathbf{j})2^{g(\mathbf{j})},$$

where the summation extends over all partitions $\mathbf{j} = (j_1, j_2, \ldots)$ of n in which each $j_{2k} = 0$, and where

$$g(\mathbf{j}) = \tfrac{1}{2}\left\{\sum_k \sum_i \gcd(i, k) j_i j_k - \sum_k j_k\right\},$$

and

$$h(\mathbf{j}) = \frac{n!}{\prod_k k^{j_k} j_k!}. \quad \|$$

We illustrate the theorem for the case $n = 7$. There are five choices for $\mathbf{j}$:

$\mathbf{j}$	$g(\mathbf{j})$
(7, 0, 0, 0, 0, 0, 0)	21
(4, 0, 1, 0, 0, 0, 0)	11
(1, 0, 2, 0, 0, 0, 0)	7
(2, 0, 0, 0, 1, 0, 0)	5
(0, 0, 0, 0, 0, 0, 1)	3

Hence,

$$T(7) = \frac{2^{21}}{5040} + \frac{2^{11}}{72} + \frac{2^7}{18} + \frac{2^5}{10} + \frac{2^3}{7} = 456.$$

In Table 1 we reproduce the values found in Moon [67, p. 87] for the numbers of n-tournaments with $n \leqslant 12$. Further values, attributed to Paul Stein, can be found in Harary and Palmer [51, p. 245].

Table 1 Tournaments

n	1	2	3	4	5	6	7	8	9	10	11	12
$T(n)$	1	1	2	4	12	56	456	6880	191 536	9 733 056	903 753 248	154 108 311 168

In the sum for $T(7)$, the first term, $2^{21}/7!$, is clearly dominant. In fact, the estimate mentioned above for $T(n)$ gets better as n gets larger. The values of this dominant term for $n \leqslant 30$ may be found in Harary and Palmer [51, pp. 245–246]. For a proof of the general statement, see Moon [67, p. 88]:

Theorem 5.2.

$$\lim_{n\to\infty} \frac{n!\,T(n)}{2^{\binom{n}{2}}} = 1. \;\|$$

One specific class of tournaments which have been enumerated are those with a unique spanning circuit. (Recall that a tournament is strong if and only if it has a spanning circuit.) One example is the tournament obtained from a transitive tournament by reversing the arcs in its spanning path. Camion [22] and Douglas [28] gave several characterizations of such tournaments, and Douglas found an imposing formula for their enumeration. Later, Garey [38] observed that the numbers involved formed part of a familiar set:

Theorem 5.3. *If $n \geqslant 4$, the number of unlabeled n-tournaments with exactly one spanning circuit is F_{2n-6}, the $(2n-6)$th Fibonacci number.* ‖

The concept of a "random tournament of order n" also has two possible interpretations. We choose to treat each of the labeled n-tournaments as equally likely in this context; this is equivalent to taking each pair of arcs vw and wv as being equally likely. We note, however that statements of the form—"almost all tournaments . . ." apply to both the labeled and unlabeled cases, by Theorem 5.2.

Theorem 5.4. *Almost all tournaments are strong—that is, the probability that a random n-tournament is strong approaches 1 as $n \to \infty$.*

Sketch of Proof (Moon and Moser [71]). Let T be a random n-tournament, and for $1 \leqslant j \leqslant n-1$, let $P(j)$ denote the probability that the unique strong component of T which dominates no other strong component has order j. Then

$$P(j) = \frac{\binom{n}{j}P(j)2^{\binom{j}{2}}2^{\binom{n-j}{2}}}{2^{\binom{n}{2}}} = \frac{\binom{n}{j}P(j)}{2^{j(n-j)}},$$

so that

$$P(n) = 1 - \sum_{j=1}^{n-1} \binom{n}{j} \frac{P(j)}{2^{j(n-j)}}.$$

Hence

$$1 - \frac{2n}{2^{n-1}} - \sum_{j=2}^{n-2} \binom{n}{j} \frac{1}{2^{j(n-j)}} < P(n) < 1 - \frac{n}{2^{n-1}} - \frac{nP(n-1)}{2^{n-1}},$$

since each $P(j) \leqslant 1$. For $n \geqslant 13$ and $2 \leqslant j \leqslant n-2$, one can show that

$$\binom{n}{j} \frac{1}{2^{j(n-j)}} < \frac{1}{n2^{n-1}},$$

so that, for $n > 12$,

$$1 - \frac{2n+1}{2^{n-1}} < P(n) < 1 - \frac{2n-1}{2^{n-1}},$$

and the theorem follows. ||

We can also give an intuitive argument for a stronger result. If vw is an arc and u another vertex in a random tournament, then the probability that both arcs uv and wu occur (so that u, v, w, u is a 3-circuit) is only $\frac{1}{4}$. However, if x is a fourth vertex, then the probability that neither u nor x yields a 3-circuit is $\frac{9}{16}$, and as n increases, the probability that the arc vw lies on a 3-circuit approaches 1.

Theorem 5.5. *In almost all tournaments, every pair of vertices lies on a 3-circuit.* ||

In their monograph on probabilistic techniques in combinatorial mathematics, Erdős and Spencer [**33**] give a number of results on tournaments, some of which we shall mention later.

6. Extremal Problems

In this section we consider, for several types of structures, the maximum or minimum number of these structures which can occur in a tournament of given order. The structures we consider are strong and transitive subtournaments, circuits, and spanning paths.

Strong Subtournaments

From Theorem 4.5 one can determine the maximum number of cyclic triples in an n-tournament by taking the scores to be as nearly equal as possible—that is, by taking a tournament which is regular or near-regular. This result has been extended to strong subtournaments of all orders by Beineke and Harary [**13**]:

Theorem 6.1. *If $S(n, k)$ denotes the maximum number of strong k-tournaments in any n-tournament, then, for $3 \leqslant k \leqslant n$,*

$$S(n, k) = \begin{cases} \binom{n}{k} - n\binom{m}{k-1}, & \text{if } n = 2m+1, \\ \binom{n}{k} - m\left(\binom{m}{k-1} + \binom{m-1}{k-1}\right), & \text{if } n = 2m. \end{cases}$$

Proof. Consider an n-tournament T. Any k-tournament containing a vertex of score $k-1$ is reducible, and each vertex of score s_i in T is such a vertex in $\binom{s_i}{k-1}$ subtournaments of T. Therefore, T has at most $\binom{n}{k} - \sum \binom{s_i}{k-1}$ strong subtournaments of order k, where the summation extends over the scores of all of the vertices in T. This expression is a maximum when the scores are as nearly equal as possible.

To complete the proof, we construct suitable tournaments. First consider the rotational n-tournament $R(S)$, where $n = 2m+1$ and $S = \{1, 2, \ldots, m\}$. We claim that each of its subtournaments is either strong or transitive. Suppose U is neither, so that $n \geqslant 5$. Since U is non-transitive, it contains a circuit C, and since U is reducible, some vertex v either dominates all the vertices of C or is dominated by all of them. But in $R(S)$, the set of vertices dominated by any one vertex forms a transitive subtournament, as does the set dominating one vertex, and so this is impossible. Hence, the number of strong k-tournaments in $R(S)$ is

$$\binom{n}{k} - \sum_{v} \binom{m}{k-1} = \binom{n}{k} - n\binom{m}{k-1}, \qquad \text{for} \qquad k = 3, 4, \ldots, n.$$

For $n = 2m$, we delete a vertex from $R(S)$ above to get an n-tournament with

$$\binom{n}{k} - m\left(\binom{m}{k-1} + \binom{m-1}{k-1}\right)$$

strong k-tournaments. This completes the proof. ||

The tournaments used in this proof have the property that every subtournament is either transitive or strong. Moon [**66**] has found a characterization of all such tournaments. The following corollary is obtained by summing over k the results in the previous theorem:

Corollary 6.2. *The maximum number of non-trivial strong subtournaments in any n-tournament is*

$$\begin{cases} 2^n - n2^{\frac{1}{2}(n-1)} - 1, & \text{if } n \text{ is odd,} \\ 2^n - 3n2^{\frac{1}{2}(n-4)} - 1, & \text{if } n \text{ is even.} \end{cases} \quad \|$$

Obviously a tournament need not contain any non-trivial strong subtournaments, but as we have seen, a strong tournament contains strong subtournaments of every possible order. The minimum numbers of such were found by Moon **[66]**:

Theorem 6.3. *If $s(n, k)$ denotes the minimum number of strong k-tournaments in any strong n-tournament, then, for $3 \leqslant k \leqslant n$,*

$$s(n, k) = n-k+1.$$

Proof. Let T be a strong n-tournament. By Corollary 3.4, T has a strong $(n-1)$-tournament S. Let v be the vertex not in S. By Theorem 3.3, there is a k-tournament containing v, which therefore is not contained in S. Hence,

$$s(n, k) \geqslant s(n-1, k)+1, \qquad \text{for} \qquad 3 \leqslant k \leqslant n-1.$$

Since $s(k, k) = 1$, it follows by induction on n that $s(n, k) \geqslant n-k+1$. The tournament obtained by reversing the arcs in the spanning path of the transitive tournament TT_n is seen to have precisely $n-k+1$ strong k-tournaments. ||

Corollary 6.4. *The minimum number of non-trivial strong subtournaments in any strong n-tournament is* $\binom{n-1}{2}$. ||

Transitive Subtournaments

A sort of dual problem to the one just discussed is that of determining the maximum number of transitive subtournaments in a strong tournament. The following result is due to Moon **[66]**:

Theorem 6.5. *If $t(n, k)$ denotes the maximum number of transitive k-tournaments in a strong n-tournament, then, for $3 \leqslant k \leqslant n$,*

$$t(n, k) = \binom{n}{k} - \binom{n-2}{k-2}.$$

Proof. We first show by induction that $t(n, k) \leqslant \binom{n}{k} - \binom{n-2}{k-2}$. This result is clearly true for $n = k$, and since every 3-tournament is either cyclic or transitive, it holds for $k = 3$ and all n by Theorem 6.3. Now fix $n \geqslant 4$, and assume that the inequality holds for all k with $n \geqslant k$. Let T be a strong $(n+1)$-tournament, and let v be a vertex for which the tournament $S = T-\{v\}$ is strong. Then for $k \geqslant 4$, S has at most $t(n, k)$ transitive k-tournaments, and T has at most $t(n, k-1)$ such tournaments which contain v. It follows that

$$\begin{aligned} t(n+1, k) &\leqslant t(n, k)+t(n, k-1) \\ &\leqslant \binom{n+1}{k} - \binom{n-1}{k-2}, \end{aligned}$$

using the induction hypothesis.

Now consider the tournament obtained from the transitive tournament TT_n by reversing the arc from v_1 to v_n, where v_1 and v_n are the vertices of scores $n-1$ and 0 respectively. The result is a strong tournament in which all subtournaments, other than those containing both v_1 and v_n, are transitive, so that it has $\binom{n}{k}-\binom{n-2}{k-2}$ transitive k-tournaments. ||

Many of the transitive subtournaments in the example given in the preceding proof are contained in others. The problem of determining the maximum number of maximal transitive subtournaments in any n-tournament seems to be quite difficult (see Moon [67, p. 18]). Another apparently difficult problem is that of determining the minimum number of transitive k-tournaments in an n-tournament. In fact, determining whether an n-tournament contains a transitive k-tournament is quite difficult for certain values of n and k (see Section 9).

Spanning Paths

Although it is known that every n-tournament has an odd number of spanning paths, it is not known how large this number can be. For small values of n, Szele [91] established the maximum number $H(n)$ of such paths, and these results are given in Table 2.

Table 2 Spanning paths

n	3	4	5	6	7
$H(n)$	3	5	15	45	189

Szele also established the results given in the following theorem, and conjectured that the value of the limit is $\frac{1}{2}$:

Theorem 6.6. *If $H(n)$ denotes the maximum number of spanning paths in a n-tournament, then*

$$H(n) \leqslant \frac{(n+1)!}{2^{\frac{3}{4}n-3}}.$$

Furthermore,

$$\lim_{n \to \infty} \left(\frac{H(n)}{n!} \right)^{1/n}$$

exists, and lies between $\frac{1}{2}$ *and* $2^{-\frac{3}{4}}$. ||

Subsequently, Moon **[69]** considered the minimum number of spanning paths:

Theorem 6.7. *If $h(n)$ denotes the minimum number of spanning paths in any strong n-tournament, then*

$$6^{\frac{1}{4}(n-1)} \leqslant h(n) \leqslant \begin{cases} 3.5^{\frac{1}{3}(n-3)}, & \text{if } n \equiv 0 \text{ (modulo 3)}, \\ 5^{\frac{1}{3}(n-1)}, & \text{if } n \equiv 1 \text{ (modulo 3)}, \\ 9.5^{\frac{1}{3}(n-5)}, & \text{if } n \equiv 2 \text{ (modulo 3)}. \end{cases} \quad ||$$

Circuits

Recall that, in a strong tournament, each vertex lies on circuits of all possible lengths, and in an r-strong tournament, each vertex lies on at least r such circuits. We now turn to problems involving the total number of circuits of various lengths. It is not surprising that some of the results are closely related to results on the number of strong subtournaments. For example, the next theorem follows from the proof of Theorem 6.3, and from the observation that each strong k-tournament in the example given there has just one k-circuit. Incidentally, Las Vergnas **[64]** has proved that if $k \geqslant 4$, then the extremal tournament is unique.

Theorem 6.8. *If $c(n, k)$ denotes the minimum number of k-circuits in a strong n-tournament, then for* $3 \leqslant k \leqslant n$,

$$c(n, k) = n-k+1. \quad ||$$

Corollary 6.9. *The minimum number of circuits in a strong n-tournament is* $\binom{n-1}{2}$. ||

Not a great deal is known about the *maximum* possible number of k-circuits; the following bounds are due to Korvin **[57]**:

Theorem 6.10. *If $C(n, k)$ denotes the maximum number of k-circuits in an n-tournament, then*

$$\binom{n}{k}\frac{(k-1)!}{2^k} \leqslant C(n, k) \leqslant (k+1)\binom{n}{k}\frac{(k-1)!}{2^{\frac{3}{4}k-3}}.$$

Proof. In a random n-tournament, the probability that a set of k vertices lies on a k-circuit is

$$(k-1)!\,\frac{2^{\binom{k}{2}-k}}{2^{\binom{k}{2}}} = \frac{(k-1)!}{2^k},$$

so that the expected number of k-circuits is $\binom{n}{k}(k-1)!/2^k$, and the lower bound follows. In order to obtain the upper bound, we observe that $kC(k, k) \leqslant H(k)$, the maximum number of spanning paths in a k-tournament. Furthermore, $C(n, k) \leqslant \binom{n}{k}C(k, k)$, so that the required bound follows from Theorem 6.6. ‖

A strong 3-tournament or 4-tournament has just one spanning circuit, so that we have the following result as a consequence of Theorem 6.1:

Theorem 6.11. *The maximum numbers of 3-circuits and 4-circuits in any n-tournament are given by*

$$C(n, 3) = \begin{cases} \frac{1}{24}n(n^2-1), & \text{if } n \text{ is odd,} \\ \frac{1}{24}n(n^2-4), & \text{if } n \text{ is even,} \end{cases}$$

and

$$C(n, 4) = \tfrac{1}{2}(n-3)C(n, 3). \quad \|$$

An example of a tournament for which this maximum is achieved was given earlier, but as we shall see later (see Section 8), others exist whenever certain matrices exist. The lower bound, $\frac{1}{4}\binom{n}{3}$ for $C(n, 3)$ in Theorem 6.10, is the mean or expected number of cyclic triples in an n-tournament, and is in fact asymptotically equal to the correct value.

We observed earlier (see Theorem 4.5) that the number of 3-circuits in a tournament depends only on the scores. In particular, all regular tournaments of a given order have the same number of 3-circuits, and all vertices lie in the same number. This led Chvátal and Kotzig **[25]** to consider arc-disjoint 3-circuits containing a fixed vertex, and they proved the following result:

Theorem 6.12. *Let T be a regular $(2r+1)$-tournament, and for each vertex v, let $c(T, v)$ denote the maximum number of arc-disjoint 3-circuits containing v. Then, for $r \geqslant 2$,*

$$[\tfrac{1}{3}(2r+4)] \leqslant c(T, v) \leqslant r,$$

and there exist tournaments attaining each bound. ‖

The two 5-tournaments in Fig. 3 show that the scores above do not determine the number of 4-circuits; they have the same score list, but the first has four 4-circuits, and the second only three. Beineke and Harary **[13]** (see

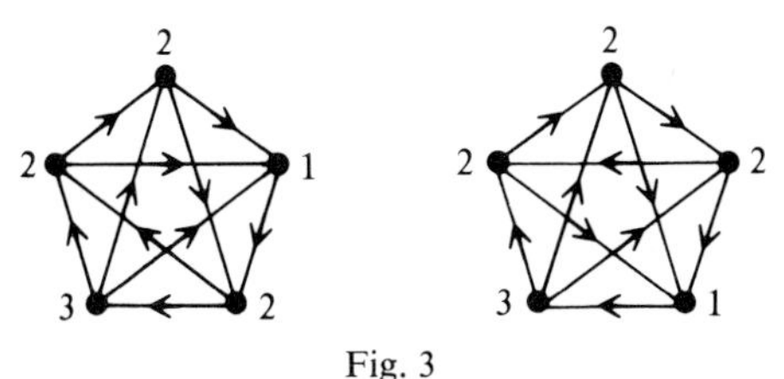

Fig. 3

also Kotzig **[59]**) observed that if, in addition to knowing the scores s_i, one knows the number s_{ij} of vertices dominated by the pair of vertices v_i and v_j, then one can show that the number of 4-circuits is

$$\binom{n}{4} - \sum_i \left\{\binom{s_i}{3} + \binom{n-s_i-1}{3}\right\} + \sum_{i<j} \binom{s_{ij}}{2}.$$

For circuits of length greater than 4, little is known beyond the bounds in Theorem 6.10. We observe that even the two tournaments in Fig. 3 have different numbers of spanning circuits.

It was noted by Beineke (see **[48]**) that if an n-tournament has more 3-circuits than any $(n-1)$-tournament, then it has a spanning circuit, and hence circuits of all possible lengths. Reid **[78]** has shown that for arbitrary j and k, one can determine the maximum number of j-circuits in a tournament with no k-circuits, in terms of the numbers $C(n, k)$ of Theorem 6.10:

Theorem 6.13. *Let $3 \leqslant j < k \leqslant n$, and let $n = q(k-1)+r$, where $0 \leqslant r \leqslant k-2$. If the n-tournament T has more than $qC(k-1, j)+C(r, j)$ j-circuits, then it has a k-circuit. There is an n-tournament with $qC(k-1, j)+C(r, j)$ j-circuits, but no k-circuit.* ||

7. Automorphisms of Tournaments

An **automorphism** of a tournament is a permutation of the vertices which preserves the dominance relation. As is the case for other structures, the set of automorphisms of a tournament T forms a group $\Gamma(T)$, called the **automorphism group** (or, simply, the **group**) of T.

As an example, consider the 4-tournament in Fig. 4. The permutation $(uvw)(x)$ is clearly dominance-preserving, as are its inverse and the identity permutation, and there are no others. Thus, the group is isomorphic to the cyclic group of order 3. As for the other three 4-tournaments, the converse of the one in Fig. 4 has the same group, whereas the transitive and strong 4-tournaments have only the identity automorphism.

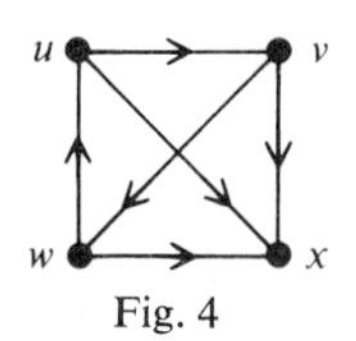

Fig. 4

We note first that an automorphism α of a tournament cannot have order 2, since if α were of order 2, then it would interchange some pairs of vertices, and hence could not preserve dominance. Since every group of even order must have an element of order 2, it follows that the group of a tournament must have odd order. Moon **[65]** showed that this is the only restriction on groups of tournaments. We omit his proof, which is constructive in nature.

Theorem 7.1. *A finite group Γ is isomorphic to the automorphism group of some tournament if and only if Γ has odd order.* ||

A related problem is to determine the maximum order $\gamma(n)$ of the group of any n-tournament, a problem which, by Theorem 7.1, is equivalent to determining the maximum order among the subgroups of odd order of the symmetric group S_n. In establishing some partial results, Goldberg and Moon (see **[67**, p. 81**]**) conjectured that $\gamma(n) \leqslant 3^{\frac{1}{2}(n-1)}$, with equality holding if and only if n is a power of 3. Dixon **[27]** first proved this; later, Alspach **[2]** gave a simpler combinatorial proof, and another proof involving ternary representations was given by Alspach and Berggren **[4]**.

Theorem 7.2. *The maximum order $\gamma(n)$ of the group of any n-tournament satisfies the inequality $\gamma(n) \leqslant 3^{\frac{1}{2}(n-1)}$, with equality if and only if n is a power of 3. Furthermore, if the ternary representation of n has r 1s and no 2s, then $\gamma(n) = 3^{\frac{1}{2}(n-r)}$.* ||

Composition of Tournaments

Let S be an m-tournament with vertex-set $U = \{u_1, u_2, \ldots, u_m\}$, and let T be an n-tournament with vertex-set $V = \{v_1, v_2, \ldots, v_n\}$. The **composition** $S \circ T$ is defined to be the tournament with vertex-set $U \times V$ in which (u_i, v_j) dominates (u_h, v_k) if either $h = i$ and v_j dominates v_k in T, or $h \neq i$ and u_i dominates u_h in S. Intuitively, each vertex of S is replaced by a copy of T, and arcs between copies of T are oriented according to the dominance in S.

In order to describe the group of the composition of two tournaments, we require the concept of group composition, due to Pólya [**75**]. To this end, let Γ and Δ be permutation groups with object sets $U = \{u_1, u_2, \ldots, u_m\}$ and $V = \{v_1, v_2, \ldots, v_n\}$, respectively. For each permutation γ in Γ, and each m-tuple $(\delta_1, \delta_2, \ldots, \delta_m)$ of permutations in Δ, there is a permutation α in $\Gamma \circ \Delta$ given by

$$\alpha(u_i, v_j) = (\gamma(u_i), \delta_i(v_j)).$$

Alspach, Goldberg and Moon [**5**] proved that these two types of composition correspond:

Theorem 7.3. *The group of the composition of tournaments S and T is the composition of their groups: $\Gamma(S \circ T) = \Gamma(S) \circ \Gamma(T)$.* ||

We mention in passing that Goldberg and Moon [**40**] studied the semigroup of tournaments under composition, and proved that two tournaments commute if and only if both are powers of the same tournament.

Vertex-homogeneity

Because of the use of the terms "symmetric" and "transitive" in describing relations, we shall generally avoid phrases such as "vertex-transitive" for automorphism groups, and follow the lead of Fried [**35**] by using the term "homogeneous".

A tournament T is called **vertex-homogeneous** if, for every pair of vertices v and w, there is an automorphism which sends v to w. Such a tournament must clearly be regular, and hence of odd order.

Alspach [**3**] observed that an n-tournament is rotational if and only if it has an automorphism which is an n-cycle. It follows that every rotational tournament is vertex-homogeneous, but the converse is not true. The smallest counterexample has order 21, and was found by Alspach, who also proved the following result:

Theorem 7.4. *A tournament of prime order is vertex-homogeneous if and only if it is rotational.* ||

Alspach also found an expression for the number of vertex-homogeneous tournaments of prime order; his result was later refined by Astie [**10**]:

Theorem 7.5. *The number of vertex-homogeneous tournaments of prime order p is*

$$\frac{1}{p-1} \sum_k \phi(k) 2^{(p-1)/2k},$$

where ϕ is the Euler ϕ-function, and the summation extends over all odd divisors of $p-1$. ‖

Quadratic Residue Tournaments and Arc-homogeneity

One family of tournaments for which all automorphisms have been found are the quadratic residue tournaments. The following result was first found by Goldberg **[39]**, and subsequently by Berggren **[16]**:

Theorem 7.6. *The group of the quadratic residue tournament QT_{p^k}, where k is odd and $p \equiv 3$ (modulo 4) is a prime, consists of all permutations π of the form*

$$\pi(x) = a^2\alpha(x)+c,$$

where α is an automorphism of the field $GF(p^k)$, and a and c are elements of $GF(p^k)$, with $a \neq 0$. ‖

A tournament is called **arc-homogeneous** if, for every pair of arcs vw and xy, there is an automorphism taking v to x and w to y. It is not difficult to see that every arc-homogeneous n-tournament is also vertex-homogeneous, except for the 2-tournament. Furthermore, for any vertex v, the subtournament $\langle N(v)\rangle$ is also vertex-homogeneous. Hence, both n and $\frac{1}{2}(n-1)$ must be odd that is, $n \equiv 3$ (modulo 4).

Fried **[35]** observed that the quadratic residue tournaments are arc-homogeneous. For, if ij and hk are arcs in QT_{p^k}, then the mapping

$$\pi(x) = \left(\frac{k-h}{j-i}\right)x+\left(h-\frac{k-h}{j-i}\,i\right)$$

is a permutation sending vertex i to h, and vertex j to k. It follows from Theorem 7.6 that π is an automorphism of the tournament. That these are essentially the only arc-homogeneous tournaments was established by Berggren **[16]**:

Theorem 7.7. *An n-tournament ($n \geqslant 3$) is arc-homogeneous if and only if it is a quadratic residue tournament.* ‖

Some special cases of arc-homogeneous tournaments were studied by Fried **[35]**, who also proved that the only tournaments which have exactly one automorphism sending one given arc to another are those of prime order.

Astie **[12]** extended Berggren's result by determining all vertex-homogeneous n-tournaments which, for fixed n, have the minimum number of arc-orbits, and what these numbers are. She has also investigated those tournaments whose automorphism group acts primitively on the vertices.

We conclude this section with one last result on arc-homogeneity. It is clear that if an n-tournament is arc-homogeneous, then all of its $\binom{n}{2}$ subtournaments of order $n-2$ are isomorphic. Jean **[54]** showed that these are the only strong tournaments with this property:

Theorem 7.8. *For $n \geqslant 5$, an n-tournament T has the property that all its subtournaments of order $n-2$ are isomorphic if and only if T is transitive or arc-homogeneous.* ||

8. Regularity in Tournaments

In the preceding section we observed that if $n \geqslant 3$, then an arc-homogeneous n-tournament T has the property that T, and every subtournament induced by the vertices dominated by a given vertex, are vertex-homogeneous, and hence regular. We call a tournament **doubly-regular** if all pairs of vertices jointly dominate the same number of vertices. We first prove that doubly-regular tournaments are regular.

Theorem 8.1. *If T is a doubly-regular n-tournament, then T is regular and $n \equiv 3$ (modulo 4).*

Proof. Let k be the number of vertices jointly dominated by each pair of vertices, and let v be a vertex and s its score. Then in the subtournament $\langle N(v) \rangle$, each vertex has score k, so that $\binom{s}{2} = sk$, and hence $s = 2k+1$. This means that T is regular, and $n = 4k+3$. ||

We observe that double-regularity is an analog of strong regularity for graphs (see Chapter 12). Also, we note that some authors use the term "homogeneous" instead of "doubly-regular".

Although, as noted above, all arc-homogeneous tournaments are doubly-regular, the converse is not true. Let T be the 15-tournament with vertices v, and u_i and w_i $(i = 0, 1, \ldots, 6)$, in which

$$N(v) = \{w_0, w_1, w_2, w_3, w_4, w_5, w_6\},$$
$$N(u_i) = \{u_{i+3}, u_{i+5}, u_{i+6}, w_{i+1}, w_{i+2}, w_{i+4}, v\},$$

and

$$N(w_i) = \{u_i, u_{i+1}, u_{i+2}, u_{i+4}, w_{i+1}, w_{i+2}, w_{i+4}\},$$

where addition is modulo 7. We note that both $\langle N(v) \rangle$ and $\langle N'(v) \rangle$ are isomorphic to QT_7, and judicious use of this fact makes it quite easy to verify that T is doubly-regular. To see that it is not arc-homogeneous, we observe that there is no automorphism which sends u_0u_3 to u_0v, since $\langle u_0, w_1, u_5, w_2, u_3 \rangle$ (in that order) is transitive; but there is no intermediate set of three vertices between u_0 and v corresponding to w_1, u_5, w_2.

An interesting property of a doubly-regular tournament of order $4k+3$ is that every arc lies on $k+1$ cyclic triples. For, given an arc vw, $k+1$ of the $2k+1$ vertices dominated by v are also dominated by w, and hence the other $k+1$ vertices dominate w. This was a property studied earlier by Kotzig [**61**], and Brown and Reid [**20**] showed that it is a characterizing property of doubly-regular tournaments. Another property is that the score list of every subtournament of order $4k+1$ must consist of k $(2k+1)$s, $2k+1$ $(2k)$s and k $(2k-1)$s; Müller and Pelant [**73**] proved that this too is a characterizing property. (They also proved a result relating double-regularity to simple tournaments—see Section 9.)

Theorem 8.2. *The following statements are equivalent for a non-transitive tournament T of order $n \geqslant 5$:*

(*i*) *T is doubly-regular;*

(*ii*) *every arc of T lies on the same number of cyclic triples;*

(*iii*) *every $(n-2)$-subtournament has the same list of scores.* ||

A natural question is whether doubly-regular n-tournaments exist for all values of $n \equiv 3$ (modulo 4). Since, as we have observed, arc-homogeneous tournaments are doubly-regular, we know that they exist for all appropriate prime power orders. It turns out that the existence of doubly-regular tournaments is equivalent to the existence of skew-Hadamard matrices. Each entry of a skew-Hadamard matrix $\mathbf{H}$ of order m is $+1$ or -1, with

$$\mathbf{H}\mathbf{H}^{\mathrm{T}} = m\mathbf{I} \qquad \text{and} \qquad \mathbf{H}+\mathbf{H}^{\mathrm{T}} = 2\mathbf{I}.$$

The connection between these matrices and doubly-regular tournaments was implicitly contained in work by Szekeres and Szekeres [**89**] on "Shütte's problem" (see Section 9), but Johnsen [**55**] was apparently the first to establish the equivalence. It was also found independently by Seidel (unpublished), and by Brown and Reid [**20**].

We now indicate how one obtains a doubly-regular tournament from a skew-Hadamard matrix $\mathbf{H}$ of order $4k+4$. We may assume that every entry in the last row of $\mathbf{H}$ is $+1$, and that every entry in the last column (except the last entry) is -1. In the remaining submatrix of order $4k+3$, we replace each diagonal entry and each entry of -1 by 0. The result turns out to be the adjacency (dominance) matrix of a doubly-regular tournament. This construction is then reversed to obtain a skew-Hadamard matrix from a doubly-regular tournament.

Theorem 8.3. *There exists a doubly-regular $(4k+3)$-tournament if and only if there exists a skew-Hadamard matrix of order $4k+4$.* ||

This result is an excellent example of an inter-relationship between two areas of combinatorial mathematics, both of which have yielded information about the other. On the one hand, Szekeres **[90]** was able to construct some new skew-Hadamard matrices using doubly-regular tournaments; and on the other hand, the following corollary is a consequence of work on the matrices:

Corollary 8.4. *There exists a doubly-regular tournament of order n for every n of the form $2^k r_1 r_2 \ldots r_m - 1$, where each r_i is a multiple of 4 and of the form $p_i^{t_i} + 1$, where p_i is a prime.* ||

There are no known exceptions to the existence of these structures. Stated in terms of doubly-regular n-tournaments, the smallest unresolved values are $n = 4t - 1$, where $t = 29, 37, 39, 43, 47, 49, 59, 65, 67, 69$ and 73, as noted by Seberry (unpublished).

It is interesting that although regular tournaments are common, and there are no known values of $n \equiv 3$ (modulo 4) for which doubly-regular tournaments fail to exist, there are only seven degenerate tournaments which can be called "triply-regular". This was established by Brown and Reid **[20]** (see also **[73]** and **[53]**):

Theorem 8.5. *If T has the property that all triples of vertices jointly dominate the same number j of vertices, then $j = 0$ and T is one of the seven tournaments of order $n \leqslant 5$ in which no vertex is dominated by three others.* ||

Herzog and Reid **[53]** have called a tournament T **nearly-triply-regular** (*NTR*) if it is doubly-regular and if there are two integers j_1 and j_2 such that every triple of vertices jointly dominates either j_1 or j_2 others. The tournaments QT_7 and QT_{11} are the only known *NTR* tournaments, and Herzog and Reid found that, for $p = 19, 23$ and 31, QT_p is not an *NTR* tournament. They also proved that QT_7 is the only *NTR* tournament of order 7, that QT_{11} is the only one in which the vertices jointly dominated by some pair of vertices do not induce a strong subtournament, and that these two are the only *NTR* tournaments with $j_2 > j_1 = 0$. Thus, there are a considerable number of restrictions on other *NTR* tournaments, if such tournaments exist.

9. A Variety of Structural Problems

In the final section of this chapter, we discuss various other aspects of tournaments—primarily problems involving substructures, paths, and circuits.

Transitive Subtournaments

The first problem we discuss is that of finding the smallest number $f(n)$, such that every tournament of order $f(n)$ must contain a transitive subtournament of order n.

Obviously $f(1) = 1$, $f(2) = 2$, and $f(3) = 4$. Furthermore, every tournament of order $2f(n)$ must contain TT_{n+1}, since it has a vertex v of score at least $f(n)$, and so the vertices dominated by v include a copy of TT_n. Hence $f(n+1) \leqslant 2f(n)$. One can also show that QT_7 contains no TT_4, so that if $n \leqslant 4$, then $f(n) = 2^{n-1}$. These facts led Erdős and Moser [**32**] to conjecture that this formula holds for all n. However, Parker and Reid [**74**] found that every 14-tournament contains TT_5, so that $f(n) < 2^{n-1}$ for $n > 4$. They also verified that the 13-tournament $R(1, 2, 3, 5, 6, 9)$ contains no TT_5, and that the 27-tournament QT_{27} contains no TT_6; this gives $f(5) = 14$ and $f(6) = 28$.

For a general lower bound, we use a probabilistic argument. Suppose that every m-tournament contains a transitive n-tournament. A given transitive n-tournament can be labeled in $n!$ ways, and is contained in $2^{\binom{m}{2}-\binom{n}{2}}$ labeled m-tournaments. As there are $2^{\binom{m}{2}}$ labeled m-tournaments, and each contains $\binom{m}{n}$ n-tournaments, it follows that

$$\binom{m}{n} n!\, 2^{\binom{m}{2}-\binom{n}{2}} \geqslant 2^{\binom{m}{2}}.$$

Using the crude inequality $m^n \geqslant 2^{\binom{n}{2}}$, we see that $m \geqslant 2^{\frac{1}{2}(n-1)}$. The only known improvements on this, other than those already mentioned, are the inequality $f(7) \geqslant 30$, and its implications for other values. We summarize these results on $f(n)$ in the following theorem:

Theorem 9.1. *Let $f(n)$ denote the minimum number for which every tournament of order $f(n)$ contains a transitive n-tournament. Then*

$$\begin{aligned} f(n) &= 2^{n-1}, && \text{if } n \leqslant 4,\\ f(n) &= 7.2^{n-4}, && \text{if } n = 5 \text{ or } 6,\\ 2^{\frac{1}{2}(n-1)} \leqslant f(n) &\leqslant 7.2^{n-4}, && \text{if } n \geqslant 7. \;\|\end{aligned}$$

Circuit-free Subgraphs

An alternative view of the preceding problem is that of finding the transitive tournament of maximum order contained in every tournament of given order. We now consider a more general question: what is the maximum number $g(n)$ for which every n-tournament contains a set (called a "consistent set") of $g(n)$ arcs yielding no circuits. The known values of $g(n)$ are given in Table 3; for $n \leqslant 9$, the results are due to Reid [**77**], for $n = 10$ and 11 to Bermond [**17**] and for $n = 12$ and 13 to Hardouin Duparc [**52**].

Table 3

n	2	3	4	5	6	7	8	9	10	11	12	13
$g(n)$	1	2	4	6	9	12	20	24	30	35	44	50

Erdős and Moon [**31**] first considered this problem, and obtained general bounds. The latest lower bound is due to Spencer [**85**] (see also [**33**]):

Theorem 9.2. *The maximum number of consistent arcs in every n-tournament satisfies the inequalities*

$$\tfrac{1}{2}\binom{n}{2}+cn^{\frac{3}{2}} \leqslant g(n) \leqslant \tfrac{1}{2}\binom{n}{2}+\left(\frac{1}{2\sqrt{2}}+o(1)\right)n^{\frac{3}{2}}\sqrt{\log n},$$

where c is some positive constant. ||

(Spencer [**86**] has also reported similar types of bounds for the maximum number of arcs in transitive sub-digraphs of every n-tournament.)

Note that $\binom{n}{2}-g(n)$ is the largest number of arcs that need to be reversed in order to convert every n-tournament into a transitive tournament—that is, the largest number that need to be deleted in order to get a transitive subgraph. Kotzig [**62**] conjectured that for $n \geqslant 10$, this number is greater than $[\frac{1}{3}n[\frac{1}{2}(n-1)]]$, the maximum number of edge-disjoint circuits in the complete n-graph. Bermond [**18**] proved this conjecture, and thereby established that one cannot always destroy all circuits by reversing (or deleting) one arc from each circuit in a maximum collection of arc-disjoint circuits (see also [**19**]).

Circuit Preservation

One of Whitney's graph theorems [**95**] concerns the preservation of incidence of edges in triangles via edge mappings between graphs. Goldberg and Moon [**41**] investigated similar mappings for tournaments, and proved the following result:

Theorem 9.3. *If two strong tournaments have a one-to-one correspondence between their arcs, which maps 3-circuits into 3-circuits and 4-circuits into 4-circuits, then they are isomorphic or converse tournaments.* ||

It can be shown that the conclusion need not hold if the tournaments are not strong, or if the hypothesis on 4-circuits is omitted. It is not known whether the hypothesis on 3-circuits is necessary.

Bypasses

For $k \geqslant 2$, a ***k*-bypass** of an arc vw in a tournament is a path of length k from v to w. Of special interest are the 2-bypasses, in that they form parts of transitive triples, and we define an arc to be **good** if it has a 2-bypass. For example, only five of the ten arcs in the regular 5-tournament are good, whereas all of the arcs of QT_7 are good.

It is straightforward to show that if each arc of a tournament T is good, then the order of T is at least 7, and each arc has a k-bypass for $k = 3, 4, 5$ and 6. Alspach, Reid, and Roselle [**9**] showed that for $n \geqslant 7$, every regular n-tournament has a k-bypass for $k = 3, 4, \ldots, n-1$, and if, in addition, every arc is good, then for every pair of distinct vertices v and w, there is a k-path from v to w for $k = 2, 3, \ldots, n-1$. They also proved that every tournament score list belongs to a tournament with at least one arc that is not good, but (on the other hand) that almost all tournaments have only good arcs.

Finally, in terms of bypasses, a result of Grünbaum (see [**47**, p. 211]) states that only two tournaments do not have long bypasses: except for the 3-circuit and one 5-tournament, every n-tournament ($n \geqslant 3$) has an arc with an $(n-1)$-bypass.

Arc-reversals

Results related to reversing the direction of certain arcs in tournaments arise in several contexts. One approach to proving Rédei's theorem (Theorem 3.1) that every tournament has an odd number of spanning paths is to use a result of Szele [**91**] that the reversal of one arc in a tournament does not change the parity of the number of spanning paths. Camion [**23**], however, showed that the parity of the number of spanning circuits can always be changed by reversing one well-chosen arc. A related open problem (for digraphs in general) is given in a conjecture of Ádám [**1**]:

Conjecture. *Every non-transitive tournament T has an arc whose reversal results in a tournament with fewer circuits than T.*

Many of the results on arc-reversals give tournaments which can be transformed into one another by successively reversing arcs in certain sets. For example, Reid [**79**] showed that for any n and k with $k < n$, any n-tournament can be transformed into any other n-tournament by a succession of reversals of k-paths.

A theorem of Ryser **[84]** on 0–1 matrices can be given as an arc-reversal result; for tournaments this result states that any tournament can be transformed into any other tournament with the same score list by successively reversing the arcs in 3-circuits. Waldrop (unpublished) has extended this result to 4-circuits and 5-circuits, and has given an example of two strong 6-tournaments with the same score list which cannot be transformed into each other by reversing 6-circuits. However, he has also shown that for any $k \geqslant 3$, there exists an integer N such that for $n \geqslant N$, and for any pair of strong n-tournaments S and T with the same score list, S can be transformed into T by successively reversing the arcs of k-circuits.

Waldrop **[93]** has also considered reversing the arcs of transitive triples. Although this can result in a change of scores, the parity of each score remains unchanged. Waldrop used Ryser's result to show that if S and T are n-tournaments with the same number of even scores, then S can be transformed into T by successively reversing the arcs of transitive triples.

Path Coverings

A collection of paths is said to **cover** a tournament if each arc lies in one of the paths, and the **path-covering number** $p(T)$ of T is the smallest cardinality of a set of arc-disjoint paths covering T. For example, the path-covering number of the regular 5-tournament is 3, since no two paths can cover it, and $\{12340, 013, 30241\}$ forms a cover of $R(1, 2)$ with three paths.

Alspach, Mason and Pullman **[6]**, **[7]** proved that the path-covering number of an n-tournament lies between $[\frac{1}{2}(n+1)]$ and $[\frac{1}{4}n^2]$, inclusively. They also established that for any n, and for any k in the closed interval $[\frac{1}{2}n, \frac{1}{4}n^2]$, there is an n-tournament with path-covering number k, unless nk is odd and $k \geqslant n$. It is conjectured that this gives all the path-covering numbers.

Self-converse Subtournaments

In solving a problem of Trotter (unpublished), Reid and Thomassen **[81]** determined which tournaments have the property that all of its subtournaments are self-converse. In addition to the transitive tournaments, and those obtained from a transitive tournament by reversing the arc from the vertex of highest score to that of score 0, there are just five such tournaments—the 6-tournament with dominance

$$0\colon 1, 4; \quad 1\colon 2, 5; \quad 2\colon 0, 3; \quad 3\colon 0, 1, 4; \quad 4\colon 1, 2, 5; \quad 5\colon 2, 0, 3,$$

the 7-tournament $R(1, 2, 3)$, and their subtournaments of order 5 and 6. This problem of Trotter involves pairs of tournaments on the same vertex-set, in which corresponding subtournaments are isomorphic.

Simple Tournaments

Intuitively, a simple tournament is one which cannot be obtained (in a non-trivial way) by replacing vertices of some tournament by other tournaments, and adding the induced arcs. Formally, we define the **composition** $T(T_1, T_2, \ldots, T_n)$ of an n-tournament T (with vertices $v_1, v_2, \ldots, v_n$) and n tournaments T_i, for $i = 1, 2, \ldots, n$ (with vertices $u_{i1}, \ldots, u_{in_i}$), as that tournament with vertices (v_i, u_{ij}) in which (v_i, u_{ij}) dominates (v_h, u_{hk}) if and only if

$$h = i, \qquad \text{and} \qquad u_{ij} \text{ dominates } u_{ik} \text{ in } T_i,$$

or

$$h \neq i, \qquad \text{and} \qquad v_i \text{ dominates } v_h \text{ in } T.$$

An example is shown in Fig. 5. A tournament T^* is called **simple** if $T^* = T(T_1, T_2, \ldots, T_n)$ implies that T is either trivial or isomorphic to T^*.

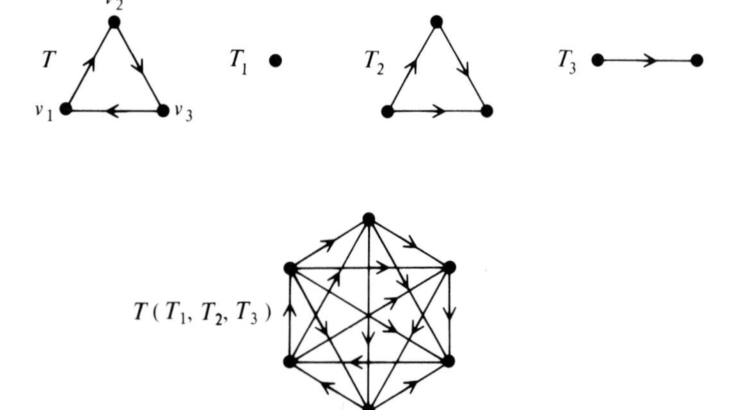

Fig. 5

Clearly, every simple tournament must be strong, and the regular tournaments of orders 3 and 5 are examples of simple tournaments.

Simple tournaments were introduced by Fried and Laskar [**36**], who proved that every n-tournament is contained in some simple $(2n+1)$-tournament. Erdős, Fried, Hajnal and Milner [**30**] improved the number $2n+1$ to $n+2$, for $n \neq 2$, and at about the same time Moon [**68**] improved it to $n+1$ for most tournaments, as stated in the following result:

Theorem 9.4. *Except for the cyclic triple and the transitive tournaments of odd order, every n-tournament is contained in a simple* $(n+1)$*-tournament, and the exceptions are contained in simple* $(n+2)$*-tournaments.* ‖

Other interesting results on simple tournaments were proved by Müller, Nešetřil and Pelant [**72**]. They showed that a tournament score list belongs to some simple tournament if and only if it belongs to some strong tournament, and also that almost all tournaments are simple.

Schütte's Problem

A tournament T is said to have **property S_k** if, for every set S of k vertices, some vertex of T dominates all vertices in S. For example, if no vertex in an n-tournament T has score $n-1$, then T has property S_1, and the tournament QT_7 has property S_2. Erdős [**29**] has used probabilistic methods to show that for each k there exist tournaments with property S_k, and later Graham and Spencer [**43**] used results on group characters to construct further examples. One such example is the following:

Theorem 9.5. *If* $p \equiv 3$ (*modulo* 4) *is a prime number, and if* $p > k^2 2^{2k-2}$, *then* QT_p *has property* S_k. ‖

The problem of determining the smallest order $m(k)$ of a tournament with property S_k remains open. Erdős [**29**] showed that $2^{k+1}-1 \leqslant m(k) \leqslant c2^k k^2$, for some constant c. The lower bound is in fact equality for $k = 1$ and 2, but Szekeres and Szekeres [**89**] showed, by considering a generalization of property S_k, that the inequality is strict for $k \geqslant 3$. An n-tournament is said to have **property $S_{k,r}$** if every set of k vertices is completely dominated by at least r vertices. Thus, property $S_{k,1}$ is identical with property S_k. Bermond (unpublished) has proved that the order of a tournament with property $S_{k,r}$ is at least $2^{k-1}(k+r-1)-1$. Moreover, a tournament with property $S_{k,1}$ for $k \geqslant 2$ also has property $S_{k-1,k+1}$, and it follows that $m(k) > 2^{k-1}(k+2)-1$. Szekeres and Szekeres also showed that equality holds here for $k = 3$ (as well as for $k = 2$). However, Wallis [**94**] showed that for $k > 3$ equality would hold if and only if certain "highly-regular" tournaments were to exist, and Brown and Reid [**20**] showed that they did not exist (see Section 8). Therefore, other than the general upper bound of Erdős, all that is known is that $m(1) = 3$, $m(2) = 7$, $m(3) = 19$, and $m(k) \geqslant 2^{k-1}(k+2)$ for $k > 3$. The Szekereses have suggested that perhaps $m(k) = 2^k(k-1)+3$.

The Reconstruction Problem

As the analog for tournaments of the reconstruction problem for graphs, the "reconstruction problem for tournaments" was to determine whether, for $n \geqslant 5$, all n-tournaments are reconstructible, in the sense that each n-tournament is determined (up to isomorphism) by its n subtournaments of order $n-1$. Chapter 8 is devoted to the general problem for graphs, and so our remarks will be confined to the basic results on tournaments.

Clearly both 3-tournaments have pairwise-isomorphic subtournaments of order 2, and the 4-tournaments with score lists (1, 1, 1, 3) and (0, 2, 2, 2) have pairwise-isomorphic subtournaments of order 3. Harary and Palmer [**51**] proved that reducible tournaments of order $n \geqslant 5$ are reconstructible. Beineke and Parker [**14**] found counter-examples for orders 5 and 6 (see Chapter 8); there is one pair of order 5 and four pairs of order 6. In a computer search, Stockmeyer [**87**] found that all 7-tournaments are in fact reconstructible, but that there are two pairs of 8-tournaments which are not. In some of the most interesting and important recent work in the theory of tournaments, Stockmeyer has found infinitely many non-reconstructible tournaments. It remains to be seen what impact, if any, his results will have on the corresponding conjecture for graphs. Stockmeyer's result is as follows:

Theorem 9.6. *If $n = 2^r + 2^s$, for any integers $r > 1$ and $s \geqslant 1$, then there exist non-isomorphic n-tournaments which have pairwise isomorphic subtournaments of order $n-1$.* ‖

Further results on the reconstruction problem will be found in Chapter 8.

References

1. A. Ádám, Bemerkungen zum graphentheoretischen Satze von I. Fidrich, *Acta. Math. Acad. Sci. Hungar.* **16** (1965), 9–11, *MR*32#7437.
2. B. Alspach, A combinatorial proof of a conjecture of Goldberg and Moon, *Canad. Math. Bull.* **11** (1968), 655–661; *MR*39#2662.
3. B. Alspach, On point-symmetric tournaments, *Canad. Math. Bull.* **13** (1970), 317–323; *MR*42#7558.
4. B. Alspach and J. L. Berggren, On the determination of the maximum order of the group of a tournament, *Canad. Math. Bull.* **16** (1973), 11–14; *MR*48#1969.
5. B. Alspach, M. Goldberg and J. W. Moon, The group of the composition of two tournaments, *Math. Mag.* **41** (1968), 77–80; *MR*37#3964.
6. B. Alspach, D. Mason and N. Pullman, Path numbers of tournaments, *J. Combinatorial Theory* (*B*) **20** (1976), 222–228; *MR*54#10065.
7. B. Alspach and N. Pullman, Path decompositions of digraphs, *Bull. Austral. Math. Soc.* **10** (1974), 421–427; *MR*50#169.
8. B. Alspach and K. B. Reid, Degree frequencies in digraphs and tournaments, *J. Graph Theory* **2** (1978), 241–249.

9. B. Alspach, K. B. Reid and D. P. Roselle, Bypasses in asymmetric digraphs, *J. Combinatorial Theory* (*B*) **17** (1974), 11–18; *MR***52**#168.
10. A. Astie, Groupes d'automorphismes des tournois sommet-symétriques d'ordre premier et dénombrement de ces tournois, *C.R. Acad. Sci. Paris* (*A*) **275** (1972), 167–169; *MR***46** #8888.
11. A. Astie, Groupes de permutations primitifs d'ordre impair et dénombrement de tournois sommet-primitifs, *Discrete Math.* **14** (1976), 1–15; *MR***52**#7958.
12. A. Astie, Vertex-symmetric tournaments of order n with the minimum number of arc orbits, in *Recent Advances in Graph Theory* (ed. M. Fielder), Academia, Prague, 1975, pp. 13–16; *MR***52**#10486.
13. L. W. Beineke and F. Harary, The maximum number of strongly connected subtournaments, *Canad. Math. Bull.* **8** (1965), 491–498; *MR***31**#5810.
14. L. W. Beineke and E. T. Parker, On nonreconstructable tournaments, *J. Combinatorial Theory* **9** (1970), 324–326; *MR***43**#6135.
15. D. H. Bent and T. V. Narayana, Computation of the number of score sequences in round-robin tournaments, *Canad. Math. Bull.* **7** (1964), 133–135.
16. J. L. Berggren, An algebraic characterization of finite symmetric tournaments, *Bull. Austral. Math. Soc.* **6** (1972), 53–59; *MR***45**#118.
17. J.-C. Bermond, Ordres à distance minimum d'un tournoi et graphes partiels sans circuits maximaux, *Math. Sci. Humaines* **37** (1972), 5–25; *MR***46**#87.
18. J.-C. Bermond, The circuit-hypergraph of a tournament, in *Infinite and Finite Sets*, Vol. I (ed. A. Hajnal *et al.*), North-Holland, Amsterdam, 1975, pp. 165–180; *MR***53**#187.
19. J.-C. Bermond and Y. Kodratoff, Une heuristique pour le calcul de l'indice de transitive d'un tournoi (to appear).
20. E. Brown and K. B. Reid, Doubly regular tournaments are equivalent to skew-Hadamard matrices, *J. Combinatorial Theory* (*A*) **12** (1972), 332–338; *MR***45**#8579.
21. P. Camion, Chemins et circuits hamiltoniens des graphes complets, *C.R. Acad. Sci. Paris* (*A*) **249** (1959), 2151–2152; *MR***23**#A75.
22. P. Camion, Quelques propriétés des chemins et circuits hamiltoniens dans la théorie des graphes, *Cahiers Centre Études Recherche Opér.* **2** (1960), 5–36; *MR***22**#4645.
23. P. Camion, Une propriété des circuits hamiltoniens des graphes complets antisymétriques, *Cahiers Centre Études Recherche Opér.* **15** (1973), 225–228; *MR***50**#170.
24. V. Chvátal, Monochromatic paths in edge-colored graphs, *J. Combinatorial Theory* (*B*) **13** (1972), 69–70; *MR***46**#8890.
25. V. Chvátal and A. Kotzig, On 3-cycles in regular tournaments, Centre de Recherches Mathematiques CRM-252, Université de Montreal (1972).
26. R. L. Davis, Structures of dominance relations, *Bull. Math. Biophys.* **16** (1954), 131–140; *MR***16**-57.
27. J. D. Dixon, The maximum order of the group of a tournament, *Canad. Math. Bull.* **10** (1967), 503–505; *MR***37**#90.
28. R. J. Douglas, Tournaments that admit exactly one Hamiltonian circuit, *Proc. London Math. Soc.* (*3*) **3** (1970), 716–730; *MR***43**#86.
29. P. Erdős, On Schütte's problem, *Math. Gaz.* **47** (1963), 220–222; *MR***28**#2566.
30. P. Erdős, E. Fried, A. Hajnal and E. C. Milner, Some remarks on simple tournaments, *Mathematika* **19** (1972), 57–62; *MR***46**#5161.
31. P. Erdős and J. W. Moon, On sets of consistent arcs in a tournament, *Canad. Math. Bull.* **8** (1965), 269–271; *MR***32**#57.
32. P. Erdős and L. Moser, On the representation of directed graphs as unions of orderings, *Publ. Math. Inst. Hungar. Acad. Sci.* **9** (1964), 125–132; *MR***29**#5756.
33. P. Erdős and J. Spencer, *Probabilistic Methods in Combinatorics*, Academic Press, New York, 1974; *MR***52**#2895.
34. R. Forcade, Parity of paths and circuits in tournaments, *Discrete Math.* **6** (1973), 115–118; *MR***47**#8352.
35. E. Fried, On homogeneous tournaments, in *Combinatorial Theory and its Applications*, Vol. II (ed. P. Erdős *et al.*), North-Holland, Amsterdam, 1970, pp. 467–476; *MR***45** #6688.

36. E. Fried and H. Laskar, Simple tournaments, *Notices Amer. Math. Soc.* **18** (1971), 395.
37. T. Gallai, On directed paths and circuits, in *Theory of Graphs* (ed. P. Erdős and G. Katona), Academic Press, New York, 1968, pp. 115–118; *MR***38**#2054.
38. M. R. Garey, On enumerating tournaments that admit exactly one Hamiltonian circuit, *J. Combinatorial Theory* (*B*) **13** (1972), 266–269; *MR***46**#7076.
39. M. Goldberg, The group of the quadratic residue tournament, *Canad. Math. Bull.* **13** (1970), 51–53; *MR***41**#6818.
40. M. Goldberg and J. W. Moon, On the composition of two tournaments, *Duke Math. J.* **37** (1970), 323–332; *MR***41**#1563.
41. M. Goldberg and J. W. Moon, Arc mappings and tournament isomorphisms, *J. London Math. Soc.* (2) **3** (1971), 378–384; *MR***43**#4725.
42. M. Goldberg and J. W. Moon, Cycles in k-strong tournaments, *Pacific J. Math.* **40** (1972), 89–96; *MR* **46**#3363.
43. R. L. Graham and J. H. Spencer, A constructive solution to a tournament problem, *Canad. Math. Bull.* **14** (1971), 45–48; *MR***45**#1798.
44. B. Grünbaum, Antidirected Hamiltonian paths in tournaments, *J. Combinatorial Theory* (*B*) **11** (1971), 249–257; *MR***45**#116.
45. A. Gyárfas and J. Lehel, A Ramsey-type problem in directed and bipartite graphs, *Periodica Math. Hungar.* **3** (1973), 299–304; *MR***49**#112.
46. M. Hall, Jr., *Combinatorial Theory*, Blaisdell, Waltham, Mass., 1967; *MR***37**#80.
47. F. Harary, *Graph Theory*, Addison-Wesley, Reading, Mass., 1969; *MR***41**#1566.
48. F. Harary and L. Moser, The theory of round robin tournaments, *Amer. Math. Monthly* **73** (1966), 231–246; *MR***33**#5512.
49. F. Harary, R. Z. Norman and D. Cartwright, *Structural Models: An Introduction to the Theory of Directed Graphs*, John Wiley & Sons, New York, 1965; *MR***32**#2345.
50. F. Harary and E. M. Palmer, On the problem of reconstructing a tournament from sub-tournaments, *Monatsh. Math.* **71** (1967), 14–23; *MR* **35**#86.
51. F. Harary and E. M. Palmer, *Graphical Enumeration*, Academic Press, New York, 1973; *MR***50**#9682.
52. J. Hardouin Duparc, Quelques résultats sur l'indice de transitivité de certains tournois, *Math. Sci. Humaines* **51** (1975), 34–41; *MR***53**#188.
53. M. Herzog and K. B. Reid, Regularity in tournaments, in *Theory and Applications of Graphs*, Lecture Notes in Mathematics **642** (ed. Y. Alavi and D. R. Lick), Springer-Verlag, Berlin, Heidelberg and New York, 1978, pp. 442–453.
54. M. Jean, Line symmetric tournaments, in *Recent Progress in Combinatorics* (ed. W. T. Tutte), Academic Press, New York, 1969, pp. 265–271; *MR***40**#5482.
55. E. J. Johnsen, Integral solutions to the incidence equation for finite projective plane cases of orders $n \equiv 2 \pmod 4$, *Pacific J. Math.* **17** (1966), 97–120; *MR***32**#7428.
56. D. Kleitman, The number of tournament score sequences for a large number of players, in *Combinatorial Structures and Their Applications* (ed. R. Guy *et al.*), Gordon and Breach, New York, 1970, pp. 209–213; *MR***42**#5815.
57. G. Korvin, Some combinatorial problems on complete directed graphs, in *Theory in Graphs* (ed. P. Rosenstiehl), Gordon and Breach, New York, and Dunod, Paris, 1967, pp. 197–213; *MR***36**#3677.
58. A. Kotzig, Des cycles dans les tournois, in *Theory of Graphs* (ed. P. Rosenstiehl), Gordon and Breach, New York, and Dunod, Paris, 1967, pp. 203–208.
59. A. Kotzig, Sur le nombre des 4-cycles dans un tournoi, *Mat. Časopis Sloven. Akad. Vied.* **18** (1968), 247–254; *MR***40**#4155.
60. A. Kotzig, The decomposition of a directed graph into quadratic factors consisting of cycles, *Acta. Fac. Rerum Natur. Univ. Comenian Math. Publ.* **22** (1969), 27–29; *MR***44**#117.
61. A. Kotzig, Sur les tournois avec des 3-cycles régulièrement placés, *Mat. Časopis Sloven. Akad. Vied.* **19** (1969), 126–134; *MR***46**#1634.
62. A. Kotzig, On the maximal order of cyclicity of antisymmetric directed graphs, *Discrete Math.* **12** (1975), 17–25; *MR***52**#10483.

63. H. G. Landau, On dominance relations and the structure of animal societies, III: The condition for a score structure, *Bull. Math. Biophys.* **15** (1953), 143–148; *MR***14**#1000.
64. M. Las Vergnas, Sur le nombre de circuits dans un tournoi fortement connexe, *Cahiers Centre Études Recherche Opér.* **17** (1975), 261–265; *MR***53**#2734.
65. J. W. Moon, Tournaments with given automorphism group, *Canad. J. Math.* **16** (1964), 485–489; *MR***29**#603.
66. J. W. Moon, On subtournaments of a tournament, *Canad. Math. Bull.* **9** (1966), 297–301; *MR***34**#95.
67. J. W. Moon, *Topics on Tournaments*, Holt, Rinehart and Winston, New York, 1968; *MR***41**#1574.
68. J. W. Moon, Embedding tournaments in simple tournaments, *Discrete Math.* **2** (1972), 389–395; *MR***46**#1635.
69. J. W. Moon, The minimum number of spanning paths in a strong tournament, *Publ. Math. Debrecen* **19** (1972), 101–104.
70. J. W. Moon, Tournaments whose subtournaments are irreducible or transitive, *Canad. Math. Bull.* (to appear).
71. J. W. Moon and L. Moser, Almost all tournaments are irreducible, *Canad. Math. Bull.* **5** (1962), 61–65; *MR***24**#A3079.
72. V. Müller, J. Nešetřil and J. Pelant, Either tournaments or algebras, *Discrete Math.* **11** (1975), 37–66; *MR***50**#9675.
73. V. Müller and J. Pelant, On strongly homogeneous tournaments, *Czechoslovak Math. J.* **24** (1974), 378–391; *MR***50**#174.
74. E. Parker and K. B. Reid, Disproof of a conjecture of Erdős and Moser on tournaments, *J. Combinatorial Theory* **9** (1970), 225–238; *MR***43**#93.
75. G. Pólya, Kombinatorische Anzahlbestimmungen für Gruppen, Graphen und chemische Verbindungen, *Acta Math.* **68** (1937), 145–254.
76. L. Rédei, Ein kombinatorischer Satz, *Acta Litt. Sci. Szeged* **7** (1934), 39–43.
77. K. B. Reid, On sets of arcs containing no cycles in a tournament, *Canad. Math. Bull.* **12** (1969), 261–264; *MR***40**#4158.
78. K. B. Reid, l-cycles in n-tournaments having no k-cycles, in *Proceedings of the Second Louisiana Conference on Combinatorics, Graph Theory and Computing* (ed. R. Mullin *et al.*), Congressus Numerantium **III**, Utilitas Mathematica, Winnipeg, 1971, pp. 473–482; *MR***47**#8342.
79. K. B. Reid, Equivalence of n-tournaments via k-path reversals, *Discrete Math.* **6** (1973), 263–280; *MR***48**#1965.
80. K. B. Reid, Score sets of tournaments, in *Proceedings of the Ninth Southeastern Conference on Combinatorics, Graph Theory and Computing* (ed. R. Mullin *et al.*) (to appear).
81. K. B. Reid and C. Thomassen, Strongly self-complementary and hereditarily isomorphic tournaments, *Monatsh. Math.* **81** (1976), 291–304.
82. M. Rosenfeld, Antidirected Hamiltonian paths in tournaments, *J. Combinatorial Theory* (*B*) **12** (1972), 93–99; *MR***44**#2670.
83. M. Rosenfeld, Antidirected Hamiltonian circuits in tournaments, *J. Combinatorial Theory* (*B*) **16** (1974), 234–242; *MR***49**#4857.
84. H. Ryser, Matrices of zeros and ones in combinatorial mathematics, in *Recent Advances in Matrix Theory*, University of Wisconsin Press, Madison, Wisc., 1964, pp. 103–124; *MR***29**#2196.
85. J. H. Spencer, Optimal ranking of tournaments, *Networks* **1** (1971), 135–138; *MR***45**#8580.
86. J. H. Spencer, Random regular tournaments, *Periodica Math. Hungar.* **5** (1974), 105–120; *MR***53**#5359.
87. P. K. Stockmeyer, The reconstruction conjecture for tournaments, in *Proceedings of the Sixth Southeastern Conference on Combinatorics, Graph Theory and Computing* (ed. F. Hoffman *et al.*), Congressus Numerantium **XIV**, Utilitas Mathematica, Winnipeg, 1975, pp. 561–566; *MR***52**#13475.
88. P. K. Stockmeyer, The falsity of the reconstruction conjecture for tournaments, *J. Graph Theory* **1** (1977), 19–26.

89. E. Szekeres and G. Szekeres, On a problem of Schütte and Erdős, *Math. Gaz.* **49** (1965), 290–293; *MR***32**#4025.
90. G. Szekeres, Tournaments and Hadamard matrices, *Enseignement Math.* **15** (1969), 269–278; *MR***40**#56.
91. T. Szele, Kombinatorikai vizsgálatok az irányitott teljes gráffal kapcscolatban, *Mat. Fiz. Lapok* **50** (1943), 223–256 = Kombinatorische Untersuchungen über gerichteten vollständigen Graphen, *Publ. Math. Debrecen* **13** (1966), 145–168; *MR***34**#7406.
92. C. Thomassen, Antidirected Hamiltonian circuits and paths in tournaments, *Math. Ann.* **201** (1973), 231–238; *MR***50**#1960.
93. C. Waldrop, Jr., An arc-reversal theorem for tournaments, in *Proceedings of the Seventh Southeastern Conference on Combinatorics, Graph Theory and Computing* (ed. F. Hoffman *et al.*), Congressus Numerantium **XVII**, Utilitas Mathematica, Winnipeg, 1976, pp. 501–508.
94. W. D. Wallis, On extreme tournaments, in *International Conference on Combinatorial Mathematics* (ed. A. Gewirtz and L. Quintas), *Annals N.Y. Acad. Sci.* **175** (1970), 403–404; *MR***42**#2974.
95. H. Whitney, Congruent graphs and the connectivity of graphs, *Amer. J. Math.* **54** (1932), 150–168.

8
The Reconstruction Problem

C. St. J. A. NASH-WILLIAMS

1. Introduction

If v is a vertex of a graph G, then $G-v$ is the graph obtained from G by deleting the vertex v and its incident edges. We shall call $G-v$ a **vertex-deleted subgraph** of G. One of the most well-known unsolved problems of graph theory asks whether a graph can be reconstructed up to isomorphism if we know all of its vertex-deleted subgraphs up to isomorphism. More precisely, it asks whether the following conjecture is true:

The Reconstruction Conjecture (First Version). *If G and H are graphs with at least three vertices, and if there exists a bijection $\sigma: V(G) \to V(H)$ such that $G-v \cong H-\sigma(v)$ for every $v \in V(G)$, then $G \cong H$.*

We must necessarily exclude graphs with two vertices, because K_2 and N_2, the complete and null graphs on two vertices, are non-isomorphic, but if $\sigma: V(K_2) \to V(N_2)$ is a bijection, then $K_2 - v \cong N_2 - \sigma(v)$ for each $v \in V(K_2)$.

The above conjecture was formulated by P. J. Kelly and S. M. Ulam in 1942, and the first published work [**17**] on it proved Kelly's well-known theorem that the conjecture is true for trees. (A proof of this will be outlined in Section 4.)

An informal, but very useful, formulation of the problem has been suggested by Harary [**12**]. Suppose that (as illustrated in Figs. 1 and 2) a graph G

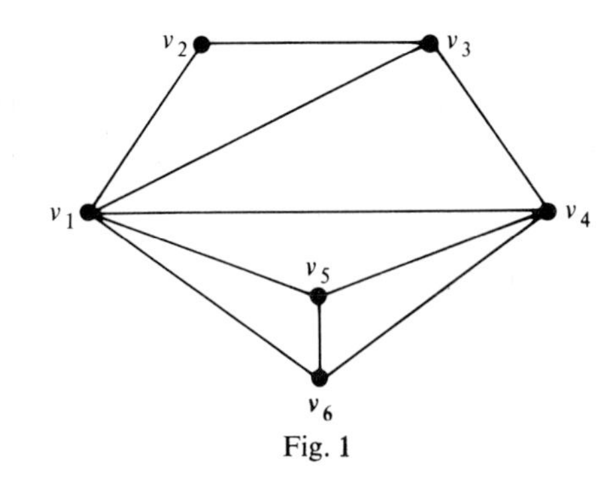

Fig. 1

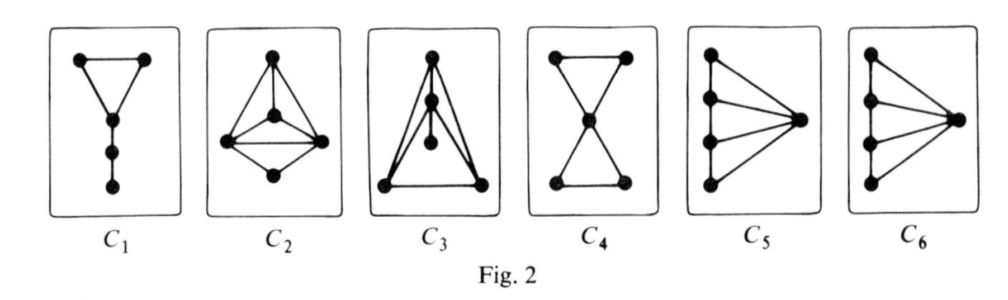

Fig. 2

has vertices $v_1, \ldots, v_p$, and we are presented with p cards $C_1, \ldots, C_p$ such that a graph isomorphic to $G-v_i$ is drawn on C_i for $i = 1, \ldots, p$. Is the information on the cards sufficient to enable us to draw a graph isomorphic to G (assuming that we have unlimited time and intelligence!)? If so, we might say that the graph G is "reconstructible". Thus a reconstructible graph is a graph which can be reconstructed up to isomorphism when we know its vertex-deleted subgraphs up to isomorphism. We shall rephrase this definition more carefully in Section 2, and the following alternative formulation of the reconstruction conjecture is then equivalent to the one given above:

The Reconstruction Conjecture (Second Version). *Every graph with at least three vertices is reconstructible.*

Many of the known partial results concerning the reconstruction conjecture fall into one of two categories:

(i) results stating that graphs of some specified kind are reconstructible;

(ii) results stating that, if there exists a bijection $\sigma: V(G) \to V(H)$ such that $G-v \cong H-\sigma(v)$ for every $v \in V(G)$, then the graphs G and H must resemble each other to some specified extent (less than the ultimate resemblance of being isomorphic).

As illustrations of (*i*), we shall outline proofs of the reconstructibility of trees and disconnected graphs with at least three vertices, and graphs in which the valencies of the vertices satisfy certain conditions. As illustrations of (*ii*), we shall see that, if $|V(G)|$, $|V(H)| \geqslant 3$ and if a bijection σ of the specified kind exists, then G and H have equal numbers of vertices and of edges, and have the same number of vertices of each valency. Moreover, the "valency-sequences" of the vertices of G are the same as those of the vertices of H, where the "valency-sequence" of a vertex v is the sequence of valencies of the vertices adjacent to v.

One can obviously formulate a conjecture concerning digraphs which is analogous to the reconstruction conjecture, and a somewhat unsettling recent discovery of Stockmeyer [**27**] is that this conjecture is false, not just for a few small digraphs (which might be regarded as analogous to the failure of the reconstruction conjecture for graphs with two vertices), but for some arbitrarily large digraphs. This will be discussed in Section 6. Stockmeyer's result might tend to suggest that the reconstruction conjecture itself is likely to be false, and that perhaps more effort should be concentrated on a search for counter-examples. However, against this one might weigh a relatively strong positive result of Müller [**23**] concerning another analog of the reconstruction conjecture.

Specifically, if e is an edge of a graph G, then the graph $G-e$ obtained by deleting e from G will be called an **edge-deleted subgraph** of G. Thus $V(G-e) = V(G)$ and $E(G-e) = E(G)-\{e\}$. A natural analog [**12**] of the reconstruction conjecture is the conjecture that a simple graph can be reconstructed up to isomorphism if we know its edge-deleted subgraphs up to isomorphism, in the following precise sense:

The Edge-reconstruction Conjecture. *If G and H are graphs with at least four edges, and if there exists a bijection $\sigma: E(G) \to E(H)$ such that $G-e \cong H-\sigma(e)$ for every $e \in E(G)$, then $G \cong H$.*

(If graphs with fewer than four edges were allowed, the graphs G and H of Fig. 3 would constitute a counter-example to this conjecture, and so would the graphs G and H of Fig. 4.)

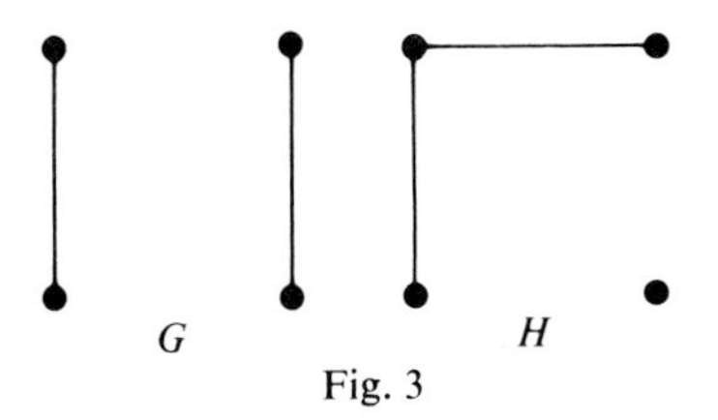

Fig. 3

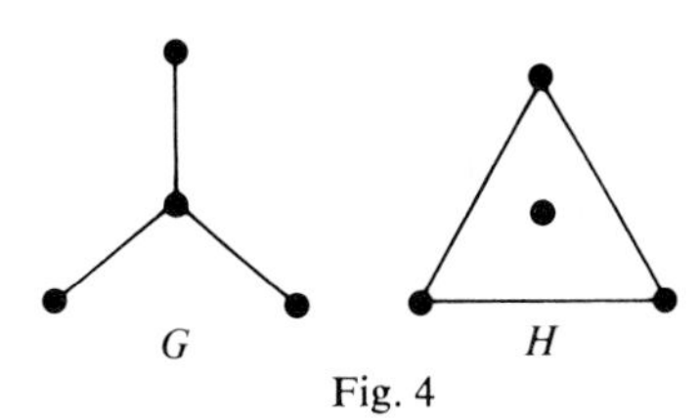

Fig. 4

Müller [**23**] has recently proved that the edge-reconstruction conjecture is—in a certain sense—true for almost all graphs. The proof, which involves a somewhat unexpected type of trick, is surprisingly short and will be given in Section 5.

A very useful and extensive survey and bibliography of work on the reconstruction problem has been written by Bondy and Hemminger [**6**], and should surely be read or consulted by every serious investigator of the subject. To reduce duplication and conform to space limitations, this chapter will be somewhat less comprehensive than [**6**], but will endeavor to highlight some ideas, results and questions which seem particularly important or interesting to the present author; inevitably this selection will involve some element of personal bias.

I am indebted to the authors of [**6**] for a pre-publication copy which has greatly helped the writing of this chapter. I am also indebted to J. A. Bondy, S. Fiorini and B. Manvel for reading a preliminary draft of this chapter and making a number of very helpful suggestions.

2. Definitions and Simple Results

Although we do not wish to overload the subject with jargon, a few definitions will make our discussion much easier. Since we have seen that graphs with just two vertices constitute exceptions to the reconstruction conjecture, we shall commonly restrict our discussion to graphs with at least three vertices: accordingly, such graphs will be called **ordinary graphs**. A **hypomorphism** of a graph G onto a graph H is a bijection $\sigma: V(G) \to V(H)$ such that $G-v \cong H-\sigma(v)$ for every $v \in V(G)$. Two graphs G and H are **hypomorphic** if there exists a hypomorphism of G onto H. A graph G is **reconstructible** if every graph hypomorphic to G is isomorphic to G.

As suggested in Section 1, we can imagine a graph G with vertices $v_1, \ldots, v_p$ giving rise to a deck of p cards $C_1, \ldots, C_p$ such that a graph isomorphic to $G-v_i$ is drawn on the card C_i for $i = 1, \ldots, p$. Hypomorphic graphs are graphs which can give rise to the same deck of cards. Consequently, Theorem 2.1 below says in effect that the number p of vertices of G and (if $p \geqslant 3$) the number of edges of G can be determined from the cards $C_1, \ldots, C_p$. Indeed, $p = |V(G)|$ can be immediately found either by counting the cards or by counting the vertices on one card and adding 1. Then $|E(G)|$ can be found by adding up the numbers of edges on all the cards and dividing by $p-2$, since each edge of G is represented on $p-2$ cards. We now translate these intuitive ideas into formal proofs.

Theorem 2.1. *(i) If G and H are hypomorphic graphs, then $|V(G)| = |V(H)|$; (ii) if G and H are hypomorphic ordinary graphs, then $|E(G)| = |E(H)|$.*

Proof. (*i*) Since there exists a hypomorphism of G onto H, and since any such hypomorphism is a bijection of $V(G)$ onto $V(H)$, it follows that $|V(G)| = |V(H)|$.

(*ii*) Let σ be a hypomorphism of G onto H. Since each edge of G belongs to $|V(G)|-2$ vertex-deleted subgraphs of G, and similarly for H, and since $G-v \cong H-\sigma(v)$ for every $v \in V(G)$, it follows that

$$\begin{aligned} |E(G)|(|V(G)|-2) &= \sum_{v \in V(G)} |E(G-v)| \\ &= \sum_{v \in V(G)} |E(H-\sigma(v))| = \sum_{w \in V(H)} |E(H-w)| \\ &= |E(H)|(|V(H)|-2), \end{aligned}$$

from which we infer that $|E(G)| = |E(H)|$, since $2 \neq |V(G)| = |V(H)|$. ‖

Informally speaking, our next theorem states that if $p \geqslant 3$, then the information on the cards $C_1, \ldots, C_p$ enables us to determine whether the graph G from which they arose is connected or disconnected. In fact, G is connected if and only if the graphs on at least two of the cards are connected.

Theorem 2.2. *If two ordinary graphs are hypomorphic, then they are both connected or both disconnected.*

Proof. A connected ordinary graph has at least two vertices which are not cut-vertices (for example, two end-vertices of the same spanning tree of the graph), and consequently has at least two connected vertex-deleted subgraphs. On the other hand, a disconnected ordinary graph G can have at most one connected vertex-deleted subgraph, because such a subgraph must be a component of G which includes all but one of its vertices. Since two hypomorphic ordinary graphs have the same number of connected vertex-deleted subgraphs, they are both connected if this number is at least two, and both disconnected if it is at most one. ‖

With the terminology introduced at the beginning of this section, the reconstruction conjecture can be restated in either of the following forms:

(RC1). *If two ordinary graphs are hypomorphic, then they are isomorphic.*

(RC2). *Every ordinary graph is reconstructible.*

We shall next prove that disconnected ordinary graphs are reconstructible, by an argument based on the following intuitive idea. Suppose that G is a disconnected ordinary graph with vertices $v_1, \ldots, v_p$, and that we are presented with cards $C_1, \ldots, C_p$ as previously described. How can we reconstruct G up to isomorphism from the information on the cards? First identify (up to isomorphism) one of the components of G with the largest number of vertices by selecting, from among all the components of all the graphs on the cards, one with as many vertices as possible. If (to conform with the notation of the formal proof below) M' is this component of G, and if L' is one of its connected vertex-deleted subgraphs, select now a card C_i such that the graph on C_i has as few components isomorphic to M' as possible and, subject to this prior requirement, has as many components isomorphic to L' as possible. Then we know that the graph on C_i must have been obtained by deleting from G a vertex v_i of a component M'' of G such that $M'' \cong M'$ and $M''-v_i \cong L'$. Thus an isomorphic copy of G can be reconstructed from the graph on C_i by replacing a component isomorphic to L' by one isomorphic to M'.

Although this type of informal discussion often provides the best way of understanding such proofs intuitively, it can lead to confused thinking and writing unless we translate the argument into fairly careful mathematical language based on proper mathematical definitions. "Deck of cards" is not mathematically defined, nor is ability to deduce something from information on cards, which appears to refer to some sort of human capability and not to any abstract mathematical concept. Moreover, a sentence like "Suppose that G is disconnected, and that we are presented with cards $C_1, \ldots, C_p$" tends to conceal the fact that the person attempting to reconstruct G is given the information on the cards but must find out that G is disconnected. Even if such confusion is not disastrous in a comparatively simple proof, it could become increasingly serious in more complicated arguments. We shall therefore pay some attention to translating informal proofs, including the one just given, into formal ones, even though unfortunately this may make them considerably more cumbersome. However, a fairly precise and rigorous language for reconstruction theory (if *reinforced* by good intuitions) may actually increase our power to obtain results, as well as meeting more pedantic requirements.

Theorem 2.3. *Every disconnected ordinary graph is reconstructible.*

Proof. Let G be a disconnected ordinary graph, and let H be a graph hypomorphic to G. We shall show that $H \cong G$.

Since G is ordinary and disconnected, we infer from Theorem 2.1(i) that H is an ordinary graph, and from Theorem 2.2 that H is disconnected.

There exists a hypomorphism σ of G onto H, and since $G-v \cong H-\sigma(v)$

for every $v \in V(G)$, it follows that the maximum of the orders of the components of the vertex-deleted subgraphs of G is equal to the corresponding maximum for H: we denote this common maximum by μ.

If G had a component with more than μ vertices, then, since G is disconnected, this component would be a component of at least one vertex-deleted subgraph of G, contradicting the definition of μ. Therefore, no component of G can have more than μ vertices. Similarly, since H is disconnected, no component of H can have more than μ vertices. In particular, if $\mu = 1$, then each component of G or H consists of just one vertex, and so $G \cong H$ by Theorem 2.1(i). We therefore assume that $\mu \geqslant 2$.

For any two graphs Γ, Z, let $c_Z(\Gamma)$ denote the number of components of Γ which are isomorphic to Z.

The definition of μ implies that we can select a connected graph M with μ vertices such that at least one vertex-deleted subgraph of G has a component isomorphic to M, and therefore at least one vertex-deleted subgraph of H has a component isomorphic to M. Select a connected vertex-deleted subgraph $L = M-u$ of M (for example, by taking u to be an end-vertex of a spanning tree of M). Consider a vertex-deleted subgraph $G-w$ of G such that

(i) $c_M(G-w) \leqslant c_M(G-v)$, for every $v \in V(G)$;

(ii) $c_L(G-w) \geqslant c_L(G-v)$, for every $v \in V(G)$ such that

$$c_M(G-w) = c_M(G-v).$$

Let G_w be the component of G which includes w.

Some vertex-deleted subgraph $G-z$ of G has a component M' isomorphic to M. If M' were not a component of G, it would be contained in a component M'' of G such that $V(M') \cup \{z\} \subseteq V(M'')$, and therefore

$$|V(M'')| > |V(M')| = |V(M)| = \mu,$$

whereas we have proved that no component of G has more than μ vertices. Therefore, M' is a component of G. From this and (i), it follows that $G_w \cong M$, because otherwise we should have $c_M(G-w) \geqslant c_M(G) > c_M(G-v)$ for each $v \in V(M')$. It now follows from (ii) that $G_w - w \cong L$, because otherwise, since $G_w \cong M$, we could select $v \in V(G_w)$ such that $G_w - v \cong L$, giving

$$c_M(G-w) = c_M(G-v) \qquad \text{and} \qquad c_L(G-v) > c_L(G-w).$$

Since $G_w \cong M$ and $G_w - w \cong L$, it follows that G is isomorphic to a graph obtained from $G-w$ on replacing a component of $G-w$ isomorphic to L by a component isomorphic to M (that is,

$$c_M(G) = c_M(G-w)+1,\ c_L(G) = c_L(G-w)-1,$$

and

$$c_Z(G) = c_Z(G-w),$$

for every graph Z not isomorphic to either L or M).

Let $\sigma(w) = x$. Since $G-v \cong H-\sigma(v)$ for every $v \in V(G)$, we infer from (i) and (ii) that

(i') $c_M(H-x) \leqslant c_M(H-y)$, for every $y \in V(H)$,

(ii') $c_L(H-x) \geqslant c_L(H-y)$, for every $y \in V(H)$ such that

$$c_M(H-x) = c_M(H-y).$$

By repeating the argument of the preceding paragraph with H, x replacing G, w, and with (i'), (ii') used in place of (i), (ii) respectively, we find that H is isomorphic to a graph obtained from $H-x$ on replacing a component of $H-x$ isomorphic to L by a component isomorphic to M. Since $G-w \cong H-x$, and G and H are isomorphic to graphs obtained from $G-w$ and $H-x$ respectively, on replacing a component isomorphic to L by one isomorphic to M, it follows that $G \cong H$. This establishes the result. ‖

The class of known reconstructible graphs can be almost doubled by using the obvious fact that the complement of a reconstructible graph is reconstructible. For example, Theorem 2.3 implies that a graph is reconstructible if its complement is disconnected, and some other results in this chapter have similar corollaries.

If G and Q are graphs, let $s_Q(G)$ denote the number of subgraphs of G which are isomorphic to Q. The following result, known as **Kelly's lemma** **[17]** is one of the basic tools of reconstruction theory.

Theorem 2.4 (Kelly's Lemma). *If G and H are hypomorphic graphs, and if Q is a graph such that $|V(Q)| < |V(G)|$, then $s_Q(G) = s_Q(H)$.*

Proof. There exists a hypomorphism σ of G onto H. Since each subgraph of G isomorphic to Q is contained in $|V(G)|-|V(Q)|$ vertex-deleted subgraphs of G, and a similar remark applies to H, and since $G-v \cong H-\sigma(v)$ for every $v \in V(G)$, it follows that

$$\begin{aligned} s_Q(G)(|V(G)|-|V(Q)|) &= \sum_{v \in V(G)} s_Q(G-v) \\ &= \sum_{v \in V(G)} s_Q(H-\sigma(v)) = \sum_{w \in V(H)} s_Q(H-w) \\ &= s_Q(H)(|V(H)|-|V(Q)|), \end{aligned}$$

from which we infer that $s_Q(G) = s_Q(H)$, since $|V(Q)| < |V(G)| = |V(H)|$ by Theorem 2.1(i). ‖

The similarity between the proofs of Theorems 2.1(*ii*) and 2.4 is not accidental; the former is essentially the special case of the latter in which $Q = K_2$.

Theorem 2.4 yields a neat alternative proof of Theorem 2.3. To see this, begin with the first, second and fifth paragraphs of the previous proof of Theorem 2.3. Then observe that, if $c_Z(G)$ were unequal to $c_Z(H)$ for some connected graph Z, then it would be fairly easy to derive a contradiction from Theorem 2.4 by taking Q to be a Z with $|E(Z)|$ maximized subject to the condition that $c_Z(G) \neq c_Z(H)$. We conclude that $c_Z(G) = c_Z(H)$ for all Z, and so $G \cong H$.

A similar type of argument shows that, if G and H are hypomorphic separable graphs, then for every graph Z, G and H have the same number of blocks isomorphic to Z; but this does not imply that $G \cong H$, and so does not provide a proof that all separable graphs are reconstructible. However, Bondy **[3]** has proved that separable graphs with no end-vertices are reconstructible, and some further kinds of separable graphs have been proved to be reconstructible in [**3**], [**11**] and [**18**]. Of course, ordinary trees (and certain similar graphs [**9**], [**22**]) constitute a further known class of reconstructible separable graphs.

Kelly's lemma, which requires the hypothesis $|V(Q)| < |V(G)|$, may be supplemented to some extent by the following theorem of Tutte [**32**].

Theorem 2.5. *Let G and H be hypomorphic ordinary graphs, and let Q be a graph such that $|V(Q)| = |V(G)|$. Then*

(*i*) $s_Q(G) = s_Q(H)$, *if Q is disconnected*;

(*ii*) $s_Q(G) = s_Q(H)$, *if Q is a circuit*;

(*iii*) $s_Q(G) = s_Q(H)$, *if Q is a path.* ||

The proof uses an algebraic technique on which we shall comment briefly in Section 8. Part (*i*) of Theorem 2.5 could be regarded as subsuming Theorem 2.3 because it tells us that, if G is a disconnected ordinary graph and if H is hypomorphic to G, then $s_G(H) = s_G(G) = 1$, and consequently $H \cong G$ by Theorem 2.1. Parts (*ii*) and (*iii*) of Theorem 2.5 say (in view of Theorem 2.1 (*i*)) that hypomorphic ordinary graphs have the same number of Hamiltonian circuits and the same number of Hamiltonian paths.

3. Reconstruction and Valencies

We recall from Chapter 1 that the **valency-sequence of a graph** is the sequence obtained by listing the valencies of its vertices in non-decreasing order. For example, the graph of Fig. 1 has valency-sequence (2, 3, 3, 3, 4, 5).

Let us once again suppose that an ordinary graph G has vertices $v_1, \ldots, v_p$, and that we are presented with the cards $C_1, \ldots, C_p$ as previously described. The information on the cards can be used to determine the valencies of the vertices of G. To do this, first determine $|E(G)|$ from the information on the cards, as explained in Section 2. From this number, subtract the number of edges on any card C_i to discover the valency $\rho(v_i)$ of the corresponding vertex v_i. The following theorem and proof express this idea more formally:

Theorem 3.1. *If σ is a hypomorphism of an ordinary graph G onto an ordinary graph H, then $\rho_G(v) = \rho_H(\sigma(v))$ for every $v \in V(G)$.*

Proof. By Theorem 2.1 (*ii*) and the definition of hypomorphism, we have $|E(G)| = |E(H)|$, and $G - v \cong H - \sigma(v)$ for every $v \in V(G)$. Therefore

$$\rho_G(v) = |E(G)| - |E(G-v)| = |E(H)| - |E(H-\sigma(v))| = \rho_H(\sigma(v)),$$

for every $v \in V(G)$. ||

Corollary 3.2. *Hypomorphic ordinary graphs have the same valency-sequence.* ||

We now define the **valency-sequence of a vertex v** (denoted by $\text{vs}(v)$ or $\text{vs}_G(v)$) of a graph G to be the sequence obtained by listing the valencies of the neighbors of v in non-decreasing order. For example, in the graph of Fig. 1, $\text{vs}(v_3) = (2, 4, 5)$, and $\text{vs}(v_4) = (3, 3, 3, 5)$. The **valency-sequence sequence** of a graph G is then defined to be the sequence of sequences obtained by listing the valency-sequences of the vertices of G in "dictionary order". For example, the valency-sequence sequence of the graph of Fig. 1 is

$$\text{vs}(v_1), \text{vs}(v_3), \text{vs}(v_4), \text{vs}(v_5), \text{vs}(v_6), \text{vs}(v_2),$$

which may be more fully written as

$$(2, 3, 3, 3, 4), \quad (2, 4, 5), \quad (3, 3, 3, 5), \quad (3, 4, 5), \quad (3, 4, 5), \quad (3, 5).$$

(The choice of dictionary order is made only for the sake of definiteness, and has no special significance.)

The information on the cards $C_1, \ldots, C_p$ corresponding to an ordinary graph G suffices to determine not only the valencies of its vertices $v_1, \ldots, v_p$, but also their valency-sequences. For, having ascertained the valencies of $v_1, \ldots, v_p$, we can determine, for any non-negative integer k, the number of k-valent neighbors of a vertex v_i. To do so, count those vertices which have valency less than k in the graph depicted on C_i, and subtract the number of vertices other than v_i whose valency in G is less than k (the valency of v_i itself being irrelevant at this point). By doing this for $k = 0, 1, \ldots, p-1$, we discover the valency-sequence of v_i. We leave it as an exercise to translate this informal argument into a proof of the following lemma, which contains the essence of Theorem 3.4 below.

Lemma 3.3. *If G and H are hypomorphic ordinary graphs, $v \in V(G)$, $w \in V(H)$, and $G-v \cong H-w$, then* $\text{vs}_G(v) = \text{vs}_H(w)$. ||

From this lemma we immediately deduce the following result:

Theorem 3.4. *If σ is a hypomorphism of an ordinary graph G onto an ordinary graph H, then* $\text{vs}_G(v) = \text{vs}_H(\sigma(v))$, *for every* $v \in V(G)$. ||

Corollary 3.5. *Hypomorphic ordinary graphs have the same valency-sequence sequence.* ||

Theorem 3.1 is very well known. Theorem 3.4, although a natural refinement of Theorem 3.1, seems to have received less explicit publicity, but clearly some investigators have been aware of it (see, for example, [**20**], [**33**]).

Returning once more to our cards $C_1, \ldots, C_p$, we remark that G can obviously be reconstructed up to isomorphism if, on some card C_i, we can identify the G-neighbors of v_i (or, more precisely, the vertices corresponding to those neighbors in G under an isomorphism between $G-v_i$ and the graph on C_i). However, a G-neighbor w of v_i such that $\rho_G(w) = k$ can be identified on C_i if G has no $(k-1)$-valent vertex, because w will appear to have valency $k-1$ on C_i, which is only explainable if one of the edges missing from C_i joins v_i to w. This suggests calling a vertex w of G "good" if G has no vertex of valency $\rho_G(w)-1$; good neighbors of v_i are identifiable on C_i, and so G is reconstructible if it has a vertex v_i all of whose neighbors are good. We shall give a precise proof of this below.

We shall approach this proof by means of two preliminary lemmas, because these seem useful in formalizing many arguments of this type, and will in fact be used again in Section 4. One may perhaps see intuitively that a certain graph G is reconstructible, by thinking how to reconstruct G from its deck of cards or something of the sort, but it can sometimes be a cumbersome process to translate this type of loose argument into a precise one which starts with an arbitrary graph H hypomorphic to G and proves that $H \cong G$. Lemma 3.6 can sometimes substantially simplify this formalization, and perhaps bring the language of the formal proof closer to that of our intuition, by showing that we need consider only those graphs H hypomorphic to G which can be obtained from G by adding and removing edges incident to some particular vertex z—intuitively, this is because, when trying to reconstruct G, we are supplied with a copy of $G-z$ on a card, but might guess incorrectly which of its vertices we should join by edges to z. Lemma 3.7 increases the power of this method by telling us more about which edges incident with z may be added and removed. This lemma is based on the intuitive idea that valency considerations can limit the possibilities of error in guessing which vertices of $G-z$ should be joined to z.

Let z be a vertex of a graph G. Then a **z-reconstruction** of G is a graph H such that $V(G) = V(H)$, $G-z = H-z$, and G is hypomorphic to H. The conditions $V(G) = V(H)$ and $G-z = H-z$ mean that G is obtainable from H by removing a (possibly empty) set L of edges incident to z and adding a (possibly empty) set M of edges incident to z; in symbols, $H = G-L+M$.

Lemma 3.6. *A graph G is reconstructible if it has a vertex z such that every z-reconstruction of G is isomorphic to G.* ‖

Lemma 3.6 is proved by showing that, if a graph is hypomorphic to G, then it is isomorphic to some z-reconstruction of G. We omit the details of this proof, which are fairly easy to supply.

A vertex v of a graph G is **bad** or **good** according as G does, or does not, possess a vertex of valency $\rho_G(v)-1$. The **neighborhood** of a vertex v in G, denoted by $N_G(v)$, is the set of vertices of G adjacent to v. We can now state our next lemma:

Lemma 3.7. *Let z be a vertex of an ordinary graph G, and let H be a z-reconstruction of G. Then, for some non-negative integer r, there exist r distinct bad neighbors $v_1, \ldots, v_r$ of z in G, and r distinct vertices $w_1, \ldots, w_r$ of $V(G)-N_G(z)$, such that $\rho_G(w_i) = \rho_G(v_i)-1$ for $i = 1, \ldots, r$, and*

$$N_H(z) = (N_G(z)-\{v_1, \ldots, v_r\}) \cup \{w_1, \ldots, w_r\}.$$

Proof. Since H is hypomorphic to G, and $H-z = G-z$, it follows by Lemma 3.3 that $\mathrm{vs}_G(z) = \mathrm{vs}_H(z)$. Therefore, for each non-negative integer k, the number of vertices in $N_H(z)-N_G(z)$ with valency k in H must equal the number of vertices in $N_G(z)-N_H(z)$ with valency k in G. We can, therefore, write $N_G(z)-N_H(z) = \{v_1, \ldots, v_r\}$ and $N_H(z)-N_G(z) = \{w_1, \ldots, w_r\}$, where $\rho_G(v_i) = \rho_H(w_i) = \rho_G(w_i)+1$ for $i = 1, \ldots, r$, so that $v_1, \ldots, v_r$ are bad in G, and the result follows. ‖

Theorem 3.8. *An ordinary graph is reconstructible if it has a vertex all of whose neighbors are good.*

Proof. Suppose that an ordinary graph G has a vertex z all of whose neighbors are good. If H is a z-reconstruction of G, then $N_H(z) = N_G(z)$ by Lemma 3.7, and therefore $H \cong G$. It follows from Lemma 3.6 that G is reconstructible. ‖

In particular, if no two vertices of an ordinary graph G have valencies which differ by 1, then every vertex of G is good, and so G is reconstructible by Theorem 3.8. For example, every regular ordinary graph is reconstructible and so is every graph with valency-sequence 2, 2, 5, 5, 5, 7, 7, 7, 7, 7. In fact, we can say a little more. If the sum of the valencies of the bad vertices of an ordinary graph G is less than $|V(G)|$, then the union of the neighborhoods of these vertices cannot be the whole of $V(G)$, and so G has a vertex all of whose neighbors are good; therefore, G is reconstructible, by Theorem 3.8. Thus a graph with valency-sequence

$$2, 2, 2, 2, 2, 2, 2, 2, 2, 2, 3, 3, 3, 4, 6, 6, 6, 6, 6, 6, 6, 6, 7$$

is necessarily reconstructible.

Theorem 3.8 is a superficial observation; but there seems to be scope for somewhat more subtle investigations based on Theorems 3.1 and 3.4, Lemmas 3.6 and 3.7, and related ideas, to increase the class of known reconstructible graphs, and possibly also guide the search for a counter-example to the reconstruction conjecture if one thinks it to be false. The author must however confess that, in some tentative explorations in this direction (some of which have involved lengthy and painstaking arguments), he has hitherto obtained only a few slender results, such as the following:

Theorem 3.9. *An ordinary graph G is reconstructible if it has a vertex z such that all vertices connected to z by paths of length two in G are good (that is, all neighbors of neighbors of z, with the possible exception of z itself, are good).* ‖

The proof of Theorem 3.9, although not unduly difficult, is less trivial than that of Theorem 3.8. From these two theorems one can fairly easily deduce (for what it is worth) that an ordinary graph G is reconstructible if the sum of the valencies of its bad vertices is less than $|V(G)| + \frac{1}{2}b$, where b is the number of bad vertices in G.

4. The Reconstruction of Trees

When a problem concerning graphs is difficult to solve in general, one commonly begins by trying to solve it for some limited class of graphs with a fairly simple type of structure, and the class of trees might seem a natural choice. The reconstructibility of ordinary trees was proved by P. J. Kelly [**17**]. In this section we shall outline a proof of this theorem; limitations of space make it necessary to summarize some of the more elaborate parts of the argument.

We refer the reader to Chapter 1 for the necessary elementary definitions relating to trees. In particular, we shall use the concept of a rooted tree. The **vertices** and **order** of a rooted tree (T, t) are defined to be the vertices and order of the tree T, respectively. A **rooted path** is a rooted tree (T, t) such that T is a path and $\rho(t) \leqslant 1$ (that is, t is either an end-vertex of T or its only vertex).

Let T be a tree with more than one vertex. Vertices whose valency in T is at least 3 will be called **junctions** of T, and the set of such vertices will be denoted by $J(T)$. The set of end-vertices of T will be denoted by $V_1(T)$. A vertex of T is **peripheral** if it is an end-vertex of one of the longest paths (that is, paths of greatest length) in T. The set of peripheral vertices of T will be denoted by $\Pi(T)$. Clearly $\Pi(T) \subseteq V_1(T)$. A **vertex-deleted subtree** of T is a vertex-deleted subgraph of T which is a tree; that is, it is a vertex-deleted subgraph $T-v$ of T such that $v \in V_1(T)$. If T is bicentral and has center e, a **branch** of T is a rooted tree (B, b) such that B is a component of $T-e$ and b is

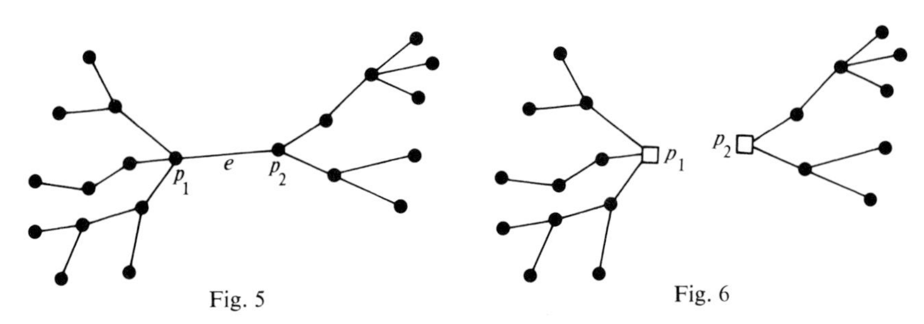

Fig. 5 Fig. 6

the bicenter of T which belongs to $V(B)$. If T is central, a **branch** of T is a rooted tree (B, c) such that c is the center of T and B is a subtree of T consisting of c, a single edge g incident with c, and the whole of the component of $T-g$ which does not include c. Thus T has exactly two branches if it is bicentral, and exactly $\rho(c)$ branches if it is central with center c. The branches of the trees in Figs. 5 and 7 are depicted in Figs. 6 and 8 respectively, using the convention that the root of a rooted tree is shown as a square vertex. A branch (B, b) of T is **peripheral** if $V(B)$ includes at least one peripheral vertex of T. Thus both branches of T are peripheral if T is bicentral, but the central tree of Fig. 7 has 3 peripheral and 2 non-peripheral branches.

Our proof that an ordinary tree T is reconstructible is based on the idea of trying to reconstruct T by identifying isomorphic copies of its branches in isomorphic copies of its vertex-deleted subtrees, in much the same way as our proof of Theorem 2.3 involved reconstructing a disconnected graph G by identifying isomorphic copies of its components in isomorphic copies of its vertex-deleted subgraphs. If the diameter of T is a, the recognition of a branch (B, b) of T in at least one vertex-deleted subtree of T is not too hard, provided

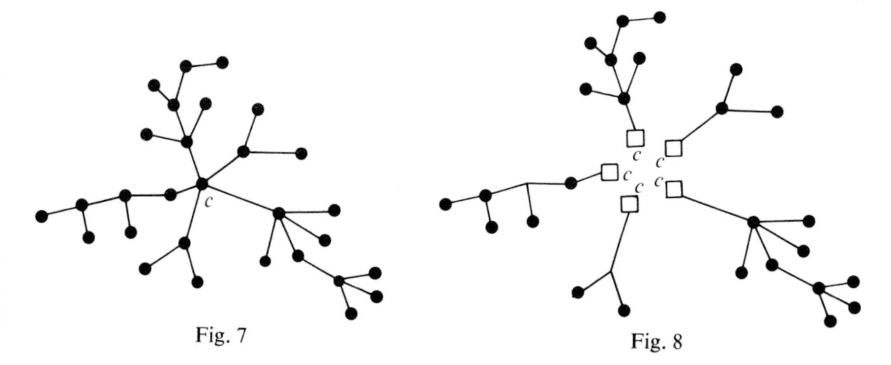

Fig. 7

Fig. 8

that (B, b) is a branch of at least one vertex-deleted subtree of diameter a; but the fact that this may sometimes not occur introduces some awkwardness into the proof, since this special case must be examined separately. (Perhaps some reader may feel challenged to find a more elegant way round this difficulty.)

A vertex-deleted subtree $T-v$ has diameter a, unless T has just two peripheral branches and v is the sole peripheral vertex of T in one of these branches. It is, therefore, fairly easily seen that a branch (B, b) of T will be a branch of at least one vertex-deleted subtree of T of diameter a except in the following cases (illustrated by Figs. 9 and 10, respectively):

(*i*) T has exactly two branches, and the branch other than (B, b) is a rooted path;

(*ii*) (B, b) is not peripheral (so that T is necessarily central), and T has exactly two other branches, both of which are peripheral and are rooted paths.

We shall, therefore, indicate first a somewhat *ad hoc* proof of the reconstructibility of ordinary trees in which the situations (*i*) and (*ii*) can arise (Lemma 4.4 below), and thereafter outline a general argument, on the lines of the proof of Theorem 2.3, for all remaining ordinary trees. (Actually, Lemma 4.4 covers some other ordinary trees in addition to those in which (*i*) or (*ii*) occurs.)

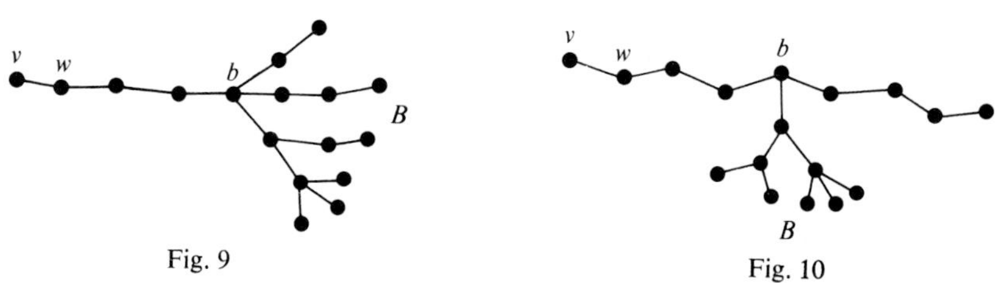

Fig. 9

Fig. 10

Lemma 4.1. *Any graph hypomorphic to an ordinary tree is itself an ordinary tree.*

Proof. Suppose that a graph G is hypomorphic to an ordinary tree T. Then $|V(G)| = |V(T)|$ by Theorem 2.1, and so G is ordinary. From this, and the facts that T is connected and $|E(T)| = |V(T)|-1$, it follows by Theorems 2.1 and 2.2 that G is connected, and that $|E(G)| = |V(G)|-1$; therefore, G is a tree. ‖

Lemma 4.2. *Every ordinary tree with at most one junction is reconstructible.*

Proof. Let T be an ordinary tree with at most one junction, and let G be a graph which is hypomorphic to T. Then there exists a hypomorphism σ of T onto G. By Lemma 4.1, G is a tree. If $|J(T)| = 0$, then $|J(G)|$ is also 0, by Corollary 3.2, so that the trees G and T are both paths, and are therefore isomorphic by Theorem 2.1. If $|J(T)| = 1$, let $J(T) = \{v\}$. Then $J(G) = \{\sigma(v)\}$ by Theorem 3.1. In this case, the facts that G and T are trees, $J(T) = \{v\}$, $J(G) = \{\sigma(v)\}$ and $T-v \cong G-\sigma(v)$ clearly imply that $G \cong T$. We have now proved that $G \cong T$ for every graph G which is hypomorphic to T, and so T is reconstructible. ‖

If G, Q are graphs and $v \in V(G)$, then $s_Q(G, v)$ will denote the number of subgraphs of G which include v and are isomorphic to Q. We can now deduce the following lemma:

Lemma 4.3. *If v is a vertex of a finite graph G, if H is a v-reconstruction of G, and if Q is a graph such that $|V(Q)| < |V(G)|$, then $s_Q(G, v) = s_Q(H, v)$.*

Proof. Since $G-v = H-v$, we have, by Kelly's lemma (Theorem 2.4),

$$s_Q(G, v) = s_Q(G)-s_Q(G-v) = s_Q(H)-s_Q(H-v) = s_Q(H, v). \quad \|$$

The distance in a tree T between two vertices v and w will be denoted by $d_T(v, w)$, and if W is a non-empty subset of $V(T)$, then $d_T(v, W)$ will denote $\min\{d_T(v, w): w \in W\}$. For any positive integer α, Y_α will denote a tree with exactly one junction and exactly three end-vertices, in which the junction is adjacent to two of the end-vertices and at distance α from the third.

Lemma 4.4. *Let T be an ordinary tree such that*

either (i) T has exactly two branches, and at least one of them is a rooted path,

or (ii) T is central, and all of its peripheral branches are rooted paths.

Then T is reconstructible.

Indication of Proof. By Lemma 4.2, we may assume that $|J(T)| \geqslant 2$. Let the diameter of T be a. By (*i*) or (*ii*), T has at least one peripheral branch which is a rooted path. Let v be the peripheral vertex of T in one such peripheral branch, and let w be the neighbor of v in T (as illustrated in Figs. 9 and 10). Let S be a v-reconstruction of T; according to Lemma 3.6, it suffices to prove that $S \cong T$.

Since $|J(T)| \geqslant 2$, it follows that $\rho(w) = 2$. Therefore, by Lemma 3.7, either $N_S(v) = \{w\}$, in which case clearly $S \cong T$, or $N_S(v) = \{z\}$ for some $z \in V_1(T) - \{v\}$. In the latter case, S is clearly a tree such that $J(S) = J(T)$; denote this set by J. By Lemma 4.3, the smallest α such that v belongs to a subtree of T isomorphic to Y_α is equal to the smallest α such that v belongs to a subtree of S isomorphic to Y_α—that is,

$$d_T(v, J) = d_S(v, J) = d_T(z, J) + 1. \qquad (1)$$

Moreover, $d_T(v, J) \geqslant \frac{1}{2}a$, by the definition of v, and $d_T(v, z) \leqslant \operatorname{diam} T = a$. From these inequalities, (1), and condition (*i*) or (*ii*), it is not hard to deduce that the vz-path in T includes exactly one element (u, say) of J. By (1), $d_T(v, u) = d_T(z, u) + 1$, and therefore $S \cong T$. ‖

We can now give the main result of this section:

Theorem 4.5. *Every ordinary tree is reconstructible.*

Indication of Proof. Let S, T be hypomorphic ordinary trees. By Lemma 4.1, it suffices to prove that in these circumstances $S \cong T$.

If either S or T is shown to be reconstructible by Lemma 4.4, then the desired conclusion $S \cong T$ follows from the fact that S, T are hypomorphic and that one of them is reconstructible. We therefore assume that neither S nor T satisfies the hypotheses of Lemma 4.4.

By Kelly's lemma (Theorem 2.4), the largest value of l such that S contains a path of length l is equal to the largest value of l such that T contains a path of length l; that is, S and T have the same diameter (a, say). Let $\mathscr{D}(S)$ be the set of all vertex-deleted subtrees of S of diameter a, and define $\mathscr{D}(T)$ analogously. Since S and T are hypomorphic, each tree in $\mathscr{D}(S)$ or $\mathscr{D}(T)$ is isomorphic to a tree in $\mathscr{D}(T)$ or $\mathscr{D}(S)$, respectively. Therefore, the maximum of the orders of the branches of the trees in $\mathscr{D}(S)$ is equal to the corresponding maximum for $\mathscr{D}(T)$; we shall denote this common maximum by μ.

For any ordinary tree Y, and any rooted tree (R, r), let $b_{(R,r)}(Y)$ denote the number of branches of Y which are isomorphic to (R, r).

The definition of μ implies that we can select a rooted tree (M, m) of order μ, such that at least one member of $\mathscr{D}(S)$ has a branch isomorphic to (M, m). It is easily shown that (M, m) is not a rooted path, and so we can select a vertex $u \in V_1(M)-\{m\}$ such that $d_M(m, v) = [\frac{1}{2}a]$ for at least one vertex $v \in V_1(M)-\{u\}$. Let $L = M-u$, and consider an element $S-w$ of $\mathscr{D}(S)$ such that

(*i*) $b_{(M,m)}(S-w) \leqslant b_{(M,m)}(S-v)$, for every $S-v \in \mathscr{D}(S)$;

(*ii*) $b_{(L,m)}(S-w) \geqslant b_{(L,m)}(S-v)$, for every $S-v \in \mathscr{D}(S)$ such that $b_{(M,m)}(S-w) = b_{(M,m)}(S-v)$.

Let $T-x$ be a member of $\mathscr{D}(T)$ which is isomorphic to $S-w$.

The fact that S does not satisfy the hypotheses of Lemma 4.4 ensures that each branch of S is a branch of at least one member of $\mathscr{D}(S)$, and a similar remark applies to T. This enables us to argue along the lines of the proof of Theorem 2.3 (with S, T replacing G, H, members of $\mathscr{D}(S)$ and $\mathscr{D}(T)$ replacing vertex-deleted subgraphs of G and H, branches replacing components, and (M, m), (L, m), $S-w$, $T-x$ replacing M, L, $G-w$, $H-x$), to show that S is isomorphic to a tree obtained from $S-w$ by replacing a branch isomorphic to (L, m) by one isomorphic to (M, m), and that T is isomorphic to a tree obtained from $T-x$ by a similar replacement. From this and the fact that $S-w \cong T-x$, it follows that $S \cong T$. ‖

Related Results. Bondy [**2**] has proved that if S and T are ordinary trees, and if there exists a bijection $\sigma: \Pi(S) \rightarrow \Pi(T)$ such that $S-v \cong T-\sigma(v)$ for every $v \in \Pi(S)$, then $S \cong T$. This result is stronger than Theorem 4.5. In informal language, it says that in order to reconstruct an ordinary tree T, we do not need isomorphic copies of all of its vertex-deleted subgraphs, but only isomorphic copies of those obtained by deleting peripheral vertices; that is, a certain subdeck of the deck of cards corresponding to T contains enough information to reconstruct T.

Geller and Manvel [**9**] have proved that every ordinary cactus (that is, every ordinary graph in which no two circuits have a common edge—see Fig. 11) is reconstructible. Manvel and Weinstein [**22**] have proved that an ordinary graph is reconstructible if at least one of its vertex-deleted subgraphs

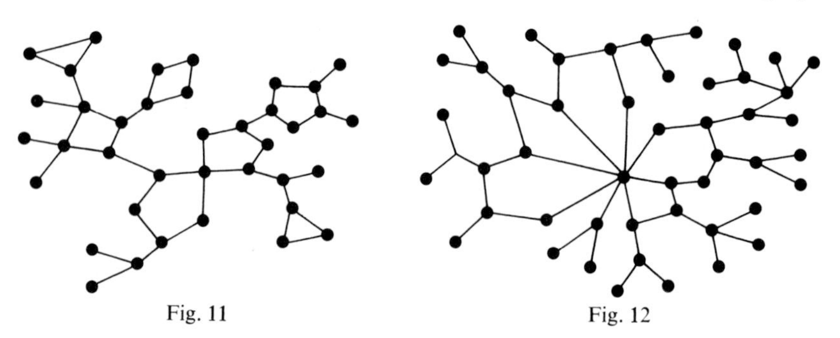

Fig. 11

Fig. 12

is a forest (see Fig. 12). It is easy to see that these results and Theorem 2.3 together imply that an ordinary graph G is reconstructible if

$$|E(G)| \leqslant |V(G)|+1.$$

This perhaps challenges us to try to prove reconstructibility of a graph G if $|E(G)| = |V(G)|+2$, and then if $|E(G)| = |V(G)|+3$, and so forth, as far as we can go. Possibly such an investigation might not justify its existence if it merely results in increasingly long and cumbersome arguments to hack out successive cases, but it might have some merit if it helps to develop and sharpen reconstruction-theoretic techniques in a useful way.

The proof of reconstructibility of ordinary trees depends on their having, in some sense, recognizable centers. Since the interrelation of the blocks and cut-vertices of a finite separable graph can be described by a tree, which again has a center, some of the work on reconstructibility of separable graphs mentioned in Section 2 makes use of this center, and thus has features in common with the proof of Theorem 4.5.

5. Edge-reconstruction

We define an **edge-hypomorphism** of a graph G onto a graph H to be a bijection $\sigma: E(G) \rightarrow E(H)$ such that $G-e \cong H-\sigma(e)$ for every $e \in E(G)$. Two graphs G and H are called **edge-hypomorphic** if there exists an edge-hypomorphism of G onto H, and a graph G is **edge-reconstructible** if every graph which is edge-hypomorphic to G is isomorphic to G. Thus the four graphs in Figs. 3 and 4 are not edge-reconstructible. The edge-reconstruction conjecture, previously stated in Section 1, can now be restated in the equivalent form: *every graph with at least four edges is edge-reconstructible.*

Intuitively, the edge-reconstruction conjecture seems even more likely to be true than the reconstruction conjecture, since a "typical" graph has more edge-deleted than vertex-deleted subgraphs, and the edge-deleted subgraphs tend to be more nearly the whole graph. For example, the edge-reconstruction conjecture envisages reconstructing the graph of Fig. 1 from the deck of cards in Fig. 13, which contains more information than the one in Fig. 2. In fact, the reconstruction conjecture can be shown to imply the edge-reconstruction conjecture by using either of the following facts:

(*i*) *a graph G with at least four edges is edge-reconstructible if and only if its line-graph L(G) is reconstructible* (see **[13]**, **[16]**);

(*ii*) *every reconstructible graph with at least four edges and no isolated vertices is edge-reconstructible* (see **[6]**, **[10]**).

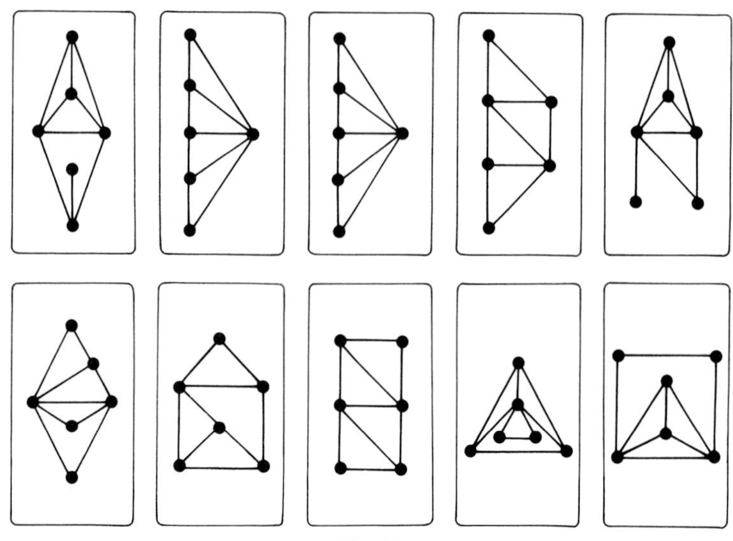

Fig. 13

The proof of (*i*) uses the theorem of Whitney [**34**] that the isomorphism class of $L(G)$ determines that of G, if G is assumed to be connected and $|E(G)| \geqslant 4$ (see Section 3 of Chapter 10). The proof of (*ii*) involves showing how to determine up to isomorphism the vertex-deleted subgraphs of the graph when we know its edge-deleted subgraphs up to isomorphism.

Striking progress on the edge-reconstruction conjecture has recently been made by Lovász [**19**] and Müller [**23**]; the former paper proved the edge-reconstruction conjecture true for graphs with p vertices and more than $\frac{1}{4}p(p-1)$ edges, and the latter paper refined the argument to prove the conjecture for graphs with p vertices and more than $p \log p/\log 2$ edges. (This result is Corollary 5.3 below.) Since the maximum possible number of edges of a graph with p vertices is $\frac{1}{2}p(p-1)$, we might say that "almost all" graphs with p vertices have more than $p \log p/\log 2$ edges if p is large; in this sense, Müller has proved the edge-reconstruction conjecture for almost all graphs. Moreover, the proof is comparatively short and we shall now present it in full.

The group of automorphisms of a graph G will be denoted by Aut G. We shall be specifically concerned with Aut K_p, the symmetric group on p elements.

Lemma 5.1. *Suppose that a spanning subgraph G of the complete graph K_p is not edge-reconstructible. Then, for every subset A of $E(G)$ such that $|A| \equiv |E(G)|$ (modulo 2), there exists an automorphism ϕ of K_p such that $E(G \cap \phi(G)) = A$.*

(As a simple exercise, verify Lemma 5.1 directly when $p = 4$ in each of the four cases in which G is isomorphic to one of the four graphs in Figs. 3 and 4.)

Proof. The conclusion of Lemma 5.1 is clearly true if $E(G) = \varnothing$; we shall therefore assume that $E(G) \neq \varnothing$.

Since G is not edge-reconstructible, there exists a graph H' which is edge-hypomorphic, but not isomorphic, to G. Since any edge-deleted subgraph of H' is isomorphic to an edge-deleted subgraph of G, it follows that $|V(H')| = |V(G)| = p$, and so we can select a spanning subgraph H of K_p such that $H \cong H'$. Clearly H is edge-hypomorphic, but not isomorphic, to G.

Let σ be an edge-hypomorphism of G onto H, and let $|E(G)| = q$; then $|E(H)| = q$ also, since $\sigma: E(G) \to E(H)$ is bijective. Let P be a spanning subgraph of K_p such that $|E(P)| = q$, and let D be a subset of $E(P)$ such that $|D| \equiv q$ (modulo 2).

Let $\mathscr{T}_G$ denote the set of all ordered pairs (ϕ, S) such that ϕ is an automorphism of K_p, S is a set of edges, and $D \subseteq S \subseteq E(P \cap \phi(G))$. We shall name various subsets of $\mathscr{T}_G$ as follows. For any fixed set S_0 such that $D \subseteq S_0 \subseteq E(P)$, let $\mathscr{T}_G(-, S_0)$ denote the set of all ordered pairs belonging to $\mathscr{T}_G$ whose second component is S_0. Let $\mathscr{T}_G(-, \equiv q)$ denote the set of all elements (ϕ, S) of $\mathscr{T}_G$ such that $|S| \equiv q$ (modulo 2), and let $\mathscr{T}_G(-, \not\equiv q)$ denote the set of all elements (ϕ, S) of $\mathscr{T}_G$ such that $|S| \not\equiv q$ (modulo 2). For any fixed automorphism ϕ_0 of K_p, let $\mathscr{T}_G(\phi_0, -)$ denote the set of all ordered pairs belonging to $\mathscr{T}_G$ whose first component is ϕ_0, let $\mathscr{T}_G(\phi_0, \equiv q)$ denote the set of all elements (ϕ_0, S) of $\mathscr{T}_G(\phi_0, -)$ such that $|S| \equiv q$ (modulo 2), and let $\mathscr{T}_G(\phi_0, \not\equiv q)$ denote the set of all elements (ϕ_0, S) of $\mathscr{T}_G(\phi_0, -)$ such that $|S| \not\equiv q$ (modulo 2). Symbols such as $\mathscr{T}_H$, $\mathscr{T}_H(-, S_0)$, etc. are defined analogously.

Let $\mu(G, P)$ denote the number of automorphisms ϕ of K_p such that $\phi(G) = P$, and let $\nu(G, P, D)$ denote the number of automorphisms ϕ of K_p such that $E(P \cap \phi(G)) = D$. Let $\mu(H, P)$ and $\nu(H, P, D)$ be analogously defined.

Consider a fixed set S_0 such that

$$D \subseteq S_0 \subsetneqq E(P). \tag{2}$$

For any spanning subgraph J of K_p, let $\gamma(J)$ denote the number of automorphisms ψ of K_p such that $\psi(S_0) \subseteq E(J)$. For any automorphism ϕ of K_p, $(\phi, S_0) \in \mathscr{T}_G$ if and only if $D \subseteq S_0 \subseteq E(P \cap \phi(G))$, which, in view of (2), occurs if and only if $S_0 \subseteq E(\phi(G))$—that is, if and only if $\phi^{-1}(S_0) \subseteq E(G)$. Hence $|\mathscr{T}_G(-, S_0)|$ is equal to the number of automorphisms ϕ of K_p such that $\phi^{-1}(S_0) \subseteq E(G)$; that is,

$$|\mathscr{T}_G(-, S_0)| = \gamma(G). \tag{3}$$

Similarly,

$$|\mathscr{T}_H(-, S_0)| = \gamma(H). \tag{4}$$

If $\psi \in \text{Aut } K_p$, and if $\psi(S_0) \subseteq E(G)$, then $\psi(S_0) \subseteq E(G-e)$ for exactly $q-|S_0|$ edges e of G; and so

$$\gamma(G) = \frac{1}{q-|S_0|} \sum_{e \in E(G)} \gamma(G-e). \tag{5}$$

Similarly,

$$\gamma(H) = \frac{1}{q-|S_0|} \sum_{f \in E(H)} \gamma(H-f). \tag{6}$$

It is easily seen that $\gamma(J) = \gamma(J')$ if J, J' are isomorphic spanning subgraphs of K_p. (For, select any isomorphism of J onto J', and extend it to an automorphism β of K_p. Then $\psi \to \beta \circ \psi$ is clearly a bijection of

$$\{\psi \in \text{Aut } K_p : \psi(S_0) \subseteq E(J)\} \text{ onto } \{\psi \in \text{Aut } K_p : \psi(S_0) \subseteq E(J')\},$$

and so these two sets have the same cardinality.) Therefore $\gamma(G-e) = \gamma(H-\sigma(e))$ for each $e \in E(G)$, and so

$$\sum_{e \in E(G)} \gamma(G-e) = \sum_{e \in E(G)} \gamma(H-\sigma(e)) = \sum_{f \in E(H)} \gamma(H-f). \tag{7}$$

By (3), (4), (5), (6) and (7), we have

$$|\mathscr{T}_G(-, S_0)| = |\mathscr{T}_H(-, S_0)|. \tag{8}$$

We note that the proof of (8) becomes invalid when $S_0 = E(P)$ because it involves division by $q-|S_0| = |E(P)|-|S_0|$. However, since $|E(P)| = q$, we may sum (8) over all sets S_0 such that $D \subseteq S_0 \subseteq E(P)$ and $|S_0| \not\equiv q$ modulo 2), to give

$$|\mathscr{T}_G(-, \not\equiv q)| = |\mathscr{T}_H(-, \not\equiv q)|; \tag{9}$$

and summing (8) over sets S_0 such that $D \subseteq S_0 \subsetneqq E(P)$ and $|S_0| \equiv q$ (modulo 2) gives

$$|\mathscr{T}_G(-, \equiv q)| - |\mathscr{T}_G(-, E(P))| = |\mathscr{T}_H(-, \equiv q)| - |\mathscr{T}_H(-, E(P))|. \tag{10}$$

For any automorphism ϕ of K_p, $(\phi, E(P)) \in \mathscr{T}_G$ if and only if

$$D \subseteq E(P) \subseteq E(P \cap \phi(G)),$$

which occurs if and only if $\phi(G) = P$, in view of the hypotheses that

$D \subseteq E(P)$ and $|E(P)| = q = |E(G)|$. Therefore $|\mathcal{T}_G(-, E(P))| = \mu(G, P)$. Similarly $|\mathcal{T}_H(-, E(P))| = \mu(H, P)$, and so (10) may be written as

$$|\mathcal{T}_G(-, \equiv q)| - \mu(G, P) = |\mathcal{T}_H(-, \equiv q)| - \mu(H, P). \tag{11}$$

Now consider a fixed automorphism ϕ_0 of K_p. Suppose first that $E(P \cap \phi_0(G)) \neq D$. Then it is an easy exercise to see that the number of sets S of even cardinality such that $D \subseteq S \subseteq E(P \cap \phi_0(G))$ is equal to the number of sets S of odd cardinality such that $D \subseteq S \subseteq E(P \cap \phi_0(G))$. (In particular, both of these numbers are zero if $D \nsubseteq E(P \cap \phi_0(G))$.) Therefore the number of sets S of even cardinality such that $(\phi_0, S) \in \mathcal{T}_G$ is equal to the number of sets S of odd cardinality such that $(\phi_0, S) \in \mathcal{T}_G$; that is,

$$|\mathcal{T}_G(\phi_0, \equiv q)| = |\mathcal{T}_G(\phi_0, \not\equiv q)|. \tag{12}$$

Now suppose that $E(P \cap \phi_0(G)) = D$. Then D is the only set S such that $D \subseteq S \subseteq E(P \cap \phi_0(G))$, and so (ϕ_0, D) is the only element of $\mathcal{T}_G(\phi_0, -)$. Since $|D| \equiv q$ (modulo 2), it follows in this case that

$$|\mathcal{T}_G(\phi_0, \equiv q)| = 1, \ |\mathcal{T}_G(\phi_0, \not\equiv q)| = 0. \tag{13}$$

Since (13) holds for the $v(G, P, D)$ automorphisms ϕ_0 of K_p such that $E(P \cap \phi_0(G)) = D$, and since (12) holds for all other automorphisms ϕ_0 of K_p, it follows that

$$\sum_{\phi_0 \in \operatorname{Aut} K_p} |\mathcal{T}_G(\phi_0, \equiv q)| = v(G, P, D) + \sum_{\phi_0 \in \operatorname{Aut} K_p} |\mathcal{T}_G(\phi_0, \not\equiv q)|;$$

that is,

$$|\mathcal{T}_G(-, \equiv q)| = v(G, P, D) + |\mathcal{T}_G(-, \not\equiv q)|. \tag{14}$$

A similar argument shows that

$$|\mathcal{T}_H(-, \equiv q)| = v(H, P, D) + |\mathcal{T}_H(-, \not\equiv q)|. \tag{15}$$

It follows from (9), (11), (14) and (15) that

$$v(G, P, D) - v(H, P, D) = \mu(G, P) - \mu(H, P). \tag{16}$$

We have proved (16) for any spanning subgraph P of K_p with q edges, and any subset D of $E(P)$ such that $|D| \equiv q$ (modulo 2). Hence, if A is any subset of $E(G)$ such that $|A| \equiv q$ (modulo 2), then taking $P = G$, $D = A$ in (16) gives

$$v(G, G, A) - v(H, G, A) = \mu(G, G) - \mu(H, G). \tag{17}$$

Moreover, $\mu(G, G) > 0$, since $\phi(G) = G$ when ϕ is the identity automorphism of K_p, and $\mu(H, G) = 0$, since $G \ncong H$. Therefore (17) implies that

$\nu(G, G, A) > 0$; that is, there is an automorphism ϕ of K_p such that $E(G \cap \phi(G)) = A$. ‖

Note that we could have taken $P = G$ and $D = A$ throughout this proof, but this would have destroyed some symmetry between G and H, and might have made the argument harder to grasp.

We can now prove Müller's results:

Theorem 5.2 (Müller). *A graph G is edge-reconstructible if*

$$2^{|E(G)|-1} > |V(G)|!.$$

Proof. Let $|V(G)| = p$, $|E(G)| = q$, and let K_p be a complete graph such that G is a spanning subgraph of K_p (that is, K_p is obtained by adding $\binom{p}{2}-q$ edges to G). Assume that $2^{q-1} > p!$. Then $q > 0$, and so $E(G)$ has 2^{q-1} subsets A such that $|A| \equiv q$ (modulo 2). On the other hand, K_p has only $p!$ automorphisms, since these automorphisms are in one-to-one correspondence with the permutations of $V(K_p)$. Since $p! < 2^{q-1}$, it cannot be true that, for every $A \subseteq E(G)$ such that $|A| \equiv q$ (modulo 2), there exists an automorphism ϕ of K_p such that $E(G \cap \phi(G)) = A$. Therefore, G is edge-reconstructible by Lemma 5.1. ‖

Corollary 5.3. *A graph with p vertices and q edges is edge-reconstructible if* $q > (p \log p)/(\log 2)$.

Proof. Since $q > (p \log p)/(\log 2) \geqslant 0$, the graph has at least one edge, and therefore $p \geqslant 2$. Since $q \log 2 > p \log p$, it follows that $2^q > p^p \geqslant 2(p!)$, so that the graph is edge-reconstructible, by Theorem 5.2. ‖

It is also fairly easy to deduce from Lemma 5.1 that a graph G with p (>0) vertices and q edges is edge-reconstructible if $q \geqslant p+(p \log \rho_{\max}/\log 2)$, where $\rho_{\max}$ is the maximum valency of G. This is sharper than Corollary 5.3 when $\rho_{\max} \leqslant \frac{1}{2}p$.

The foregoing proof of Theorem 5.2 did not depend very heavily on the fact that G and H were graphs; the same arguments would work if, for example, they were digraphs, or if each of them was a hypergraph in which each edge is incident with exactly 3 vertices. It would not be hard to formulate and prove a generalization of Theorem 5.2 which contains these and many other variants of the theorem as special cases, but to preserve some transparency, the treatment was not thus generalized above. It might be of interest to enquire how near the inequality corresponding to $2^{|E(G)|-1} > |V(G)|!$ in such a generalization of Theorem 5.2 comes to being best possible.

6. The Reconstruction of Digraphs

This section will partially describe Stockmeyer's proof **[27]** that no close analog of the reconstruction conjecture can be true for digraphs. Let us say that two digraphs D and D' constitute a "counter-example to the reconstructibility of digraphs" if $D \not\cong D'$ but there exists a bijection $\sigma: V(D) \to V(D')$ such that $D - v \cong D' - \sigma(v)$ for every $v \in V(D)$. It is an easy exercise to find counter-examples of order p to the reconstructibility of digraphs for $p = 2$, 3, 4, and counter-examples for $p = 5, 6$ found by Beineke and Parker **[1]** are depicted in Figs. 14 and 15. More recently, Stockmeyer **[26]** found a counter-example on 8 vertices, but one could still have propounded a reconstruction conjecture for digraphs with sufficiently many vertices, until Stockmeyer **[27]** discovered, for each integer p of the form $2^k + 1$ or $2^k + 2$, a pair of tournaments which constitute a counter-example of order p to the reconstructibility of digraphs. We shall briefly describe these counter-examples.

Let k be a positive integer, and let $K = 2^k$. Observe that every integer $i \neq 0$ can be uniquely expressed as a product $2^{\text{pow}(i)}\text{odd}(i)$, where $\text{pow}(i)$ is a non-negative integer, and $\text{odd}(i)$ is an odd integer; for example,

$$\text{pow}(56) = 3,\ \text{odd}(56) = 7; \qquad \text{pow}(64) = 6,\ \text{odd}(64) = 1.$$

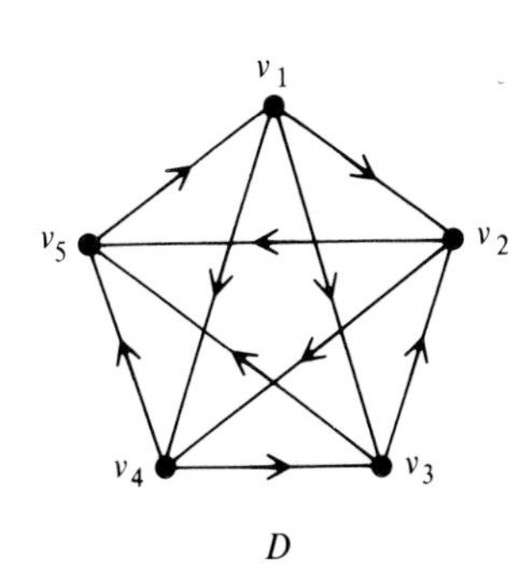

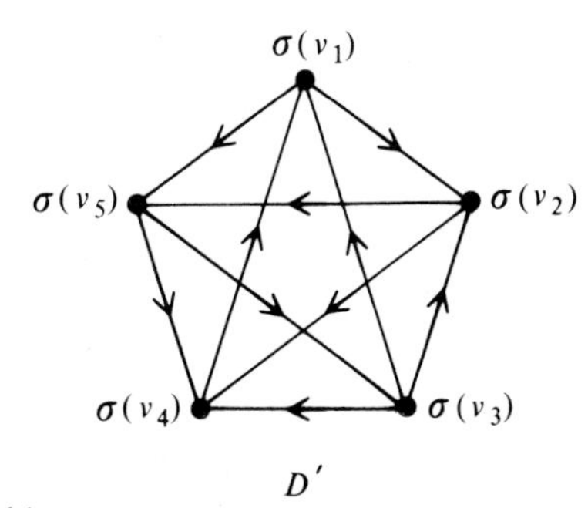

Fig. 14

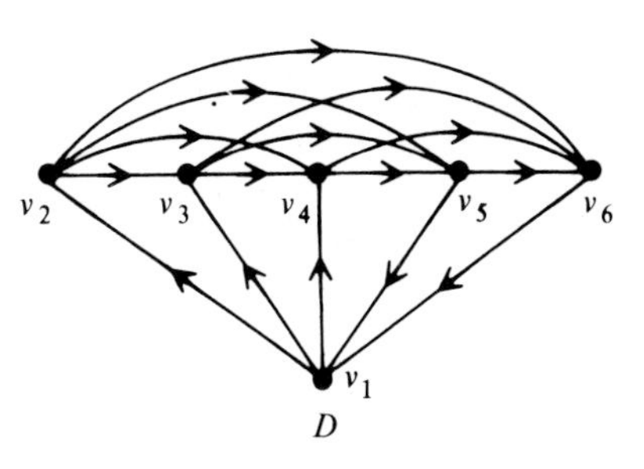

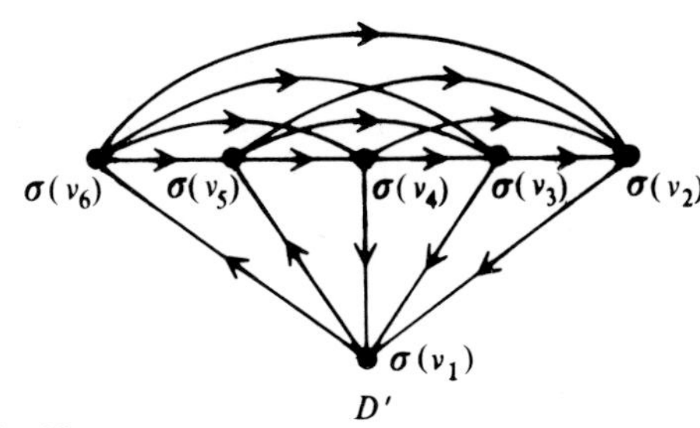

Fig. 15

Let T_{K+1} be a tournament with $K+1$ vertices $v_1, v_2, \ldots, v_K, w$ such that

(i) v_i dominates v_j (that is, T_{K+1} has an v_iv_j-arc) if and only if $\text{odd}(i-j) \equiv 1$ (modulo 4), for $i, j = 1, 2, \ldots, K$ $(i \neq j)$;

(ii) w dominates v_i if i is odd, and v_i dominates w if i is even, for $i = 1, 2, \ldots, K$.

Since $\text{odd}(p-q) = -\text{odd}(q-p)$, rule ($i$) determines exactly one direction of orientation for each arc joining two of $v_1, v_2, \ldots, v_K$. Let U_{K+1} be a tournament with $K+1$ vertices $v_1, v_2, \ldots, v_K, z$, and with arcs specified by the rules (i) and

(iii) z dominates v_i if i is even, and v_i dominates z if i is odd, for $i = 1, 2, \ldots, K$.

Then clearly $T_{K+1} - w \cong U_{K+1} - z$; and Stockmeyer [**27**] showed that $T_{K+1} - v_i \cong U_{K+1} - v_{K+1-i}$, for $i = 1, 2, \ldots, K$, and $T_{K+1} \ncong U_{K+1}$, thus proving that T_{K+1}, U_{K+1} constitute a counter-example to the reconstructibility of digraphs. Let R_{K+2} be a tournament with $K+2$ vertices $v_1, v_2, \ldots, v_K, w, z$, in which w dominates z and the remaining arcs are specified by the rules (i), (ii), (iii). Let S_{K+2} be a tournament with the same $K+2$ vertices in which z dominates w and the remaining arcs are specified by (i), (ii), (iii). Then clearly $R_{K+2} - w \cong S_{K+2} - w$ and $R_{K+2} - z \cong S_{K+2} - z$; and Stockmeyer showed that $R_{K+2} - v_i \cong S_{K+2} - v_{K+1-i}$, for $i = 1, 2, \ldots, K$, and $R_{K+2} \ncong S_{K+2}$, thus proving that R_{K+2}, S_{K+2} constitute a counter-example to the reconstructibility of digraphs. (The isomorphisms between $T_{K+1} - v_i$ and $U_{K+1} - v_{K+1-i}$, and between $R_{K+2} - v_i$ and $S_{K+2} - v_{K+1-i}$, have a somewhat intricate and non-obvious appearance; we omit the full details.)

All of the counter-examples of [**1**], [**26**] and [**27**] are pairs of tournaments; but more recently Stockmeyer [**28**] reported the discovery of further counter-examples to the reconstructibility of digraphs, including, for infinitely many values of p, counter-examples of order p which are not pairs of tournaments.

Define the **valency-pair** of a vertex v of a digraph to be the ordered pair $(\rho_{\text{out}}(v), \rho_{\text{in}}(v))$, where $\rho_{\text{out}}(v)$ and $\rho_{\text{in}}(v)$ are the out-valency and in-valency of v respectively. Define the **valency-pair sequence** of a digraph D to be the sequence of $|V(D)|$ ordered pairs of numbers obtained by listing the valency-pairs of the vertices of D in dictionary order (for the sake of definiteness). Manvel [**21**] has proved that, if D, D' are digraphs with at least five vertices, and if there exists a bijection $\sigma: V(D) \rightarrow V(D')$ such that $D - v \cong D' - \sigma(v)$ for each $v \in V(D)$, then D and D' have the same valency-pair sequence. This provides an analog for digraphs of Corollary 3.2, although its proof is more difficult. Informally it says that, although a digraph cannot necessarily be reconstructed from its vertex-deleted subgraphs, its valency-pair sequence *can* be reconstructed therefrom.

There may likewise be scope for investigating the possible existence of analogs of Theorems 3.1 and 3.4 and Corollary 3.5 for digraphs with more than some appropriate number of vertices. Presumably a first step might be to check any conjecture of this sort against Stockmeyer's counter-examples.

Harary and Palmer [**14**] have shown that a tournament with at least five vertices is reconstructible if it is not strongly connected. By analogy with Theorem 2.3, one might enquire whether, more generally, a digraph with sufficiently many vertices is necessarily reconstructible if it is not strongly connected.

7. The Reconstruction of Infinite Graphs

Two hypomorphic infinite graphs need not be isomorphic, as the following counter-example of Fisher, Graham and Harary [**8**], and Nešetřil [**24**] shows. Let T_α be a tree in which the valency of every vertex is the same infinite cardinal number α. For any cardinal number β, let βT_α denote a graph which is the union of β disjoint subgraphs isomorphic to T_α. Since

$$|V(T_\alpha)| = \alpha = |V(2T_\alpha)|,$$

there exist bijections of $V(T_\alpha)$ onto $V(2T_\alpha)$. Moreover, *every* such bijection is a hypomorphism of T_α onto $2T_\alpha$, because $T_\alpha - v \cong \alpha T_\alpha \cong (2T_\alpha) - w$, for every $v \in V(T_\alpha)$ and every $w \in V(2T_\alpha)$. Hence T_α, $2T_\alpha$ are hypomorphic infinite graphs which are obviously not isomorphic.

This counter-example seems to depend fairly heavily on the fact that $2T_\alpha$ is disconnected and that the graphs concerned are not locally finite. One is, therefore, led to wonder whether two hypomorphic infinite graphs may have to be isomorphic if they satisfy suitable restrictions, particularly restrictions involving connectedness and/or local finiteness. For instance, the following questions might naturally come to mind:

(*i*) are two connected hypomorphic infinite graphs necessarily isomorphic?

(*ii*) are two locally-finite hypomorphic infinite graphs necessarily isomorphic?

(*iii*) are two locally-finite connected hypomorphic infinite graphs necessarily isomorphic?

(*iv*) are two hypomorphic infinite trees necessarily isomorphic?

Although T_α is both connected and a tree, $2T_\alpha$ is neither of these things, and so the pair T_α, $2T_\alpha$ does not yield a negative answer to (*i*) or (*iv*). However, since hypomorphic graphs clearly have hypomorphic complements, the complements of T_α and $2T_\alpha$ yield a quick negative answer to (*i*). A more complicated counter-example of Fisher **[7]** shows that, even if two hypomorphic infinite graphs G and H *and their complements* are all connected, G need not be isomorphic to H. A counter-example of Harary, Schwenk and Scott **[15]** shows that the answer to (*ii*) is negative, even if we require both of the graphs concerned to be forests. On the other hand, (*iii*) and (*iv*) seem to be open questions. Results of Bondy and Hemminger **[4]** and Thomassen **[30]**, when taken together, say that the answer to (*iv*) is affirmative for locally-finite trees which do not contain an infinite number of disjoint infinite paths.

There is often more than one way of extending to infinite graphs a problem about finite graphs. For example, if G and H denote finite graphs, the statements "$G \cong H$" and "each of G, H is isomorphic to a subgraph of the other" are clearly equivalent. Consequently, the formulation (RC1) of the reconstruction conjecture can be translated into the equivalent statement:

(RC1*). *If two ordinary (finite) graphs are hypomorphic then each of them is isomorphic to a subgraph of the other.*

On the other hand, the analogs of (RC1) and (RC1*) for infinite graphs are essentially different statements, because one can easily find two non-isomorphic infinite graphs, each of which is isomorphic to a subgraph of the other. Two such graphs are depicted in Figs. 16 and 17. Thus the following is an interesting variant of the reconstruction conjecture which might be true for both finite and infinite graphs.

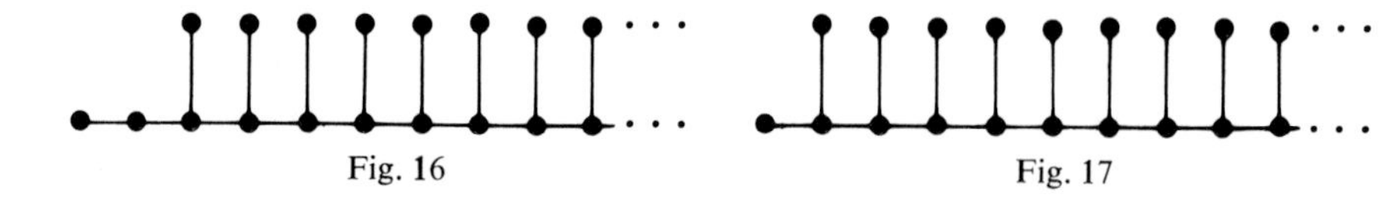

Fig. 16 Fig. 17

Halin's Conjecture. *If G and H are hypomorphic graphs with at least three vertices, then G is isomorphic to a subgraph of H, and H is isomorphic to a subgraph of G.*

No counter-examples to this conjecture are known. Every known counter-example to the more obvious analog of the reconstruction conjecture for infinite graphs involves a pair of graphs, each of which is isomorphic to a subgraph of the other. For example, each of T_α, $2T_\alpha$ is isomorphic to a subgraph of the other, since T_α is isomorphic to either component of $2T_\alpha$, and $2T_\alpha$ is isomorphic to any edge-deleted subgraph of T_α.

Bondy and Hemminger [**5**] have proved Halin's conjecture for the case in which G and H are both disconnected, thereby providing an analog of Theorem 2.3 for infinite graphs.

The analog for infinite graphs of the edge-reconstruction conjecture seems to escape the more obvious types of counter-example (for example, T_α and $2T_\alpha$ do not constitute a counter-example to it), but Thomassen [**29**] has recently established its falsity by an ingenious class of counter-examples.

8. Further Results

Some of the proofs in this chapter show that counting arguments play a large part in the theory of reconstruction of (finite) graphs, and enumerative combinatorics often leads to the use of polynomials or power series in one or more indeterminates, whose coefficients count appropriate objects.

Tutte [**32**] has recently made powerful and wide-ranging use of such techniques in reconstruction theory, thus introducing what may be one of the most exciting current lines of development in the subject. Among the concepts featuring in it are the chromatic and dichromatic polynomials of a finite graph, which have played a large part elsewhere in graph theory, especially in work related to graph-coloring and the four-color problem. (Actually, in place of the dichromatic polynomial, [**32**] uses a related polynomial called the "rank polynomial" of a graph G, which is equal to $x^{|V(G)|}D_G(x^{-1}, y)$ where $D_G(x, y)$ is the dichromatic polynomial of G.) The chromatic polynomial $P_G(x)$ of a graph G was defined in Chapter 1. The dichromatic polynomial $D_G(x, y)$ of G is the polynomial

$$y^{-|V(G)|}\sum_F x^{c(F)}y^{|E(F)|+c(F)}$$

in two indeterminates x and y, where the summation is over all spanning subgraphs F of G, and $c(F)$ denotes the number of components of F. It is easily proved that

$$P_G(x) = (-1)^{|V(G)|}D_G(-x, -1). \tag{18}$$

In [**32**], Tutte proved that, if G and H are hypomorphic ordinary graphs, then $D_G(x, y) = D_H(x, y)$, and therefore, by (18), $P_G(x) = P_H(x)$. Informally this says that the chromatic and dichromatic polynomials of an ordinary graph can be determined from its deck of cards. An earlier proof of this fact was also (in effect) given by Tutte in [**31**]. From $D_G(x, y) = D_H(x, y)$, Tutte [**32**] deduced that, for every integer $k \geqslant |V(G)| = |V(H)|$, G and H have equal numbers of non-separable spanning subgraphs with exactly k edges. In particular, by taking $k = |V(G)| = |V(H)|$, we see that they have equal numbers of Hamiltonian circuits, which is Theorem 2.5 (*ii*). The proofs of the other two parts of Theorem 2.5 also use polynomials associated with graphs. From parts (*i*) and (*ii*) of Theorem 2.5, it can be deduced that the adjacency matrices of G and H have the same determinant and the same characteristic polynomial. (This implication was first observed by Pouzet [**25**], but a proof may also be found in [**32**]; see also Section 12 of Chapter 11.) Tutte further showed that a finite graph is reconstructible if its characteristic polynomial is irreducible (that is, not a product of two polynomials of non-zero degree with integer coefficients).

Some other reconstruction-theoretic topics which have been considered to varying extents include the following:

(*i*) The reconstruction of a graph G from information other than a knowledge, up to isomorphism, of all of its vertex-deleted subgraphs or all of its edge-deleted subgraphs. For instance, we might only be allowed to know which isomorphism classes of graphs include at least one vertex-deleted subgraph of G—that is, informally speaking, the deck of cards corresponding to G is reduced by omitting duplicates until no two cards show isomorphic graphs, and we are asked to reconstruct G from this reduced deck. Alternatively, we might be given, up to isomorphism, all the subgraphs obtained from G by deleting r vertices, for some specified number $r > 1$. See also the reference to [**2**] in Section 4.

(*ii*) The reconstruction of graphs, some or all of whose vertices and/or edges are colored, and its application to the reconstruction of uncolored graphs.

(*iii*) Given a deck of cards each of which has a graph drawn on it, how can we decide whether this could have arisen from some graph in the manner described in Section 1?

(*iv*) The reconstruction of objects other than graphs and digraphs.

Some further discussion of these topics, references to relevant literature, and a list of unsolved problems can be found in [**6**].

References

1. L. W. Beineke and E. T. Parker, On nonreconstructable tournaments, *J. Combinatorial Theory* **9** (1970), 324–326; *MR*43#6135.
2. J. A. Bondy, On Kelly's congruence theorem for trees, *Proc. Cambridge Phil. Soc.* **65** (1969), 387–397; *MR*41#5238.

3. J. A. Bondy, On Ulam's conjecture for separable graphs, *Pacific J. Math.* **31** (1969), 281–288; *MR***41**#6708.
4. J. A. Bondy and R. L. Hemminger, Reconstructing infinite graphs, *Pacific J. Math.* **52** (1974), 331–340; *MR***50**#9646.
5. J. A. Bondy and R. L. Hemminger, Almost reconstructing infinite graphs, in *Recent Advances in Graph Theory* (ed. M. Fiedler), Academia, Prague, 1975, pp. 69–73.
6. J. A. Bondy and R. L. Hemminger, Graph reconstruction—a survey, *J. Graph Theory* **1** (1977), 227–268.
7. J. Fisher, A counterexample to the countable version of a conjecture of Ulam, *J. Combinatorial Theory* **7** (1969), 364–365; *MR* **41**#6712.
8. J. Fisher, R. L. Graham and F. Harary, A simpler counterexample to the reconstruction conjecture for denumerable graphs, *J. Combinatorial Theory* (*B*) **12** (1972), 203–204; *MR***45**#5007.
9. D. Geller and B. Manvel, Reconstruction of cacti, *Canad. J. Math.* **21** (1969), 1354–1360; *MR***40**#5476.
10. D. L. Greenwell, Reconstructing graphs, *Proc. Amer. Math. Soc.* **30** (1971), 431–433; *MR***44**#3908.
11. D. L. Greenwell and R. L. Hemminger, Reconstructing graphs, in *The Many Facets of Graph Theory*, Lecture Notes in Mathematics **110** (ed. G. Chartrand and S. F. Kapoor), Springer-Verlag, Berlin, Heidelberg and New York, 1969, pp. 91–114; *MR***40**#5479.
12. F. Harary, On the reconstruction of a graph from a collection of subgraphs, in *Theory of Graphs and its Applications* (ed. M. Fiedler), Czechoslovak Academy of Sciences, Prague, 1964, pp. 47–52; reprinted by Academic Press, New York; *MR***30**#5296.
13. F. Harary and E. Palmer, A note on similar points and similar lines of a graph, *Rev. Roumaine Math. Pures et Appl.* **10** (1965), 1489–1492; *MR***33**#5511.
14. F. Harary and E. Palmer, On the problem of reconstructing a tournament from sub-tournaments, *Monatsh. Math.* **71** (1967), 14–23; *MR***35**#86.
15. F. Harary, A. J. Schwenk and R. L. Scott, On the reconstruction of countable forests, *Publ. Math. Inst.* (*Beograd*) **13** (1972), 39–42; *MR***48**#5894.
16. R. L. Hemminger, On reconstructing a graph, *Proc. Amer. Math. Soc.* **20** (1969), 185–187; *MR***38**#1019.
17. P. J. Kelly, A congruence theorem for trees, *Pacific J. Math.* **7** (1957), 961–968; *MR***19**–442.
18. V. Krishnamoorthy and K. R. Parasarathy, On the reconstruction conjecture for separable graphs (preprint).
19. L. Lovász, A note on the line reconstruction problem, *J. Combinatorial Theory* (*B*) **13** (1972), 309–310; *MR***46**#8913.
20. B. Manvel, On reconstruction of graphs, Ph.D. thesis, University of Michigan, 1970.
21. B. Manvel, Reconstructing the degree-pair sequence of a digraph, *J. Combinatorial Theory* (*B*) **15** (1973), 18–31; *MR***48**#1963.
22. B. Manvel and J. M. Weinstein, Nearly acyclic graphs are reconstructible, *J. Graph Theory* **2** (1978), 25–39.
23. V. Müller, The edge reconstruction hypothesis is true for graphs with more than $n \log_2 n$ edges, *J. Combinatorial Theory* (*B*) **22** (1977), 281–283.
24. J. Nešetřil, On reconstructing of infinite forests, *Comment. Univ. Math. Carolinae* **13** (1972), 503–510; *MR***51**#5413.
25. M. Pouzet, Note sur le problème de Ulam, *J. Combinatorial Theory* (*B*) (to appear).
26. P. K. Stockmeyer, The reconstruction conjecture for tournaments, in *Proceedings of the Sixth Southeastern Conference on Combinatorics, Graph Theory and Computing* (ed. F. Hoffman *et al.*), Congressus Numerantium **XIV**, Utilitas Mathematica, Winnipeg, 1975, pp. 561–566; *MR***52**#13475.
27. P. K. Stockmeyer, The falsity of the reconstruction conjecture for tournaments, *J. Graph Theory* **1** (1977), 19–25.
28. P. K. Stockmeyer, New counterexamples to the digraph reconstruction conjecture, *Notices Amer. Math. Soc.* **23** (1976), Abstract A-654.

29. C. Thomassen, Counterexamples to the edge reconstruction conjecture for infinite graphs, University of Waterloo Research Report CORR 76-19 (1976).
30. C. Thomassen, Reconstructing 1-coherent locally finite trees. Aarhus Universitet Matematisk Institute Preprint Series 1976/77, No. 19 (1977).
31. W. T. Tutte, On dichromatic polynomials, *J. Combinatorial Theory* **2** (1967), 301–320; *MR*36#6320.
32. W. T. Tutte, All the king's horses, in *Graph Theory and Related Topics* (ed. J. A. Bondy and U. S. R. Murty), Academic Press, New York (to appear).
33. J. M. Weinstein, Reconstructing colored graphs, *Pacific J. Math.* **57** (1975), 307–314; *MR*52#5465.
34. H. Whitney, Congruent graphs and the connectivity of graphs, *Amer. J. Math.* **54** (1932), 150–168.

9
Minimax Theorems in Graph Theory

D. R. WOODALL

1. Introduction

It is arguable that the most important theorem in graph theory is Menger's theorem, which, in one of its many variant forms, may be stated as follows:

Menger's Theorem. *Let X and Y be sets of vertices in a graph G, and let k be a positive integer; then there exist k disjoint XY-paths in G if and only if every XY-separating set of vertices of G contains at least k vertices.*

(We shall give precise definitions of these terms in the next section.) As it stands, this gives a necessary and sufficient condition for the existence of k disjoint XY-paths. But it is clear that we can restate Menger's theorem in the form: *the largest number of disjoint XY-paths in G is equal to the smallest number of vertices of G in an XY-separating set*. In this form, it is a "minimax theorem": a theorem which states that the minimum of one thing is equal to the maximum of another. Many of the necessary and sufficient conditions of graph theory can be rephrased as minimax theorems. And every minimax theorem can be rephrased as a necessary and sufficient condition. For the theorem *the maximum value of x is equal to the minimum value of y* is the same as *for every real number k, there exists an x that is at least k if and only if every y is at least k*.

There are frequently several advantages to be obtained from rephrasing theorems as minimax theorems. For example, there is often a very close connection between a minimax theorem "max x = min y" and an algorithm for constructing max x. Typically, the algorithm starts by choosing an x. There is then an iterative step consisting of a construction that will lead to a larger x *or* to a y with $x = y$. The theorem shows that in the latter case we have found the common value of max x and min y. If (as is usual) x and y are restricted to integer values, this situation will be reached after a finite number of steps. Thus the theorem shows that the algorithm terminates. But, conversely, the algorithm can usually be used to give a proof of the theorem. For applying the construction of the algorithm to a *maximal* x must yield a y with $x = y$, so that max $x \geqslant$ min y; and in most of these theorems it is obvious that max $x \leqslant$ min y. (However, a proof obtained in this way from a good algorithm may not be particularly direct or simple; conversely, a direct and simple proof will almost certainly yield an algorithm of sorts, but this algorithm may not be particularly good or easy to apply.) As illustrations, we shall exhibit "proofs by algorithm" for Menger's theorem (in Section 2), the max-flow min-cut theorem (Section 3), and the maximal-matching and optimal-assignment theorems for a bipartite graph (Section 4).

Another advantage of rephrasing a theorem in minimax form is that it then fits into a pattern with other theorems. I believe that every minimax theorem in graph theory can be derived as a consequence of the duality theorem of linear programming. As in the case of a "proof by algorithm", this is not necessarily a very sensible way of obtaining the result: the deduction from the duality theorem may be long and indirect, whereas much shorter and more direct proofs are often available. However, this does show that all of these theorems fit into some sort of pattern, and thereby provides the subject with some degree of unity. We shall discuss the duality theorem of linear programming in Section 6, where, as an illustration of the method, we shall use it to prove two minimax theorems about bipartite graphs.

Finally, there is another pattern that many minimax theorems fall into. Many of these theorems say essentially that a certain family of sets has one or other of two properties, which I have called elsewhere the *Menger property* and the *König property*, since Menger's theorem and König's theorem are early theorems of these two types. In many cases it turns out that not only the original family of sets, but many other families derived from it, have the relevant property. In the case of the König property, this phenomenon is partly explained by the perfect-graph theorem, and there exists a conjecture (the strong version of Berge's perfect-graph conjecture) that, if true, will almost completely explain it. In the case of the Menger property, matters are not so far advanced. Sections 7–9 of this chapter are devoted to a discussion of these two properties, with many examples drawn from graph theory.

So the rest of this chapter divides naturally into three parts. The first (and longest) part, consisting of Sections 2–5, is devoted to the classic minimax theorems of graph theory (Menger's theorem, the max-flow min-cut theorem, König's theorem and Tutte's factor theorems, and related results), with a strong emphasis on algorithmic proofs. The second part, consisting just of Section 6, is about applications of the duality theorem of linear programming. And the final part, Sections 7–9, is about the Menger and König properties.

I should like to express my thanks to C. J. H. McDiarmid and P. D. Seymour for their useful comments on an earlier draft of this chapter.

2. Menger's Theorem

Let G be a graph or digraph. If x and y are vertices of G, an ***xy*-path** is a path (or directed path) from x to y in G. If X and Y are sets of vertices, an ***XY*-path** is a path (or directed path) from some $x \in X$ to some $y \in Y$ passing through no other vertex of X or Y. (If $v \in X \cap Y$ then v itself is an XY-path.) If X, Y and Z are sets of vertices, then Z **separates** Y from X, or is an ***XY*-separating set**, if every XY-path contains a vertex of Z. (In particular, X and Y are both XY-separating sets.) The version of Menger's theorem stated in the previous section was the following, proved by Menger [**29**] in 1927:

Theorem 2.1. *Let k be a positive integer, and let X and Y be sets of vertices in a graph or digraph G. Then there are k disjoint XY-paths in G if and only if every XY-separating set contains at least k vertices—that is, if and only if there remains an XY-path in G whenever fewer than k vertices of G are removed.*

Proof. "Only if" is clear: if there exist k disjoint XY-paths, then every separating set must contain at least one vertex of each of these paths.

It remains to prove "if". We shall give two proofs of this. The first is the shortest and simplest proof that I know, but would not yield a good algorithm for constructing the paths; the second proof is algorithmic.

First Proof (Pym [**34**], 1969). We prove the result by induction on $a(G) := |V(G) \cup E(G)|$.† We may clearly suppose without loss of generality that there is an XY-separating set of k vertices (increasing the value of k if necessary to make this true), but no such set of $k-1$ vertices. There are two cases to consider:

† Throughout this chapter the symbol := indicates that the equation in which it occurs acts as the definition of the expression on the left-hand side of the equation.

Case 1. There is an XY-separating set $Z \neq X$ or Y such that $|Z| = k$. Let G_{XZ} (respectively, G_{ZY}) be the subgraph of G consisting of all the vertices and edges of all the XZ-paths (ZY-paths) in G. Clearly $G_{XZ} \cap G_{ZY} \subseteq Z$, since Z is a separating set. Since Z is a *minimal* separating set, and $Z \neq Y$, we have $Z \not\supseteq Y$; thus there is a vertex in $Y - Z$. Since this vertex cannot be in G_{XZ}, we have $a(G_{XZ}) < a(G)$. Moreover, Z is not separated from X by any set of fewer than k vertices in G_{XZ}, since an XZ-separating set in G_{XZ} is an XY-separating set in G. So, by the induction hypothesis, there are k disjoint XZ-paths in G_{XZ}. Similarly, there are k disjoint ZY-paths in G_{ZY}. But since $|Z| = k$, these paths can be joined together in a one–one manner on the vertices in Z to form k disjoint XY-paths in G. This completes the discussion of Case 1.

Case 2. Every XY-separating set of k vertices coincides with X or Y. Suppose without loss of generality that one coincides with X, so that $|X| = k$. If $X \subseteq Y$, the result is obvious; so suppose that there is a vertex $x \in X - Y$. The set $X - \{x\}$ is not a separating set, so there is an edge xy with $y \notin X$. Delete such an edge (but no vertices) from G to form a graph G' with $a(G') < a(G)$.

If every XY-separating set in G' has at least k vertices, then the result follows by the induction hypothesis. Otherwise, let Z be a separating set in G' with $k-1$ vertices. Then $Z \cup \{x\}$ and $Z \cup \{y\}$ are separating sets in G. By the hypotheses of the theorem and of Case 2,

$$|Z \cup \{x\}| = |Z \cup \{y\}| = k, \quad Z \cup \{x\} = X \quad \text{and} \quad Z \cup \{y\} = Y;$$

the result is now immediate. ||

Second proof (Ore [**31**], 1962). We prove the result when $X \cap Y = \varnothing$; it clearly then follows when $X \cap Y \neq \varnothing$, since every vertex of $X \cap Y$ is an XY-path on its own.

We present the proof in the form of an algorithm. Readers familiar with the max-flow min-cut algorithm, which we shall present in the next section, will recognize the similarity between these two algorithms. The present algorithm starts with a set Π_l of l disjoint XY-paths $P_1, \ldots, P_l$ for some non-negative integer l (such as $l = 0$).

Iterative step. Form two sets S and T of vertices recursively, as follows:

(*i*) put all vertices of X into both S and T;

(*ii*) if $v \in S$ and $vw \in E(G)$, and if vw is not in any of the paths in Π_l and w is not in any of the paths in Π_l, then put w into both S and T;

(*iii*) if $v \in S$ and $vw \in E(G)$, and if vw is not in any of the paths in Π_l and w is in a path in Π_l, then put w into T only (but leave it in S if it is in S already);

(*iv*) if $v \in T$ and wv is in one of the paths in Π_l, so that w precedes v when the path is traversed from X to Y, then put w into S and T.

The construction of S and T continues until it terminates, as it must do after a finite number of steps. There are then two cases to consider:

Case 1. S contains some vertex in Y. Then, by the construction of S, G contains a *reverse cross-sequence* for Π_l—that is, an XY-trial Q (allowing repeated vertices, but not repeated edges) whose only intersection with the paths in Π_l consists of segments of at least one edge that it traverses in the opposite direction from the XY-direction in Π_l (see Fig. 1). (Note that, if G is a digraph, then Q is not a directed trail, since, although it traverses edges outside Π_l in the correct direction (the direction of the arrow), it traverses edges in Π_l in the reverse direction.) In this case there is an obvious way of enlarging Π_l into Π_{l+1} (see Fig. 2). Do this, and go back to the iterative step, with Π_{l+1} in place of Π_l.

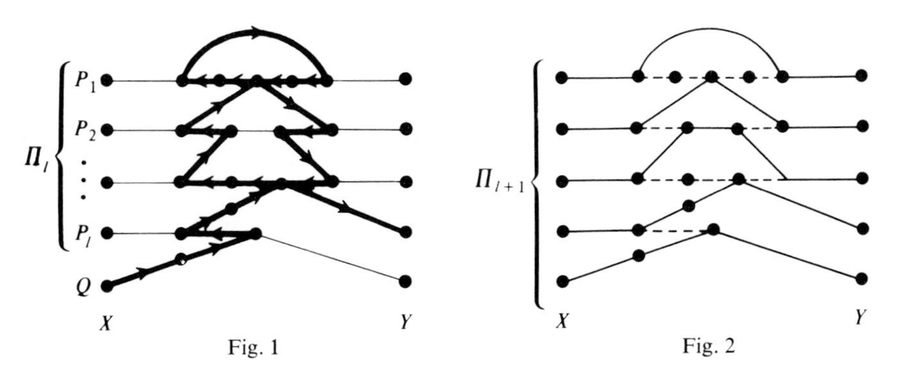

Fig. 1

Fig. 2

Case 2. S does not contain any vertex in Y. In this case, let v_i be the last vertex along P_i that is in T $(i = 1, \ldots, l)$. Then $v_1, \ldots, v_l$ is an XY-separating set of l vertices, which shows that there cannot exist $l+1$ disjoint XY-paths. It follows from the hypotheses of the theorem that $l = k$, and the algorithm has terminated by constructing k disjoint XY-paths. ||

We now briefly abandon our convention that graphs are finite, and present Erdős's elegant proof of Menger's theorem, still with k finite, for two arbitrary sets X and Y of vertices in an infinite graph, assuming the truth of the result for any finite graph. This is not necessary if one adopts the second proof above, since this proof (although not, of course, the algorithm) can be modified easily to prove the result for an infinite graph directly. However, Pym's inductive proof clearly works only for a finite graph, and Erdős's argument (produced in 1931 or 1932, when he was about 19, and published by König [**20**]) is sufficiently pleasing to be worth including.

Proof of Menger's Theorem for an Infinite Graph G. "Only if" clearly holds as before. We prove "if". Suppose the result is false, so that there are at most $l < k$ disjoint XY-paths in G, say $P_1, \ldots, P_l$. Note that an XY-path, like any path with two ends, must be finite in length, even in an infinite graph. Suppose that we delete vertices v_1 on P_1, v_2 on P_2, . . ., v_l on P_l. Since $l < k$, there is still an XY-path left in G: call it $P(v_1, \ldots, v_l)$. There are *finitely* many possible choices for each of the vertices $v_1, \ldots, v_l$, and so the subgraph G' of G consisting of all of the vertices and edges in $P_1, \ldots, P_l$ and all of the paths $P(v_1, \ldots, v_l)$ is *finite*. But Y is not separated from X in G' by any set of l or fewer vertices: for if we delete one vertex, say v_i, from each P_i, then $P(v_1, \ldots, v_l)$ remains; and if we delete l (or fewer) vertices in any other way, then one of the paths P_i remains. So, by Menger's theorem for finite graphs, there are $l+1$ disjoint paths in G', and hence in G. This contradiction completes the proof. ‖

It is almost trivial to prove Menger's theorem for an infinite cardinal k. However, the result is of little interest. The extension that one would like to prove is contained in the following classic conjecture of Erdős:

Conjecture. *Let X and Y be arbitrary sets of vertices in an infinite graph G. Then G contains a set Π of XY-paths and an XY-separating set Z that are in one-to-one correspondence, each vertex of Z lying on a unique path in Π, and each path in Π containing a unique vertex of Z.*

As far as I know, virtually nothing is known about this problem (however, see McDiarmid [**27**]).

We conclude this section with a list of related theorems on (finite) graphs which we present without proof; the interested reader may care to regard these as exercises:

Theorem 2.2. *Let v and w be two vertices of a graph or digraph G such that v is not joined to w by an edge (or arc). Then there exist k vw-paths in G that are pairwise disjoint (except for the vertices v and w themselves) if and only if there remains a vw-path in G whenever fewer than k vertices of $G-\{v, w\}$ are removed.* ‖

Remark. This theorem is essentially the original "Menger's theorem" [**29**], and can easily be proved by adapting the second proof of Theorem 2.1.

Alternatively (and better), it is easily shown by elementary graph-theoretical constructions that each of Theorems 2.1 and 2.2. implies the other.

Theorem 2.3. (Dilworth's Chain-Decomposition Theorem). *The number of elements in the largest anti-chain* (*set of mutually incomparable elements*) *in a finite partially ordered set P is equal to the minimum number of chains* (*totally ordered subsets*) *whose union is P.* ‖

Remark. This theorem, proved by Dilworth [**4**] in 1950, can be obtained by applying Menger's theorem to a bipartite graph constructed in a suitable way from P. However, it is easier to prove Dilworth's theorem directly, copying the first proof of Theorem 2.1. (This is the reverse of the historical order; Perles [**33**] presented such a proof of Dilworth's theorem in 1963, and Pym [**34**] modified it in 1969 to prove Menger's theorem.)

In order to state our next theorem, we define a graph to be ***k*-connected** if it has at least $k+1$ vertices and if each two distinct vertices v and w are connected by at least k vw-paths which are pairwise disjoint except for the vertices v and w.

Theorem 2.4 (The Menger–Whitney Theorem). *A graph G with at least $k+1$ vertices is k-connected if and only if it cannot be disconnected by the removal of any subset of $k-1$ or fewer vertices.* ‖

Remark. This was proved by Whitney [**42**] in 1932, and is a fairly straightforward consequence of Theorem 2.2; in proving "if", one must not forget that v and w might be joined by an edge of G.

Theorem 2.5. *If v and w are vertices in a graph or digraph G, then v is connected to w by k edge-disjoint paths if and only if v remains connected to w by a path whenever $k-1$ or fewer edges of G are removed.* ‖

Remark. This result can be proved by adapting either proof of Theorem 2.1. It can also be shown by fairly simple graph-theoretical constructions to be equivalent to both Menger's theorem and the max-flow min-cut theorem, thereby demonstrating that these two theorems are equivalent (in the sense that each can be used to prove the other).

We now define a graph to be ***k*-edge-connected** if each two distinct vertices are connected by at least k edge-disjoint paths.

Theorem 2.6. *A graph G is k-edge-connected if and only if it cannot be disconnected by the removal of any set of $k-1$ or fewer edges.* ‖

Remark. This follows from Theorem 2.5 even more easily than Theorem 2.4 follows from Theorem 2.2.

Theorem 2.7. *Suppose that X, Y and Z are (not necessarily disjoint) sets of vertices in a graph or digraph G, such that Z separates Y from X in G. Suppose further that, whenever fewer than k vertices of Z are deleted, there remains an XY-path in the graph. Then there are k XY-paths in G that are pairwise disjoint over Z (that is, no two of them have a vertex of Z in common).* ‖

Remark. This was proved by Dirac [5] in 1963, and follows by applying Menger's theorem to a new graph $\tilde{G}$ obtained from G by taking k copies of each vertex not in Z and one copy of each vertex in Z, with edges multiplied appropriately. It is clear that Theorem 2.7 includes Menger's theorem as the special case $Z = V(G)$. Thus, by applying Menger's theorem to a modified graph, we can obtain a result that is strictly stronger than Menger's theorem. This capacity of a theorem to yield results stronger than itself is sometimes expressed by the statement that the theorem is "self-refining". We have just seen that, in this sense, Menger's theorem is self-refining, and we shall see in later sections that so are the theorems of Hall and Tutte.

3. The Max-Flow Min-Cut Theorem

Let D be a digraph. We follow Ford and Fulkerson [**11**] in writing $\mathscr{A}(x)$ ("after x") to denote the set of vertices to which x is joined by an arc, and $\mathscr{B}(x)$ ("before x") to denote the set of vertices that are joined to x by an arc. Suppose that to each arc xy there is assigned a non-negative real number $c(xy)$ called its **capacity**, so that c can be thought of as a real-valued function on $A(D)$, the arc-set of D. If s and t are two vertices of D, called respectively the **source** and the **sink**, a **flow** f of **value** $v(f)$ from s to t is a function $f: A(D) \to \mathbf{R}^+$ such that, for each vertex x of D,

$$F(x) := \sum_{y \in \mathscr{A}(x)} f(xy) - \sum_{y \in \mathscr{B}(x)} f(yx) = \begin{cases} v(f), & \text{if } x = s, \\ 0, & \text{if } x \neq s \text{ or } t, \\ -v(f), & \text{if } x = t, \end{cases}$$

and $0 \leqslant f(xy) \leqslant c(xy)$ for each arc xy of D. $F(x)$ is called the **net flow out of the vertex x**. We call f (or c) **integral** if $f(xy)$ (or $c(xy)$) is an integer for each arc xy of D.

Let X and Y be sets of vertices of D such that $X \cup Y = V(D)$, $X \cap Y = \varnothing$, $s \in X$ and $t \in Y$. Let $\langle X, Y \rangle$ denote the set of arcs joining a vertex in X to a vertex in Y. The set $\langle X, Y \rangle$ is called a **cut** in D **separating** s and t, and its **capacity** is

$$c(X, Y) := \sum_{xy \in \langle X, Y \rangle} c(xy).$$

If S and T are any sets of vertices, we define $f(S, T) := \sum f(xy)$, where the summation extends over all $x \in S$ and $y \in T$ for which $xy \in A(D)$.

Theorem 3.1 (The Max-Flow Min-Cut Theorem). *The maximum flow value $v(f)$, taken over all flows f from s to t, is equal to the minimum cut capacity $c(X, Y)$, taken over all cuts $\langle X, Y \rangle$ separating s and t.*

Proof (Ford and Fulkerson [9], 1956). We remarked in the previous section that this result can be obtained from Menger's theorem. However, the derivation is somewhat involved, and we here give Ford and Fulkerson's algorithmic proof.

We show first that $v(f) \leqslant c(X, Y)$ for each flow f and each cut $\langle X, Y \rangle$. Let f be a flow and $\langle X, Y \rangle$ a cut. It is intuitively obvious that

$$v(f) = f(X, Y) - f(Y, X).$$

To prove this formally, note that

$$\begin{aligned} v(f) = \sum_{x \in X} F(x) &= \sum_{x \in X} \{f(\{x\}, V(D)) - f(V(D), \{x\})\} \\ &= f(X, V(D)) - f(V(D), X) \\ &= f(X, X) + f(X, Y) - f(X, X) - f(Y, X) \\ &= f(X, Y) - f(Y, X). \end{aligned}$$

But $f(X, Y) \leqslant c(X, Y)$ and $f(Y, X) \geqslant 0$, so that $v(f) \leqslant c(X, Y)$, as required.

Thus it suffices to exhibit a flow f and a cut $\langle X, Y \rangle$ such that $v(f) = c(X, Y)$. We shall give an algorithm for constructing such a flow and a cut when the capacity function is integral. When some of the capacities are non-integral rational numbers, one can simply multiply all the capacities by the lowest common denominator of all these fractions, and then proceed with the integral algorithm. This covers all cases of practical importance. (After all, no computer has ever heard of irrational numbers.) To prove the result for irrational capacities requires some analysis. One can prove the existence of a flow of maximum value, by using the fact that a continuous (in fact, linear) function on a closed bounded subset of $\mathbf{R}^n$ attains its bounds; a modified version of the following proof will then go through. Alternatively, one can derive the result from the rational version by taking a sequence of rational approximations to the capacities, using the fact that an infinite sequence of points in a closed bounded subset of $\mathbf{R}^n$ has a convergent subsequence.

So, let us suppose that the capacities of the arcs are all integers. The algorithm starts by choosing a flow f_v of value v (for example, the flow of value 0 that is zero on every arc).

Iterative step. Form a set X of vertices recursively, as follows:

(*i*) put the source s into X;

(*ii*) if $x \in X$, and x is joined to y by an arc, and $f_v(xy) < c(xy)$, then put y into X;

(*iii*) if $x \in X$, and y is joined to x by an arc, and $f_v(yx) > 0$, then put y into X.

The construction of X continues until it terminates, which it must do after a finite number of steps. There are then two cases to consider:

Case 1. The sink t is in X. Then, by the construction of X, there is an undirected path P from s to t, say

$$s = x_0, e_1, x_1, e_2, x_2, \ldots, e_n, x_n = t,$$

where each edge e_i is either a *forward edge* of P (that is, e_i is an $x_{i-1}x_i$-arc in D) when $f_v(e_i) < c(e_i)$, or a *reverse edge* of P (that is, e_i is an x_ix_{i-1}-arc in D) when $f_v(e_i) > 0$. Thus we can increase f_v by 1 on all forward edges of P, and decrease it by 1 on all reverse edges, to obtain a flow f_{v+1} of value $v+1$. Do this, and go back to the iterative step with f_{v+1} in place of f_v.

Case 2. The sink t is not in X. Define $Y := V(D) - X$. Then $t \in Y$, and so $\langle X, Y \rangle$ is a cut separating s and t. If $x \in X$ and $y \in Y$, then clearly $f_v(xy) = c(xy)$ if $xy \in A(D)$, and $f_v(yx) = 0$ if $yx \in A(D)$, since otherwise y would have been put into X; it follows that $f_v(X, Y) = c(X, Y)$ and $f_v(Y, X) = 0$. Thus $v = v(f_v) = f_v(X, Y) - f_v(Y, X) = c(X, Y)$, and the algorithm has terminated by exhibiting a "maximum flow" f_v and a "minimum cut" $\langle X, Y \rangle$. ‖

We call an undirected path P from s to t a *flow-augmenting path* with respect to f if $f < c$ on all forward edges of P and $f > 0$ on all reverse edges.

Corollary 3.2. *A flow f has maximum value if and only if there is no flow-augmenting path with respect to f.* ‖

Corollary 3.3 (The Integrity Theorem). *If c is integral, then there is a flow f of maximum value that is also integral.* ‖

4. Matchings in a Bipartite Graph

In order to state the results in this section, we shall need some elementary definitions. If G is a graph, the **vertex-covering number** of G, denoted by $\tau(G)$, is the smallest number of vertices of G that cover all the edges—that is, $\tau(G)$ is the smallest number of vertices of G such that each edge of G is incident with at least one of them. The **edge-covering number** of G is the smallest number of edges of G that cover all the vertices. The **vertex-independence number** of G is the largest number of pairwise non-adjacent vertices of G. A **matching** in G is a set of pairwise non-adjacent edges of G, and the **edge-independence number** of G, denoted by $\nu(G)$, is the largest number of edges in a matching.

We can now present König's theorem and various related results:

Theorem 4.1 (König's Theorem). *If G is a bipartite graph, then its edge-independence number $\nu(G)$ is equal to its vertex-covering number $\tau(G)$.*

Remark. This result, proved by König [**19**] in 1931, is just Menger's theorem applied to the two partite sets X and Y of the bipartite graph. In Section 6 we shall give an alternative derivation of it, using the duality theorem of linear programming. Here we shall give Lovász's direct proof, which is based on Rado's elegant proof [**35**] of Hall's theorem in 1967.

Proof (Lovász [**23**], 1975). It is clear that the largest number of edges in a matching cannot exceed the vertex-covering number, so that $\nu(G) \leqslant \tau(G)$. In order to prove equality, we use induction on the number of edges in G. If the maximum valency of G is 1, then the result obviously holds. So suppose some vertex x of G has at least two edges incident with it, e joining it to v, and f joining it to w. If $\tau(G-e) = \tau(G)$ or $\tau(G-f) = \tau(G)$, then the conclusion follows by the induction hypothesis. So suppose that there are sets E and F of $\tau(G)-1$ vertices that cover all the edges of $G-e$ and $G-f$, respectively. Evidently $x \notin E$, $x \notin F$, $v \in F-E$ and $w \in E-F$, or else E or F would cover all the edges of G.

Let the two partite sets of G be X and Y, with $x \in X$, and note that

$$|((E \cap F) \cap X) \cup ((E \cup F) \cap Y)| \geqslant \tau(G),$$

and

$$|((E \cup F \cup \{x\}) \cap X) \cup ((E \cap F) \cap Y)| \geqslant \tau(G),$$

since both of the sets on the left-hand side cover all the edges of G. Adding, we obtain

$$|E \cap F| + |E \cup F| + 1 \geqslant 2\tau(G),$$

so that

$$|E| + |F| \geqslant 2\tau(G) - 1,$$

which plainly contradicts the choice of E and F. This contradiction completes the proof of Theorem 4.1. ||

We can now deduce the König–Egerváry theorem, proved by König [**19**] and Egerváry [**8**] in 1931:

Theorem 4.2 (The König–Egerváry Theorem). *In any matrix, the maximum number of non-zero entries such that no two are in the same row or column is equal to the minimum number of rows and columns that between them contain all the non-zero entries.*

Proof. This is just a restatement of Theorem 4.1. The rows and columns of the matrix correspond to the two partite sets in the bipartite graph, and a non-zero entry in the matrix corresponds to an edge in the graph joining the vertices representing the appropriate row and column of the matrix. ‖

Another simple consequence of König's theorem is the following result; in Section 6 we shall show how to derive it using the duality theorem of linear programming:

Theorem 4.3. *The edge-covering number of a bipartite graph is equal to its vertex-independence number.*

Proof. By Theorem 4.1, the edge-independence number of a bipartite graph is equal to its vertex-covering number. However, Gallai (see [**18**], Theorem 10.1) proved in 1959 that in any graph, not necessarily bipartite, the sum of the edge-covering and edge-independence numbers equals the sum of the vertex-covering and vertex-independence numbers (this common sum being just the number of vertices in the graph). The result follows immediately. ‖

If x is a vertex of a graph or digraph G, let $N(x)$ denote the neighborhood of x, and if X is a set of vertices, let $N(X) := \bigcup_{x \in X} N(x)$. So $N(X) = \{y : xy \text{ is an edge (arc) of } G \text{ for some } x \in X\}$.

Theorem 4.4. (The König–Hall Theorem). *Let G be a bipartite graph on two partite sets X and Y. Then G has a matching that covers all the vertices in X if and only if $|N(S)| \geqslant |S|$ for every subset S of X.* ‖

Remark. This theorem is just a translation into graphical language of Hall's theorem [**16**] on distinct representatives of a family of sets. As its name implies, it is an easy consequence of König's theorem (Theorem 4.1). Since König's theorem is just Menger's theorem for a bipartite graph, one can equally well deduce Theorem 4.4 from Menger's theorem. (Indeed, many results on systems of representatives can be deduced very elegantly from Menger's theorem, as shown by Perfect [**32**] in 1968, or from the max-flow min-cut theorem or related theorems, as shown by Ford and Fulkerson [**11**] in 1962. All of these results in transversal theory can be regarded as minimax theorems in graph theory.) Among the many direct proofs of Theorem 4.4, two in particular stand out: that of Halmos and Vaughan [**17**] in 1950, and that of Rado [**35**] in 1967 already referred to. Here we shall do no more than observe that Theorem 4.4 is a trivial consequence of the "defect" form of Hall's theorem, which is the minimax version and which translates into graph theory as follows:

Thereom 4.5. *Let G be a bipartite graph on two partite sets X and Y. Then its edge-independence number $\nu(G)$ is equal to*

$$\min_{S \subseteq X} (|X| + |N(S)| - |S|).$$

Remark. This result, proved by Ore [30] in 1955, both implies Hall's theorem, and is easily obtained from Hall's theorem by adding sufficiently many extra vertices to G joined to all the vertices in X, until Hall's theorem applies. (This illustrates one of the many self-refining aspects of Hall's theorem.) It can also be proved by minor modifications of most of the proofs of Hall's theorem. However, to illustrate the minimax nature of the theorem, we shall give an algorithmic proof (the **Hungarian method**) which is similar in form to the two algorithms that we have met already.

Proof. It is easy to see that

$$\nu(G) \leqslant \min_{S \subseteq X} (|X| + |N(S)| - |S|);$$

we shall use the algorithm to prove equality.

The algorithm starts with a matching M_l of l edges for some non-negative integer l (say $l = 0$).

Iterative step. Form two sets $S \subseteq X$ and $T \subseteq Y$ of vertices recursively, as follows:

(i) put into S all the vertices of X that are not covered by M_l;

(ii) if $v \in S$, and v is joined to w by an edge of G not in M_l, then put w into T;

(iii) if $w \in T$, and v is joined to w by an edge of M_l, then put v into S.

The construction of S and T continues until it terminates, as it must do after a finite number of steps. There are then two cases to consider:

Case 1. T contains some vertex y of Y that is not covered by M_l. Then there are a vertex x of X that is not covered by M_l, and a path P, connecting x to y, consisting of an odd number of edges that are alternately in M_l and not in M_l. Interchange the edges of $P \cap M_l$ and $P - M_l$ to form a new matching M_{l+1} with $l+1$ edges. Go back to the iterative step with M_{l+1} in place of M_l.

Case 2. All the vertices of T are covered by M_l. Then S consists of the $|T|$

vertices of X that are matched with vertices in T, together with the $|X|-l$ vertices of X that are not covered by M_l; and $N(S) = T$. So

$$|X|+|N(S)|-|S| = l.$$

Thus the algorithm has terminated with the construction of a largest matching. ||

Our next theorem (Theorem 4.6) is a minimax theorem that can be regarded as a quantitative version of König's theorem, in much the same way that the max-flow min-cut theorem can be regarded as a quantitative version of Menger's theorem. The main point of Theorem 4.6 is that its proof gives an effective algorithm for solving the optimal-assignment problem.

Suppose that we have n jobs $J_1, \ldots, J_n$ to fill, and n "individuals" $I_1, \ldots, I_n$ to fill them, and suppose that the "rating" of I_i for the job J_j is given by the real number α_{ij}. The **optimal-assignment problem** is to match the n individuals onto the n jobs in such a way as to maximize the sum of the ratings over the matching—that is, to find non-negative integers x_{ij} $(i, j = 1, \ldots, n)$ that maximize the expression

$$\sum_{i=1}^{n} \sum_{j=1}^{n} \alpha_{ij} x_{ij},$$

subject to the constraints

$$\sum_{i=1}^{n} x_{ij} = 1 \ (j = 1, \ldots, n), \qquad \text{and} \qquad \sum_{j=1}^{n} x_{ij} = 1 \ (i = 1, \ldots, n).$$

(These constraints force the integers x_{ij} to be 0 or 1, and the n integers that are equal to 1 correspond to n edges that match the n individuals to the n jobs.)

Theorem 4.6. *In the above optimal-assignment problem, the maximum sum of the ratings over all matchings is equal to the minimum value of*

$$\sum_{i=1}^{n} y_i + \sum_{j=1}^{n} z_j, \tag{1}$$

taken over all sets of real numbers y_i and z_j that satisfy

$$y_i + z_j \geqslant \alpha_{ij}, \qquad \text{for all } i \text{ and } j. \tag{2}$$

Remark. We shall see at the end of Section 6 how this theorem can be deduced from the duality theorem of linear programming. However, we here give the direct algorithmic proof, since this is the main value of the theorem.

Proof. Note first that, given any numbers x_{ij}, y_i and z_j satisfying the constraints—and such numbers clearly exist—, we have

$$\sum_{i,j} \alpha_{ij}x_{ij} \leqslant \sum_{i,j} (y_i+z_j)x_{ij} = \sum_i y_i + \sum_j z_j. \tag{3}$$

So it suffices to exhibit a set of numbers for which equality holds. As in the proof of the max-flow min-cut theorem, we shall confine our attention to the case when the ratings α_{ij} are all integers. The algorithm starts by choosing sets of numbers y_i and z_j for which (2) holds: say $y_i := \max_j \alpha_{ij}$ for each i, and $z_j := 0$ for each j.

Iterative step. Consider the bipartite graph G with

$$V(G) := \{I_1, \ldots, I_n\} \cup \{J_1, \ldots, J_n\},$$

in which there is an edge joining I_i and J_j if and only if $y_i+z_j = \alpha_{ij}$. Apply the Hungarian algorithm of Theorem 4.5 to G. There are two cases to consider:

Case 1. The Hungarian algorithm terminates with a set $S \subseteq \{I_1, \ldots, I_n\}$ such that $|N(S)| < |S|$. Then we can decrease the value of y_i by 1 for each $I_i \in S$, and increase the value of J_j by 1 for each $J_j \in N(S)$, without violating (2), and in so doing we decrease (1) by $|S|-|N(S)| > 0$. Do this, and go back to the iterative step. (Since (3) shows that (1) is bounded below, Case 1 can arise at most a finite number of times.)

Case 2. The Hungarian algorithm terminates with a "complete matching" of G (that is, one that covers all the vertices). Define $x_{ij} := 1$ if I_i is matched with J_j, and 0 otherwise. Then equality holds in (3). So the algorithm has terminated by finding an optimal assignment. ||

A complete matching of a graph G is often called a **1-factor**, and two 1-factors are called "disjoint" when they have no edges in common. For a bipartite graph to have a 1-factor, it is evidently necessary that the two partite sets should have the same number of vertices. When that happens, Theorem 4.4 gives a necessary and sufficient condition for the graph to have a 1-factor. Theorem 4.5 is a minimax extension of this, giving a formula for $\nu(G)$, the largest number of edges in a matching in G. One might ask for a different type of minimax extension, giving a formula for the largest number of disjoint 1-factors. This is provided by the following theorem, which (with $k = 1$) gives a slightly different condition for a bipartite graph to have a 1-factor.

Theorem 4.7. *Let G be a bipartite graph on two partite sets X and Y of equal cardinality n. Then G contains k disjoint 1-factors if and only if, for each pair* *of subsets $S \subseteq X$ and $T \subseteq Y$, the number of edges of G joining a vertex in S to one in T is at least $k(|S|+|T|-n)$.* ‖

Remark. This theorem is a consequence of the necessary and sufficient condition for two families of sets to have a system of common distinct representatives, which was first proved by Ford and Fulkerson [**10**] in 1958. It illustrates the extent to which the theory of factors in bipartite graphs is more advanced than in arbitrary graphs. If an analog of Theorem 4.7 could be found for an arbitrary graph, even for $k = 2$, it would probably yield a proof of the four-color theorem.

5. Factors in an Arbitrary Graph

If G is any graph, not necessarily bipartite, then (as we have just mentioned) a 1-factor, or complete matching (or perfect matching) of G is a matching that covers all the vertices of G. If S is a set of vertices of G, let $\theta(S)$ denote the number of odd components (that is, components with an odd number of vertices) of $G-S$. The most important result on 1-factors is the following theorem, first proved by Tutte [**38**] in 1947:

Theorem 5.1 (Tutte's 1-Factor Theorem). *A graph G has a 1-factor if and only if $\theta(S) \leqslant |S|$ for each subset S of $V(G)$.* ‖

Remark. There are several neat proofs of this result, of which two in particular stand out: those of Anderson [**1**] in 1971 using Hall's theorem, and of Lovász [**23**] in 1975. Here, as with the König–Hall theorem, we shall refrain from proving the result directly, and shall instead proceed straight to the minimax version, obtained by Berge [**2**] in 1958.

Theorem 5.2. *The maximum number of edges in a matching of a graph G is $\Theta(G)$, where*

$$\Theta(G) := \min_{S \subseteq V(G)} \tfrac{1}{2}(|V(G)| - \theta(S) + |S|).$$

Remark. This result both implies Tutte's theorem, and is easily obtained from it by adding to G sufficiently many extra vertices, joined to all the vertices of G and to each other, until Tutte's theorem applies; see the elegant proof by McCarthy [**26**] in 1973. (This illustrates one aspect of the self-refining nature of Tutte's theorem.) It may also be obtained by modifying Anderson's proof of Tutte's theorem, using Ore's theorem (Theorem 4.5) where Anderson used Hall's theorem; proofs of this type have been produced by Mader [**28**] and myself [**43**], both in 1973. It is however very difficult to give an algorithmic proof, since the only algorithms I know for finding a largest matching in a general graph are very complicated: see Edmonds **[6]** (1965). We prove Theorem 5.2 here by adapting Lovász's elegant proof **[23]** of Tutte's theorem.

Proof. It is easy to see that the number of edges in a matching cannot exceed $\Theta(G)$, and so it suffices to prove equality. Note that $|V(G)|-\theta(S)+|S|$ is even for each $S \subseteq V(G)$, so that $\Theta(G)$ is an integer. Note also that adding an extra edge to G cannot decrease $\Theta(G)$. Let G' be a maximal graph on $V(G)$ containing all the edges of G, with $\Theta(G') = \Theta(G)$, such that G' has no matching of cardinality $\Theta(G)$. We shall obtain a contradiction.

Let V_1 be the set of those vertices of G' that are joined by edges to every other vertex, and let $V_2 := V(G)-V_1$. Let G_2 be the subgraph of G induced by V_2. If G_2 consists of disjoint complete subgraphs, then (taking $S := V_1$) the number of odd complete graphs among these cannot exceed

$$|V_1|+|V(G)|-2\Theta(G).$$

There is then clearly a matching containing all but $|V(G)|-2\Theta(G)$ of the vertices of G, and this matching contains $\Theta(G)$ edges, which is the required contradiction. So we may assume that G_2 does not consist of disjoint complete graphs. Thus there exist vertices a, b and c in V_2 such that ab and $bc \in G'$, but $ac \notin G'$. Since $b \in V_2$, there exists a vertex d such that $bd \notin G'$.

By the maximality of G', the graphs $G' \cup \{ac\}$ and $G' \cup \{bd\}$ have matchings M_1 and M_2 (respectively) of cardinality $\Theta(G)$. Clearly, $ac \in M_1 - M_2$, and $bd \in M_2 - M_1$. Also, $M_1 \cup M_2$ consists of isolated edges (the edges of $M_1 \cap M_2$) and disjoint circuits and paths. Let C be the circuit or path of $M_1 \cup M_2$ containing ac.

If $bd \notin C$, interchange the edges of M_1 and M_2 in C to form new matchings M_1' (of G') and M_2' (of $G' \cup \{ac, bd\}$). If $|M_1'| < \Theta(G)$, then $|M_2'| \geqslant \Theta(G)+1$, and $M_2' \cup \{ab\}-\{ac, bd\}$ is a matching of G' of cardinality $\Theta(G)$. Otherwise, M_1' is a matching of G' with cardinality at least $\Theta(G)$. In either case we have a contradiction, and this shows that $bd \in C$.

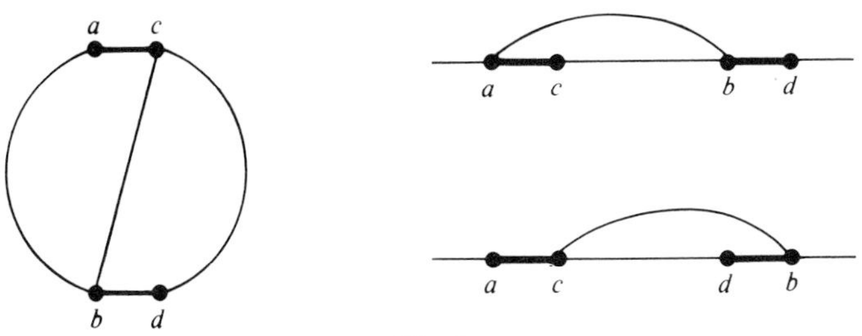

Fig. 3

It is now easy to see that $C \cup \{ab\}$ or $C \cup \{bc\}$ contains a circuit C' that alternates with respect to M_1 or M_2 (see the three cases in Fig. 3; the three remaining cases, obtained by interchanging a and c, are treated similarly). Interchanging the edges of $C' \cap M_1$ and $C' - M_1$ (or $C' \cap M_2$ and $C' - M_2$, if appropriate), we form a new matching M_1' (or M_2') of G' with cardinality $\Theta(G)$. This final contradiction completes the proof. ‖

We conclude this section with the statements of some more general factor theorems. Suppose we are given a graph G and a mapping $f: V(G) \to \mathbf{Z}^+$ (the non-negative integers). An ***f*-factor** of G is a subgraph in which each vertex v has valency $f(v)$. The first general factor theorem was proved by Tutte [**39**] in 1952:

Theorem 5.3. *If G is a graph, and if $f: V(G) \to \mathbf{Z}^+$ is such that $0 \leqslant f(v) \leqslant \rho(v)$ (the valency of v) for each vertex v of G, then G has an f-factor if and only if*

$$\psi(S, T) \leqslant \sum_{v \in S} f(v) - \sum_{v \in T} f(v) + \sum_{v \in T} \rho_{G-S}(v)$$

for every pair (S, T) of disjoint sets of vertices of G, where ρ_{G-S} denotes valency in $G-S$, and $\psi(S, T)$ denotes the number of components C of $G-(S \cup T)$ with the property that the number of edges joining them to T has different parity from $\sum_{v \in C} f(v)$. ‖

Remark. The neatest proof of this result is that found in 1954 by Tutte himself [**40**]. We form a new graph $\tilde{G}$ from G by replacing each vertex v of G by $\rho(v)$ copies of itself, each receiving exactly one edge of G, and then joining all of these vertices to $\rho(v) - f(v)$ new independent vertices (see Fig. 4); this is done for each vertex of G simultaneously. Then it is easy to see that G has an f-factor if and only if the new graph $\tilde{G}$ has a 1-factor. Applying the condition of Theorem 5.1 to $\tilde{G}$ yields the required result.

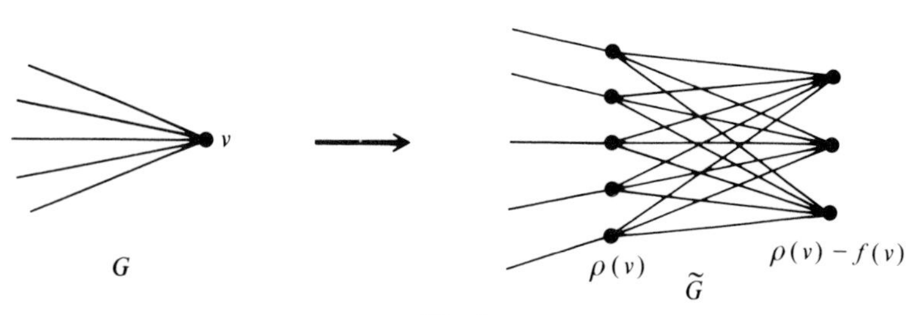

Fig. 4

Theorem 5.4. *With the terminology of Theorem 5.3, the largest number of edges of G in a subgraph in which each vertex v has valency at most f(v) is*

$$\min \tfrac{1}{2}\left(\sum_{v \in V(G)} f(v) - \Psi(S, T)\right),$$

where the minimum is taken over all pairs (S, T) of disjoint sets of vertices of G, and

$$\Psi(S, T) := \psi(S, T) - \sum_{v \in S} f(v) + \sum_{v \in T} f(v) - \sum_{v \in T} \rho_{G-S}(v). \; \|$$

Remark. This is the minimax version of Theorem 5.3. It can be proved by mimicking the proof of Theorem 5.3, using Theorem 5.2 instead of Theorem 5.1. Alternatively, it can be derived from Theorem 5.3 by adding sufficiently many extra vertices to G, joined to all the vertices of G and to each other, until Theorem 5.3 applies.

Finally, we turn our attention to digraphs. A **factor** in a digraph is a subgraph in which each vertex v has $\rho_{\text{out}}(v) = \rho_{\text{in}}(v) = 1$.

Theorem 5.5. *A digraph D has a factor if and only if $|N(X)| \geqslant |X|$ for each subset X of V(D).* $\|$

Remark. This looks very much like the König–Hall theorem (Theorem 4.4), and indeed it is practically the same result. It can be proved by applying the König–Hall theorem to a bipartite graph $\tilde{G}$ constructed from G in a simple way that the reader can no doubt work out for himself. In an exactly similar way, one can prove a general factor theorem for digraphs by applying a general factor theorem for bipartite graphs to the same graph $\tilde{G}$. In order to obtain a general factor theorem for bipartite graphs, one can either specialize Theorem 5.3 to bipartite graphs, or else (and more easily) use the condition for two families of sets to have a system of common distinct representatives, which will yield results like Theorem 4.7.

6. The Duality Theorem of Linear Programming

The duality theorem of linear programming is a very powerful tool that can be used to prove many results in graph theory and other branches of combinatorics. In this section we state (without proof) the duality theorem, and the integrity theorem for totally unimodular matrices, and show how they can be used to prove two results about bipartite graphs—namely, König's theorem (Theorem 4.1), and Theorem 4.3.

If $\mathbf{A} = (a_{ij})$ and $\mathbf{B} = (b_{ij})$ are real matrices of the same size, we shall write

$\mathbf{A} \geqslant \mathbf{B}$ if $a_{ij} \geqslant b_{ij}$ for each i and j. Suppose that we are given a real $m \times n$ matrix $\mathbf{A}$, a real m-vector $\mathbf{b}$, and a real n-vector $\mathbf{c}$. The standard linear-programming minimization problem is to find the minimum value of $\mathbf{c}^{\mathrm{T}}\mathbf{x}$ over all n-vectors $\mathbf{x}$ satisfying the constraints $\mathbf{Ax} \geqslant \mathbf{b}$, and $\mathbf{x} \geqslant \mathbf{0}$. If this is called the *primal problem*, its *dual problem* is to find the maximum value of $\mathbf{y}^{\mathrm{T}}\mathbf{b}$ over all m-vectors $\mathbf{y}$ satisfying the constraints $\mathbf{y}^{\mathrm{T}}\mathbf{A} \leqslant \mathbf{c}^{\mathrm{T}}$, and $\mathbf{y} \geqslant \mathbf{0}$. We summarize the problems as follows:

$$\textit{Primal problem} \quad \begin{cases} \mathbf{Ax} \geqslant \mathbf{b}, \\ \mathbf{x} \geqslant \mathbf{0}, \end{cases} \quad \text{find min } \mathbf{c}^{\mathrm{T}}\mathbf{x}. \tag{4}$$

$$\textit{Dual problem} \quad \begin{cases} \mathbf{y}^{\mathrm{T}}\mathbf{A} \leqslant \mathbf{c}^{\mathrm{T}}, \\ \mathbf{y} \geqslant \mathbf{0}, \end{cases} \quad \text{find max } \mathbf{y}^{\mathrm{T}}\mathbf{b}. \tag{5}$$

Theorem 6.1 (The Duality Theorem for the Standard Linear Program). *If either of the above problems has a solution, or if the systems of constraints* (4) *and* (5) *are both realizable, then both of the problems are solvable and the two solutions are equal.* ||

A proof of this result will be found in almost any book on linear programming.

In order to apply this result to combinatorial problems, we need one of the so-called "integrity theorems". A matrix $\mathbf{A}$ is called **totally unimodular** if every square submatrix has determinant 0, 1 or -1. (In particular, every element of $\mathbf{A}$ must be 0, 1 or -1.) It is easy to see that $\mathbf{A}^{\mathrm{T}}$ is totally unimodular if $\mathbf{A}$ is. An **integer vector** is a vector all of whose coordinates are integers. The required integrity theorem follows by a simple application of Cramer's rule:

Theorem 6.2. *If* $\mathbf{A}$ *is a totally unimodular matrix, and if* $\mathbf{b}$ *is an integer vector, then* min $\mathbf{c}^{\mathrm{T}}\mathbf{x}$ *in the primal problem is attained* (*if at all*) *at an integer vector* $\mathbf{x}$. *If* $\mathbf{A}$ *is totally unimodular, and if* $\mathbf{c}$ *is an integer vector, then* max $\mathbf{y}^{\mathrm{T}}\mathbf{b}$ *in the dual problem is attained* (*if at all*) *at an integer vector* $\mathbf{y}$. ||

As an illustration of the way in which linear programming can be used to prove theorems about graphs, we shall use these results to prove Theorem 4.1 (König's theorem) and Theorem 4.3. The **incidence matrix** of a graph G is the matrix in which the rows are indexed by the vertices of G, and the columns by the edges of G, with an entry 1 in the matrix if the corresponding vertex and edge are incident, and an entry 0 if they are not. Since we are restricting our attention to bipartite graphs, we can make use of Theorem 6.2, as shown by the following lemma:

Lemma 6.3. *The incidence matrix* $\mathbf{B}$ *of a bipartite graph* G *is totally unimodular.*

Proof. Each column of $\mathbf{B}$ has exactly two 1s, the rest of the entries being 0. Let $\mathbf{M}$ be a square submatrix of $\mathbf{B}$ of minimal order such that det $\mathbf{M}$ is not 0, 1 or -1. Clearly no column of $\mathbf{M}$ is all zeros. Equally clearly, no column has exactly one 1. Thus every column has exactly two 1s. The sum of the rows of $\mathbf{M}$ corresponding to vertices in one partite set of G is thus equal to the sum of all the other rows, which correspond to vertices in the other partite set of G (both of these sums being a row of 1s). Thus det $\mathbf{M} = 0$, contradicting the choice of $\mathbf{M}$. ||

We are now in a position to prove Theorems 4.1 and 4.3.

Theorem 6.4. *If G is a bipartite graph, then*

(*i*) *its edge-covering number is equal to its vertex-independence number;*

(*ii*) *its edge-independence number is equal to its vertex-covering number.*

Proof. We prove (*i*); (*ii*) is similar. Let $\mathbf{B}$ be the incidence matrix of G. (To prove (*ii*), we should work with $\mathbf{B}^T$.) Let $\mathbf{1}$ be the column vector of 1s (of any size, just as $\mathbf{0}$ is used for the zero vector of any size), and consider the two mutually dual problems:

$$\textit{Primal problem} \qquad \begin{cases} \mathbf{Bx} \geqslant \mathbf{1}, \\ \mathbf{x} \geqslant \mathbf{0}, \end{cases} \qquad \text{find min } \mathbf{1}^T\mathbf{x},$$

and

$$\textit{Dual problem} \qquad \begin{cases} \mathbf{y}^T\mathbf{B} \leqslant \mathbf{1}^T, \\ \mathbf{y} \geqslant \mathbf{0}, \end{cases} \qquad \text{find max } \mathbf{y}^T\mathbf{1}.$$

Since $\mathbf{B}$ is totally unimodular, the extremal values are both realized at integer vectors, $\mathbf{x}$ and $\mathbf{y}$ respectively. The solution to the primal problem is thus the smallest sum of the non-negative integers (which we can clearly take to be 0s and 1s) that we can assign to the edges of G in such a way that at each vertex the sum of the numbers on all of the incident edges is at least 1; that is, it is the smallest number of edges covering all the vertices (the edge-covering number). And the solution to the dual problem is the largest sum of the non-negative integers (which again we can clearly take to be 0s and 1s) that we can assign to the vertices of G in such a way that the sum of the numbers at the ends of each edge is at most 1; that is, it is the largest number of pairwise non-adjacent vertices (the vertex-independence number). The equality of these two numbers now follows from the duality theorem (Theorem 6.1). ||

Note that the optimal-assignment theorem (Theorem 4.6) follows in a similar way from the mutually dual linear programs

$$\begin{cases} \mathbf{Bx} \geqslant \boldsymbol{\alpha}, \\ \mathbf{x} \geqslant \mathbf{0}, \end{cases} \qquad \text{find min } \mathbf{1}^T\mathbf{x},$$

and

$$\begin{cases} \mathbf{y}^{\mathrm{T}}\mathbf{B} \leqslant \mathbf{1}^{\mathrm{T}}, \\ \quad \mathbf{y} \geqslant \mathbf{0}, \end{cases} \qquad \text{find max } \mathbf{y}^{\mathrm{T}}\boldsymbol{\alpha},$$

where, in order to agree with the terminology of Theorem 4.6, we should write $\mathbf{x}$ as

$$\left[\frac{\mathbf{y}}{\mathbf{z}}\right],$$

and $\mathbf{y}$ as $\mathbf{x}$.

7. The Menger and König Properties

The purpose of this and the next two sections is to exhibit a large number of minimax theorems on graphs, and to show how they all fit into a pattern (or, more accurately, into two closely-related patterns). The treatment follows closely that in [**44**]. This section is mainly concerned with the definitions.

Let $(S, \mathscr{F})$, or just $\mathscr{F}$, be a finite **set-system**—that is, a collection $\mathscr{F}$ of distinct subsets of a finite set S (elsewhere called a *configuration*, *design*, *hypergraph*, etc.). The elements of S will be called **points**, and the sets of $\mathscr{F}$ will be called **blocks**. For the moment, suppose that the blocks of $\mathscr{F}$ are mutually incomparable as sets ($\mathscr{F}$ is a *Sperner family*, or *clutter*). Then the **Menger dual** (or *blocking clutter*, or *blocker*) $(S, \mathscr{F}_{\geqslant 1})$ of $\mathscr{F}$ consists of all the minimal subsets of S that have at least one element in common with each block of $\mathscr{F}$. (We adopt the convention that $\{\varnothing\}_{\geqslant 1} = \varnothing$ and $\varnothing_{\geqslant 1} = \{\varnothing\}$.) It is easy to check that this is a genuine dual.

Theorem 7.1. $(\mathscr{F}_{\geqslant 1})_{\geqslant 1} = \mathscr{F}$. ||

If $\mathscr{F}$ contains k disjoint non-empty blocks, then clearly every block in $\mathscr{F}_{\geqslant 1}$ has cardinality at least k. We say that $\mathscr{F}$ is a **Menger system**, or has the **Menger property** (or *packs*, or has the *Boolean max-flow min-cut property*) if the maximum number $\nu(\mathscr{F})$ of disjoint blocks in $\mathscr{F}$ is equal to the cardinality $\tau(\mathscr{F})$ of the smallest block in $\mathscr{F}_{\geqslant 1}$. (We adopt the convention that $\{\varnothing\}$ and $\varnothing$ both have the Menger property.) The reason for the name "Menger property" is that Menger's theorem and its edge-separation analog (Theorems 2.1 and 2.5) state that the XY-paths in a graph or digraph have the Menger property, the paths being regarded either as sets of vertices or as sets of edges (or arcs); in the latter case, we must modify Theorem 2.5 to sets X and Y of vertices, and we must specify that these sets be disjoint.

In a similar way, the **König dual** (or *antiblocking clutter*, or *antiblocker*) $(S, \mathscr{F}_{\leqslant 1})$ of $\mathscr{F}$ consists of all the maximal subsets of S that have at most one element in common with each block of $\mathscr{F}$. However $(\mathscr{F}_{\leqslant 1})_{\leqslant 1} \neq \mathscr{F}$ in general. $\mathscr{F}$ is **clique-complete** (or *conformal*, as in Berge [**3**]) if it has the following property, possessed by the maximal cliques (vertex-sets of complete subgraphs) of any graph:

> if $S' \subseteq S$, and every pair of elements of S' is contained in some block of $\mathscr{F}$, then S' is contained in some block of $\mathscr{F}$.

(Equivalently, every point of S is contained in some block of $\mathscr{F}$, and, if $B_1, B_2, B_3 \in \mathscr{F}$, then there is a block of $\mathscr{F}$ that contains all points that are in at least two of B_1, B_2 and B_3.) The **clique-completion** $\mathscr{F}_{cc}$ of $\mathscr{F}$ consists of all the maximal sets among those subsets of S whose pairs of elements are all contained in blocks of $\mathscr{F}$. It is easy to check that the analog of Proposition 7.1 now becomes:

Theorem 7.2. $(\mathscr{F}_{\leqslant 1})_{\leqslant 1} = \mathscr{F}$ *if and only if* $\mathscr{F}$ *is clique-complete. In general,* $(\mathscr{F}_{\leqslant 1})_{\leqslant 1} = \mathscr{F}_{cc}$. *But* $((\mathscr{F}_{\leqslant 1})_{\leqslant 1})_{\leqslant 1} = \mathscr{F}_{\leqslant 1}$ *always* (since $\mathscr{F}_{\leqslant 1}$ *is clique-complete*). ‖

We note that, in practice, all of the most interesting examples tend to be clique-complete.

Suppose that S is the union of all the blocks of $\mathscr{F}$. If $\mathscr{F}_{\leqslant 1}$ contains a block of cardinality k, then clearly S cannot be expressed as the union of fewer than k blocks of $\mathscr{F}$. $\mathscr{F}$ is a **König system**, or has the **König property**, if the smallest number $\rho(\mathscr{F})$ of blocks of $\mathscr{F}$ whose union is S is equal to the cardinality $\alpha(\mathscr{F})$ of the largest block in $\mathscr{F}_{\leqslant 1}$. (We adopt the convention that, if S is not the union of all the blocks of $\mathscr{F}$, then $\mathscr{F}$ automatically has the König property.) $\mathscr{F}$ has the **dual König property** if the smallest number $\gamma(\mathscr{F}) = \rho(\mathscr{F}_{\leqslant 1})$ of blocks of $\mathscr{F}_{\leqslant 1}$ whose union is S is equal to the cardinality $\mu(\mathscr{F})$ of the largest block in $\mathscr{F}$. If $\mathscr{F}$ is clique-complete, this is the same as saying that $\mathscr{F}_{\leqslant 1}$ has the König property. The reason for the name "König property" is that König's theorem (Theorem 4.1) states that the system of vertex-coboundaries of a bipartite graph (regarded as subsets of the set of all its edges) has the König property, where a **vertex-coboundary** is the set of all edges incident with a given vertex.

If $(S, \mathscr{F})$ is any finite set system (not necessarily a clutter), and if $\mathscr{F}_{\min}$ denotes the set of minimal blocks of $\mathscr{F}$, then we define $\mathscr{F}_{\geqslant 1} := (\mathscr{F}_{\min})_{\geqslant 1}$, and say that $\mathscr{F}$ has the Menger property if $\mathscr{F}_{\min}$ has. (Note the consequence that $\mathscr{F}$ automatically has the Menger property if it contains an empty block.) Similarly, if $\mathscr{F}_{\max}$ denotes the set of maximal blocks of $\mathscr{F}$, then we define $\mathscr{F}_{\leqslant 1} := (\mathscr{F}_{\max})_{\leqslant 1}$, and say that $\mathscr{F}$ has the König property, or dual König property, if $\mathscr{F}_{\max}$ has.

We remarked in Section 4 that König's theorem is just Menger's theorem for a bipartite graph. This suggests that there may be a connection between the Menger and König properties. In order to exhibit this connection, we let $\mathscr{F}_x$ be the collection of blocks of $\mathscr{F}$ containing x, for each $x \in S$, and let $\mathscr{F}^{\mathrm{T}}$, the **transpose** (or *design dual*, or *hypergraph dual*) of $\mathscr{F}$, be the collection $(\mathscr{F}_x : x \in S)$ of subsets of $\mathscr{F}$. The **incidence matrix** of $\mathscr{F}$ has a row for each block B of $\mathscr{F}$ and a column for each point $x \in S$, with 1 in the (B, x)-position if $x \in B$, and 0 otherwise; so the incidence matrix of $\mathscr{F}^{\mathrm{T}}$ is just the transpose of that of $\mathscr{F}$. Clearly $(\mathscr{F}^{\mathrm{T}})^{\mathrm{T}} = \mathscr{F}$.

Theorem 7.3. *$\mathscr{F}$ has the Menger property if and only if $\mathscr{F}^{\mathrm{T}}$ has the König property, and vice versa.*

Proof. If $X \subseteq S$, $X \in \mathscr{F}_{\geqslant 1}$ means the same as $\bigcup_{x \in X} \mathscr{F}_x = \mathscr{F}$; similarly, the largest number of disjoint blocks in $\mathscr{F}$ is equal to the cardinality of the largest block in $(\mathscr{F}^{\mathrm{T}})_{\leqslant 1}$. Thus $\mathscr{F}$ has the Menger property if and only if $\mathscr{F}^{\mathrm{T}}$ has the König property. The "vice versa" follows on taking transposes. (What we have shown is that $\tau(\mathscr{F}) = \rho(\mathscr{F}^{\mathrm{T}})$ and $\nu(\mathscr{F}) = \alpha(\mathscr{F}^{\mathrm{T}})$; hence $\tau(\mathscr{F}^{\mathrm{T}}) = \rho(\mathscr{F})$ and $\nu(\mathscr{F}^{\mathrm{T}}) = \alpha(\mathscr{F})$.) $\|$

In order to discuss the Menger and König properties in more detail, we need a large number of rather specialized definitions, which we now give. The reader may prefer to skip to the next section, and refer back to these definitions as they are needed.

If $(S, \mathscr{F})$ is a finite set-system, and if $S' \subseteq S$, we define the **weak point-restriction** $\mathscr{F}\backslash S'$ of $\mathscr{F}$ to $S - S'$ (or *subhypergraph*, as in Berge [3], or *contraction* $\mathscr{F}/S'$, as in Seymour [37]) to be the collection $(B - S' : B \in \mathscr{F})$ of subsets of $S - S'$, and the **strong point-restriction** $\mathscr{F}/S'$ (or *section hypergraph*, as in Lovász [22], or *deletion* $\mathscr{F}\backslash S'$, as in Seymour [37]) to be $(B \in \mathscr{F} : B \cap S' = \varnothing)$. So a weak point-restriction of $\mathscr{F}$ is obtained by deleting a set of columns of the incidence matrix—that is, by deleting a set of points from S and from any block containing them—while a strong point-restriction is obtained by deleting a set of points from S and deleting all blocks containing any of them. A **minor** is anything obtained by a sequence of weak and strong point-restrictions; note that these operations commute. It is not difficult to see that $(\mathscr{F}\backslash S')_{\leqslant 1} = (\mathscr{F}_{\leqslant 1}\backslash S')_{\max}$, $(\mathscr{F}\backslash S')_{\geqslant 1} = \mathscr{F}_{\geqslant 1}/S'$, and $(\mathscr{F}/S')_{\geqslant 1} = (\mathscr{F}_{\geqslant 1}\backslash S')_{\min}$.

If $x \in S$, the system obtained from $\mathscr{F}$ by **weakly replicating x r times** is

$$(S \cup \{x_1, \ldots, x_r\} - \{x\}, \mathscr{F} \cup (A \cup \{x_1, \ldots, x_r\} - \{x\} : x \in A \in \mathscr{F}) \\ - (A : x \in A \in \mathscr{F})),$$

where $x_1, \ldots, x_r$ are new elements not in S. (Here r is a non-negative integer;

if we take $r = 0$, we just get the weak point-restriction $\mathscr{F}\backslash\{x\}$, and if $r = 1$, we leave $\mathscr{F}$ essentially unchanged.) If $c: S \to \mathbf{Z}^+$ is any function, the **weak c-replication** of $\mathscr{F}$ is the system obtained by weakly replicating each element $x \in S$ $c(x)$ times (the result being independent of the order in which these operations are carried out). The **weak point-replications** of $\mathscr{F}$ are the weak c-replications for all $c: S \to \mathbf{Z}^+$; these include all the weak point-restrictions.

Similarly, the system obtained from $\mathscr{F}$ by **strongly replicating x r times** is

$$(S \cup \{x_1, \ldots, x_r\} - \{x\}, \mathscr{F} \cup (A \cup \{x_i\} - \{x\} : 1 \leqslant i \leqslant r, x \in A \in \mathscr{F}) - (A : x \in A \in \mathscr{F})),$$

where $x_1, \ldots, x_r$ are new elements not in S. (If $r = 0$, we just get the strong point-restriction $\mathscr{F}/\{x\}$, and, as before, if $r = 1$, we leave $\mathscr{F}$ essentially unchanged.) If $c: S \to \mathbf{Z}^+$ is as before, the **strong c-replication** of $\mathscr{F}$ is defined analogously to the weak c-replication, and the **strong point-replications** of $\mathscr{F}$ are the strong c-replications, for all c; these include all the strong point-restrictions. Note that the strong c-replication of $\mathscr{F}_{\geqslant 1}$ is the Menger dual of the system of minimal blocks of the weak c-replication of $\mathscr{F}$, and vice versa. And, if c is nowhere zero, then the strong c-replication of $\mathscr{F}_{\leqslant 1}$ is the König dual of the system of maximal blocks of the weak c-replication of $\mathscr{F}$, and vice versa.

Following Seymour [**37**], we call a set-system $(S, \mathscr{F})$ **Mengerian** if $\varnothing \in \mathscr{F}$ or, for each function $c: S \to \mathbf{Z}^+$, there is a function $f: \mathscr{F} \to \mathbf{Z}^+$ such that,

$$\text{for each } x \in S, \qquad \sum_{\substack{B \\ x \in B \in \mathscr{F}}} f(B) \leqslant c(x), \tag{6}$$

and

$$\sum_{B \in \mathscr{F}} f(B) = \min_{C \in \mathscr{F}_{\geqslant 1}} \sum_{x \in C} c(x). \tag{7}$$

It is not difficult to see that this implies the Menger property (take $c(x) = 1$, for each x), but is not implied by it (consider, for example, the system of three sets (wxy, wxz, wyz), with $c(w) = 2$, and $c(x) = c(y) = c(z) = 1$). The assertion that the arc-sets of xy-paths in a digraph are Mengerian is equivalent to the max-flow min-cut theorem together with the integrity theorem (Theorem 3.1 and Corollary 3.3).

Theorem 7.4. *(i) $\mathscr{F}$ is Mengerian if and only if $\mathscr{F}$ and all its strong point-replications have the Menger property;*

(ii) if a clutter $\mathscr{F}$ is Mengerian, then so are all its minors. ‖

Remark. These results are not difficult to prove: *(i)* is almost tautological, and (*ii*) is proved in [37]. It is not known whether or not the statements *a clutter $\mathscr{F}$ is Mengerian* and *$\mathscr{F}$ and all its minors are Menger systems* are equivalent, although from the two parts of Theorem 7.4 the forward implication clearly holds.

8. König Systems

We remarked in the previous section that König's theorem states simply that the vertex-coboundaries of a bipartite graph have the König property. To check this, note that the system of vertex-coboundaries of any graph, and the system of maximal matchings (maximal by inclusion), are König duals of each other, so that both of these systems are clique-complete. To say that the vertex-coboundaries of a bipartite graph G have the König property is to say that the smallest number of vertices that cover all of the edges of G is equal to the largest number of edges in a matching of G—which is just König's theorem. To say that the maximal matchings have the König property is to say that the smallest number of maximal matchings that cover all the edges of G (that is, the chromatic index of G) is equal to the maximum number of edges incident to a vertex of G (that is, the maximum valency of G)—which is another well-known theorem (Theorem 2.2 of Chapter 5). Thus the vertex-coboundaries and the maximal matchings of a bipartite graph G form two König dual systems of sets, both of which have the König property. Moreover, every weak point-restriction of either of these systems has as its maximal blocks the vertex-coboundaries or maximal matchings of a subgraph of G, and so it too has the König property.

An exactly similar situation holds with the edges (regarded as pairs of vertices) and the maximal independent sets of vertices in a bipartite graph. These two systems are König duals of each other, and so both are clique-complete. To say that the former is a König system is to say that the edge-covering number of a bipartite graph is equal to its vertex-independence number, which is Theorem 4.3. To say that the latter is a König system is to say that the chromatic number of a bipartite graph with at least one edge is 2, which is certainly true. So both of these systems are König systems, and so are all their weak point-restrictions, since the maximal blocks of the latter are obtained just by deleting vertices from the graph.

Yet another example is afforded by the maximal chains and anti-chains of a finite partially ordered set P. It is easy to see that the system of maximal chains and the system of maximal anti-chains are König duals of each other, and so both of them are clique-complete. Dilworth's chain-decomposition theorem (Theorem 2.3) states that the smallest number of chains that cover

P is equal to the size of the largest anti-chain—that is, the maximal chains have the König property. It is easy to see that the smallest number of anti-chains that cover P is equal to the length of the longest chain—that is, the maximal anti-chains have the König property. So we have a third example of a pair of König dual systems of sets, both of which have the König property. Moreover, every weak point-restriction of either of these families has as its maximal blocks the maximal chains or anti-chains of a partially ordered subset of P, and so it too has the König property.

Now, if a set-system $(S, \mathscr{F})$ is clique-complete, it can be represented as the system of (vertex-sets of) cliques in a graph G (taking $V(G) = S$, and joining two vertices by an edge if there is a block of $\mathscr{F}$ containing both of them). The König dual system is then the system of maximal independent sets of vertices of G. In 1961 Berge (see [3, p. 360]) tried to generalize the results of the previous three paragraphs by making his **weak perfect-graph conjecture**: that for a graph G, the following two statements are equivalent:

(*i*) every weak point-restriction of the system of maximal cliques is a König system;

(*ii*) every weak point-restriction of the system of maximal independent sets of vertices is a König system.

This result—in fact, a slightly stronger one—was proved in transpose form in 1972 by Fulkerson and Lovász (see [**12**], [**21**], and [**3**, Theorem 7 on p. 459]):

Theorem 8.1. *Every weak point-restriction of a set-system $\mathscr{F}$ has the König property if and only if every weak point-restriction of $\mathscr{F}$ has the dual König property. This in turn implies (and, if $\mathscr{F}$ is clique-complete, is implied by) the fact that every weak point-restriction of $\mathscr{F}_{\leqslant 1}$ has the König property.* ‖

The truth of the weak perfect-graph conjecture follows immediately from this. A graph G that satisfies the equivalent conditions (*i*) and (*ii*) above is called a **perfect graph**. This gives rise to the final pair of König-dual König systems in Table 1. In view of our previous remarks, it is clear that the first three pairs of König systems in Table 1 are special cases of this last pair, although this is only obvious in the case of the second pair. (A bipartite graph is perfect.)

Theorem 8.1 goes some way towards explaining why these examples of König systems occur in König-dual pairs: whenever all weak point-restrictions of $\mathscr{F}$ have the König property, then $\mathscr{F}_{\leqslant 1}$ will also have the König property. What Theorem 7.1 does not explain is why the particular systems in Table 1, and all weak point-restrictions of them, should have the König property.

Table 1 König systems

{Vertex-coboundaries of a bipartite graph (König's theorem) Maximal matchings of a bipartite graph
{Edges (regarded as pairs of vertices) of a bipartite graph Maximal independent sets of vertices of a bipartite graph
{Maximal chains of a partially ordered set (Dilworth's theorem) Maximal anti-chains of a partially ordered set
{Maximal cliques (vertex-sets) in a perfect graph Maximal independent sets of vertices in a perfect graph

Berge has attempted to answer this question by making his *strong perfect-graph conjecture* [**3**, p. 361], which remains unsettled, and is undoubtedly the most important unsolved problem on the König property.

Strong Perfect-Graph Conjecture. *A graph is perfect if and only if neither it, nor its complement, contains a chordless odd circuit of length greater than* 3.

9. Menger Systems and Mengerian Systems

In the previous section we gave some examples of König systems that occur in König-dual pairs, and stated a theorem and a conjecture to explain this phenomenon. In the case of Menger systems, the empirical evidence is even more striking. There are many examples of Menger systems that occur in Menger-dual pairs (as we shall see in Table 2). However, as far as I know, there is no theorem analogous to Theorem 8.1 to explain this phenomenon—only a conjecture of Seymour [**37**] reproduced below—, and there is no conjecture analogous to the strong perfect-graph conjecture.

There are certainly Menger systems whose Menger duals are not Menger systems. For example, in 1966 Rothschild and Whinston [**36**] proved that if v, v', w and w' are four distinct vertices of an Eulerian graph, then the set of all vv'-paths and ww'-paths (regarded as sets of edges) has the Menger property. But the Menger dual of this system need not have the Menger property (take v, w, v', w' in order round a square C_4 in the octahedral graph). It is interesting to note that although all weak point-restrictions of Rothschild and Whinston's system have the Menger property (by contracting edges), not all strong point-restrictions do (or their theorem would hold in a non-Eulerian graph, which it needn't). So this system is not Mengerian.

All of the Menger systems listed in Table 2 are actually Mengerian; and all of them (with one possible exception, queried in the table) occur in Menger-dual pairs. This might lead one to conjecture that the appropriate analog of

Theorem 8.1 would be that, if $\mathscr{F}$ is Mengerian, then $\mathscr{F}_{\geqslant 1}$ has the Menger property. Seymour [37] has pointed out that this is false—the system of seven sets (*abc*, *cde*, *bdf*, *aef*, *cf*, *be*, *ad*), with Menger dual (*abc*, *cde*, *bdf*, *aef*), is Mengerian, but its Menger dual does not have the Menger property. (The same example was cited by Lovász [22]). However, Seymour made the following conjecture, which is an analog for Menger systems of Theorem 8.1 for König systems.

Conjecture. *If* $\mathscr{F}$ *is Mengerian and does not contain* (*abc*, *cde*, *bdf*, *aef*, *cf*, *be*, *ad*) *as a minor, then* $\mathscr{F}_{\geqslant 1}$ *has the Menger property* (*and so is Mengerian*).

If *all minors of* $\mathscr{F}$ *are Menger systems* does not imply that $\mathscr{F}$ *is Mengerian*, then a stronger conjecture would be obtained by replacing the latter by the former in the above conjecture.

The remainder of this section is taken up with a brief description of the Menger systems in Table 2. First we need some terminology. If (X, Y) is a partition of the vertices of a graph or digraph G into two disjoint sets, then (as in Section 3) $\langle X, Y\rangle$ denotes the set of edges (arcs) joining a vertex in X to a vertex in Y. If G is a graph, then $\langle X, Y\rangle = \langle Y, X\rangle$, and is called a **coboundary**; if X consists of just one vertex, then $\langle X, Y\rangle$ is a **vertex-coboundary**. If G is a digraph, then $\langle X, Y\rangle$ and $\langle Y, X\rangle$ are disjoint sets, and are called **semi-directed coboundaries**; if $\langle Y, X\rangle = \varnothing$, then $\langle X, Y\rangle$ is a **directed coboundary**. Similarly if C is a circuit (elementary) in a digraph, then the sets C_- and C_+ of "clockwise" and "counter-clockwise" oriented edges round C

Table 2 Mengerian systems

XY-paths (vertices or edges) in a graph or digraph (Menger's theorem) XY-separating sets (vertices or edges) in a graph or digraph
Vertex-coboundaries of a bipartite graph Minimal covering sets of edges of a bipartite graph (Gupta's theorem)
Edges (regarded as pairs of vertices) of a bipartite graph Minimal covering sets of vertices of a bipartite graph
Spanning out-trees with fixed root in a digraph (Edmonds's theorem) The Menger dual thereof (Fulkerson's theorem)
Directed coboundaries of a digraph (Lucchesi and Younger's theorem) ?The Menger dual thereof?
Semi-directed coboundaries in a digraph without directed circuits The Menger dual thereof
Semi-directed circuits in a digraph without directed coboundaries The Menger dual thereof
$(C-\{x\}: x \in C \in \mathscr{C})$, where $\mathscr{C}$ is the system of circuits of a regular matroid $(D-\{x\}: x \in D \in \mathscr{D})$, where $\mathscr{D}$ is the system of cocircuits of a regular matroid

are called **semi-directed circuits** (the choice of positive direction being arbitrary); if C_+ or $C_- = \varnothing$, then C_- or C_+ is a **directed circuit**. Finally, it should be clear (in context) what we mean by "replicating edges of G in series" or "in parallel"; we count the contraction of an edge as a special case of the former, and deletion as a special case of the latter, regarding these as "replicating 0 times".

The first entry in Table 2 states that the XY-paths in a graph or digraph have the Menger property, the paths being regarded either as sets of vertices or as sets of edges (arcs). It is easy to see that the Menger dual of this system has the Menger property. In fact, all strong point-replications of these systems have the Menger property (although several different constructions are needed to show this for the different cases: graph or digraph, paths or separating sets, edge-sets or vertex-sets); so these systems are all Mengerian.

Since, as we have seen, the edges of a bipartite graph (regarded as pairs of vertices) have the König property, Theorem 7.3 shows that the transposed system, the system of vertex-coboundaries of a bipartite graph, has the Menger property. The Menger dual of this system, the system of minimal covering sets of edges of a bipartite graph, also has the Menger property, as was proved in 1967 by Gupta [**15**] (see also [**3**, p. 455]). The incidence matrix of the former system is totally unimodular, from which it can be deduced that both of these systems are Mengerian.

In a similar way, since the vertex-coboundaries of a bipartite graph have the König property, the transposed system, the system of edges (regarded as pairs of vertices) must have the Menger property. The Menger dual of this system, the system of minimal covering sets of vertices, also has the Menger property: this just says that the largest number of disjoint covering sets of vertices in a bipartite graph is 2. Again, we have a totally unimodular matrix, so that both of these systems are Mengerian.

Edmonds [7] (see also [**22**, p. 113], [**24**]) proved in 1973 that the spanning "out-trees" with fixed root in a digraph D, when regarded as sets of arcs, have the Menger property. All strong point-replications of this system also have the Menger property (by applying the theorem to the results of replicating arcs of D in parallel), so this system is Mengerian. The Menger dual of this example has the Menger property (consider the sets of arcs entering the vertices other than the root), and Fulkerson [**13**] proved in 1974 that this Menger dual is also Mengerian.

Lucchesi and Younger [**25**] (see also [**24**]) have recently proved that the directed coboundaries of a digraph have the Menger property. All strong point-replications of them also have the Menger property (by replication of arcs in series), so this system is Mengerian. The situation for the Menger dual of this system seems completely open: I strongly suspect that it does have the Menger property, but this seems quite a hard unsolved problem. An affirmative answer would follow from the truth of Seymour's conjecture (stated earlier in this section).

By dualizing (graphically) the result of Lucchesi and Younger, we see that (as they themselves pointed out) the directed circuits in a planar digraph have the Menger property—but not in an arbitrary digraph (take $K_{3,3}$ with its three "vertical" edges directed upwards and its six "sloping" edges directed downwards).

The semi-directed coboundaries in a Hamiltonian digraph have the Menger property (since the Hamiltonian circuit of p arcs intersects every semi-directed coboundary, and the p sets of arcs entering the p vertices are disjoint semi-directed coboundaries), but the same does not necessarily hold for their point-restrictions (weak or strong), strong point-replications, or Menger dual. The result does not hold in an arbitrary digraph: consider the Petersen graph with every edge replaced by two oppositely-oriented arcs in parallel. The semi-directed circuits of an arbitrary digraph also do not have the Menger property: consider $K_{3,3}$ with every edge replaced by two oppositely-oriented arcs in series. However, I proved recently [**44**] that if a digraph D contains no directed circuits (that is, every arc is in a directed coboundary), then the minimal semi-directed coboundaries of D are pairwise disjoint, and so this system and its Menger dual are both Mengerian. Similarly, if a digraph D contains no directed coboundaries (that is, every arc is in a directed circuit), then the minimal semi-directed circuits of D are pairwise disjoint.

Finally we consider an example from matroid theory; the terms used here are defined in [**41**]. Let M be a matroid on a set S with systems $\mathscr{C}$ of circuits and $\mathscr{D}$ of cocircuits. Gallai [**14**] proved in 1959 (in a more general form) that if M is a regular matroid (that is, representable over every field), then, for each $x \in S$, the set systems

$$\mathscr{C}_x := (C - \{x\} : x \in C \in \mathscr{C}) \qquad \text{and} \qquad \mathscr{D}_x := (D - \{x\} : x \in D \in \mathscr{D}),$$

which are Menger duals, both have the Menger property. Seymour [**37**] extended this by proving that $\mathscr{C}_x$ and all of its minors have the Menger property if and only if M does not have a minor U_4^2 or F^* containing x, and $\mathscr{D}_x$ and all its minors have the Menger property if and only if M does not have a minor U_4^2 or F containing x (the matroid U_4^2 is the 2-uniform matroid on a set of four elements, and F and F^* are the Fano matroid and its dual); moreover, when these systems have the Menger property, they are actually Mengerian.

References

1. I. Anderson, Perfect matchings of a graph, *J. Combinatorial Theory* (*B*) **10** (1971), 183–186; *MR***43**#1853.
2. C. Berge, Sur le couplage maximum d'un graphe, *C.R. Acad. Sci. Paris* (*A*) **247** (1958), 258–259; *MR***20**#7278.
3. C. Berge, *Graphs and Hypergraphs*, North-Holland, Amsterdam, 1973; *MR***50**#9640.
4. R. P. Dilworth, A decomposition theorem for partially ordered sets, *Ann. of Math.* (2) **51** (1950), 161–166; *MR***11**–309.
5. G. A. Dirac, Extensions of Menger's theorem, *J. London Math. Soc.* **38** (1963), 148–161; *MR***27**#1939.
6. J. Edmonds, Paths, trees and flowers, *Canad. J. Math.* **17** (1965), 449–467; *MR***31**#2165.
7. J. Edmonds, Edge-disjoint branchings, in *Combinatorial Algorithms* (ed. R. Rustin), Algorithmic Press, New York, 1973, pp. 91–96; *MR***50**#4377.
8. E. Egerváry, Matrixok kombinatorius tulajdonságairól, *Mat. Fiz. Lapok* **38** (1931), 16–28.
9. L. R. Ford and D. R. Fulkerson, Maximal flow through a network, *Canad. J. Math.* **8** (1956), 399–404; *MR***18**–56.
10. L. R. Ford and D. R. Fulkerson, Network flow and systems of representatives, *Canad. J. Math.* **10** (1958), 78–84; *MR***20**#4502.
11. L. R. Ford and D. R. Fulkerson, *Flows in Networks*, Princeton University Press, Princeton, 1962; *MR***28**#2917.
12. D. R. Fulkerson, Antiblocking polyhedra, *J. Combinatorial Theory* (*B*) **12** (1972), 50–71; *MR***44**#2629.
13. D. R. Fulkerson, Packing rooted directed cuts in a weighted directed graph, *Math. Programming* **6** (1974), 1–13; *MR***52**#7953.
14. T. Gallai, Über reguläre Kettengruppen, *Acta Math. Acad. Sci. Hungar.* **10** (1959), 227–240; *MR***23**#A1553.
15. R. P. Gupta, A decomposition theorem for bipartite graphs, in *Theory of Graphs* (ed. P. Rosenstiehl), Gordon and Breach, New York, and Dunod, Paris, 1967, pp. 135–138.
16. P. Hall, On representatives of subsets, *J. London Math. Soc.* **10** (1935), 26–30.
17. P. R. Halmos and H. E. Vaughan, The marriage problem, *Amer. J. Math.* **72** (1950), 214–215; *MR***11**–423.
18. F. Harary, *Graph Theory*, Addison-Wesley, Reading, Mass., 1969; *MR***41**#1566.
19. D. König, Graphok és matrixok, *Mat. Fiz. Lapok* **38** (1931), 116–119.
20. D. König, Über trennende Knotenpunkte in Graphen (nebst Anwendungen auf Determinanten und Matrizen), *Acta Litt. Sci. Szeged* **6** (1933), 155–179.
21. L. Lovász, Normal hypergraphs and the perfect graph conjecture, *Discrete Math.* **2** (1972), 253–267; *MR***46**#1624.
22. L. Lovász, Minimax theorems for hypergraphs, in *Hypergraph Seminar*, Lecture Notes in Mathematics **411** (ed. C. Berge and D. K. Ray-Chaudhuri), Springer-Verlag, Berlin, Heidelberg and New York, 1974, pp. 111–126; *MR***53**#10648.
23. L. Lovász, Three short proofs in graph theory, *J. Combinatorial Theory* (*B*) **19** (1975), 269–271; *MR***53**#211.
24. L. Lovász, On two minimax theorems in graph [*sic*], *J. Combinatorial Theory* (*B*) **21** (1976), 96–103; *MR***55**#174.
25. C. L. Lucchesi and D. H. Younger, A minimax theorem for directed graphs, *J. London Math. Soc.* (2) **17** (1978), 369–374.
26. P. J. McCarthy, Matchings in graphs, *Bull. Austral. Math. Soc.* **9** (1973), 141–143; *MR***48**#2001.
27. C. McDiarmid, On separated separating sets and Menger's theorem, in *Proceedings of the Fifth British Combinatorial Conference* (ed. C. St. J. A. Nash-Williams and J. Sheehan), Congressus Numerantium **XV**, Utilitas Mathematica, Winnipeg, 1975, pp. 455–459.
28. W. Mader, 1-Faktoren von Graphen, *Math. Ann.* **201** (1973), 269–282; *MR***50**#12807.

29. K. Menger, Zur allgemeinen Kurventheorie, *Fund. Math.* **10** (1927), 96–115.
30. O. Ore, Graphs and matching theorems, *Duke Math. J.* **22** (1955), 625–639; *MR* **17**–394.
31. O. Ore, *Theory of Graphs*, Amer. Math. Soc. Colloq. Publ. XXXVIII, American Mathematical Society, Providence, Rhode Island, 1962; *MR* **27**#740.
32. H. Perfect, Applications of Menger's graph theorem, *J. Math. Anal. Appl.* **22** (1968), 96–111; *MR* **37**#93.
33. M. A. Perles, A proof of Dilworth's decomposition theorem for partially ordered sets, *Israel J. Math.* **1** (1963), 105–107; *MR* **29**#5758.
34. J. S. Pym, A proof of Menger's theorem, *Monatsh. Math.* **73** (1969), 81–83; *MR* **39**#1352.
35. R. Rado, Note on the transfinite case of Hall's theorem on representatives, *J. London Math. Soc.* **42** (1967), 321–324; *MR* **35**#2758.
36. B. Rothschild and A. Whinston, On two commodity network flows, *Operations Res.* **14** (1966), 377–387.
37. P. D. Seymour, The matroids with the max-flow min-cut property, *J. Combinatorial Theory* (*B*) **23** (1977), 189–222.
38. W. T. Tutte, The factorization of linear graphs, *J. London Math. Soc.* **22** (1947), 107–111; *MR* **9**–297.
39. W. T. Tutte, The factors of graphs, *Canad. J. Math.* **4** (1952), 314–328; *MR* **14**–67.
40. W. T. Tutte, A short proof of the factor theorem for finite graphs, *Canad. J. Math.* **6** (1954), 347–352; *MR* **16**–57.
41. D. J. A. Welsh, *Matroid Theory*, Academic Press, London, 1976; *MR* **55**#148.
42. H. Whitney, Congruent graphs and the connectivity of graphs, *Amer. J. Math.* **54** (1932), 150–168.
43. D. R. Woodall, The binding number of a graph and its Anderson number, *J. Combinatorial Theory* (*B*) **15** (1973), 225–255; *MR* **48**#5915.
44. D. R. Woodall, Menger and König systems, in *Theory and Applications of Graphs*, Lecture Notes in Mathematics **642** (ed. Y. Alavi and D. R. Lick), Springer-Verlag, Berlin, Heidelberg and New York, 1978, pp. 620–635.

10
Line Graphs and Line Digraphs

ROBERT L. HEMMINGER AND LOWELL W. BEINEKE

1. Introduction

The line graph transformation is probably the most interesting of all graph transformations, and it is certainly the most widely studied (there are over 250 journal articles dealing, at least in part, with line graphs). Much of this activity was stimulated by Ore's discussion of line graphs, and problems about them, in **[50]**.

The line graph concept is quite natural, and has been introduced in several ways. One view of the concept is as an alternative way of describing a graph—the edges and their adjacencies are given without reference to the vertex-set. Of course, the question of whether this is well defined immediately arises: if the graphs G_1 and G_2 are edge-isomorphic (that is, if there is a bijection of $E(G_1)$ onto $E(G_2)$ that preserves the adjacencies of edges), are they necessarily isomorphic? In the founding paper on this subject, Whitney **[64]** showed that for connected graphs the answer to this question is almost always "yes", the only exception being the pair K_3 and $K_{1,3}$.

A significant portion of the material that we shall cover is concerned with two problems that Grünbaum **[19]** formulated for any graphical transformation. He refers to them as the determination and characterization problems—we state them here for line graphs:

Determination Problem. Determine which graphs have a given graph as their line graph.

Characterization Problem. Characterize those graphs that are line graphs of some graph.

A proof of Whitney's theorem, which solves the determination problem, will be given in Section 3. As a by-product we show that, with just four small exceptions, the group of edge-isomorphisms of a connected graph is isomorphic to its group of isomorphisms.

Solutions of the characterization problem will be given in Section 4. One of these solutions is used as the basis of an efficient algorithm for detecting whether or not a given graph H is a line graph, and if so, for producing a "root graph" for H—that is, a solution G of the equation $L(G) = H$. In Section 5 the line graphs of special families of graphs are discussed, and in some cases characterizations are given, while in Section 6 we examine various properties of line graphs and the relationships between the properties of line graphs and their root graphs. For example, when is a line graph Eulerian, Hamiltonian, or planar? And what can be said about the connectivity, genus, or chromatic number of a line graph in terms of its root graph?

It is quite natural to try to extend the line graph concept to digraphs. However, because of the different types of adjacencies between arcs (head-to-head, head-to-tail, and tail-to-tail), it is not obvious how this should be done. The definition used in nearly all of the literature, and the one that we shall use here, reflects only the head-to-tail adjacencies. Because of this, the results for line digraphs often differ considerably from the corresponding ones for line graphs. For example, the line digraph of a connected digraph can be disconnected, and digraphs isomorphic to their own line digraphs are much more complicated than in the line graph case. A significant portion of the literature is devoted to characterizing this latter class of digraphs, and that material will be discussed in Section 9. Section 8 is devoted to the determination and characterization problems for digraphs, and Section 10 to special properties of line digraphs. For the sake of simplicity we shall assume that, unless otherwise specified, digraphs may have loops and multiple arcs; this is not the case for graphs which (again, unless it is specified otherwise) are assumed to be simple. Many of the results mentioned here can be extended to the infinite case in a straightforward way (see Hemminger [**28**]).

2. Elementary Results on Line Graphs

With each graph G we associate another graph $L(G)$, called the **line graph** of G, in which $V(L(G)) = E(G)$, and where two vertices are adjacent if and only if they are adjacent as edges of G. (For convenience, we allow here the empty graph, which has neither vertices nor edges.) Some examples are given in Fig. 1.

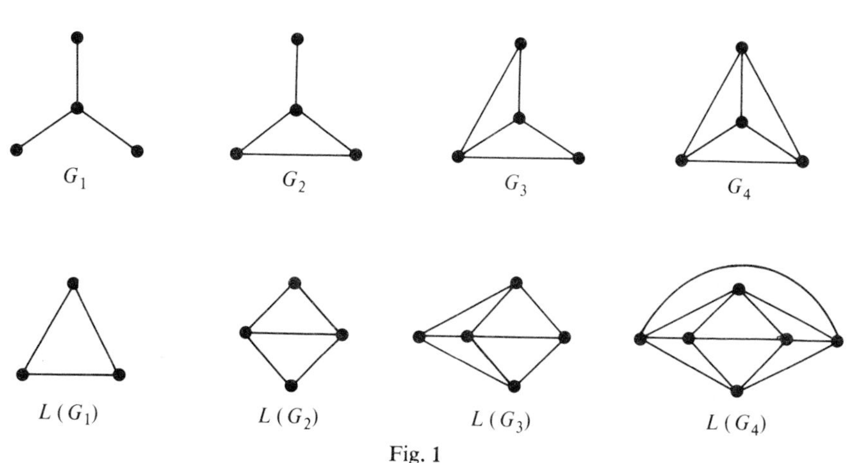

Fig. 1

Technically we should use the name "edge graph", but the name "line graph", introduced in 1960 by Harary and Norman **[23]** (who used "lines" instead of "edges"), is now almost universally accepted. It has not always been so: Kotzig **[40]** called it the "ϑ-obrazom", Fisher **[16]** the "covering graph", Sabidussi **[57]** the "derivative", Seshu and Reed **[60]** the "edge-to-vertex dual", Ore **[50]** the "interchange graph", Menon **[45]** the "adjoint", Vizing **[63]** the "conjugate", Beineke **[5]** the "derived graph", and Berge **[8**, Chapter 17] the "representative graph". Interestingly, neither of the first two to study the concept, Whitney **[64]** and Krausz **[41]**, gave it a name.

The following theorem contains a number of basic observations about line graphs; as usual, the path and circuit of length p are denoted by P_p and C_p, respectively, and the valency of the vertex v is denoted by $\rho(v)$:

Theorem 2.1. *Let G be a graph with p vertices and q edges. Then*

(*i*) *$L(G)$ has q vertices and $\Sigma_v \frac{1}{2}\{\rho(v)\}^2 - q$ edges;*

(*ii*) *the valency in $L(G)$ of an edge vw of G is $\rho(v)+\rho(w)-2$;*

(*iii*) *G and $L(G)$ have the same number of connected components if G has no isolated vertices;*

(*iv*) $L(P_p) \cong P_{p-1}$, *for* $p \geqslant 1$;

(*v*) $L(C_p) \cong C_p$, *for* $p \geqslant 3$,

(*vi*) $L(K_3) \cong L(K_{1,3})$;

(*vii*) *If G is connected and* $p \geqslant 4$, *then* $L(G) \cong K_p$ *if and only if* $G \cong K_{1,p}$. ‖

In Fig. 1, we observe that $L(G_2) \cong G_3$, so that $L(G_3) \cong L(L(G_2))$. Thus we are led to the concept of the *n*th iterated line graph $L^n(G)$, defined by $L^1(G) = L(G)$, and $L^n(G) = L(L^{n-1}(G))$ for $n > 1$. In their investigations of $L^n(G)$, van Rooij and Wilf [**54**] proved our next theorem. One method of proof relies on two facts:

(*i*) if G is connected and not a path, circuit, or $K_{1,3}$, then $L^3(G)$ contains two circuits with at most one vertex in common;

(*ii*) if a graph has two circuits joined by a path of length $k \geqslant 0$, then its line graph has two circuits joined by a path of length $k+1$.

Theorem 2.2. *Let G be a connected graph which is not isomorphic to* $K_{1,3}$ *or a path or circuit, and let* p_n *denote the order of* $L^n(G)$. *Then* $\lim_{n\to\infty} p_n = \infty$. ‖

Corollary 2.3. *If G is connected and* $L^n(G) \cong G$ *for some n, then* $L(G) \cong G$ *and G is a circuit.* ‖

Much of the study of line graphs is related to mappings from one graph to another, and in particular to isomorphisms. We denote the set of all isomorphisms of a graph G into a graph H by $\Gamma(G, H)$. (Thus, $\Gamma(G, G)$ is equal to the automorphism group $\Gamma(G)$.)

An edge-isomorphism from G onto H is a bijection $\sigma: E(G) \to E(H)$ such that two edges are adjacent in G if and only if their images are adjacent in H. We let $\Gamma'(G, H)$ denote the set of all edge-isomorphisms of G onto H, and it is readily seen that $\Gamma(L(G), L(H)) = \Gamma'(G, H)$.

We close this section with some results on edge-mappings induced by (vertex-)isomorphisms. For $\sigma \in \Gamma(G, H)$, define $\sigma^*: E(G) \to E(H)$ by $\sigma^*(vw) = \sigma(v)\sigma(w)$, and let $\Gamma^*(G, H) = \{\sigma^* : \sigma \in \Gamma(G, H)\}$.

Theorem 2.4. *If G and H are graphs, then*

(*i*) $\Gamma^*(G, H) \subseteq \Gamma'(G, H)$;

(*ii*) *the mapping* $T: \Gamma(G, H) \to \Gamma^*(G, H)$ *given by* $T(\sigma) = \sigma^*$ *is one-to-one if and only if G has at most one isolated vertex and no isolated edges.* ‖

3. Isomorphisms and Whitney's Theorem

Whitney's 1932 paper [**64**] was a pioneering work on connectivity and planarity of graphs, and in that setting it is fairly clear that what we shall refer to as "Whitney's theorem" is just a lemma used to show that 3-connected planar graphs have unique duals. Be that as it may, his theorem still marks the beginning of our subject.

Our proof is basically that given by Jung [36], and is also valid for infinite graphs. Before giving it, we require some definitions and a lemma. Any subset of the set of edges incident to a vertex v of a graph G is called a **star** in G, and we let $S(v)$ denote the set of all edges incident to v. A mapping $\sigma: E(G) \to E(H)$ is called **star-preserving** if the set $\sigma(S)$ is a star in H whenever S is a star in G.

Lemma 3.1. *If G and H are connected graphs and $\sigma: E(G) \to E(H)$ is a bijection, then σ is induced by an isomorphism of G onto H if and only if σ and σ^{-1} preserve stars.*

Proof. The condition is clearly necessary, so we assume that σ and σ^{-1} preserve stars. Hence for each vertex v in G, there is at least one vertex v' in H such that $\sigma(S(v)) \subseteq S(v')$. Moreover, v' is uniquely determined by v if $\rho(v) > 1$, since $S(v') \cap S(v'')$ is a singleton set if $v' \neq v''$. Thus, if $\rho(v) > 1$, we have $\rho(v') \geqslant \rho(v) > 1$; so, as with σ, we must have $\sigma^{-1}(S(v')) \subseteq S(v)$. We conclude that the function σ determines a well-defined function $\tilde{\sigma}: \{v \in V(G): \rho(v) > 1\} \to \{v' \in V(H): \rho(v') > 1\}$ which is onto (since σ^{-1} enjoys the same properties as σ) and for which $\sigma(S(v)) = S(\tilde{\sigma}(v))$.

We now assume that G has at least 3 vertices, since otherwise the result is trivial. Thus, if $vw \in E(G)$ and $\rho(w) = 1$, then $\rho(v) > 1$, and so

$$\sigma(vw) \in \sigma(S(v)) = S(\tilde{\sigma}(v)).$$

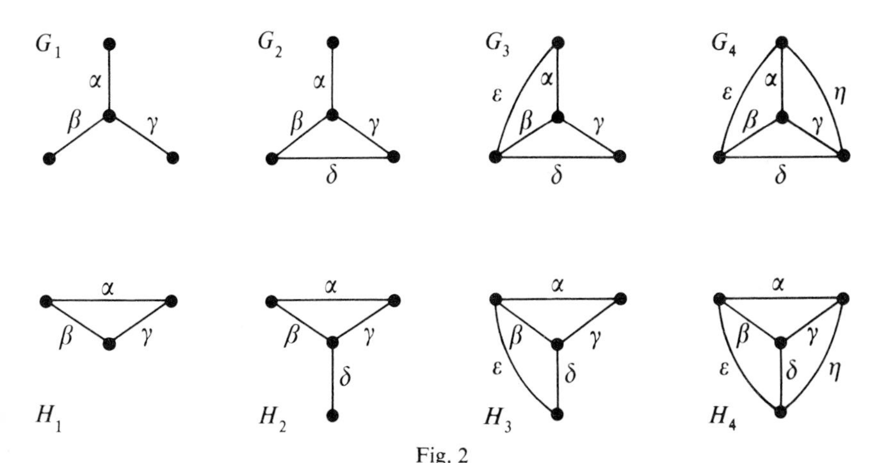

Fig. 2

By the results of the preceding paragraph, we must have $\sigma(vw) = v'w'$, where $v' = \tilde{\sigma}(v)$ and $\rho(w') = 1$. If we extend $\tilde{\sigma}$ by defining $\tilde{\sigma}(w) = w'$ (and still denote the resulting function by $\tilde{\sigma}$), then we conclude that σ determines a function $\tilde{\sigma}: V(G) \to V(H)$ such that $\sigma(S(v)) = S(\tilde{\sigma}(v))$ for all $v \in V(G)$. However, $S(v) = S(w)$ if and only if $v = w$ (since $p \geqslant 3$), and so $\tilde{\sigma}$ is a bijection. The proof is now complete, since $\tilde{\sigma}$ obviously preserves adjacency (and non-adjacency) and induces σ. ‖

Figure 2 gives four examples of edge-isomorphisms between graphs which are not induced by isomorphisms between the graphs. Whitney proved that these are the only such examples.

Theorem 3.2. *If G and H are connected graphs, then, except for the four cases shown in Fig. 2, each edge-isomorphism of G onto H is induced by an isomorphism of G onto H.*

Proof. If σ is an edge-isomorphism of G onto H that is not induced by an isomorphism, then (by the lemma) either σ or σ^{-1} fails to preserve stars. Without loss of generality, we may assume that there is a vertex v in G such that $\sigma(S(v))$ is not a star. Then $\rho(v) = 3$, and $\sigma(S(v))$ is the edge-set of a triangle (the only way four or more edges can be pairwise adjacent is in a star—this is the content of Theorem 2.1(*vii*)).

Let $S(v) = \{\alpha, \beta, \gamma\}$. If $S(v) = E(G)$, then we have the first and basic exceptional case. It is illustrated in Fig. 2 by the pair (G_1, H_1). If $S(v) \neq E(G)$, then there is an edge δ in G that is adjacent to one of the edges in the set $S(v)$, since G is connected. But δ must be adjacent to two of the edges in $S(v)$ since $\sigma(\delta)$ is adjacent to two edges in $\sigma(S(v))$. There are exactly three such edges possible in G. The pairs (G_2, H_2), (G_3, H_3) and (G_4, H_4) illustrate the remaining exceptions to Whitney's theorem; they correspond, respectively, to the existence in G of one, two or three such edges. ‖

Corollary 3.3. *If G and H are connected graphs, then $L(G) \cong L(H)$ if and only if either $G \cong H$ or $\{G, H\}$ is, up to isomorphism, the unordered pair $\{K_3, K_{1,3}\}$.* ‖

In one form, Whitney's theorem tells us when $\Gamma^*(G, H) = \Gamma'(G, H)$ for connected graphs G and H. And, by Theorem 2.4, the natural mapping T from $\Gamma(G, H)$ onto $\Gamma^*(G, H)$ is usually one-to-one. Thus the sets $\Gamma(G, H)$, $\Gamma^*(G, H)$ and $\Gamma'(G, H)$ are closely related—all the more so if we take $G = H$. In that case they are groups (which we denote by $\Gamma(G)$, $\Gamma^*(G)$ and $\Gamma'(G)$, respectively), and T is a homomorphism (since

$$(\tau\sigma)^*(vw) = \tau\sigma(v)\tau\sigma(w) = \tau^*(\sigma(v)\sigma(w)) = \tau^*\sigma^*(vw)$$

for all $vw \in E(G)$). Since isomorphisms and edge-isomorphisms preserve connected components, combining these observations, Whitney's theorem, and Theorem 2.5, we have the following result:

Theorem 3.4. *Let G be a graph. Then*

(i) $\Gamma^*(G)$ *is a subgroup of* $\Gamma'(G)$;

(ii) $\Gamma(G) \cong \Gamma^*(G)$ *if and only if G has at most one isolated vertex and no isolated edges;*

(iii) $\Gamma^*(G) = \Gamma'(G)$ *if and only if G has no connected component isomorphic to* G_2, G_3 *or* G_4 *(of Fig. 2) nor a pair of connected components isomorphic to* K_3 *and* $K_{1,3}$;

(iv) for G connected, $\Gamma(G) \cong \Gamma'(G)$ *if and only if G is not isomorphic to* G_2, G_3 *or* G_4 *(of Fig. 2).* ||

Although this theorem (or portions of it) has been discovered several times, Sabidussi [**56**] was the first to point out the connection between these groups. Harary and Palmer [**24**] also characterized graphs for which $\Gamma(G)$ and $\Gamma^*(G)$ are isomorphic as permutation groups.

Inspired by Whitney's theorem, Nešetřil [**49**] considered edge-functions induced by homomorphisms rather than just isomorphisms. For this purpose we let $\Pi(G, H)$ denote the set of homomorphisms of G into H. As before, a function $\sigma \in \Pi(G, H)$ induces, in the natural way, the function

$$\sigma^{\#} : E(G) \to E(H)$$

given by $\sigma^{\#}(vw) = \sigma(v)\sigma(w)$. But now $\sigma^{\#}$ need not be in $\Pi(L(G), L(H))$. Nešetřil gave necessary and sufficient conditions for this to be the case, as well as giving a characterization of the homomorphisms of $L(G)$ into $L(H)$ that are induced by homomorphisms of G into H. Finally, he showed that $\Pi(G)(=\Pi(G, G))$ and $\Pi(L(G))$ are isomorphic via the #-function if and only if $\Pi(G)$ and $\Pi(L(G))$ are groups.

Nebeský (unpublished) has observed that by use of Krausz's result (see Section 4), Whitney's theorem can also be extended in the following way: *if G and H are connected graphs different from* $K_{1,3}$, *and if* $L(G)$ *is homeomorphic to* $L(H)$, *then G is homeomorphic to H.*

A different generalization was initiated by Whitney himself. A ***k*-skein** in a graph is a collection of k paths with just their terminal vertices in common, and a ***k*-skein isomorphism** from a graph G onto a graph H is a bijection $\sigma: E(G) \to E(H)$ such that a set N of edges forms a k-skein in G if and only if $\sigma(N)$ forms a k-skein in H. A 1-skein isomorphism is an edge-isomorphism, and a 2-skein isomorphism is called a "cycle-isomorphism". As an intermediate step in his proof of the uniqueness of duals, Whitney showed that cycle-isomorphisms of 3-connected graphs are induced by isomorphisms. In a subsequent paper [**65**] he showed that the assumption could be reduced to 2-connectedness. This problem lay dormant until Halin and Jung [**20**], in response to a question of Ore [**50**, p. 249], proved the following general result:

Theorem 3.5. *Let $k > 3$, and let G and H be $(k+1)$-connected graphs. Then any k-skein isomorphism from G to H is induced by an isomorphism.* ‖

For $k = 3$, the hypotheses cannot be weakened in the way Whitney did it for $k = 2$, as is shown by K_4. The existence of other such examples for $k \geqslant 3$ is undetermined.

4. Characterizations of Line Graphs

In this section we discuss characterizations of the class of line graphs, and a number of these have been found—by Krausz [**41**], van Rooij and Wilf [**54**], A. R. Rao (unpublished), Beineke [**6**], and Nebeský [**48**]. In the next section we discuss characterizations of some special classes.

Because not all graphs are line graphs, the characterization problem is not vacuous. For example, $K_{1,3}$ is not a line graph; for, if it were the line graph of G, then G would have each of three edges adjacent to a fourth, and hence two of the three would also be adjacent and $L(G)$ would contain a triangle. Eight other minimal graphs that are not line graphs are shown in Fig. 3. We shall see that in one characterization these nine graphs are forbidden induced subgraphs.

One does not have to consider many examples of line graphs before realizing that direct reliance on the line graph definition is not the easiest method of drawing them. Rather, one uses the complete subgraphs of the line graphs induced by the stars at the vertices of the given graph. Indeed, these subgraphs are quite special, and a graph which is a line graph must have such complete subgraphs. We define a collection $\mathscr{K}$ of subgraphs of a graph H to be a **Krausz partition** of H if it has the following three properties:

(*i*) each member of $\mathscr{K}$ is a complete graph;

(*ii*) every edge of H is in exactly one member of $\mathscr{K}$;

(*iii*) every vertex of H is in exactly two members of $\mathscr{K}$.

It is clear that if H is the line graph of G, then the family of subgraphs $\langle S(v) \rangle$ of H induced by the stars at the vertices of G form a Krausz partition.

Cliques (maximal complete subgraphs) generally serve as members of a

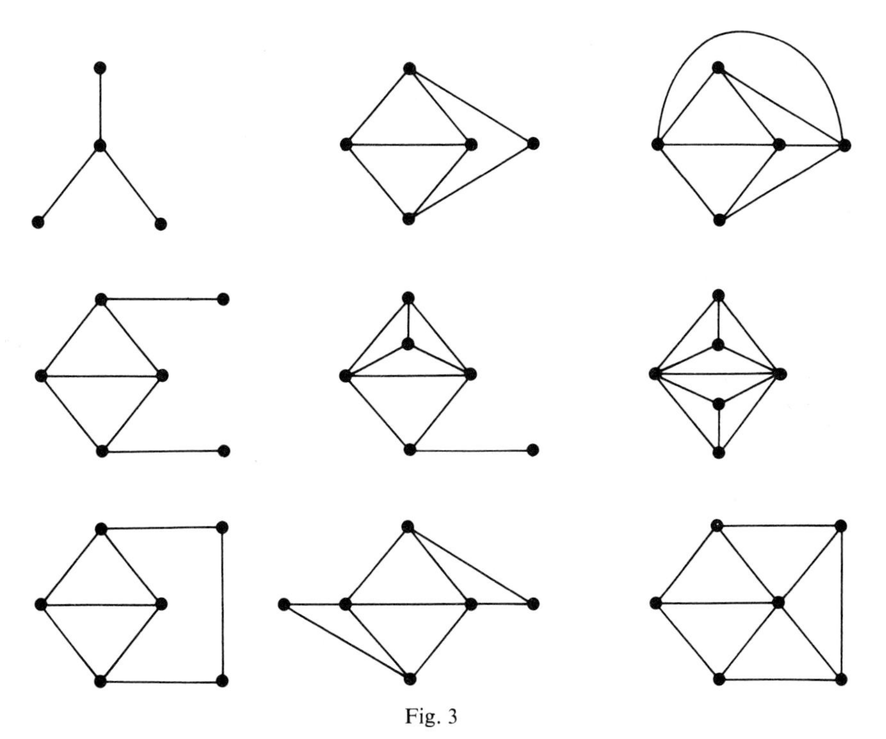

Fig. 3

Krausz partition, with triangles playing a special role (as one might expect from Whitney's theorem). We shall call a triangle **odd** if there is some vertex adjacent to an odd number of its vertices; otherwise we call it **even**.

Lemma 4.1. *Let G be a connected graph other than one of the graphs G_i of Fig. 1, let H be its line graph, let $\mathcal{K}$ be the Krausz partition of H induced by the stars in G, and let $A \in \mathcal{K}$. Then*

(*i*) *if A has at least three vertices, it is a clique of H;*

(*ii*) *if A has exactly three vertices, it is an odd triangle;*

(*iii*) *if A has exactly two vertices, it is either a clique or part of an even triangle;*

(*iv*) *if A has just one vertex, then it and all its neighbors form a clique which is not an even triangle;* .

(*v*) *a clique of H is in $\mathcal{K}$ if and only if it is not an even triangle;*

(*vi*) *if t triangles share a common edge in H, then at least $t-1$ of them are odd, and all the odd ones are contained in a member of $\mathcal{K}$.* ‖

The main observation to be made in the proof of this result is that a 3-clique of H is in $\mathscr{K}$ if and only if it is odd. This is because triangles in G go to even triangles of H, while 3-stars in G go to odd triangles. The latter statement is not true for the graphs in Fig. 1; in fact, in those line graphs all triangles are even.

Before giving the characterization theorem, we observe that the family $\mathscr{K}$ in the lemma is in fact the only Krausz partition of H. For if $\mathscr{K}$ is a Krausz partition, then the members of $\mathscr{K}$ with four or more vertices are precisely the cliques of those orders, the 3-vertex members are precisely the odd 3-cliques, the 2-vertex members are the 2-cliques and the edges of even triangles not in a member of $\mathscr{K}$ already specified, and the singleton members are those vertices in only one of the members of $\mathscr{K}$ already specified. We state this formally:

Corollary 4.2. *If H is a connected line graph other than one of the graphs $L(G)$ in Fig.* 1, *then H has one and only one Krausz partition.* ‖

In the following theorem, the first of the characterizations of a line graph is due to Krausz [**41**], the second to van Rooij and Wilf [**54**], and the third to Beineke [**6**] and Robertson (unpublished), independently. (The graph $K_4 - K_2$ is the graph obtained from K_4 by deleting an edge.)

Theorem 4.3. *Let H be a graph. Then the following statements are equivalent*:

(*i*) *H is a line graph*;

(*ii*) *H has a Krausz partition*;

(*iii*) *H does not have $K_{1,3}$ as an induced subgraph, and any induced subgraph isomorphic to $K_4 - K_2$ has one of its triangles even*;

(*iv*) *H does not contain an induced subgraph isomorphic to any of the graphs of Fig.* 3.

Sketch of Proof. We assume that H is connected, and we have already seen that each of the conditions (*ii*), (*iii*) and (*iv*) is necessary for H to be a line graph (for (*iii*) and (*iv*), we note that a vertex-induced subgraph of a line graph is a line graph).

(*ii*) $\Rightarrow$ (*i*). Suppose that (*ii*) holds, and let $\mathscr{K}$ be the given Krausz partition. Define the graph G by taking $V(G) = \mathscr{K}$ and, for distinct members A, B in $\mathscr{K}$, putting $AB \in E(G)$ if and only if $A \cap B \neq \varnothing$. By properties (*i*) and (*ii*) of a Krausz partition, $A \cap B$ is a singleton set whenever $AB \in V(L(G)) = E(G)$, so that letting $\sigma(AB)$ be the element of $A \cap B$ gives a well-defined function $\sigma: V(L(G)) \rightarrow V(H)$. The proof is completed by showing that σ is a one-to-one and onto function that preserves adjacencies; these properties all follow directly from the definition of a Krausz partition, and their verification is left as an exercise.

(*iii*) $\Rightarrow$ (*ii*). Suppose that (*iii*) holds, and assume that H has two even triangles T_1 and T_2 sharing an edge. To be explicit, let T_1 and T_2 be as in Fig. 4(a). If H is not the graph in Fig. 4(a), then there is some vertex y adjacent to exactly two vertices of T_1. But then y is adjacent to T_2, and so it must also be adjacent to exactly two vertices of T_2. This pair cannot be $\{u, v\}$ since H has no induced subgraph isomorphic to $K_{1,3}$. Thus y is as in Fig. 4(b); moreover, H has no other vertex adjacent to v, x and w, since such a vertex would be adjacent to y (there is no $K_{1,3}$), and we would then have two odd triangles sharing an edge. Likewise, H can have at most one more vertex, and then H must be as in Fig. 4(c). Note that the graphs of Fig. 4 are line graphs (of three of the graphs in Fig. 1). Thus we can assume that at least one of the two triangles sharing an edge is odd. Under these conditions, the set $\mathscr{K}$ of subgraphs of H (as described immediately preceding Corollary 4.2) is easily seen to be a Krausz partition of H, as required.

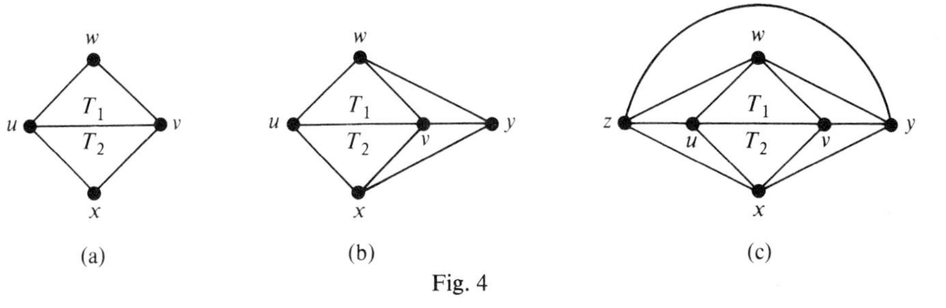

Fig. 4

(*iv*) $\Rightarrow$ (*iii*). Finally, suppose that (*iv*) holds. Then it can be shown that the eight subgraphs of Fig. 3 different from $K_{1,3}$ represent the different ways in which H can contain an induced subgraph isomorphic to $K_4 - K_2$ in which both triangles are odd. We omit the details. ||

We now sketch an alternative proof of Corollary 4.2 that is of some interest, since it uses the theorems of both Krausz and Whitney. Suppose that $\mathscr{K}$ and $\mathscr{K}'$ are two Krausz partitions for a connected graph $H \not\cong L(G_i)$, $i = 1, 2, 3, 4$ (of Fig. 1). Then, as in the proof of Krausz's theorem, there are graphs G and G' with $V(G) = \mathscr{K}$, $V(G') = \mathscr{K}'$, and isomorphisms $\sigma : L(G) \cong H$ and $\sigma' : L(G') \cong H$. Thus $(\sigma')^{-1}\sigma$ is an edge-isomorphism of G onto G', and so, by Whitney's theorem, it is induced by an isomorphism $\tau : G \cong G'$. Let $A \in \mathscr{K}$, and let v be a vertex of A. Then there is another element B of $\mathscr{K}$ containing v, so that AB is an edge of G and $\sigma(AB) = v$. Since τ is induced by $(\sigma')^{-1}\sigma$, we must have that $(\sigma')^{-1}\sigma(AB) = (\sigma')^{-1}(v) = \tau(A)\tau(B)$ is an edge of G', and so $\sigma'(\tau(A)\tau(B)) = v$. Thus $v \in \tau(A)$, and so $A \subseteq \tau(A)$. By symmetry, we have $A = \tau(A)$. But A was arbitrary, and so we have (again by symmetry) that $\mathscr{K} = \mathscr{K}'$. By an appropriate modification of this argument, we can also show that Whitney's theorem follows from Krausz's result and Corollary 4.2.

Since the graphs in Beineke's characterization involve at most six vertices, we can decide in $O(p^6)$ operations whether or not a given graph H of order p is a line graph. Of course, if H is a line graph we should also like to be able to find its root graph efficiently. Roussopoulos **[55]** has shown that for a line graph with p vertices and q edges this can be done in $O(\max\{p, q\})$ steps by an algorithm that also decides whether or not H is a line graph. The algorithm constructs a Krausz partition $\mathscr{K}$ and is based on the following lemma, whose proof is obvious:

Lemma 4.4. *Let $\mathscr{K}$ be a Krausz partition for the line graph H, let $A \in \mathscr{K}$, and let $A_1 = \{v \in A : \{v\} \in \mathscr{K}\}$. Then*

$$\mathscr{K}_A = \mathscr{K} - \{A\} \cup (\{\{v\} : v \in A - A_1\} - \{\{v\} : v \in A_1\})$$

is a Krausz partition for the graph H_A obtained from H by deleting the edges of A and the vertices of A_1. ||

The singleton sets in a Krausz partition cause no trouble, and so we ignore them until the end of the algorithm. Throughout this algorithm, the phrase "put the subgraph A into $\mathscr{K}$" means also that we delete the edges of A from H. The algorithm is as follows:

Step 1. If there are no edges, go to Step 3; otherwise, pick an edge e, and find the number t of triangles of H that contain e:

(a) if $t = 0$, put this edge subgraph into $\mathscr{K}$ and return to Step 1;
(b) if $t = 1$, and if another edge f on this triangle is contained in more than one triangle, then replace e by f and go to Step 2; otherwise, put this triangle into $\mathscr{K}$ and return to Step 1;
(c) if $t \geqslant 2$, go to Step 2.

Step 2. Find the number s of odd triangles in H that contain e:

(a) if $t = 2$ and $s = 0$, then either $H \cong L(G_2)$, $L(G_3)$ or $L(G_4)$ (of Fig. 1), or H is not a line graph. Put either of the two triangles into $\mathscr{K}$ and return to Step 1 (the non-line graphs will be detected eventually);
(b) if $s = t$ or $s = t-1$, and if the s odd triangles form a complete

subgraph of H, then put this complete subgraph into $\mathscr{K}$ and return to Step 1;

(c) if none of the above happens, then H is not a line graph.

Step 3. If a vertex v is in only one member of $\mathscr{K}$, then add $\{v\}$ to $\mathscr{K}$.

Step 4. STOP.

If at any time a vertex has to appear in more than two of the members of $\mathscr{K}$ produced up to that point, then G is not a line graph. Otherwise, the algorithm will clearly produce a Krausz partition $\mathscr{K}$, and G can be constructed from $\mathscr{K}$ as in the proof of Krausz's result.

The proof that the algorithm works follows from the properties enunciated in Lemma 4.1. In order to obtain the stated efficiency, a depth first search was used in Step 2.

The only other characterizations for the class of all line graphs were given by Rao (unpublished) and Nebeský **[48]**, and their characterizations depend on whether or not certain associated graphs are line graphs.

Ray-Chaudhuri **[52]** gave a partial eigenvalue characterization of line graphs. It is easy to see that if H is a line graph in which every vertex has valency at least 4, then (*i*) its minimum eigenvalue is -2, and (*ii*) any two adjacent vertices v and w are mutually adjacent to fewer than $\min(\rho(v), \rho(w))-2$ vertices. Ray-Chaudhuri showed that any graph with these two properties and minimum valency greater than 43 is a line graph. The bound on the minimum valency was needed to deduce from the eigenvalue condition that the graph has no subgraph isomorphic to $K_{1,3}$. Some restriction such as this is necessary since Hoffman **[33]** had constructed a graph of order 28 and minimum valency 12 which satisfies conditions (*i*) and (*ii*), but is not a line graph. Extensions of these results have been obtained by Hoffman **[34]**, **[35]**. In a remarkable paper, Cameron *et al.* **[12]** introduced a new approach which uses the theory of root systems from Lie algebra. Their results not only determine optimal bounds on the order and valency, but also explain the exceptional cases. This material is discussed further in Chapter 11, Section 9 and Chapter 12.

5. Line Graphs of Certain Classes of Graphs

In this section we consider characterizations of the line graphs of some restricted classes of graphs, primarily of bipartite graphs.

Since 3-cliques are the only source of ambiguity in the Krausz partition of a line graph it is not surprising that conditions have been imposed to remove this ambiguity. Of course, the line graphs lacking that ambiguity are those with no even 3-cliques; that is, the line graphs of triangle-free graphs—in this case, the Krausz partition is the set of cliques augmented by the appropriate singleton vertex-sets. Furthermore, the construction of root graphs uses the Krausz partition, so that, in this context, the clique graph transformation K is almost the inverse of the line graph transformation, failing only at those augmented singleton sets. (Here, the **clique graph** $K(H)$ is the intersection graph of the cliques of H—that is, the vertices of $K(H)$ correspond to the cliques of H, and two of these vertices are joined by an edge if and only if the corresponding cliques intersect.) Thus we have the following theorem of Hedetniemi and Slater [**26**]:

Theorem 5.1. *If G is a triangle-free connected graph with at least three vertices, then* $K(L(G)) = G - \{v : \rho(v) = 1\}$. ‖

This characterization of the line graph of triangle-free graphs is part (*ii*) of the next theorem; part (*iii*) is a previously unpublished result due to A. R. Rao:

Theorem 5.2. *The following statements are equivalent for a connected graph H with at least four vertices*:

(*i*) *H is the line graph of a triangle-free graph*;

(*ii*) *two cliques of H have at most one vertex in common, and the clique graph $K(H)$ is triangle-free*;

(*iii*) *all vertices adjacent to the two vertices of an edge of H induce a complete subgraph of H, and the set of vertices adjacent to a given vertex can be partitioned into no more than two complete subgraphs of H*;

(*iv*) *H contains no induced subgraph isomorphic to $K_{1,3}$ or $K_4 - K_2$.*

Proof. The necessity of the conditions is clear: $K(H)$ is triangle-free by Theorem 5.1; the first condition of (*iii*) follows from Lemma 4.1 (*vi*), since H has no even triangles; and a copy of $K_4 - K_2$ in a line graph comes from a copy of graph G_2 in Fig. 1. So we need prove only the sufficiency of the conditions.

(*ii*) ⇒ (*i*). Let $\mathscr{K}$ be the set of cliques of H, augmented by the singleton sets of vertices in only one clique of H. It follows that $\mathscr{K}$ is a Krausz partition of H, since a vertex in more than two cliques of H would lead to a triangle in $K(H)$. Hence H is a line graph, and (as before) it must be the line graph of a triangle-free graph.

(*iii*) ⇒ (*ii*). By the first condition of (*iii*), two cliques of H have at most one vertex in common. Suppose that $K(H)$ has a triangle, say on cliques A, B

and C. By the second condition of (*iii*), these do not intersect at a single vertex, and so their pairwise intersections occur on a set $\{u, v, w\}$. But then these three vertices are contained in a clique that intersects one of A, B or C in more than one vertex. Thus $K(H)$ is triangle-free.

(*iv*) ⇒ (*i*). Finally, suppose that (*iv*) holds. Then $H = L(G)$ for some graph G, by the van Rooij and Wilf condition of Theorem 4.3. Furthermore, G cannot contain a triangle, since if it did, then it would contain a subgraph isomorphic to the graph G_2 of Fig. 1, and then H would necessarily contain a subgraph isomorphic to $K_4 - K_2$. ||

As corollaries, we obtain characterizations of two classes of graphs of special interest. The bipartite case is due independently to Hedetniemi [**25**], Hedetniemi and Slater [**26**], and Kundu (unpublished), and part (*iii*) is due to Chartrand [**13**] (see also [**21**, p. 78]).

Corollary 5.3. *The following statements are equivalent for a graph H*:

(*i*) *H is the line graph of a bipartite graph*;

(*ii*) *two cliques of H have at most one vertex in common, and the clique graph $K(H)$ is bipartite*;

(*iii*) *H has no induced subgraph isomorphic to $K_{1,3}$, $K_4 - K_2$, or C_{2k+1} for $k \geqslant 2$.* ||

Corollary 5.4. *The following statements are equivalent for a graph H*:

(*i*) *H is the line graph of a tree*;

(*ii*) *two cliques of H have at most one vertex in common, and the clique graph $K(H)$ is a tree*;

(*iii*) *H is connected, every block of H is complete, and each cut-vertex lies in exactly two blocks.* ||

Results of another type were obtained by Balconi [**3**], [**4**]. In particular, he obtained the following characterization of line graphs in which all vertices have sufficiently large valency. We let $\rho(v, w)$ denote the number of vertices adjacent to the pair of vertices v and w.

Theorem 5.5. *A graph H whose minimum valency exceeds* 20 *is a line graph if and only if*

(*i*) $\rho(v, w) \leqslant 4$, *if v and w are not adjacent*;

(*ii*) *for each vertex v there exist numbers $\alpha(v)$ and $\beta(v)$ satisfying*

$$\alpha(v) \geqslant \beta(v) \geqslant 10,$$

and $\alpha(v)+\beta(v) = \rho(v)$, *such that, for* $\alpha(v)$ *of the vertices* w *adjacent to* v, $\alpha(v)-1 \leqslant \rho(v, w) \leqslant \alpha(v)$, *and for the other* $\beta(v)$ *of these vertices,*

$$\beta(v)-1 \leqslant \rho(v, w) \leqslant \beta(v). \; \|$$

Several characterizations have come out of work in the design of experiments. Classes involved in these association schemes include the line graphs of complete and complete bipartite graphs, projective and affine planes and other BIBD's, and most of this is presented in an excellent survey by Bose **[11]**. For the characterizations of $L(K_p)$ and $L(K_{r,s})$, see Chapters 11 and 12. Here we present only the following result of Rao **[51]**:

Theorem 5.6. *Let* m *and* n *be positive integers, and let* $k = (m-1)n$. *If* $k > 10$, *a graph* H *is the line graph of the complete* m*-partite graph* $K_{n(m)}$ *if and only if it satisfies the following conditions:*

(*i*) H *has* $\binom{m}{2}n^2$ *vertices;*

(*ii*) H *is regular of valency* $2k-2$;

(*iii*) *if* $d(v, w) = 2$, *then* $\rho(v, w) \leqslant 4$;

(*iv*) *there are* $\binom{m}{2}n^2(n-1)$ *pairs of adjacent vertices* v *and* w *for which* $\rho(v, w) = k-1$;

(*v*) *the remaining pairs of adjacent vertices* v *and* w *satisfy* $\rho(v, w) = k-2$. ‖

Any discussion of characterizations of special classes of graphs would be incomplete without a mention of infinite graphs and hypergraphs. As far as infinite graphs are concerned, locally-finite graphs which are isomorphic to their line graphs have been characterized (see **[58]**), but the general case remains open. A considerable amount of work has been done in the area of hypergraphs. The place to begin the study of hypergraphs is Berge **[8]**, with special work in the area of line graphs found in Bermond **[9]**, Heydemann and Sotteau **[32]** and Marczyk **[43]**. Topics covered include the following: generalizations of the Whitney and Krausz theorems, characterizations of the line graphs of particular hypergraphs (mostly concerned with h-uniform complete q-partite hypergraphs), the computation of parameters for the line graphs of particular hypergraphs, and the characterization of graphs that are line graphs of particular kinds of hypergraphs (in particular, h-uniform hypergraphs). This last problem seems to be very difficult in that infinite families of irreducible forbidden subgraphs are known.

6. Line Graphs with Special Properties

The line graph transformation, by the very nature of its definition, translates edge concepts to vertex concepts. Thus, the chromatic index (edge-chromatic number) of a graph is the (vertex-)chromatic number of its line graph, and sometimes results such as Vizing's theorem are stated in terms of vertex-colorings of line graphs. (For a survey of edge-colorings of graphs, see Chapter 5.) However, since the line graph transformation is not one-to-one, the translation of concepts is not always perfect (or, on occasion, it is fine, but the proof of that fact is non-trivial). An example of this is given by Hemminger **[27]** who showed that a graph G is edge-reconstructible if and only if $L(G)$ is vertex-reconstructible (see Section 5 of Chapter 8).

In the last section we considered line graphs $L(G)$ for which $G \in \mathscr{C}$ for certain classes $\mathscr{C}$ of graphs; in this section we consider classes $\mathscr{C}$ for which $L(G) \in \mathscr{C}$. The following general result shows a relationship between the two topics in some cases. It was discovered independently by Greenwell **[17]** and Nebeský **[47]**:

Theorem 6.1. *If G is a connected graph with at least one circuit, then $L(G)$ contains a spanning subgraph homeomorphic to G.* ||

In order to prove this result, one takes a spanning connected subgraph H of G with just one circuit, and extends $L(H)$ to yield the desired spanning subgraph of $L(G)$.

As applications of this theorem, we observe that the line graph of a non-planar graph must be non-planar, and that of a Hamiltonian graph must be Hamiltonian. These are two of the properties we shall discuss in this section.

Planarity

Sedláček **[59]** discovered the following result; once the conditions are known, the proof is relatively straightforward:

Theorem 6.2. *A graph G has a planar line graph if and only if the following three conditions hold*:

(*i*) *G is planar*;

(*ii*) *G has maximum valency at most* 4;

(*iii*) *every vertex of valency* 4 *is a cut-vertex.* ||

Using this result, Greenwell and Hemminger **[18]** gave a forbidden subgraph characterization of planar line graphs:

Theorem 6.3. *A graph has a planar line graph if and only if it has no subgraph homeomorphic to $K_{3,3}$, $K_{1,5}$, K_1+P_4, or K_2+K_3 (see Fig. 5).* ||

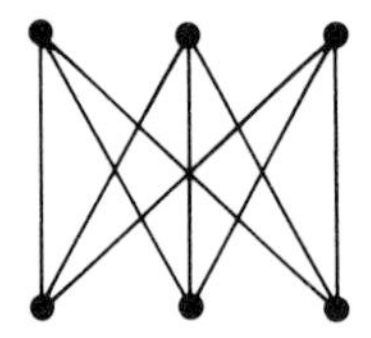
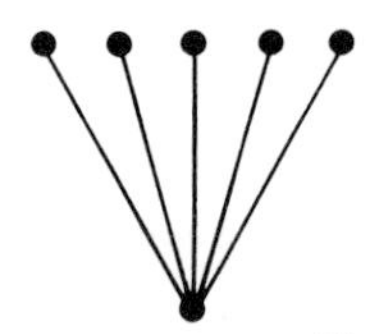
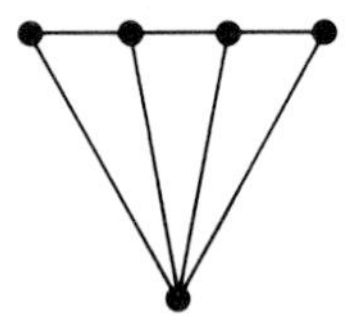
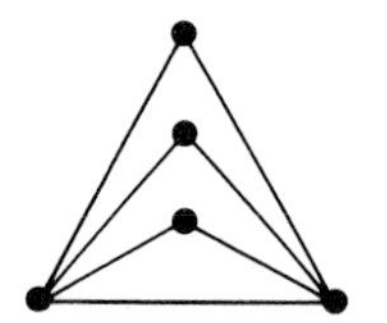

Fig. 5

D. Greenwell (unpublished) extended this by characterizing those graphs G for which $L^n(G)$ is planar, and independently Kulli and Sampathkumar [**42**] gave the corresponding Sedláček-type conditions for these graphs. For $n \geqslant 4$, the answers are all the same, and the only connected graphs satisfying the properties are paths, circuits, and $K_{1,3}$.

Another consequence of Theorem 6.1 is that the genus of $L(G)$ is at least as great as that of G. The following result of Greenwell (unpublished) gives a lower bound for the genus of $L(G)$:

Theorem 6.4. *Let G be a graph with p vertices, q edges, and valencies $\rho_1, \rho_2, \ldots, \rho_p$. Then*

$$\gamma(L(G)) > \sum_{i=1}^{p} \gamma(K_{\rho_i}) + \tfrac{1}{2}q - 2p + 1. \quad \|$$

Hamiltonian Circuits

We observed above that the line graph of a Hamiltonian graph is also Hamiltonian. Two other families of graphs which are easily recognizable and have Hamiltonian line graphs are the Eulerian graphs and stars (see also Section 5 of Chapter 6). A combination of these families describes those graphs having Hamiltonian line graphs—they are Eulerian graphs with edges added at some vertices. Our next theorem, due to Harary and Nash-Williams [**22**], states this more precisely, and gives further aspects of the relationship between Eulerian graphs and Hamiltonian line graphs. Here, the **subdivision graph** $S_1(G)$ of a graph G is obtained from G by inserting a new vertex into each edge, and the **second subdivision graph** $S_2(G)$ has two new vertices inserted into each edge.

Theorem 6.5. *Let G be a graph. Then*

(*i*) *$L(G)$ is Hamiltonian if and only if G has a closed trail incident with each edge;*

(*ii*) *$L(S_1(G))$ is Hamiltonian if and only if G has a spanning closed trail;*

(*iii*) *$L(S_2(G))$ is Hamiltonian if and only if G is Eulerian.* $\|$

The role of the second subdivision graph $S_2(G)$ is to separate the Krausz partition members in $L(G)$, and to insert a path of length two between the vertices which two of them had in common. Consequently, a Hamiltonian circuit in $L(S_2(G))$ must traverse all of these inserted paths. But then, by the properties of a Krausz partition, this forces the Hamiltonian circuit, upon entering a clique corresponding to the star of a vertex of G (that is, not one of the vertices inserted in forming $S_2(G)$), to leave that clique on the very next vertex of the cycle—this is what makes it correspond to an Eulerian trail in G. In Fig. 6 the cliques in $L(S_1(G))$ and $L(S_2(G))$ corresponding to the Krausz

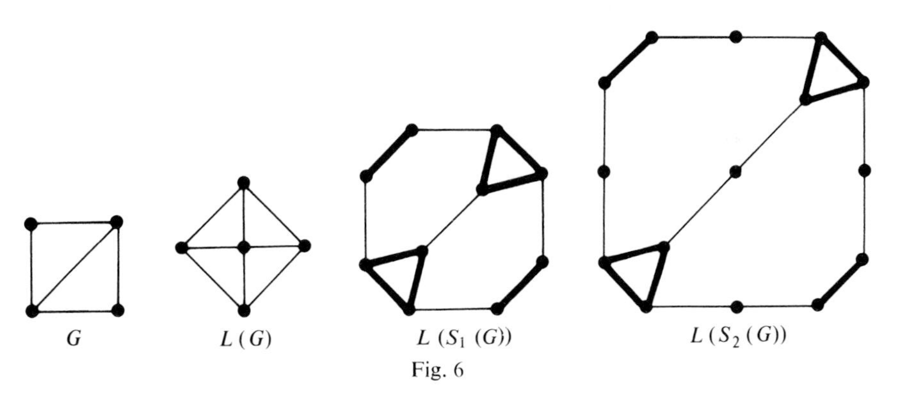

Fig. 6

partition of $L(G)$ are drawn with heavy lines; note that G is not Eulerian, but that $L(G)$ and $L(S_1(G))$ are both Hamiltonian.

Chartrand [**14**] showed that iterated line graphs also tend to be Hamiltonian:

Theorem 6.6. *If G is a connected graph of order p (other than a path), then $L^n(G)$ is Hamiltonian for all $n \geqslant p-3$.* ||

Before leaving Hamiltonian line graphs, we should like to mention three other results, the first due to Kotzig [**40**] (independently rediscovered by Martin [**44**]), the second to Nebeský [**47**], and the last to Bermond and Rosenstiehl [**10**]. Further results on Hamiltonian line graphs may be found in Chapter 6.

Theorem 6.7. *A cubic graph G is Hamiltonian if and only if L(G) has two edge-disjoint Hamiltonian circuits.* ||

Theorem 6.8. *If G is a graph of order 5 or more, then G or $\bar{G}$ has a Hamiltonian line graph.* ||

Theorem 6.9. *The square of the line graph of a connected graph is pancyclic.* ||

Connectivity

Bounds on the connectivity κ and edge-connectivity λ of line graphs have been given by Chartrand and Stewart [**15**]:

Theorem 6.10. *For any graph G,*

(*i*) $\kappa(L(G)) \geqslant \lambda(G)$;

(*ii*) $\lambda(L(G)) \geqslant 2\lambda(G)-2$;

(*iii*) *if* $\lambda(L(G)) \neq 2$, *then* $\lambda(L(G)) = 2\lambda(G)$ *if and only if G has two adjacent vertices of valency* $\lambda(G)$. ‖

Zamfirescu [**66**] proved the following result, which has parts (*ii*) and (*iii*) of the preceding theorem as corollaries:

Theorem 6.11. *If* $\lambda(L(G)) < \lambda(G)[\frac{1}{2}(\lambda(G)+1)]$, *then some edge of G is adjacent to precisely* $\lambda(L(G))$ *other edges.* ‖

Miscellany

There are a multitude of isolated results on line graphs which we have not yet included, and we conclude this section by listing some of these:

Theorem 6.12. *Let G be a connected graph. Then*

(*i*) *L(G) is Eulerian if and only if all of the valencies in G have the same parity*;

(*ii*) *L(G) is bipartite if and only if G is a path or an even circuit*;

(*iii*) *L(G) has a 1-factor if and only if G has an even number of edges*;

(*iv*) *except for* $L(K_{r,s})$, *which is isomorphic to* $K_r \times K_s$, *line graphs are irreducible with respect to the Cartesian product*;

(*v*) $L(G^2)$ *is Hamiltonian, and* $L(G)^2$ *is pancyclic*;

(*vi*) *if G is k-path Hamiltonian, then L(G) is* $(k+1)$*-path Hamiltonian*;

(*vii*) *the strong perfect graph conjecture (see Chapter 9) is true for line graphs.* ‖

Additionally, the line graphs of multigraphs have been characterized, as have minimum line graphs; the solutions of various line graph equations, such as $L(G) \cong \bar{G}$ and $L(G) \cong \overline{L(H)}$, have been found; and it has been shown that several (but not all) NP-complete problems for graphs are polynomial time problems for line graphs (see Section 4 of Chapter 15).

7. Elementary Results on Line Digraphs

The **line digraph** $L(D)$ of a digraph D has as its vertex-set the set of arcs of D; (a, b) is an arc of $L(D)$ if and only if there are vertices u, v, w in D with $a = (u, v)$ and $b = (v, w)$. Some examples of line digraphs are given in Fig. 7.

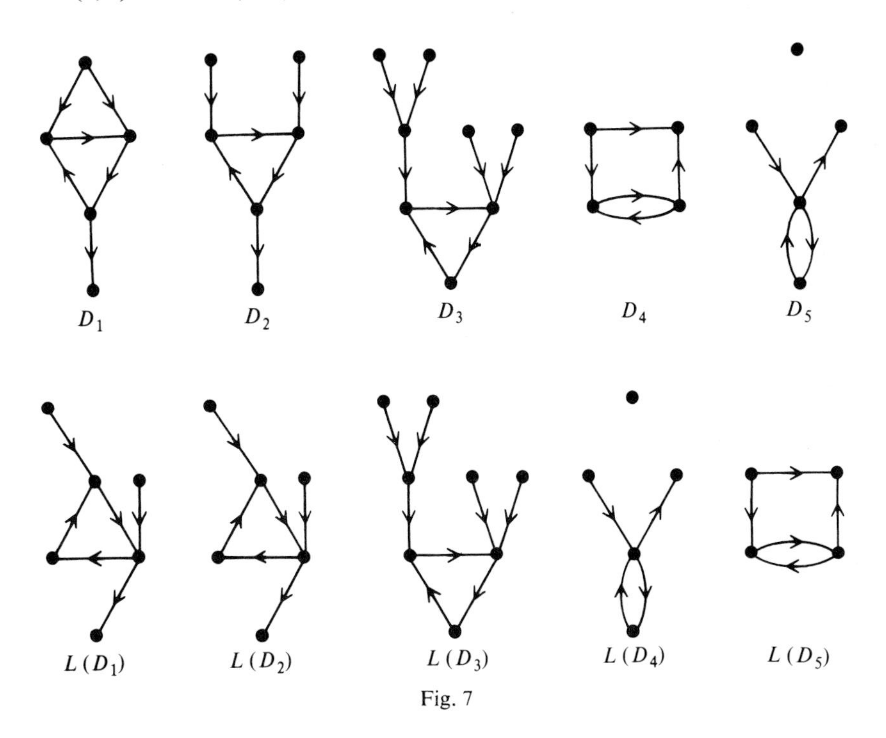

Fig. 7

This definition reflects only the head-to-tail adjacencies between arcs. It was introduced by Harary and Norman [**23**] in 1960, and since it seems to be the most natural analog of the line graph, it is the one which is most in use. (For an alternative definition which reflects all adjacencies between arcs, see Hemminger and Klerlein [**30**].) Unless otherwise specified, our digraphs will be allowed to have loops and multiple arcs. However, it is clear from the definition that a line digraph has no multiple arcs, and furthermore, that it has a loop at a vertex a if and only if a is a loop in the original digraph.

The examples in Fig. 7 have some interesting features which contrast with some of the results on line graphs: (*i*) $L(D_1) \cong L(D_2)$; (*ii*) $L(D_3) \cong D_3$; (*iii*) $L(D_4) \cong D_5$, and $L(D_5) \cong D_4$; and (*iv*) $L(D_4)$ is disconnected, although D_4 itself is not. These types of differences are part of what makes the subject of line digraphs interesting.

Our first theorem is a compilation of some of the basic results on line digraphs. For its statement we need some notation: as in Chapter 1, the out-valency $\rho_{\text{out}}(v)$ of a vertex v is the number of out-going arcs at v, and in-valency $\rho_{\text{in}}(v)$ is its converse, the number of in-coming arcs. The digraphs $\vec{P}_p$ and $\vec{C}_p$ are the directed paths and circuits of length p (for $p \geqslant 1$), and the digraph $\vec{K}_{r,s}$ is obtained from the graph $K_{r,s}$ by orienting all edges from the set of r vertices to the set of s vertices. Also, if A and B are two sets of vertices (not necessarily disjoint, but not both empty), then the digraph $\vec{K}(A, B)$ has vertex-set $A \cup B$ and arc-set $A \times B$.

Theorem 7.1. *Let D be a digraph with p vertices (none of which is isolated), and q arcs. Then*

(i) L(D) has q vertices and $\sum \rho_{\text{out}}(v)\rho_{\text{in}}(v)$ *arcs;*

(ii) the out-valency in L(D) of an arc vw in D is $\rho_{\text{out}}(w)$ *and the in-valency is* $\rho_{\text{in}}(v)$;

(iii) $L(D) \cong \vec{P}_{p-1}$ *if and only if* $D \cong \vec{P}_p$;

(iv) $L(D) \cong \vec{C}_p$ *if and only if* $D \cong \vec{C}_p$;

(v) $L(D) \cong \vec{K}_{r,s}$ *if and only if D consists of r in-coming arcs and s out-going arcs (no loops) at some vertex.* ||

Iterated line digraphs are defined as one would expect, and the next result gives some of their elementary properties:

Theorem 7.2. *Let D be a digraph. Then*

(i) $L^n(D)$ *is a null graph, for some n, if and only if D has no directed circuits;*

(ii) if D has two directed circuits joined by a directed path (possibly of length 0), then

$$\lim_{n \to \infty} p_n = \infty,$$

where p_n *is the order of* $L^n(D)$;

(iii) if no two directed circuits of D are joined by a directed path, then for all sufficiently large values of n, each connected component of $L^n(D)$ *has at most one directed circuit.* ||

We note that this result is incomplete in a sense, but in Section 9 we shall see that the converse of part (*ii*) also holds. We shall also investigate further those digraphs with just one directed circuit. The three statements in the theorem are illustrated in Fig. 8.

We now give the digraph analog of Corollary 2.3; the digraph D_3 of Fig. 7 shows the necessity of assuming that D is strongly connected.

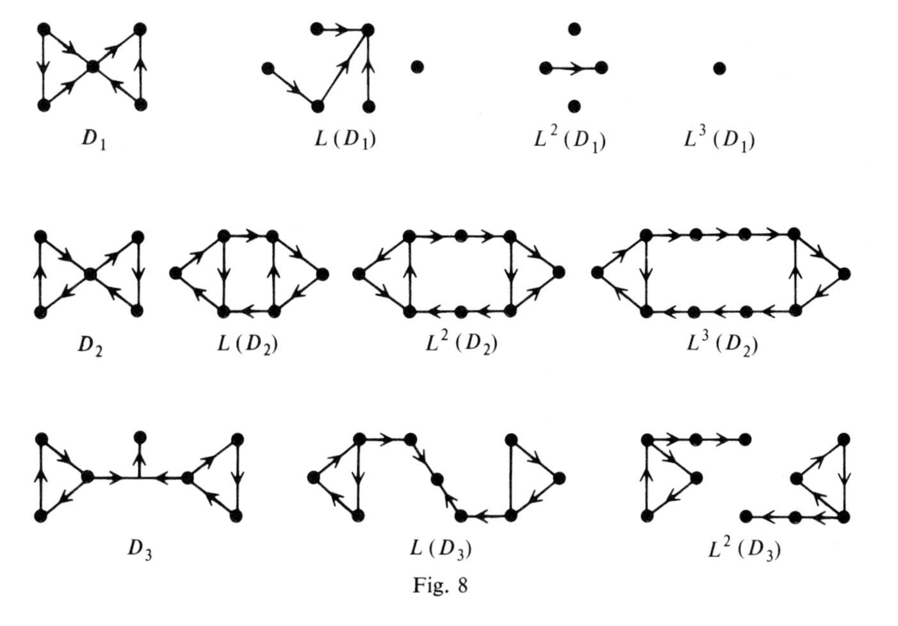

Fig. 8

Corollary 7.3. *If D is strongly connected, and if $L^n(D) \cong D$ for some n, then $L(D) \cong D$, and D is a directed circuit.* ‖

We close this section with a classification of digraphs, due primarily to Aigner [**1**]; this classification involves the type of connectedness of their line digraphs. In this theorem, a **source** is a vertex with in-valency 0, and a **sink** is a vertex with out-valency 0.

Theorem 7.4. *Let D be a digraph with at least three vertices (none of which is isolated). Then*

(*i*) *$L(D)$ is strongly connected if and only if D is strongly connected;*

(*ii*) *$L(D)$ is unilaterally connected if and only if* (a) *D is unilaterally connected, and* (b) *for each arc vw, if there are at least two directed paths from v to w, then there is also a directed path from w to v;*

(*iii*) *$L(D)$ is connected if and only if* (a) *D is connected and* (b) *there is no separating set of vertices consisting only of sources and sinks.* ‖

8. Characterizations of Line Digraphs

We begin this section with the determination problem—that is, if $F = L(D)$, to what extent is F characterized by D? Of course, we are interested only in digraphs D without isolated vertices, but even so, two root graphs can be very different, since the transformation L ignores the adjacencies of arcs at source and sinks. This is illustrated in Fig. 9 which shows the four digraphs (with no isolated vertices) whose line digraphs consist of two isolated arcs.

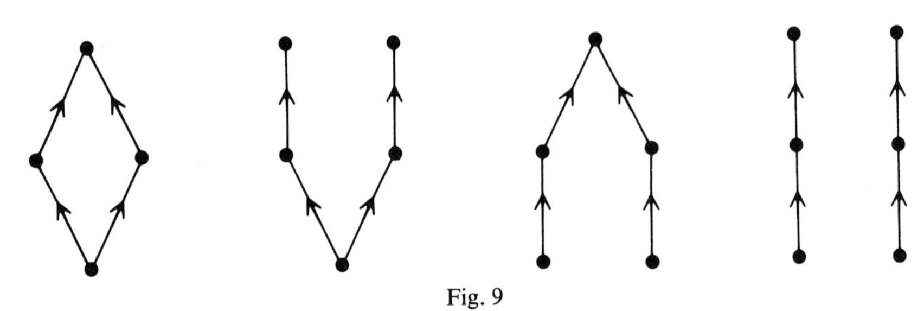

Fig. 9

However, the following theorem, due to Harary and Norman [**23**], shows that this is the only way in which two root graphs can differ. We let $\tilde{D}$ denote the subdigraph of D induced by the vertices which are neither sources nor sinks; for details of the proof of this theorem, see Hemminger [**28**]:

Theorem 8.1. *Let D and F be digraphs, and let σ be an isomorphism of $L(D)$ onto $L(F)$. Then the function σ restricted to $A(D)$ (the arc-set of D) is induced, in the natural way, by an isomorphism of $\tilde{D}$ onto $\tilde{F}$. Moreover, the arc a is incident from a source or incident to a sink in D if and only if $\sigma(a)$ has the same property in F.* ||

As an immediate corollary we have a result of Aigner [**1**] which can be viewed as the directed version of Whitney's theorem; it gives a class of digraphs $\mathscr{R}$ with the property that, up to isomorphism, the transformation L is one-to-one from $\mathscr{R}$ onto the class of line digraphs.

Corollary 8.2. *Let $\mathscr{R}$ be the class of digraphs with at most one source and at most one sink and no isolated vertices. Then for any line digraph F, there is a unique digraph D in $\mathscr{R}$ with $L(D) \cong F$.* ||

We observe that another class which could be used here is the class of those digraphs in which all sources have out-valency 1 and all sinks have in-valency 1. This would be most useful if one were considering only digraphs without multiple arcs.

A second corollary is the following result on automorphism groups:

Corollary 8.3. *If D is a connected digraph with at most one source and at most one sink, then its automorphism group is isomorphic to that of its line digraph—that is, $\Gamma(D) \cong \Gamma'(D)$.* ||

We observe that some restrictions on D (such as those given) are necessary, since $\vec{K}_{2,2}$ has just four automorphisms, whereas its line digraph has 24.

We now turn to the characterization problem—which digraphs are line digraphs? Several answers have been given to this question. Of the following characterizations, (*ii*) is due to Harary and Norman [**23**], (*iii*) to Heuchenne [**31**], and (*iv*) and (*v*) to Richards [**53**]; conditions (*ii*) and (*iii*) have each been rediscovered several times. One definition is required: a collection $\{S_i\}_{i \in I}$ of (possibly empty) subsets of a set S is called a **general partition** of S if $S = \bigcup_{i \in I} S_i$, and if $S_i \cap S_j = \varnothing$ whenever $i \neq j$. An example illustrating this theorem follows its proof:

Theorem 8.4. *Let F be a digraph, let $A(F)$ be its arc-set, and let* $\mathbf{M}$ *be its adjacency matrix. Then the following statements are equivalent*:

(*i*) *F is a line digraph*;

(*ii*) *there exist two general partitions $\{A_i\}_{i \in I}$ and $\{B_i\}_{i \in I}$ of $V(F)$ such that $A(F) = \bigcup_{i \in I} \vec{K}(A_i, B_i)$*;

(*iii*) *if vw, uw and ux are arcs of F, then so is vx*;

(*iv*) *any two rows of* $\mathbf{M}$ *are either identical or orthogonal*;

(*v*) *any two columns of* $\mathbf{M}$ *are either identical or orthogonal.*

Proof. We show in sequence that (*i*) implies (*ii*), (*ii*) implies (*iii*), (*iii*) implies (*iv*), (*iv*) is equivalent to (*v*), (*iv*) implies (*ii*), and (*ii*) implies (*i*).

(*i*) $\Rightarrow$ (*ii*). Let $F = L(D)$. For each $v_i \in V(D)$, let A_i and B_i be the sets of in-coming and out-going arcs at v_i. Then the subdigraph of $L(D)$ induced by $A_i \cup B_i$ is $\vec{K}(A_i, B_i)$ and, if ab is an arc in $L(D)$, (where, say, $a = v_h v_i$ and $b = v_i v_j$ in D), then $ab \in \vec{K}(A_i, B_i)$. The result follows.

(*ii*) $\Rightarrow$ (*iii*). If the arcs vw, uw, and ux lie in F, then there exist i, j such that $\{u, v\} \subseteq A_i$ and $\{w, x\} \subseteq B_j$. It follows that the arc vx must lie in F.

(*iii*) $\Rightarrow$ (*iv*). Let μ_i be the ith row of $\mathbf{M} = (m_{ij})$, and assume that (*iv*) does not hold—that is, that there exist i and j such that μ_i and μ_j are neither identical nor orthogonal. Then there exist h and k such that $m_{ih} = 1$, $m_{jh} = 0$ (or vice versa), $m_{ik} = 1$, and $m_{jk} = 1$. But this means that $v_i v_h$, $v_i v_k$ and $v_j v_k$ are arcs of F, whereas $v_j v_h$ is not. This contradicts (*iii*).

(*iv*) $\Rightarrow$ (*v*). Both (*iv*) and (*v*) are equivalent to the statement:

$$\text{for all } i, j, h, k, \text{ if } m_{ih} = m_{ik} = m_{jk} = 1, \qquad \text{then} \qquad m_{jh} = 1.$$

(*iv*) $\Rightarrow$ (*ii*). For each i and j with $m_{ij} = 1$, let $A_{ij} = \{v_h : m_{hj} = 1\}$ and let $B_{ij} = \{v_k : m_{ik} = 1\}$. Then, by (*iv*), A_{ij} is the set of vertices whose row vectors in $\mathbf{M}$ are identical to the ith row vector, whereas B_{ij} is the set of vertices whose column vectors in $\mathbf{M}$ are identical to the jth column vector (the latter follows from (v), which follows from (iv)). Consequently, $A_{ij} \times B_{ij} \subseteq A(F)$, and so $A(F) = \bigcup_{i,j} A_{ij} \times B_{ij}$. But, by the orthogonality condition, A_{ij} and A_{hk} are either equal or disjoint, as are B_{ij} and B_{hk}. If there is a zero row vector in $\mathbf{M}$ (say the ith row), let A_{ij} be the set of vertices whose row vector in $\mathbf{M}$ is the zero vector, and let $B_{ij} = \varnothing$. Doing the same with the zero column vectors of $\mathbf{M}$ gives us the general partitions as in (ii).

(ii) $\Rightarrow$ (i). Let D be the digraph with ordered pairs (A_i, B_i) as vertices, and with $|A_j \cap B_i|$ arcs from (A_i, B_i) to (A_j, B_j) for each i and j (including $i = j$). Let σ_{ij} be a one-to-one function from $A_j \cap B_i$ onto this set of arcs of D. Then the function σ defined on $V(F)$ by taking σ to be σ_{ij} on $A_j \cap B_i$ is a well-defined function of $V(F)$ into $V(L(D))$, since $\{A_j \cap B_i\}_{i,j \in I}$ partitions $V(F)$. Moreover, σ is one-to-one and onto since each σ_{ij} is one-to-one and onto, and it is easily seen that σ is an isomorphism from F onto $L(D)$; for example, if $vw \in A(F)$, then there exist i, j and k such that $v \in A_j \cap B_i$ and $w \in A_k \cap B_j$. Thus $\sigma(v)$ is an arc of D from (A_i, B_i) to (A_j, B_j), and $\sigma(w)$ is an arc of D from (A_j, B_j) to (A_k, B_k). Hence $(\sigma(v), \sigma(w)) \in A(L(D))$, and this completes the proof. ‖

Whereas (ii) is a very useful criterion, (iii) is certainly an elegant one, and is also pleasing in that it gives a forbidden induced subdigraph characterization. Seven of these forbidden induced subdigraphs are given in Fig. 10, where the dotted lines indicate arcs missing in Heuchenne's condition. Any other forbidden induced subdigraph can be obtained from one of the seven by adding arcs (other than the dotted one) on the same vertex-set. Conditions (iv) and (v) are easy to check if one has the adjacency matrix of F; they more or less describe on a large scale what (iii) describes locally.

Because of the relative densities of the subdigraphs involved, the partition

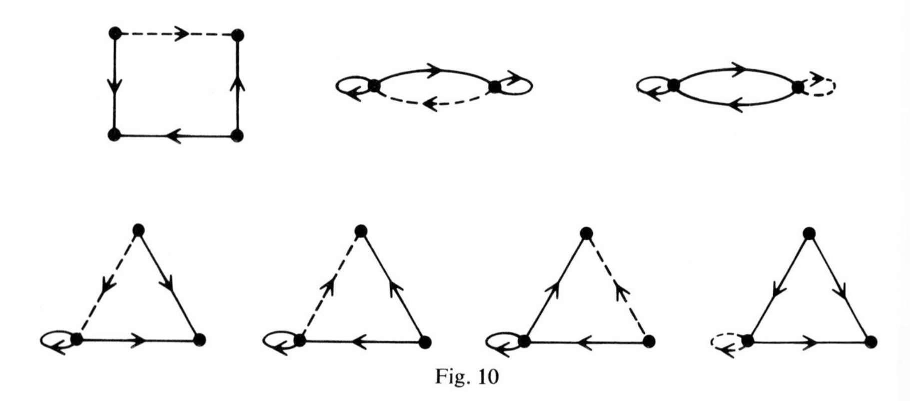

Fig. 10

criterion (*ii*) is not as recognizable in digraphs as the corresponding criterion is for graphs. We illustrate the theorem in Fig. 11, which contains a line digraph F, its adjacency matrix $\mathbf{M}$, the complete bipartite subdigraph partitioning $\vec{K}(\{1, 2\}, \{1, 5, 6\})$, $\vec{K}(\{3, 4\}, \{2\})$ and $\vec{K}(\{5\}, \{3, 4\})$, and root digraph D. Note that D is unique (up to isomorphism) since it has only one sink arc and no source arcs.

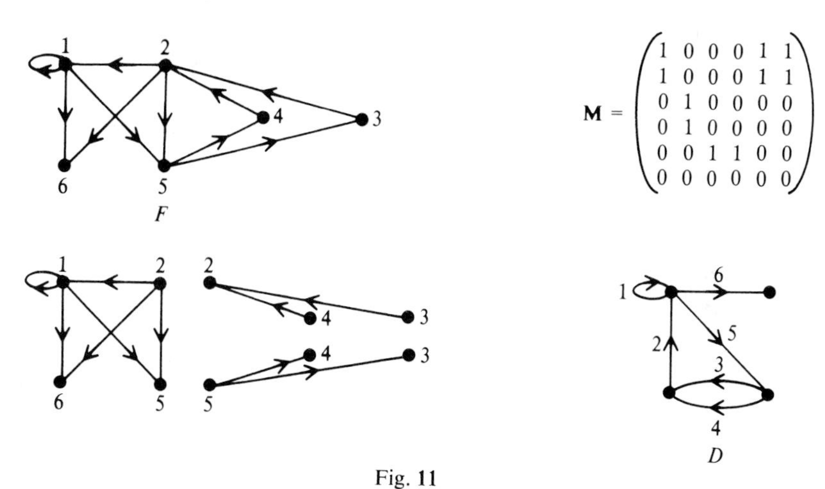

Fig. 11

We continue with the corresponding result for digraphs without loops or multiple arcs; its proof follows by eliminating those possibilities in Theorem 8.4 which give rise to loops or multiple arcs:

Theorem 8.5. *Let F be a digraph without loops or multiple arcs and let $\mathbf{M}$ be its adjacency matrix. Then the following are equivalent:*

(*i*) *F is the line digraph of a digraph without loops or multiple arcs;*

(*ii*) *there exist two general partitions $\{A_i\}_{i \in I}$ and $\{B_i\}_{i \in I}$ of $V(F)$ such that, for each i and j, $|A_j \cap B_i| \leqslant 1 - \delta_{ij}$ (where δ_{ij} is the Kronecker delta), and such that $A(F) = \bigcup_{i \in I} \vec{K}(A_i, B_i)$;*

(*iii*) *of the digraphs in Fig. 12, neither D_1 nor D_2 is a subdigraph, and*

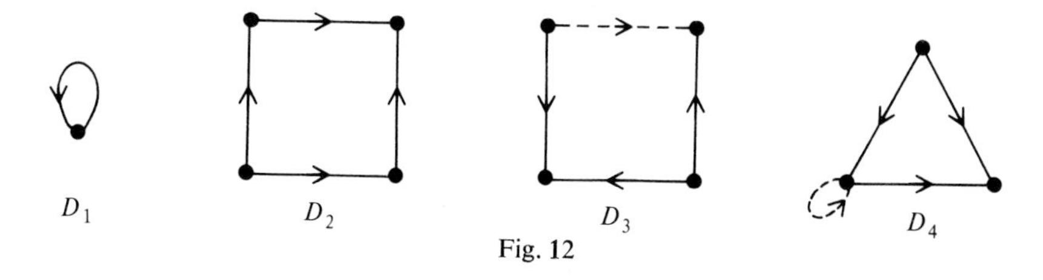

Fig. 12

neither D_3 nor D_4, nor any simple digraph obtained from them by adding arcs other than the dotted one, is an induced subdigraph;

(iv) any two rows of **M** *are identical or orthogonal, $m_{ii} = 0$ for all i, and if the ith and jth rows of* **M** *are non-zero and identical, then the ith and jth columns of* **M** *are orthogonal;*

(v) any two columns of **M** *are identical or orthogonal, $m_{ii} = 0$ for all i, and if the ith and jth columns of* **M** *are non-zero and identical, then the ith and jth rows of* **M** *are orthogonal.* ‖

The only other characterization known to us is due to Hoffman (unpublished), who proved that a (0, 1)-matrix **A** satisfies $\mathbf{A}^2 + \mathbf{A} = \mathbf{J}$ (the matrix of ones) if and only if **A** is the adjacency matrix of the line digraph of the complete symmetric digraph with no loops. He also asked whether one can determine the characteristic polynomial of $L(D)$ in terms of the characteristic polynomial of D—this is easy if D is regular.

In related work, Knuth [**39**] noted that $\mathbf{A}^2 = \mathbf{J}$, if **A** is the adjacency matrix of the line digraph of the complete symmetric digraph with loops included. However, there are other graphs with this property, and he investigated these in an algebraic context. He found five of them which have 25 vertices (it can be shown that the number of vertices is necessarily a square).

9. Periodic Iterated Line Digraphs

If, for some positive integers n and k, we have $L^{n+k}(D) \cong L^n(D)$, then we say that the digraph D is **periodic**, and the smallest value of k for which this is true is called its **period**. In this section we determine which digraphs are periodic, and study the problem of finding their periods. We do this in a rather intuitive way; to be more precise would require a notational barrage testing the stamina of the hardiest reader. Our discussion is based on work of Hemminger [**29**], but most of his results in the finite case were obtained earlier by Balconi [**2**].

Our description relies heavily on the concept of an **out-tree** which is a directed trec with one vertex (called the **root**) which can reach any other vertex by a directed path; an example is given in Fig. 13. An **in-tree** is the converse of an out-tree. (In the literature, out-trees and in-trees are sometimes called "*arborescences*" and "*counter-arborescences*".)

If D is a digraph with a single directed circuit C, we define its **basic configuration** to be that subdigraph formed by the union of all directed paths to and from vertices of C. Clearly, the line digraph of D and each of its iterates have a single directed circuit (all of the same length), and hence we can

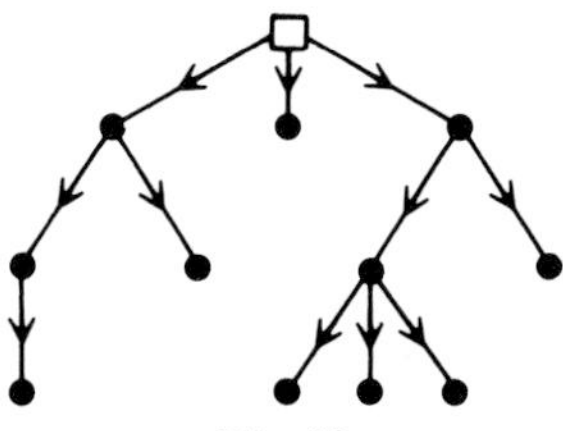

Fig. 13

speak of their basic configurations. Figure 14 shows the basic configuration of a digraph D and of its line digraphs $L(D)$, $L^2(D)$ and $L^3(D)$. It is not difficult (but somewhat messy) to show that for any digraph D and for all sufficiently large n, the basic configuration of $L^n(D)$ consists of a directed circuit and an out-tree and an in-tree (disjoint except for roots) at each vertex of the directed circuit. This result is illustrated in Fig. 14 which shows the basic configurations of $L^2(D)$ and $L^3(D)$ having the stated form.

We call such a digraph an **eddy digraph**: it consists of a directed circuit $C: v_0, v_1, \ldots, v_{k-1}, v_0$ together with an out-tree A_i and an in-tree B_i at each vertex v_i $(0 \leqslant i \leqslant k-1)$. The cyclic (modulo k) sequences $\{A_i\}$ and $\{B_i\}$ are called the **out-tree** and **in-tree sequences**. An observation crucial to what follows is this: the line digraph of an eddy digraph has the same sequence, with the in-trees advanced one vertex of the circuit relative to the out-trees. This, too, is illustrated by Fig. 14. We define the **out-tree index** of the digraph D itself to be the minimum positive integer r for which $A_{i+r} \cong A_i$ for all i; the **in-tree index** is defined similarly.

The reader may have noticed that the arcs uw and vw of $L(D)$ do not appear in Fig. 14. (Are there any others missing?) These additional arcs are necessitated by the Heuchenne condition (*iii*) of Theorem 8.4. Like Hemminger

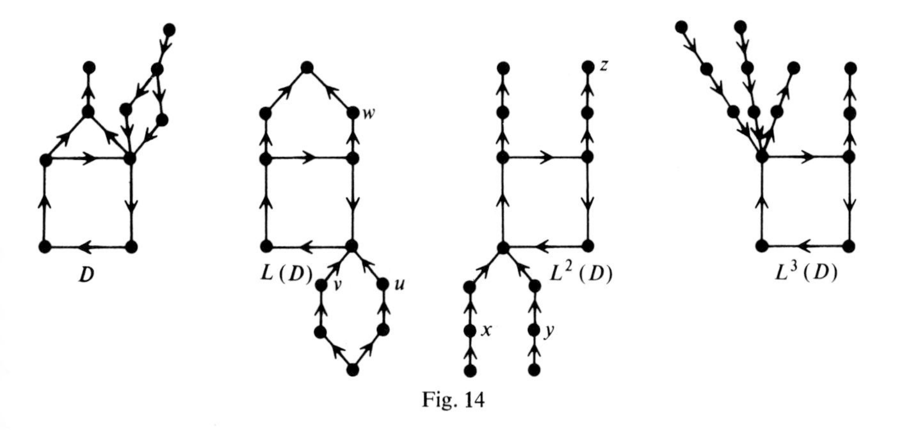

Fig. 14

[28], we define the ***n*th Heuchenne condition** as follows: whenever there are vertex-disjoint directed paths of length n from v to w, u to w, and u to x, then there is one (disjoint from the others) from v to x. It follows by induction on n that $L^n(D)$ must satisfy the first n Heuchenne conditions. Thus in $L^2(D)$ of Fig. 14, there are directed paths of length 2 from each of x and y to z. The reader will note that the image of these directed paths in $L^4(D)$ is two isolated arcs. However, it is not difficult to see that the connected component of $L^n(D)$ containing the directed circuit consists precisely of the basic configuration and the directed paths resulting from the Heuchenne conditions. It follows that if C has length k, then, for sufficiently large n, $L^{n+k}(D) \cong L^n(D)$, since the out-tree and in-tree sequences resume their relative positions, and the only other arcs are due to the Heuchenne conditions. Thus D is periodic and we get the promised converse of (*ii*) in Theorem 7.2 (and a little more):

Theorem 9.1. *Let D be a digraph. Then*

(*i*) *$L^n(D)$ is null for some n if and only if D has no directed circuits;*

(*ii*) *The order of $L^n(D)$ gets arbitrarily large if and only if D has two directed circuits joined by a directed path (possibly of length* 0);

(*iii*) *D is periodic if and only if D has directed circuits, no two of which are joined by a directed path.* ||

Thus we have characterized the periodic digraphs, and the next problem is to determine their periods. The answer is given in the next theorem:

Theorem 9.2. *Let D be a periodic digraph with a single directed circuit. Then the period of D is the greatest common divisor of its out-tree and in-tree indices.*

Proof. Let these indices be a and b, respectively. If one rotates the out-trees any multiple of a units (backward for a positive multiple, forward for a negative multiple), then the basic configuration remains the same; a similar situation holds if the in-trees are shifted through a multiple of b units. Thus, if t is a positive linear combination of a and b, and n is sufficiently large, then $L^n(D)$ and $L^{n+t}(D)$ have the same basic configuration. But gcd(a, b) is the minimum such positive linear combination, and hence the period of D is at most gcd(a, b). The reverse inequality is not hard, and is basically the elementary theory of cyclic groups (see Hemminger **[29]** for details). ||

As a corollary we have the first published result on periodic digraphs; it is due to Harary and Norman **[23]**:

Corollary 9.3. *Let D be a connected digraph. Then $L(D) \cong D$ if and only if every vertex has out-valency* 1 *or every vertex has in-valency* 1.

Proof. If every vertex has out-valency 1, then D (being connected) must consist of a directed circuit and in-trees to its vertices, and such a digraph is clearly isomorphic to its line digraph. A similar statement holds if every in-valency is 1.

Conversely, if $L(D) \cong D$, then D must be an eddy digraph; if it has both a nontrivial out-tree and in-tree, then eventually the Heuchenne paths lead to other components. ||

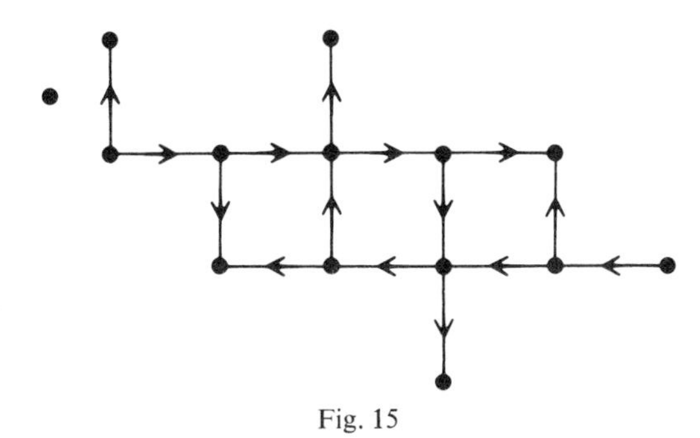

Fig. 15

In moving to the disconnected digraphs of period 1, Beineke [**5**] recognized the importance of the unicyclic digraphs. The following theorem follows immediately from Theorem 9.2, and an illustration appears in Fig. 15:

Corollary 9.4. *A digraph with a single directed circuit has period* 1 *if and only if the out-tree and in-tree indices are relatively prime.* ||

The only other way in which digraphs of period 1 can occur is for there to exist a collection of its connected components forming all iterated line digraphs of a unicyclic digraph with period greater than 1. An example of this is given in Fig. 16; for further details, see Muracchini and Ghirlanda [**46**], or Hemminger [**29**]. (The problem of determining all infinite digraphs of period 1 has been solved by Beineke and Hemminger [**7**].)

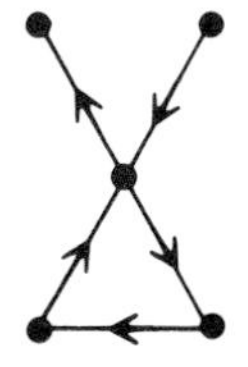
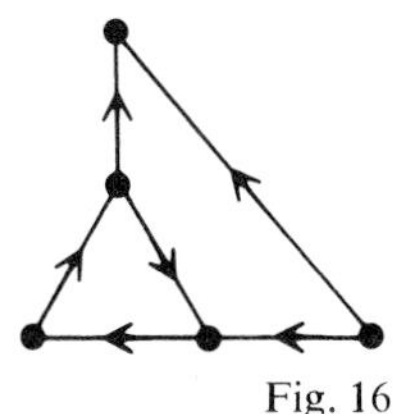
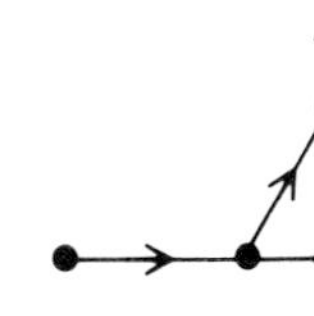

Fig. 16

10. Line Digraphs with Special Properties

We conclude the chapter with a brief discussion of some further properties of line digraphs, beginning with traversability. It is in fact quite easy to characterize both Eulerian and Hamiltonian line digraphs.

Theorem 10.1. *Let D be a strongly connected digraph. Then*

(*i*) *$L(D)$ is Eulerian if and only if $\rho_{in}(v) = \rho_{out}(w)$ for each arc vw of D;*

(*ii*) *$L(D)$ is Hamiltonian if and only if D is Eulerian.* ‖

The Hamiltonian characterization was observed by Aigner [**1**] and by Kasteleyn [**37**]. Kasteleyn also showed that the number of Hamiltonian directed circuits in $L(D)$ equals the number of Eulerian directed walks in D, and he found a formula for this number. Sysło [**61**] used this in showing that the traveling salesman problem is solvable for line digraphs satisfying certain valency conditions.

Knuth [**38**] extended to line digraphs a formula of Tutte for the number of spanning out-trees rooted at a given vertex of a digraph, and from this he obtained an extension of a result of van Ardenne-Ehrenfest enumerating Eulerian directed walks (see Section 10 of Chapter 14).

Sysło [**62**] posed the following question: given a digraph D, is there some line graph L which can be obtained from D (but is not $L(D)$ in general) such that L is Hamiltonian if and only if D is. (He found a solution for the corresponding Eulerian question.) This problem occurs by analogy with the problem of finding a minimal line digraph homeomorphic to a given digraph, a problem which is of interest in scheduling applications where optimal methods are sought for finding an "event network" from a given "activity network".

Our final results involve distances, and are due to Aigner [**1**]. In order to state them, we define the **diameter** of a strongly connected digraph D by $\text{diam}\ D = \max_{v,w} d(v, w)$; the quantities $r^+(D) = \min_v \max_w d(v, w)$ and $r^-(D) = \min_w \max_v d(v, w)$ are defined for all unilaterally connected digraphs, and are called the **out-radius** and **in-radius** respectively.

Theorem 10.2. *Let D be a strongly connected digraph. Then* $\text{diam}\ L(D) = \text{diam}\ D$ *if D is a directed circuit, and* $\text{diam}\ L(D) = \text{diam}\ D+1$ *otherwise.* ‖

Theorem 10.3. *Let D be a unilaterally connected digraph. Then*

$$r^+(D) \leqslant r^+(L(D)) \leqslant r^+(D)+1, \qquad \textit{and}$$
$$r^-(D) \leqslant r^-(L(D)) \leqslant r^-(D)+1. \ \|$$

References

1. M. Aigner, On the linegraph of a directed graph, *Math. Z.* **102** (1967), 56–61; *MR*36#76.
2. G. Balconi, Sui singrammi a commutato periodico, *Ist. Lombardo Accad. Sci. Lett. Rend.* (*A*) **102** (1968), 427–436; *MR*39#1361.
3. G. Balconi, Singrammi commutati di singrammi regolari, *Ist. Lombardo Accad. Sci. Lett. Rend.* (*A*) **106** (1972), 685–696; *MR*49#2461.
4. G. Balconi, Una caratterizzazione dei singrammi commutati, *Ist. Lombardo Accad. Sci. Lett. Rend.* (*A*) **107** (1973), 685–698; *MR*50#185.
5. L. W. Beineke, On derived graphs and digraphs, in *Beiträge zur Graphentheorie* (ed. H. Sachs *et al.*), Teubner-Verlag, Leipzig, 1968, pp. 17–23.
6. L. W. Beineke, Characterizations of derived graphs, *J. Combinatorial Theory* **9** (1970), 129–135; *MR*41#6707.
7. L. W. Beineke and R. L. Hemminger, Infinite digraphs isomorphic with their line digraphs, *J. Combinatorial Theory* (*B*) **21** (1976), 245–256.
8. C. Berge, *Graphs and Hypergraphs*, North-Holland, Amsterdam, and American Elsevier, New York, 1973; *MR*50#9640.
9. J.-C. Bermond, Graphe représentatif de l'hypergraphe *h*-parti complet, in *Hypergraph Seminar*, Lecture Notes in Mathematics **411** (ed. C. Berge and D. K. Ray-Chaudhuri), Springer-Verlag, Berlin, Heidelberg, New York, 1974, pp. 34–53; *MR*51#12611.
10. J.-C. Bermond and P. Rosenstiehl, Pancyclisme du carré du graphe aux arêtes d'un graphe, *Cahiers Centre Études Recherche Opér.* **15** (1973), 285–286; *MR*51#2980.
11. R. C. Bose, Characterization problems of combinatorial graph theory, in *A Survey of Combinatorial Theory* (ed. J. N. Srivastava), North-Holland, Amsterdam, 1973, pp. 31–51; *MR*52#115.
12. P. J. Cameron, J.-M. Goethals, J. J. Seidel and E. E. Shult, Line graphs, root systems, and elliptical goemetry, *J. Algebra* **43** (1976), 305–327.
13. G. Chartrand, Graphs and their associated line-graphs, Ph.D. thesis, Michigan State University, 1964.
14. G. Chartrand, On hamiltonian line-graphs, *Trans. Amer. Math. Soc.* **134** (1968), 559–566; *MR*38#68.
15. G. Chartrand and M. J. Stewart, The connectivity of line-graphs, *Math. Ann.* **182** (1969), 170–174; *MR*43#3161.
16. M. E. Fisher, Critical probabilities for cluster size and percolation problems, *J. Math. Phys.* **2** (1961), 620–627; *MR*23#A3602.
17. D. L. Greenwell, Forbidden subgraphs, Ph.D. thesis, Vanderbilt University, 1973.
18. D. L. Greenwell and R. L. Hemminger, Forbidden subgraphs for graphs with planar line graphs, *Discrete Math.* **2** (1972), 31–34; *MR*45#6658.
19. B. Grünbaum, Incidence patterns of graphs and complexes, in *The Many Facets of Graph Theory*, Lecture Notes in Mathematics **110** (ed. G. Chartrand and S. F. Kapoor), Springer-Verlag, Berlin, Heidelberg, New York, 1969, pp. 115–128; *MR*40#4152.
20. R. Halin and H. A. Jung, Note on isomorphisms of graphs, *J. London Math. Soc.* **42** (1967), 254–256; *MR*34#7402.
21. F. Harary, *Graph Theory*, Addison-Wesley, Reading, Mass., 1969; *MR*41#1566.
22. F. Harary and C. St. J. A. Nash-Williams, On eulerian and hamiltonian graphs and line graphs, *Canad. Math. Bull.* **8** (1965), 701–709; *MR*33#66.
23. F. Harary and R. Z. Norman, Some properties of line digraphs, *Rend. Circ. Mat. Palermo* (2) **9** (1960), 161–168; *MR*24#A693.
24. F. Harary and E. M. Palmer, On the point-group and line-group of a graph, *Acta Math. Acad. Sci. Hungar.* **19** (1968), 263–269; *MR*38#81.
25. S. T. Hedetniemi, Graphs of (0,1)-matrices, in *Recent Trends in Graph Theory*, Lecture Notes in Mathematics **186** (ed. M. Capobianco *et al.*), Springer-Verlag, Berlin, Heidelberg, New York, 1971, pp. 157–171; *MR*43#6120.
26. S. T. Hedetniemi and P. J. Slater, Line graphs of triangleless graphs and iterated clique

graphs, in *Graph Theory and Applications*, Lecture Notes in Mathematics **303** (ed. Y. Alavi *et al.*), Springer-Verlag, Berlin, Heidelberg, New York, 1972, pp. 139–147; *MR*49#151.
27. R. L. Hemminger, On reconstructing a graph, *Proc. Amer. Math. Soc.* **20** (1969), 185–187; *MR*38#1019.
28. R. L. Hemminger, Line digraphs, in *Graph Theory and Applications*, Lecture Notes in Mathematics **303** (ed. Y. Alavi *et al.*), Springer-Verlag, Berlin, Heidleberg, New York, 1972, pp. 149–163; *MR*51#243.
29. R. L. Hemminger, Digraphs with periodic line digraphs, *Studia Sci. Math. Hungar.* **9** (1974), 27–31; *MR*52#2948.
30. R. L. Hemminger and J. B. Klerlein, Line psuedodigraphs, *J. Graph Theory* **1** (1977), 365–377.
31. C. Heuchenne, Sur une certaine correspondance entre graphs, *Bull. Soc. Roy. Sci. Liège* **33** (1964), 743–753; *MR*30#5297.
32. M. C. Heydemann and D. Sotteau, Line-graphs of hypergraphs, II, in *Combinatorics* (ed. A. Hajnal and V. T. Sós), North-Holland, Amsterdam, 1978.
33. A. J. Hoffman, On the exceptional case in a characterization of the arcs of a complete graph, *IBM J. Res. Develop.* **4** (1960), 487–496; *MR*25#3861.
34. A. J. Hoffman, $-1-\sqrt{2}$?, in *Combinatorial Structures and Their Applications* (ed. R. K. Guy *et al.*), Gordon and Breach, New York, 1970, pp. 173–176.
35. A. J. Hoffman, On graphs whose least eigenvalue exceeds $-1-\sqrt{2}$, *Lin. Alg. Appl.* **16** (1977), 153–165.
36. H. A. Jung, Zu einem Isomorphiesatz von H. Whitney für Graphen, *Math. Ann.* **164** (1966), 270–271; *MR*33#5518.
37. P. W. Kasteleyn, Graph theory and crystal physics, in *Graph Theory and Theoretical Physics* (ed. F. Harary), Academic Press, London and New York, 1967, pp. 43–110; *MR*40#6903.
38. D. Knuth, Oriented subtrees of an arc digraph, *J. Combinatorial Theory* **3** (1967), 309–314; *MR*35#5361.
39. D. Knuth, Notes on central groupoids, *J. Combinatorial Theory* **8** (1970), 376–390; *MR*41#3645.
40. A. Kotzig, Aus der Theorie der endlichen regulären Graphen dritten und vierten Grades (Slovak with Russian and German summaries), *Časopis Pěst. Mat.* **82** (1957), 76–92; *MR*19-876.
41. J. Krausz, Démonstration nouvelle d'un théorème de Whitney sur les réseaux (Hungarian with French summary), *Mat. Fiz. Lapok* **50** (1943), 75–85; *MR*8-284.
42. V. R. Kulli and E. Sampathkumar. On the interchange graph of a finite planar graph, *J. Indian Math. Soc.* (*New Ser.*) **37** (1973), 339–341; *MR*51#7920.
43. A. Marczyk, On line multigraphs of hypergraphs, in *Graphs, Hypergraphs and Block Systems* (ed. M. Borowiecki *et al.*), Zielona Góra, 1976, pp. 147–154.
44. P. Martin, Cycles Hamiltoniens dans les graphes 4-réguliers 4-connexes, *Aequationes Math.* **14** (1976), 37–40; *MR*54#2544a.
45. V. V. Menon, The isomorphism between graphs and their adjoint graphs, *Canad. Math. Bull.* **8** (1965), 7–15; *MR*30#5299.
46. L. Muracchini and A. Ghirlanda, Sul grafo commutato e sul grafo opposto di un grafo orientato, *Atti Sem. Mat. Fis. Univ. Modena* **14** (1965), 87–97; *MR*33#2570.
47. L. Nebeský, A note on line graphs, *Comment. Math. Univ. Carolinae* **15** (1974), 567–570; *MR*50#4404.
48. L. Nebeský, A theorem on line graphs, in *Recent Advances in Graph Theory* (ed. M. Fiedler), Academia, Prague, 1975, pp. 399–403; *MR*52#10509.
49. J. Nešetřil, Homomorphisms of derivative graphs, *Discrete Math.* **1** (1971), 257–268; *MR*46#99.
50. O. Ore, *Theory of Graphs*, Amer. Math. Soc. Colloq. Publ. XXXVIII, American Mathematical Society, Providence, Rhode Island, 1962; *MR*27#740.
51. A. R. Rao, A characterization of a class of regular graphs, *J. Combinatorial Theory* **10** (1971), 264–274.

52. D. K. Ray-Chaudhuri, Characterization of line graphs, *J. Combinatorial Theory* **3** (1967), 201–214; MR**35**#4119.
53. P. I. Richards, Precedence constraints and arrow diagrams, *SIAM Rev.* **9** (1967), 548–553.
54. A. C. M. van Rooij and H. S. Wilf, The interchange graph of a finite graph, *Acta Math. Acad. Sci. Hungar.* **16** (1965), 263–269; *MR***33**#3959.
55. N. D. Roussopoulos, A max $\{m, n\}$ algorithm for determining the graph H from its line graph G, *Information Processing Lett.* **2** (1973), 108–112.
56. G. Sabidussi, Loewy-groupoids related to linear graphs, *Amer. J. Math.* **76** (1954), 477–487; *MR***16**-444.
57. G. Sabidussi, Graph derivatives, *Math. Z.* **76** (1961), 385–401; *MR***24**#A53.
58. B. L. Schwartz and L. W. Beineke, Locally finite self-interchange graphs, *Proc. Amer. Math. Soc.* **27** (1971), 8–12; *MR***44**#6540.
59. J. Sedláček, Some properties of interchange graphs, in *Theory of Graphs and Its Applications* (ed. M. Fiedler), Czechoslovak Academy of Sciences, Prague, and Academic Press, New York, 1964, 145–150; *MR***30**#3468.
60. S. Seshu and M. B. Reed, *Linear Graphs and Electrical Networks*, Addison-Wesley, Reading, Mass., 1961; *MR***26**#4638.
61. M. M. Sysło, A new solvable case of the traveling salesman problem, *Math. Programming* **4** (1973), 347–348; *MR***53**#13023.
62. M. M. Sysło, Remarks on line digraphs, *Bull. Acad. Polon. Sci. Ser. Sci. Math. Astronom. Phys.* **22** (1974), 5–10; *MR***52**#7957.
63. V. G. Vizing, Some unsolved problems in graph theory, *Uspekhi Mat. Nauk* **23** (1968), 117–134 = *Russian Math. Surveys* **23** (1968), 125–142; *MR***39**#1354.
64. H. Whitney, Congruent graphs and the connectivity of graphs, *Amer. J. Math.* **54** (1932), 150–168.
65. H. Whitney, 2-isomorphic graphs, *Amer. J. Math.* **55** (1933), 245–254.
66. T. Zamfirescu, On the line-connectivity of line-graphs, *Math. Ann.* **187** (1970), 305–309; *MR***44**#6541.

11
On the Eigenvalues of a Graph

ALLEN J. SCHWENK AND ROBIN J. WILSON

1. Introduction

Whenever we need a non-pictorial representation of a graph (as we may, for example, when dealing with a computer), we frequently use its adjacency matrix to encode the structure. This raises the natural question of what the highly-developed theory of matrices can tell us about graphs. Since many graphical properties are invariant under a relabeling of the vertices, we are particularly motivated to study those matrix concepts which are invariant under the simultaneous reordering of rows and columns; in particular, the eigenvalues of the matrix immediately come to mind. In this chapter we shall endeavor to find out to what extent the eigenvalues of a given graph reflect the properties of that graph.

We recall that if G is a graph (or digraph) with vertex-set $\{v_1, v_2, \ldots, v_p\}$, then its **adjacency matrix** $\mathbf{A}(G)$ is defined to be the $p \times p$ matrix (a_{ij}), in which a_{ij} denotes the number of edges (arcs) from v_i to v_j. It follows immediately that if G is a graph, then $\mathbf{A}(G)$ is a symmetric (0, 1)-matrix in which every diagonal entry is zero, and in which the sum of the entries in any row or column is equal to the valency of the corresponding vertex. We shall denote the characteristic polynomial of $\mathbf{A}(G)$ by $\phi(G; x)$, since it is uniquely determined by the graph G, and refer to it as the **characteristic polynomial** of G; thus

$$\phi(G; x) = \det(x\mathbf{I} - \mathbf{A}(G)) = \sum_{i=0}^{p} a_i x^{p-i}.$$

Since $\mathbf{A}(G)$ is a real symmetric matrix, its eigenvalues (the roots of this polynomial) must be real, and may be ordered $\lambda_1 \geqslant \lambda_2 \geqslant \ldots \geqslant \lambda_p$; these eigenvalues are called the **eigenvalues of G**, and the sequence of p eigenvalues is called the **spectrum of G**. Some examples of the eigenvalues and characteristic polynomials of graphs are given in the following table; further examples will be given in Section 6.

Graph	Adjacency matrix	Characteristic polynomial	Eigenvalues
K_2:	 $\begin{pmatrix} 0 & 1 \\ 1 & 0 \end{pmatrix}$	$x^2 - 1$	$1, -1$
P_3:	 $\begin{pmatrix} 0 & 0 & 1 \\ 0 & 0 & 1 \\ 1 & 1 & 0 \end{pmatrix}$	$x^3 - 2x$	$\sqrt{2}, -\sqrt{2}, 0$
K_4:	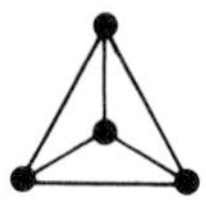 $\begin{pmatrix} 0 & 1 & 1 & 1 \\ 1 & 0 & 1 & 1 \\ 1 & 1 & 0 & 1 \\ 1 & 1 & 1 & 0 \end{pmatrix}$	$x^4 - 6x^2 - 8x - 3$	$3, -1, -1, -1$
C_4:	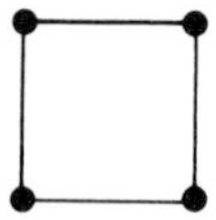 $\begin{pmatrix} 0 & 1 & 0 & 1 \\ 1 & 0 & 1 & 0 \\ 0 & 1 & 0 & 1 \\ 1 & 0 & 1 & 0 \end{pmatrix}$	$x^4 - 4x^2$	$2, -2, 0, 0$
$K_{2,3}$:	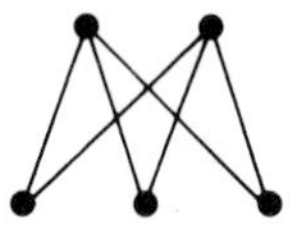 $\begin{pmatrix} 0 & 0 & 1 & 1 & 1 \\ 0 & 0 & 1 & 1 & 1 \\ 1 & 1 & 0 & 0 & 0 \\ 1 & 1 & 0 & 0 & 0 \\ 1 & 1 & 0 & 0 & 0 \end{pmatrix}$	$x^5 - 6x^3$	$\sqrt{6}, -\sqrt{6}, 0, 0, 0$

We shall see in Section 7 that the spectrum by no means specifies its graph uniquely; nevertheless, it does provide a wealth of information about the graph. Rather surprisingly, perhaps, eigenvalues also arise in a variety of applications—for example, they occur in organic chemistry, where the energy levels of certain molecules (such as polycyclic hydrocarbons) are essentially the eigenvalues of the graph of the molecule, and where the wave functions are then the corresponding eigenvectors.

So our primary concern in this chapter is to consider the following two questions:

(*i*) if we are given a graph (or digraph) of a particular type, what can we say about its spectrum?

(*ii*) if we impose restrictions on the spectrum, what restrictions does this impose on the corresponding graphs or digraphs?

In what follows, $\mathbf{I}_r$ and $\mathbf{J}_r$ will denote, respectively, the identity $r \times r$ matrix and the $r \times r$ matrix each entry of which is 1. The matrix $\mathbf{A}(G)$ will be abbreviated to $\mathbf{A}$ when there is no possibility of confusion, and $\mathbf{e}_1, \mathbf{e}_2, \ldots$ will always denote unit vectors.

We should like to thank Peter Cameron for his comments and suggestions regarding this chapter, and, in particular, for his help with Sections 9 and 11.

2. Some Simple Results

Before actually calculating the eigenvalues of some specific graphs, we shall describe some consequences of the above definitions. The reader may wish to verify these results for the examples in the above table.

Theorem 2.1. *If G is a disconnected graph, then the spectrum of G is the union of the spectra of the components of G.*

Proof. This is an immediate consequence of the properties of determinants of partitioned matrices; the characteristic polynomial of G is the product of the characteristic polynomials of the components of G. ||

Theorem 2.2. *If ψ is any polynomial, and if λ is any eigenvalue of the matrix* $\mathbf{A}$, *then $\psi(\lambda)$ is an eigenvalue of the matrix $\psi(\mathbf{A})$.*

Proof. Let $\psi(\mathbf{A})$ act on the eigenvector corresponding to λ, and use the distributive and associative laws for matrices. ||

Theorem 2.3. *The sum of the eigenvalues of a graph G is zero.*

Proof. This sum is equal to the trace of $\mathbf{A}(G)$, which is zero since every diagonal element is zero. ||

This last theorem is a simple instance of a more general result. It is well known that the *ij*-entry of $\mathbf{A}^k$ gives the number of walks in G of length k joining v_i to v_j. Consequently, the total number of closed walks of length k

is given by the trace of $\mathbf{A}^k$. Using Theorem 2.2 to find the eigenvalues of $\mathbf{A}^k$, we obtain the following result:

Theorem 2.4. *If c is the number of closed walks in G of length k, then*

$$c = \operatorname{tr} \mathbf{A}^k = \sum_{i=1}^{n} \lambda_i^k . \ \|$$

In particular, this tells us that $\sum_i \lambda_i^2$ is equal to twice the number of edges in G, and $\sum_i \lambda_i^3$ is equal to six times the number of triangles in G. For larger values of k, closed walks of length k may take several different forms. An application of Theorem 2.4 to the enumeration of walks which various chessmen can take on an $n \times n$ chessboard can be found in [**10**].

Because every symmetric matrix is similar to a diagonal matrix, the minimum polynomial of $\mathbf{A}(G)$ will be $\mu(G) = \prod_{i=1}^{m} (x - \mu_i)$, where $\mu_1, \mu_2, \ldots, \mu_m$ are the m distinct eigenvalues occurring in the spectrum. The number m is restricted by the structure of G:

Theorem 2.5. *If G is a connected graph with m distinct eigenvalues and with diameter d, then $m > d$.*

Proof. If $m \leqslant d$, then G has vertices v_i and v_j whose distance apart is m. The minimum polynomial acting on $\mathbf{A}$ has leading term $\mathbf{A}^m$. But the ij-entry of this term is positive, whereas the ij-entry of all lower terms must be zero. It follows that $\mu(\mathbf{A})$ cannot equal the zero matrix, and we conclude that $m > d$. $\|$

3. Computing the Characteristic Polynomial

There are at least three sets of variables that can be used to specify the spectrum of a graph G. The eigenvalues themselves comprise one set, the coefficients a_i of the characteristic polynomial $\phi(G; x) = \sum_i a_i x^{p-i}$ form another, and the moments $M_k = \sum_i \lambda_i^k$ for $1 \leqslant k \leqslant p$ form a third. Often, the eigenvalues themselves are not directly available, and so we determine one of the other sets first. Using Theorem 2.4, we observe that the moments M_k are equal to $\operatorname{tr} \mathbf{A}^k$, but these numbers tend to become excessively large. Thus, we often seek the coefficients a_i first. Certainly $a_0 = 1$, and using Theorems 2.3 and 2.4, we can easily verify that $a_1 = 0$, $-a_2$ is the number of edges of G, and $-a_3$ is twice the number of triangles in G (see [**9**]). These results have been extended by Sachs [**38**], Spialter [**47**], and Mowshowitz [**31**] to give the following result, usually known as "Sachs' theorem":

Theorem 3.1. *The coefficients a_i of $\phi(G; x)$ are given by*

$$a_i = \sum_H (-1)^{k(H)} 2^{c(H)},$$

where the summation extends over all subgraphs H of G on i vertices whose components are either single edges or circuits, and where $k(H)$ and $c(H)$ denote, respectively, the number of components and circuits in H. ||

Sachs' theorem can be used to compute the coefficients a_i directly, as well as providing the foundation for the following two theorems of Schwenk [**41**] (Theorems 3.2 and 3.4). To prove these theorems, it is enough to apply Sachs' theorem to each polynomial that appears, and then to verify that for each term corresponding to the subgraph H on the left-hand side of the equation, there is a corresponding equal term on the right-hand side.

Theorem 3.2. *Let v be a vertex of a graph G, let $\mathscr{C}(v)$ be the collection of circuits containing v, and let $V(Z)$ denote the set of vertices in the circuit Z. Then the characteristic polynomial $\phi(G)$ satisfies*

$$\phi(G) = x\phi(G-v) - \sum_w \phi(G-v-w) - 2\sum_Z \phi(G-V(Z)),$$

where the first summation extends over those vertices w adjacent to v, and the second summation extends over all $Z \in \mathscr{C}(v)$. ||

This theorem takes an especially simple form when v is a vertex of valency 1. In this form, it appears as Theorem 2 of Harary, King, Mowshowitz and Read [**18**]:

Corollary 3.3. *If v is an end-vertex of a graph G, and if w is the vertex adjacent to v, then $\phi(G) = x\phi(G-v) - \phi(G-v-w)$.* ||

This corollary has been used to compile the characteristic polynomials of trees [**31**], although the polynomial for a single tree is still most easily obtained directly from Sachs' theorem.

If e is any edge of G, a similar formula relates the characteristic polynomial of G to that of $G-e$:

Theorem 3.4. *Let $e = vw$ be an edge of G, and let $\mathscr{C}(e)$ be the set of all circuits containing e. Then $\phi(G)$ satisfies*

$$\phi(G) = \phi(G-e) - \phi(G-v-w) - 2\sum_Z \phi(G-V(Z)),$$

where the summation extends over all $Z \in \mathscr{C}(e)$. ||

4. Bounds on Eigenvalues

Since eigenvalues are often difficult to evaluate, it is sometimes useful to determine bounds for them; conversely, such bounds can be used to obtain spectral information. Furthermore, the nature of these bounds provides insight into the relationship between eigenvalues and graphical properties. One significant result is the specialization of a rather deep theorem on non-negative matrices proved by Perron [**33**] and Frobenius [**15**]. Since the matrices are not required to be symmetric, there is a corresponding theorem for digraphs (see Section 13). For the general version of this theorem, together with its proof, see [**16**] or [**28**].

Theorem 4.1 (Perron–Frobenius). *If G is a connected graph with at least two vertices, then*

(*i*) *its largest eigenvalue* λ_1 *is a simple root of* $\phi(G; x)$;

(*ii*) *corresponding to the eigenvalue* λ_1, *there is an eigenvector* $\mathbf{x}_1$ *all of whose coordinates are positive*;

(*iii*) *if* λ *is any other eigenvalue of G, then* $-\lambda_1 \leqslant \lambda < \lambda_1$;

(*iv*) *the deletion of any edge of G decreases the largest eigenvalue.* ‖

The largest eigenvalue λ_1 is often called the **spectral radius** of G. Since the eigenvectors corresponding to any eigenvalue other than λ_1 must be orthogonal to $\mathbf{x}_1$, we observe that the multiples of $\mathbf{x}_1$ are the only eigenvectors all of whose coordinates are positive.

It is of interest to characterize those graphs for which $\lambda_p = -\lambda_1$. This situation occurs exactly when G is a bipartite graph (see Section 6), and we present four other equivalent conditions in Theorem 4.2. The equivalence of the first and third of these are referred to as the "pairing theorem" by Coulson and Rushbrooke [**8**] and Rouvray [**36**], and first arose in a chemical context. Sachs [**38**] derived the equivalence of (*i*) and (*iv*). We have added versions (*v*) and (*vi*), which are immediate deductions from Theorem 2.4:

Theorem 4.2. *The following statements are equivalent for a connected graph G:*

(*i*) *G is a bipartite graph*;

(*ii*) $\lambda_p = -\lambda_1$;

(*iii*) $\lambda_i = -\lambda_{p+1-i}$, *for* $1 \leqslant i \leqslant \frac{1}{2}(p-1)$;

(*iv*) $a_{2i-1} = 0$, *for* $1 \leqslant i \leqslant \frac{1}{2}(p+1)$;

(*v*) $\sum_{j=1}^{p} \lambda_j^{2i-1} = 0$, *for all* $i \geqslant 1$;

(*vi*) $\sum_{j=1}^{p} \lambda_j^{2[\frac{1}{2}(p+1)]-1} = 0.$ ‖

The fact that the eigenvectors $\mathbf{x}_1, \mathbf{x}_2, \ldots, \mathbf{x}_p$ can be chosen to form an orthonormal basis in such a way that the coordinates of $\mathbf{x}_1$ are all positive provides the foundation for the next theorem:

Theorem 4.3. *Let G be a connected graph, let* $\mathbf{y}$ *be any positive vector, and let* $\mathbf{e}_j$ *denote the jth unit vector* $(1 \leqslant j \leqslant p)$. *Then*

$$\frac{(\mathbf{A}\mathbf{y}).\mathbf{y}}{\mathbf{y}.\mathbf{y}} \leqslant \lambda_1 \leqslant \max_{1 \leqslant j \leqslant p} \frac{(\mathbf{A}\mathbf{y}).\mathbf{e}_j}{\mathbf{y}.\mathbf{e}_j},$$

with equality if and only if $\mathbf{y}$ *is a scalar multiple of* $\mathbf{x}_1$.

Remark. The upper bound is Varga's improvement of Gershgorin's theorem [**50**, Theorems 1.5 and 2.2]. Because we are dealing with symmetric matrices, we can strengthen Varga's lower bound by using the "Rayleigh quotient", as described in Collatz and Sinogowitz [**7**].

Proof. To prove the lower bound, we note that if $\mathbf{y} = \sum_i c_i \mathbf{x}_i$, then the expression

$$\frac{(\mathbf{A}\mathbf{y}).\mathbf{y}}{\mathbf{y}.\mathbf{y}} = \frac{\sum c_i^2 \lambda_i}{\sum c_i^2}$$

is just a weighted average of the eigenvalues λ_i. This average is clearly less than its maximum value λ_1 unless $c_2 = c_3 = \ldots = c_p = 0$, in which case $\mathbf{y} = c_1 \mathbf{x}_1$ is an eigenvector as asserted, and $(\mathbf{A}\mathbf{y}).\mathbf{e}_j/\mathbf{y}.\mathbf{e}_j = \lambda_1$ for each value of j.

In order to prove the upper bound, we suppose that $\mathbf{y} \neq c_1 \mathbf{x}_1$, and assume that $(\mathbf{A}\mathbf{y}).\mathbf{e}_j/\mathbf{y}.\mathbf{e}_j \leqslant \lambda_1$, for each value of j. Since $\mathbf{y}$ is not an eigenvector, this inequality is strict for at least one value of j. Letting $\mathbf{x}_1 = \sum_j b_j \mathbf{e}_j$, where each $b_j > 0$, we observe that

$$\lambda_1 = \frac{c_1 \lambda_1}{c_1} = \frac{(\mathbf{A}\mathbf{y}).\mathbf{x}_1}{\mathbf{y}.\mathbf{x}_1} = \frac{\sum b_j((\mathbf{A}\mathbf{y}).\mathbf{e}_j)}{\sum b_j(\mathbf{y}.\mathbf{e}_j)} < \max_{1 \leqslant j \leqslant p} \frac{b_j((\mathbf{A}\mathbf{y}).\mathbf{e}_j)}{b_j(\mathbf{y}.\mathbf{e}_j)} \leqslant \lambda_1.$$

This contradiction completes the proof. ‖

Astute choices for $\mathbf{y}$ provide useful bounds on the spectral radius λ_1. In the following corollaries, let ρ_i be the valency of the vertex v_i, $\rho_{\max}$ the maximum valency and q the number of edges of G:

Corollary 4.4. *Let G be a connected graph. Then*

either (*i*) $2q/p < \lambda_1 < \rho_{\max}$,

or (*ii*) $2q/p = \lambda_1 = \rho_{\max}$, *G is regular, and* $(1, 1, \ldots, 1)^{\mathrm{T}}$ *is an eigenvector.*

Proof. Substitute $\mathbf{y} = (1, 1, \ldots, 1)^{\mathrm{T}}$ in Theorem 4.3, and observe that $2q/p$ is the average valency in G. ‖

Corollary 4.5. *If G is connected, but not regular, then*

$$\frac{1}{q}\sum \sqrt{\rho_i \rho_j} < \lambda_1 < \max_{1 \leqslant i \leqslant p} \frac{1}{\rho_i} \sum \sqrt{\rho_i \rho_j},$$

where the first summation extends over all edges $v_i v_j$, and the second summation extends over all vertices v_j adjacent to v_i.

Proof. Substitute $\mathbf{y} = (\sqrt{\rho_1}, \sqrt{\rho_2}, \ldots, \sqrt{\rho_p})^{\mathrm{T}}$ in Theorem 4.3. ‖

Although the second corollary is more involved, its bounds are much tighter. The Cauchy–Schwarz inequality can be used to give away a bit on the upper bound in order to obtain a form that is easier to work with—namely, $(1/\rho_i)\sum \sqrt{\rho_i \rho_j} \leqslant \sqrt{\sum \rho_j}$. The weakened upper bound becomes $(\max_{1 \leqslant i \leqslant p} \sum \rho_j)^{\frac{1}{2}}$, and is still quite tight. In particular, it is always better than the upper bound $(2q(p-1)/p)^{\frac{1}{2}}$ obtained in **[52]**.

In [7], Collatz and Sinogowitz found the range of values for the spectral radius of a connected graph:

Theorem 4.6. *If G is a connected graph of order p, then*

$$2 \cos\left(\frac{\pi}{p+1}\right) \leqslant \lambda_1 \leqslant p-1.$$

The lower bound occurs only when G is the path P_p, and the upper bound occurs only when G is the complete graph K_p. ‖

Smith **[46]** has found all connected graphs with $\lambda_1 \leqslant 2$ (see also Section 9), and he also identified those graphs for which λ_1 is the only positive eigenvalue:

Theorem 4.7. *If G is a connected graph with $\lambda_1 = 2$, then G is either $K_{1,4}$, a circuit graph C_p, or one of the trees in Fig. 1. Moreover, if G is a connected graph with $\lambda_1 < 2$, then G is a subgraph of one of these graphs.* ‖

Theorem 4.8. *A connected graph has exactly one positive eigenvalue if and only if it is a complete multipartite graph.* ‖

We conclude this section with an important result which is often alluded to, seldom referenced, and almost never proved. Our approach is loosely modeled on Gantmacher **[16]**.

Theorem 4.9 (The Interlacing Theorem). *Let G be a graph with spectrum $\lambda_1 \geqslant \lambda_2 \geqslant \ldots \geqslant \lambda_p$, and let the spectrum of $G - v_1$ be $\mu_1 \geqslant \mu_2 \geqslant \ldots \geqslant \mu_{p-1}$.*

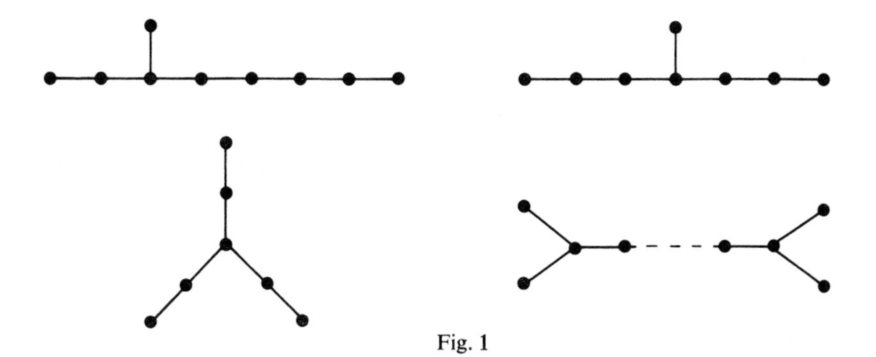

Fig. 1

Then the spectrum of $G-v_1$ is "interlaced" with the spectrum of G—that is,

$$\lambda_1 \geqslant \mu_1 \geqslant \lambda_2 \geqslant \mu_2 \geqslant \ldots \geqslant \mu_{p-1} \geqslant \lambda_p.$$

Proof. Let $\mathbf{A}^*$ be the adjacency matrix of $G-v_1$, and let $\mathbf{A}$ be written as

$$\mathbf{A} = \begin{pmatrix} \mathbf{0} & \mathbf{u} \\ \mathbf{u}^{\mathrm{T}} & \mathbf{A}^* \end{pmatrix},$$

for some $(p-1)$-vector $\mathbf{u}$. We can choose the eigenvectors of G to be orthonormal, and we can also insist that if G has a multiple eigenvalue, then at most one of its eigenvectors is not orthogonal to the unit vector $\mathbf{e}_1$. Now if $\mathbf{x}_i$ is an eigenvector orthogonal to $\mathbf{e}_1$, then $\mathbf{x}_i$ restricted to $G-v_1$ is also an eigenvector, and so $\mu_i = \lambda_i$. These equal eigenvalues can be inserted at the end of the proof, and let us relabel the remaining eigenvalues $\tilde{\lambda}_1 > \tilde{\lambda}_2 > \ldots > \tilde{\lambda}_k$. Any remaining eigenvector of $\mathbf{A}^*$ can be written as a linear combination $\mathbf{y} = \sum_{i=1}^k b_i\mathbf{x}_i$. Similarly, $\mathbf{e}_1 = \sum_{i=1}^k c_i\mathbf{x}_i$, where $c_i = \mathbf{e}_1 . \mathbf{x}_i$. Now if $\mathbf{A}^*\mathbf{y} = \mu\mathbf{y}$, then $\mathbf{A}\mathbf{y} = (\mathbf{u}.\mathbf{y})\mathbf{e}_1 + \mu\mathbf{y} = \sum_i b_i\tilde{\lambda}_i\mathbf{x}_i$. For each j, we form the inner product of this vector equation with $\mathbf{x}_j$, giving $(\mathbf{u}.\mathbf{y})c_j + \mu b_j = b_j\tilde{\lambda}_j$. That is,

$$b_j = \frac{-(\mathbf{u}.\mathbf{y})c_j}{\mu - \tilde{\lambda}_j}.$$

Since $\mathbf{y}.\mathbf{e}_1 = 0$, this implies that

$$\sum_j c_j b_j = \sum_j \frac{-(\mathbf{u}.\mathbf{y})c_j^2}{\mu - \tilde{\lambda}_j} = 0.$$

This last equation has k vertical asymptotes at $\tilde{\lambda}_1, \tilde{\lambda}_2, \ldots, \tilde{\lambda}_k$, and so by continuity there is a root lying in each interval $(\tilde{\lambda}_{i+1}, \tilde{\lambda}_i)$. It follows that $\tilde{\lambda}_1 > \tilde{\mu}_1 > \tilde{\lambda}_2 > \tilde{\mu}_2 > \ldots > \tilde{\mu}_{k-1} > \tilde{\lambda}_k$. The replacement of the identical eigenvalues removed above completes the proof. ‖

5. Regular Graphs

In Corollary 4.4 we gave a spectral characterization of a connected regular graph—namely, that $\lambda_1 = 2q/p$. We now present several further spectral results which apply only to regular graphs, starting with a result of Sachs [**37**] on the complement of a regular graph. (Cvetković [**10**] has given an analogous, but much more complicated, result for arbitrary graphs.)

Theorem 5.1. *If G is a regular ρ-valent graph, then*

$$\phi(\bar{G};\, x) = (-1)^p \left(\frac{x+\rho+1-p}{x+\rho+1}\right) \phi(G;\, -x-1).$$

Proof. The graphs G and $\bar{G}$ have the same eigenvectors, and the eigenvalues of $\bar{G}$ are readily computed. ‖

For example, taking $G = K_p$, we have $\phi(\bar{K}_p;\, x) = x^p$, so that

$$\phi(K_p;\, x) = (x+1-p)(x+1)^{p-1}.$$

If G is a regular graph, then the matrix $\mathbf{M}$ occurring in the matrix-tree theorem (see Theorem 1.10 of Chapter 1) is simply $\rho\mathbf{I}-\mathbf{A}(G)$, and it becomes a routine matter to derive the following alternative version of this theorem:

Theorem 5.2. *The number of spanning trees in a regular graph G is given by*

$$\frac{1}{p}\,\phi'(G;\, x)|_{x=\lambda_1} = \frac{1}{p}\prod_{i=2}^{p}(\lambda_1-\lambda_i). \quad ‖$$

For example, K_p has $(1/p)(p-1+1)^{p-1} = p^{p-2}$ spanning trees (a result of Cayley—see Section 5 of Chapter 14), and $K_{r,r}$ has r^{2r-2} trees. Theorem 5.2 has been generalized to non-regular graphs by Waller [**51**], but the resulting formula seems to have little advantage over the original matrix-tree theorem.

If G and H are graphs, their **join** $G+H$ can be defined by $\overline{G+H} = \bar{G} \cup \bar{H}$. Consequently, if G and H are regular, we may use Theorems 2.1 and 5.1 to obtain a result which has been rediscovered several times (see, for example, [**10**], [**14**], [**41**]):

Theorem 5.3. *If G is a regular ρ_1-valent graph of order p_1, and H is a regular ρ_2-valent graph of order p_2, then*

$$\phi(G+H;\, x) = \frac{\phi(G;\, x)\,\phi(H;\, x)}{(x-\rho_1)\,(x-\rho_2)}\{x^2-(\rho_1+\rho_2)x+(\rho_1\rho_2-p_1p_2)\}. \quad ‖$$

Cvetković [**10**] and Schwenk [**41**] have also given formulas for computing the characteristic polynomials of Cartesian products, conjunctions, strong products, and generalized compositions of graphs.

An alternative characterization of regular graphs has been found by Hoffman **[21]**:

Theorem 5.4. *The graph G is regular and connected if and only if there exists a polynomial q such that $q(\mathbf{A}(G)) = \mathbf{J}_p$, where $\mathbf{J}_p$ is the $p \times p$ matrix consisting entirely of ones.* ‖

In fact, Hoffman showed that if $\rho, \beta_1, \ldots, \beta_k$ denote the distinct eigenvalues of G, then

$$q(x) = p \prod_{i=1}^{k} \frac{(x-\beta_i)}{(\rho-\beta_i)}.$$

Moreover, since $(x-\rho)q(x)$ is evidently the minimum polynomial $\mu(G;\, x)$, the Hoffman polynomial can often be used to determine the spectrum of G, especially when there are few distinct eigenvalues. The only connected graphs with two distinct eigenvalues are the complete graphs K_p. The connected regular graphs with three distinct eigenvalues are known as "strongly regular graphs", and are closely linked to combinatorial designs (see, for example, **[1]**, **[9]**, **[26]** and **[44]**) and to a method for constructing simple groups (see Hestenes **[20]**). In Chapter 12, Peter Cameron examines these strongly regular graphs in depth, and presents an elegant application of the Hoffman polynomial to the proof of the friendship theorem.

6. Some Examples

We now have enough tools available to compute the eigenvalues of many familiar graphs. We begin by stating a simple lemma whose proof is left to the reader:

Lemma 6.1. *If $\mathbf{C} = s\mathbf{J}_p + t\mathbf{I}_p$ then $\mathbf{C}$ has eigenvalues t (with multiplicity $p-1$), and $ps+t$ (with multiplicity 1).* ‖

Complete Graphs

By applying Lemma 6.1 to the matrix $\mathbf{J}_p - \mathbf{I}_p$, we see that the eigenvalues of K_p are -1 (with multiplicity $p-1$), and $p-1$ (with multiplicity 1). The characteristic polynomial is

$$\phi(K_p;\, x) = (x-p+1)(x+1)^{p-1}.$$

Complete Bipartite Graphs

The eigenvalues of the complete bipartite graph $K_{r,s}$ are given by $\pm\sqrt{rs}$ (with multiplicity 1), and 0 (with multiplicity $r+s-2$). The characteristic polynomial is

$$\phi(K_{r,s};\, x) = (x^2-rs)x^{r+s-2}.$$

We shall give three proofs of this result to illustrate various techniques which are sometimes useful when dealing with spectra:

First Proof. Let $\mathbf{A} = \mathbf{A}(K_{r,s})$ be written in the form

$$\mathbf{A} = \begin{pmatrix} \mathbf{0} & \mathbf{B} \\ \mathbf{B}^{\mathrm{T}} & \mathbf{0} \end{pmatrix},$$

where $\mathbf{B}$ is the $r \times s$ matrix consisting entirely of ones. By subtracting the first row of $\mathbf{A}-x\mathbf{I}_{r+s}$ from each of rows 2, 3, . . ., r, and subtracting row $r+1$ from each of rows $r+2$, $r+3$, . . ., $r+s$, we see immediately that $r+s-2$ of the eigenvalues must be zero. By Theorem 2.3, the two remaining eigenvalues must be t and $-t$ for some t. The fact that $t = \sqrt{rs}$ then follows from Theorem 2.4. ‖

Second Proof. If $\mathbf{A}$ is written in the above form, then

$$\mathbf{A}^2 = \begin{pmatrix} s\mathbf{J}_r & \mathbf{0} \\ \mathbf{0} & r\mathbf{J}_s \end{pmatrix},$$

and so, by Lemma 6.1, $\mathbf{A}^2$ has eigenvalues rs (with multiplicity 2), and 0 (with multiplicity $r+s-2$), The result now follows from Theorem 2.2. ‖

Third Proof. Since $K_{r,s} = \overline{K}_r + \overline{K}_s$ is a join of two regular graphs, we have

$$\phi(K_{r,s};\, x) = (x^2-rs)x^{r+s-2},$$

by Theorem 5.3. ‖

Circuit Graphs

The easiest way to obtain the eigenvalues of the circuit graph C_p is to pull $1+[\frac{1}{2}p]$ eigenvectors out of the air, and then to verify that these actually are eigenvectors. For $0 \leqslant j \leqslant [\frac{1}{2}p]$, we let the jth eigenvector be

$$\mathbf{x}_j = \sum_{i=1}^{p} \cos\left(\frac{2\pi ij}{p}\right) \mathbf{e}_i,$$

where $\mathbf{e}_i$ denotes, as usual, the ith unit vector. Using linearity and a trigonometric identity, we can verify that

$$\begin{aligned}\mathbf{A}\mathbf{x}_j &= \sum_{i=1}^{p} \left\{\cos\left(\frac{2\pi j(i-1)}{p}\right) + \cos\left(\frac{2\pi j(i+1)}{p}\right)\right\} \mathbf{e}_i \\ &= \sum_{i=1}^{p} 2\cos\left(\frac{2\pi j}{p}\right)\cos\left(\frac{2\pi ji}{p}\right) \mathbf{e}_i \\ &= 2\cos\left(\frac{2\pi j}{p}\right) \mathbf{x}_j.\end{aligned}$$

For each $j \neq 0$ and $j \neq \frac{1}{2}p$, this eigenvector is non-symmetrical under a rotation of coordinates, so that a one-step rotation through $2\pi/p$ radians provides a new linearly independent eigenvector. Whether p is even or odd, we have found a complete set of p eigenvectors. Furthermore, since $\cos(2\pi j/p) = \cos(2\pi(p-j)/p)$, we may list the spectrum as

$$\left\{2\cos\left(\frac{2\pi j}{p}\right) : j = 0, 1, \ldots, p-1\right\}.$$

(An alternative method for evaluating the spectrum of a circuit graph is to observe that $x\mathbf{I} - \mathbf{A}(C_p)$ is a circulant matrix, and to use the formula for the determinant of such a matrix as found in, for example, **[13]**.)

Wheels

The spectrum of the wheel $W_p = K_1 + C_{p-1}$ may be found by applying Theorem 5.3, giving

$$\phi(W_p; x) = (x-1+\sqrt{p})(x-1-\sqrt{p}) \prod_{i=1}^{p-2} \left\{x - 2\cos\left(\frac{2\pi i}{p-1}\right)\right\}.$$

Path Graphs

Given any eigenvector $\mathbf{x}_j$ of the path graph P_p, we can immediately construct two eigenvectors of the circuit graph C_{2p+2}—namely $(\mathbf{x}_j, 0, -\mathbf{x}_j, 0)$ and $(0, \mathbf{x}_j, 0, -\mathbf{x}_j)$. Thus any eigenvalue of P_p is a "double eigenvalue" of C_{2p+2}, so that

$$\phi(P_p; x) = \prod_{i=1}^{p} \left\{x - 2\cos\left(\frac{\pi i}{p+1}\right)\right\}.$$

Cubes

The n-dimensional cube Q_n is the Cartesian product of n copies of K_2. Repeated use of a formula for the spectrum of a Cartesian product (see **[10]**, **[41]**) yields the characteristic polynomial

$$\phi(Q_n;\, x) = \prod_{i=0}^{n} (x+n-2i)^{\binom{n}{i}}.$$

In particular, the characteristic polynomial of the ordinary 3-dimensional cube is given by

$$\phi(Q_3;\, x) = (x-3)(x-1)^3(x+1)^3(x+3).$$

Octahedra

The m-dimensional octahedron $K_{m(2)}$ is the complete m-partite graph on m sets of size 2. Since $K_{m(2)}$ is just the complement of m disjoint copies of K_2, we may use Theorems 2.1 and 5.1 to give

$$\phi(K_{m(2)};\, x) = (x+2-2m)x^m(x+2)^{m-1}.$$

Note that

$$\mathbf{A}^2+2\mathbf{A} = (2m-2)\mathbf{J},$$

so that the Hoffman polynomial is $q(x) = (x^2+2x)/(2m-2)$. Note also that the octahedra have exactly three eigenvalues, and are therefore strongly regular; the importance of this will become apparent in Section 9.

Further Examples

The characteristic polynomials of all trees with at most ten vertices appear in **[31]**, and the polynomials of all graphs with at most five vertices may be found in Collatz and Sinogowitz [7]. This latter paper also contains the characteristic polynomials of the five Platonic graphs; only two of these are not included in the families mentioned above, and we include them here for completeness:

the dodecahedron: $(x-3)(x^2-5)^3(x-1)^5x^4(x+2)^4$;

the icosahedron: $(x-5)(x^2-5)^3(x+1)^5$.

7. Cospectral Graphs

One of the most difficult problems in graph theory is the recognition of isomorphic graphs displayed with different labelings (see Chapter 15). Since the spectrum of a graph is unchanged by a relabeling of its vertices, early investigators briefly entertained the hope that the spectrum would uniquely determine the structure of the graph, but this was soon found to be false. The smallest pair of **cospectral graphs** (that is, graphs with the same spectrum) were found to be the star graph $K_{1,4}$ and the graph $K_1 \cup C_4$. Since the second of these graphs is disconnected, we might hope that by adding connectivity or other structural restrictions we could still obtain a limited spectral characterization theorem. But the smallest connected pair of cospectral graphs have only six vertices (see **[2]**, [7]), the smallest cospectral blocks have seven vertices (**[2]**), and the smallest cospectral trees (**[2]**, [7], **[18]**) have eight vertices (see Fig. 2). In **[21]**, Hoffman constructed cospectral regular bipartite graphs

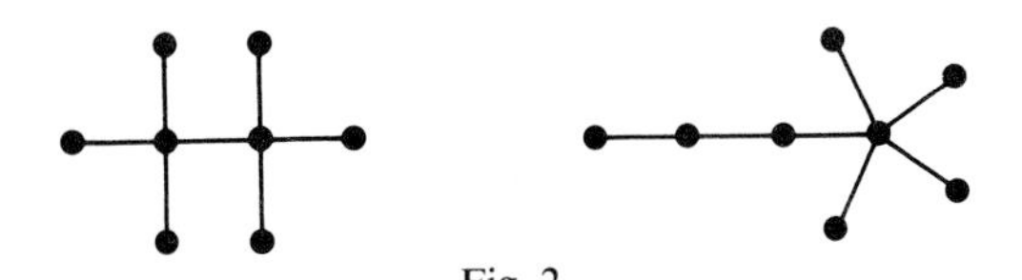

Fig. 2

of order 16, and the smallest pair of regular cospectral graphs are of order 10, and have been found by Godsil and McKay.

Hoffman has described a construction for cospectral sets of arbitrary size. The proof that follows is due to Mowshowitz **[31]**:

Theorem 7.1. *For any positive integer k, there exist k cospectral graphs which are both connected and regular.*

Proof. Let H_1 and H_2 be two given cospectral regular graphs. For $0 \leqslant i \leqslant k-1$, let G_i be the complement of the graph obtained by taking i copies of H_1 and $k-i-1$ copies of H_2. Then, by Theorems 2.1 and 5.1, the regular connected graphs $G_0, G_1, \ldots, G_{k-1}$ are cospectral. ‖

Mowshowitz **[31]** went on to construct infinitely many pairs of cospectral trees, but the construction is so special that one might still suspect that "most" trees are determined by their spectrum. Nothing could be further from the truth, as was proved by Schwenk **[40]** (see also Section 3 of Chapter 15):

Theorem 7.2. *If t_p is the number of unlabeled trees of order p, and if s_p is the number of such trees cospectral with no other tree of order p, then*

$$\lim_{p \to \infty} \frac{s_p}{t_p} = 0.$$

Sketch of Proof. A full proof of this theorem is very long, and involves asymptotics of certain generating functions. The approach is based on the concept of cospectrally-rooted graphs. Two different rooted graphs, G_1 with root-vertex r_1 and G_2 with root-vertex r_2, are called "cospectrally-rooted" if both $\phi(G_1) = \phi(G_2)$ and $\phi(G_1 - r_1) = \phi(G_2 - r_2)$. (Note that G_1 and G_2 may be different graphs, or the same graph with two different rootings.) By applying Theorem 3.2, one can easily verify that attaching G_1 to any rooted graph (H, r) by identifying the root-vertices r_1 and r produces the same spectrum as attaching G_2 to (H, r) by identifying the root-vertices r_2 and r. In [**40**], Schwenk found the smallest pair of cospectrally-rooted trees, with nine vertices, and then showed that nearly every tree of large order contains one of these two as a limb. Since this limb can be replaced by the other 9-vertex limb to form a different tree cospectral with the original tree, it follows that very few trees are spectrally unique. ‖

An exhaustive search was used to find the pair of cospectrally-rooted trees of order 9 used above, but they could have been recognized by a very nice characterization of cospectrally-rooted graphs given by Herndon and Ellzey [**19**]. In the statement of this result, we let G_1 have orthonormal eigenvectors $\mathbf{x}_1, \ldots, \mathbf{x}_p$, and G_2 have orthonormal eigenvectors $\mathbf{y}_1, \ldots, \mathbf{y}_p$, and assume that $\phi(G_1) = \phi(G_2)$ with $\mathbf{A}_1\mathbf{x}_i = \lambda_i\mathbf{x}_i$ and $\mathbf{A}_2\mathbf{y}_i = \lambda_i\mathbf{y}_i$, for each i. (If these graphs have multiple eigenvalues, then the eigenvectors for the repeated eigenvalues must be selected carefully.) We also assume that the unit vector $\mathbf{e}_1 = (1, 0, \ldots, 0)$ represents the root in both graphs.

Theorem 7.3. *Suppose that $\phi(G_1) = \phi(G_2)$, and that the eigenvectors have been selected as above. Then G_1 and G_2 are cospectrally-rooted if and only if $(\mathbf{e}_1 . \mathbf{x}_i)^2 = (\mathbf{e}_1 . \mathbf{y}_i)^2$, for each i.*

Proof. Recalling the proof of the interlacing theorem (Theorem 4.9), we observe that μ is an eigenvalue of $G_1 - r_1$ if and only if μ satisfies the equality

$$\sum_{i=1}^{p} \frac{(\mathbf{e}_1 . \mathbf{x}_i)^2}{\mu - \lambda_i} = 0,$$

and similarly μ is an eigenvalue of $G_2 - r_2$ if and only if μ satisfies the equality

$$\sum_{i=1}^{p} \frac{(\mathbf{e}_1 . \mathbf{y}_i)^2}{\mu - \lambda_i} = 0.$$

It follows that if $(\mathbf{e}_1 . \mathbf{x}_i)^2 = (\mathbf{e}_1 . \mathbf{y}_i)^2$, for each i, then we get the same roots.

Conversely, if the two summations have the same roots, then multiplying both by $\prod_{i=1}^{p} (\mu - \lambda_i)$ gives two polynomials of degree $p-1$ with identical roots. Thus, one of these polynomials must be a multiple of the other. But the leading coefficient of each polynomial is given by

$$\sum_{i=1}^{p} (\mathbf{e}_1 . \mathbf{x}_i)^2 = \mathbf{e}_1 . \mathbf{e}_1 = 1 = \sum_{i=1}^{p} (\mathbf{e}_1 . \mathbf{y}_i)^2,$$

so that the polynomials are identical. Setting $\mu = \lambda_i$ (for each i) in both polynomials gives us $(\mathbf{e}_1 . \mathbf{x}_i)^2 = (\mathbf{e}_1 . \mathbf{y}_i)^2$, as required. ||

Realizing that the spectrum fails to determine the graph, Graham and Pollak **[17]** suggested that perhaps the spectrum of the distance matrix **D** should be used. For strongly regular graphs, $\mathbf{D} = \mathbf{A} + 2(\mathbf{J} - \mathbf{I} - \mathbf{A})$ and no new theory emerges, but for other graphs, working with **D** seems to be much more difficult. Nevertheless, knowledge about the "distance-spectrum" of trees has been advanced. At first, not even a single pair of distance-cospectral trees could be found, but, by a remarkable discovery, McKay **[29]** found a 16-vertex tree with two different distance-cospectral rootings, and used it to prove the following distance analog of Theorem 7.2:

Theorem 7.4. *If t_p is the number of unlabeled trees of order p, and if d_p is the number of such trees distance-cospectral with no other tree, then*

$$\lim_{p \to \infty} \frac{d_p}{t_p} = 0. \ \|$$

Convergence in this theorem is extremely slow. By an exhaustive search, McKay has found that $d_p = t_p$ for $p \leqslant 16$, whereas $d_{17} = t_{17} - 2 = 48\,627$. The only pair not determined uniquely here are the first pair constructed from his 16-vertex trees.

8. Eigenvalues and Coloring

A well-known theorem due to Brooks (see Chapter 1) states that if G is a connected graph with chromatic number $\chi(G)$ and maximum valency ρ, then $\chi(G) \leqslant 1 + \rho$, with equality if and only if G is a complete graph or an odd circuit. The main weakness of this theorem is that if there are just a few vertices of large valency, then the upper bound is likely to be far from the true value; for example, if G is the star graph $K_{1,s}$, then $\chi(G) = 2$, whereas Brooks' theorem gives s as the upper bound.

To alleviate this difficulty, we can use the following stronger bound of Wilf **[52]**:

Theorem 8.1. *If G is a connected graph with largest eigenvalue λ_1, then $\chi(G) \leqslant 1 + \lambda_1$, with equality if and only if G is a complete graph or an odd circuit.*

Proof. Let $\mathbf{x}_1 = (c_1, c_2, \ldots, c_p)$ be the positive eigenvector corresponding to λ_1, and assume that the vertices v_i of G are labeled in such a way that $c_1 \geqslant c_2 \geqslant \ldots \geqslant c_p > 0$. For each i, we have

$$\lambda_1 c_i = \mathbf{A}\mathbf{x}_1 . \mathbf{e}_i = \sum c_j,$$

where the summation extends over all vertices v_j adjacent to v_i. In this sum, we can make every term smaller by replacing c_j by c_i, if $j < i$, and by 0, if $j > i$. So $\lambda_1 c_i \geqslant sc_i$, where s is the number of vertices adjacent to v_i with subscript less than i. Now assume that $[1+\lambda_1]$ colors are available, and color the vertices in sequence. When the vertex v_i is reached, only s of its neighbors have been colored, so at least one color is available for v_i (since $[1+\lambda_1] \geqslant 1+s$). Thus, the vertices of G can be colored with $[1+\lambda_1]$ colors. ‖

Wilf's theorem can be a substantial improvement over Brooks' theorem; for example, for the star graph we have $\chi(K_{1,s}) \leqslant 1+\sqrt{s}$, rather than Brooks' bound of s.

Hoffman **[24]** has also found a lower bound for $\chi(G)$; a good proof of this has been given by Biggs [3, p. 54]:

Theorem 8.2. $\chi(G) \geqslant 1-\lambda_1/\lambda_p$. ‖

It is worth noting that in several important cases (for example, if G is a bipartite graph, a complete graph, or an odd circuit), this lower bound for $\chi(G)$ is best possible. Note also that we can turn this theorem around to gain information on the spectrum from properties of the graph; for example, if G is a planar graph, then $\chi(G) \leqslant 4$, and so $\lambda_p \leqslant -\frac{1}{3}\lambda_1$.

9. Line Graphs

The relationship between the spectra of a graph G and its line graph $L(G)$ is based on the vertex-edge incidence matrix $\mathbf{B} = (b_{ij})$, in which $b_{ij} = 1$ if the vertex v_i is incident to the edge e_j, and $b_{ij} = 0$ otherwise. It is then a routine matter to verify the following results:

Theorem 9.1. (*i*) $\mathbf{A}(G) = \mathbf{B}\mathbf{B}^{\mathrm{T}} - \mathbf{D}$, *where* $\mathbf{D}$ *is the diagonal matrix whose entries are the vertex-valencies*;

(*ii*) $\mathbf{A}(L(G)) = \mathbf{B}^{\mathrm{T}}\mathbf{B} - 2\mathbf{I}_q$. ‖

Theorem 9.2. *If* $\mathbf{B}$ *is any real* $n \times m$ *matrix with* $n \leqslant m$, *then each eigenvalue of* $\mathbf{B}^{\mathrm{T}}\mathbf{B}$ *is non-negative, and* $\phi(\mathbf{B}^{\mathrm{T}}\mathbf{B}) = x^{m-n}\phi(\mathbf{B}\mathbf{B}^{\mathrm{T}})$. ‖

Using these results one can easily deduce that if λ_p is the smallest eigenvalue of a line graph, then $\lambda_p \geqslant -2$, and that if G is a regular ρ-valent graph, then

$$\phi(L(G); x) = \phi(G; x-\rho+2)(x+2)^{\frac{1}{2}p(\rho-2)},$$

a result due to Sachs [**39**].

The condition $\lambda_p \geqslant -2$ is a very strong condition on a graph, and it is natural to ask to what extent it characterizes line graphs. A characterization of all sufficiently large graphs with $\lambda_p \geqslant -2$ has been given by Hoffman [**24**]. However, an alternative and simpler approach to the problem has been described by Cameron, Goethals, Seidel and Shult [**5**], and we shall spend the rest of this section outlining this approach.

So suppose that $\mathbf{A}$ is the adjacency matrix of a graph G, and suppose that the smallest eigenvalue λ_p of G satisfies $\lambda_p \geqslant -2$. Then $2\mathbf{I}+\mathbf{A}$ is a positive semi-definite matrix of rank n (say), and can therefore be interpreted as the matrix of inner products of a set of p vectors $\mathbf{v}_1, \ldots, \mathbf{v}_p$ in $\mathbf{R}^n$; so

$$2\mathbf{I}+\mathbf{A} = (\mathbf{v}_i . \mathbf{v}_j).$$

Now $\mathbf{v}_i . \mathbf{v}_i = 2$, so that each vector $\mathbf{v}_i$ has length $\sqrt{2}$, and $\mathbf{v}_i . \mathbf{v}_j = 0$ or 1 if $i \neq j$, so that any two of these vectors make an angle of 90° or 60°. Conversely, if we are given any set of vectors of length $\sqrt{2}$ in which any two vectors make an angle of 90° or 60°, then these vectors represent a graph with $\lambda_p \geqslant -2$; for example, if $\{\mathbf{e}_1, \ldots, \mathbf{e}_n\}$ is an orthonormal basis, then the set

$$\{\mathbf{e}_i+\mathbf{e}_j : v_i \text{ is adjacent to } v_j\}$$

represents the line graph $L(G)$, so that $L(G)$ has smallest eigenvalue $\lambda_p \geqslant -2$. Note that the graph G is connected if and only if the set of vectors $\{\mathbf{v}_1, \ldots, \mathbf{v}_p\}$ is "indecomposable", in the sense that it is not contained in the union of a family of proper pairwise orthogonal subspaces.

Cameron *et al.* have considered the lines through the origin spanned by these vectors—and, more generally, any set of lines through the origin in R^n at angles of 90° and 60°. Such a set of lines is contained in a maximal set, which is "star-closed", in the sense that if it contains two lines at an angle of 60°, then it also contains the third line in their plane making an angle of 60° with both.

All star-closed sets can be determined. Indeed, if we take the two vectors of length $\sqrt{2}$ on each line, then the set of all such vectors is a "root system" in the sense of the classical theory of semi-simple complex Lie algebras. Conversely, given a root system all of whose vectors are of the same length, the lines spanned by these vectors make angles of 90° and 60° with each other, and the set of all such lines is star-closed. The determination of indecomposable root systems was given by Cartan and Killing (see Jacobson [27]); for a graph-theoretical proof, see [5]. A list of the various types which occur is as follows; here $\{\mathbf{e}_1, \ldots, \mathbf{e}_d\}$ denotes an orthonormal basis for $\mathbf{R}^d$:

$A_n = \{\mathbf{e}_i - \mathbf{e}_j : 1 \leqslant i, j \leqslant n+1, i \neq j\}$
(this set spans the hyperplane in $\mathbf{R}^{n+1}$ perpendicular to $\mathbf{e}_1 + \ldots + \mathbf{e}_{n+1}$);

$D_n = \{\pm\mathbf{e}_i \pm \mathbf{e}_j : 1 \leqslant i < j \leqslant n\}$;

$$E_8 = D_8 \cup \left\{\tfrac{1}{2} \sum_{i=1}^{8} \varepsilon_i \mathbf{e}_i : \varepsilon_i = \pm 1, \text{ and } \varepsilon_1 \varepsilon_2 \ldots \varepsilon_8 = 1\right\};$$

E_7 = the subset of E_8 consisting of all vectors perpendicular to a fixed vector of E_8;

E_6 = the subset of E_8 consisting of all vectors perpendicular to a fixed star of E_8.

If follows that every connected graph G with $\lambda_p \geqslant -2$ can be represented by a subset S of one of the above root systems for which all inner products are non-negative, and conversely. The vertices of G can be identified with the vectors in S, and two vertices of G are adjacent if and only if their inner product is 1. Since $A_n \subseteq D_{n+1}$, and $E_6 \subseteq E_7 \subseteq E_8$, our problem is reduced to the study of graphs represented by subsets of D_n or E_8.

Recall from Section 6 that if m is a non-negative integer, then the m-dimensional octahedron $K_{m(2)}$ is the complete multipartite graph with m partite sets of size 2. It has smallest eigenvalue -2 if $m > 1$, and is represented by the subset $\{\mathbf{e}_0 \pm \mathbf{e}_i : 1 \leqslant i \leqslant m\}$ of D_{m+1}. Given a graph G with vertex-set $\{v_1, \ldots, v_p\}$, $L(G)$ is represented by the subset $\{\mathbf{e}_i + \mathbf{e}_j : v_i \text{ is adjacent to } v_j\}$ of D_p. Hoffman [24] has defined the **generalized line graph** $L(G; m_1, \ldots, m_p)$, where $m_1, \ldots, m_p$ are non-negative integers, to consist of the disjoint union of $L(G)$ and $K_{m_i(2)}$, for $1 \leqslant i \leqslant p$, together with additional edges joining each vertex of $K_{m_i(2)}$ to each vertex $v_i v_j$ of $L(G)$, for $1 \leqslant i \leqslant p$. This generalized line graph is represented by the subset of D_n, with $n = m_1 + \ldots + m_p + p$, consisting of the vectors $\mathbf{e}_{i,0} \pm \mathbf{e}_{i,k}$, for $1 \leqslant k \leqslant m_i$, together with those vectors $\mathbf{e}_{i,0} + \mathbf{e}_{j,0}$ for which v_i and v_j are adjacent in G. We then have the following result:

Theorem 9.3. *A graph G is represented by a subset of a root system of type D_n if and only if G is a generalized line graph.* ||

From this, we can deduce Hoffman's theorem, in the form given by Cameron, Goethals, Seidel and Shult:

Theorem 9.4. *If G is a connected graph with $\lambda_p \geqslant -2$, then*

either (i) G is a generalized line graph,

or (ii) G is represented by a subset of E_8. ‖

It is easy to show that a regular generalized line graph is either a line graph or an n-dimensional octahedron, for some n. This implies the following theorem of Hoffman and Ray-Chaudhuri [**26**]:

Theorem 9.5. *If G is a regular connected graph with $\lambda_p \geqslant -2$, then*

(i) G is a line graph,

or (ii) G is an n-dimensional octahedron, for some n,

or (iii) G is represented by a subset of E_8. ‖

In this connection, Cameron *et al.* have observed that a graph G represented by a subset of E_8 has at most 36 vertices, and maximum valency at most 28; if G is regular, then it has at most 28 vertices, and valency at most 16. Furthermore, all regular graphs which are represented by subsets of E_8, but are not line graphs or octahedra have been determined by Bussemaker, Cvetković and Seidel [**4**]. There are 187 isomorphism classes of such graphs, 68 of which are cospectral with line graphs.

These theorems, and earlier versions of them, have had a number of interesting applications. Among these are Seidel's determination of all strongly regular graphs with smallest eigenvalue -2 [**43**], and several results by Hoffman and Ray-Chaudhuri [**25**], Ray-Chaudhuri [**35**], Doob [**12**] and others, characterizing certain classes of graphs by their spectra. In particular, the eigenvalues of the line graphs of finite projective planes and finite affine planes determine these graphs up to isomorphism, as also do the eigenvalues of $L(K_p)$ (except if $p = 8$, when there are three exceptions), and of $L(K_{r,r})$ (except if $r = 4$, when there is just one exception).

We conclude this section by observing that the same approach can be applied also to graphs with $\lambda_1 \leqslant 2$ (see Theorem 4.7). If G is such a graph, then $2\mathbf{I} - \mathbf{A}(G)$ is a positive semi-definite matrix, and so G can be represented by a subset of a root system in which all inner products are non-positive. If $\lambda_1 < 2$, and G is connected, then the resulting subset is called a "system of fundamental roots" in the classical theory, and we get the following result:

Theorem 9.6. *Every connected graph with $\lambda_1 < 2$ is a Dynkin diagram with no multiple bonds, and conversely.* ‖

10. Another Example

Before proceeding, let us take a particular graph and see how much we can find out about its properties simply by knowing its eigenvalues:

Problem. *What can be said about a graph G whose characteristic polynomial is* $(x-3)(x-1)^5(x+2)^4$?

Solution. The eigenvalues of G are 3, 1 (five times), and -2 (four times), and so, by the remarks following Theorem 2.4, G must have fifteen edges, no triangles, and, of course, ten vertices. By Theorem 3.1 we can also deduce that G contains a circuit of length five, and by Theorem 2.5, that G has diameter two. We also see from Theorem 4.2 that G is not bipartite, and from Theorem 4.8 that G is not a complete k-partite graph for any value of k. Theorems 8.1 and 8.2 tell us that the chromatic number of G is equal to three.

We have already discovered quite a lot about G, but we can find out much more by using Corollary 4.4, from which we deduce that G is a regular graph. We can therefore use the results of Section 5, and can deduce, for example, that G has exactly 2000 spanning trees.

We can now use Theorem 5.1 to deduce that the characteristic polynomial of $\bar{G}$ is equal to $(x-6)(x-1)^4(x+2)^5$. But we see from the remarks after Theorem 9.2 that this is precisely the characteristic polynomial of $L(K_5)$, and by a later result in the same section we know that there is no other graph with this characteristic polynomial. It follows that G must be the complement of $L(K_5)$; in other words, G is the Petersen graph.

Note that in this example G and $\bar{G}$ both have smallest eigenvalue -2, but whereas $\bar{G}$ is a line graph, G is not. The reader may like to verify that G comes under condition (*iii*) of Theorem 9.5.

11. Automorphism Groups

There are several results relating the eigenvalues of a connected graph G to its automorphism group $\Gamma(G)$. If we regard $\Gamma(G)$ as a set of permutation matrices, then a given permutation $\mathbf{P}$ lies in this group if and only if it commutes with the adjacency matrix $\mathbf{A}$ of G; in other words, if we wish to find the automorphism group of a graph G, we need only find all those permutation matrices $\mathbf{P}$ satisfying $\mathbf{PA} = \mathbf{AP}$.

Since $\mathbf{A}$ is a symmetric matrix, there is an orthogonal matrix $\mathbf{S}$ such that

$\mathbf{S}^{\mathrm{T}}\mathbf{AS}$ is a diagonal matrix $\mathbf{D}$ whose diagonal entries are the eigenvalues of G. So we can write

$$\mathbf{S}^{\mathrm{T}}\mathbf{AS} = \begin{pmatrix} \mu_1\mathbf{I} & & & 0 \\ & \mu_2\mathbf{I} & & \\ & & \ddots & \\ 0 & & & \mu_k\mathbf{I} \end{pmatrix},$$

where $\mu_1, \ldots, \mu_k$ are the distinct eigenvalues of G. It follows that $\mathbf{QD} = \mathbf{DQ}$, where $\mathbf{Q} = \mathbf{S}^{\mathrm{T}}\mathbf{PS}$, and hence that

$$\mathbf{Q} = \begin{pmatrix} \mathbf{Q}_1 & & & 0 \\ & \mathbf{Q}_2 & & \\ & & \ddots & \\ 0 & & & \mathbf{Q}_k \end{pmatrix}.$$

Since $\mathbf{Q}$ is a product of orthogonal matrices, each submatrix $\mathbf{Q}_i$ is orthogonal, and the mapping $\mathbf{P} \to \mathbf{Q}_i$ is a homomorphism from $\Gamma(G)$ to a group of orthogonal matrices $\mathbf{Q}_i$ whose size is the multiplicity of the corresponding eigenvalue μ_i.

In particular, if the eigenvalues of G are all distinct, then each matrix $\mathbf{Q}$ is a diagonal matrix, and so $\Gamma(G)$ is isomorphic to a group of diagonal matrices. It follows that $\Gamma(G)$ is an Abelian group. Moreover, each matrix $\mathbf{Q}_i$ is a 1×1 orthogonal matrix, and so $\mathbf{Q}_i = (1)$ or (-1), giving $\mathbf{Q}^2 = \mathbf{I}$. This implies the following theorem of Mowshowitz [**30**], also proved independently by Petersdorf and Sachs [**34**]. (A similar, but weaker, version of this result holds for digraphs; it will be stated in Section 13.)

Theorem 11.1. *If all the eigenvalues of a graph G are distinct, then every non-trivial automorphism of G has order* 2, *and $\Gamma(G)$ is an Abelian group.* ‖

Exactly the same argument can be used to prove the next result; here we use the fact that a finite group of orthogonal 2×2 matrices is either a cyclic group or a dihedral group:

Theorem 11.2. *If every eigenvalue of G has multiplicity* 1 *or* 2, *then $\Gamma(G)$ is a solvable group.* ‖

Considerations such as these are also very important in group theory, since many of the recently-discovered "sporadic simple groups" are defined as automorphism groups of graphs (see, for example, Section 10 of Chapter

12). In such cases the above techniques can be used to "reduce" the group into constituent groups of matrices of smaller size.

Returning to graphs, we can now state a related result concerning graphs whose characteristic polynomial is irreducible; this result is also due to Mowshowitz [**32**]:

Theorem 11.3. *If G is a graph whose characteristic polynomial is irreducible over the integers, then its automorphism group $\Gamma(G)$ is the trivial group.* ||

Note that the converse of this theorem is false, as may be seen by taking any regular graph with trivial automorphism group (or, alternatively, any bipartite graph of odd order with trivial automorphism group); in either case, the characteristic polynomial has a linear factor, contrary to hypothesis.

Petersdorf and Sachs have also obtained several powerful results concerning the eigenvalues of those graphs whose automorphism group is transitive; the reader is referred to [**34**] for the proofs:

Theorem 11.4. *Let G be a graph with largest eigenvalue λ_1. If the automorphism group of G is transitive, then G is regular of valency λ_1. If, in addition, G has odd order, then λ_1 is the only simple root of the characteristic equation. If, on the other hand, G has even order, and λ is a simple root of the characteristic equation, then $\lambda_1 - \lambda$ must be an even integer.* ||

Further results (also proved in [**34**]) include an upper bound for the number of simple eigenvalues of a graph G in terms of the cycle decomposition of any permutation in $\Gamma(G)$:

Theorem 11.5. *If there exists an automorphism of G with s odd cycles and t even cycles in its cycle decomposition, then the number of simple eigenvalues of G does not exceed $s+2t$.* ||

12. The Reconstruction Problem

As we saw in Chapter 8, the reconstruction problem remains one of the outstanding problems in graph theory: given the deck of vertex-deleted subgraphs $G_i = G - v_i$, is it possible to reconstruct the unlabeled graph G? By considering spectra we can obtain four variations of this problem, since we may be given either the deck of subgraphs G_i, or perhaps only the deck of polynomials $\phi(G_i; x)$, and from either deck, we may be asked to reconstruct either G, or perhaps only $\phi(G; x)$. Of course, reconstructing G from the deck of G_i's is the original (and still unsolved) problem, and finding G from the deck of $\phi(G_i)$'s is even harder, since a solution of this version would give *a fortiori* a solution of the original problem. But what about the more modest goal of reconstructing only $\phi(G)$? To answer this, we begin with an elementary theorem which is a special case of a result from matrix theory. The formula given here can also be used in Theorem 5.2.

Theorem 12.1. *The derivative of the characteristic polynomial is given by*

$$\phi'(G; x) = \sum_{i=1}^{p} \phi(G_i; x).$$

Proof. In $\phi(G; x) = \det(x\mathbf{I}-\mathbf{A})$, replace the x's by variables $z_1, z_2, \ldots, z_p$, and differentiate using the multivariable form of the chain rule; this gives

$$\phi'(G; x) = \sum_{i=1}^{p} \frac{\partial}{\partial z_i}\{\det(\mathrm{diag}(z_1, \ldots, z_p)-\mathbf{A})\}.\frac{dz_i}{dx}.$$

Since $z_i = x$ for each i, we can omit the factor $dz_i/dx = 1$. By expanding the determinant along row i, we see that the partial derivative with respect to z_i is simply $\phi(G_i; x)$, evaluated at $z_1 = \ldots = z_p = x$. The result follows immediately. ‖

Using this result we can easily determine $\phi'(G)$ by starting with either the deck of polynomials or the deck of subgraphs. We may then integrate term by term to obtain all but the constant term in $\phi(G)$. Tutte **[49]** has provided the *coup de grace* by showing how to reconstruct the constant term from the deck of subgraphs (see also Section 8 of Chapter 8):

Theorem 12.2. *Given the subgraphs $G_1, \ldots, G_p$, it is possible to reconstruct $\phi(G; x)$.* ‖

One might wonder whether or not the subgraphs are actually essential; perhaps the deck of $\phi(G_i)$'s would be adequate. This problem remains open.

The last, and strongest, conjecture is false! Schwenk **[42]** has given two different pairs of graphs with identical decks of $\phi(G_i)$'s. (Since each pair is cospectral, this is not a counter-example for the preceding paragraph.) One of these pairs consists of $K_4 \times K_4$ and a certain 16-vertex triangulation of the torus. (This pair of graphs has several interesting properties too involved for inclusion here—see Section 5 of Chapter 12.) Schwenk has also given analogous results for spectral versions of the edge-reconstruction problem.

13. Digraphs

Several of the spectral properties of graphs have their analogs in the study of digraphs. In this section, we shall look briefly at some of the earlier results on graphs which can be carried over to give results on the eigenvalues of a digraph.

Since the adjacency matrix $\mathbf{A}(D)$ of a digraph D is not, in general, symmetric, we cannot expect the eigenvalues of D to be real. However, we can obtain an analog of Theorem 2.4 by observing that the ij-entry of the matrix $\mathbf{A}(D)^k$ is equal to the number of directed walks of length k joining the vertices v_i and v_j; since D contains no loops or multiple arcs, it follows that the sum of the eigenvalues is zero, the sum of their squares is equal to twice the number of pairs of vertices joined by a symmetric pair of arcs, and the sum of their cubes is equal to three times the number of directed triangles.

As observed in Section 4, the Perron–Frobenius theorem can also be used to give us information about the eigenvalues of a digraph, and takes the following form (see [**16**]):

Theorem 13.1. *If D is a strongly-connected digraph with at least two vertices, then*

(*i*) *its largest eigenvalue λ_1 is real, and is a simple root of $\phi(D; x)$;*

(*ii*) *corresponding to the eigenvalue λ_1, there is an eigenvector $\mathbf{x}$ all of whose coordinates are positive;*

(*iii*) *if λ is any other eigenvalue of D, then $|\lambda| \leqslant \lambda_1$;*

(*iv*) *the deletion of any arc of D decreases the largest eigenvalue.* ||

Perron and Frobenius also proved that if there are h eigenvalues whose absolute value is equal to λ_1, then these h eigenvalues are distinct, and equal to $\lambda_1, \lambda_1\omega, \lambda_1\omega^2, \ldots, \lambda_1\omega^{h-1}$, where ω is a complex hth root of unity. In fact, in this case if λ is *any* eigenvalue of D, then so are $\lambda\omega, \lambda\omega^2, \ldots, \lambda\omega^{h-1}$. The reader should compare this result for $h = 2$ with the pairing theorem (Theorem 4.2).

There has also been a certain amount of attention paid recently to the study of cospectral digraphs. The simplest way to construct a pair of these is to take any digraph D together with its converse D' (the digraph obtained from D by reversing the direction of each arc); it is clear that if D and D' are not isomorphic, then they must form a cospectral pair. However, not all pairs of cospectral digraphs may be obtained in this way; for example, the four digraphs in Fig. 3 are cospectral. For further examples, the reader is referred to [**18**].

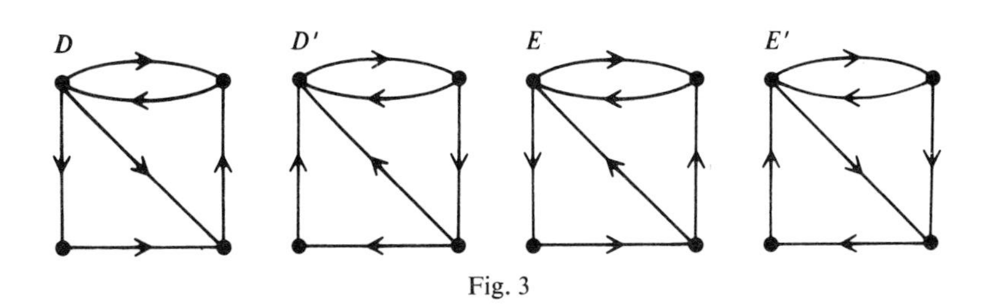

Fig. 3

There are also analogs of some of the results of Section 11; for example, Theorem 11.3 goes through without alteration. In addition, Chao [**6**] has proved the following result, which is a weaker analog of Theorem 11.1:

Theorem 13.2. *If all the eigenvalues of a digraph D are distinct, then the automorphism group of D is Abelian.* ||

Note that for a digraph it is not true, in general, that every non-trivail automorphism has order two.

14. Unsolved Problems

Although we have described a wide variety of results relating to the spectral properties of graphs, it is clear that we still have a long way to go before we can give a satisfactory answer to the questions we raised at the end of Section 1. We conclude this chapter with a few problems about which little seems to be known.

(1) For which graphs G is $\mathbf{A}(G)$ non-singular? More generally, can we say anything about the rank and signature of $\mathbf{A}(G)$ in terms of the properties of G? (Some progress has been made on these questions by Cvetković and Gutman [**11**] and others in the case in which G is bipartite, but little seems to have been done in the general case.)

(2) What can be said about the spectral properties of trees (other than those properties derivable from results on general bipartite graphs)?

(3) What can be said about the eigenvalues of a planar graph?

(4) What can be said in general about the spectral properties of tournaments? In particular are the non-reconstructible tournaments of Stockmeyer [**48**] cospectral?

(5) Are there analogs for line digraphs of the results in Section 9?

References

For a general survey which includes the eigenvalues of a large number of particular graphs, the reader is referred to the paper of L. Collatz and U. Sinogowitz [**7**]. For more complete and up-to-date studies, the reader is referred to the papers of D. M. Cvetković [**10**] and A. J. Hoffman [**22**], and to the book by N. L. Biggs [**3**].

1. M. Aigner, The uniqueness of the cubic lattice graph, *J. Combinatorial Theory* **6** (1969), 282–297; *MR***38**#5657.
2. G. A. Baker, Jr., Drum shapes and isospectral graphs, *J. Math. Phys.* **7** (1966), 2238–2242.
3. N. L. Biggs, *Algebraic Graph Theory*, Cambridge University Press, Cambridge, 1974; *MR***50**#151.
4. F. C. Bussemaker, D. M. Cvetković and J. J. Seidel, Graphs related to exceptional root systems, T. H.-Report 76-WSK-05, Department of Mathematics, Techn. Univ. Eindhoven, 1976.
5. P. J. Cameron, J. M. Goethals, J. J. Seidel and E. E. Shult, Line graphs, root systems, and elliptic geometry, *J. Algebra* **43** (1976), 305–327; *MR***56**#182.
6. C.-Y. Chao, A note on the eigenvalues of a graph, *J. Combinatorial Theory* (*B*) **10** (1971), 301–302; *MR***43**#1867.
7. L. Collatz and U. Sinogowitz, Spektren endlicher Grafen, *Abh. Math. Sem. Univ. Hamburg* **21** (1957), 63–77; *MR***19**#443.
8. C. A. Coulson and G. S. Rushbrooke, Note on the method of molecular orbitals, *Proc. Cambridge Phil. Soc.* **36** (1940), 193–200.
9. D. M. Cvetković, New characterizations of the cubic lattice graph, *Publ. Inst. Math.* (*Beograd*) **10** (24) (1970), 195–198; *MR***43**#1869.
10. D. M. Cvetković, Graphs and their spectra, *Univ. Beograd. Publ. Elektrotehn. Fak. Ser. Mat. Fiz.* **354–356** (1971), 1–50; *MR***45**#8556.
11. D. M. Cvetković and I. M. Gutman, The algebraic multiplicity of the number zero in the spectrum of a bipartite graph, *Matematicki Vesnik* (*Beograd*) **9**(**24**) (1972), 141–150; *MR***48**#1993.
12. M. Doob, A geometric interpretation of the least eigenvalue of a graph, in *Proceedings of the Second Chapel Hill Conference on Combinatorial Mathematics and its Applications* (ed. R. C. Bose *et al.*), University of North Carolina Press, Chapel Hill, N.C., 1970, pp. 126–135; *MR***42**#2959.
13. B. Elspas and J. Turner, Graphs with circulant adjacency matrices, *J. Combinatorial Theory* **9** (1970), 297–307; *MR***42**#7540.
14. H.-J. Finck and G. Grohmann, Vollständiges Produkt, chromatische Zahl und characteristisches Polynom regulärer Graphen I, *Wiss. Z. Techn. Hochsch. Ilmenau* **11** (1965), 1–3; *MR***32**#5538.
15. G. Frobenius, Über Matrizen aus nicht negativen Elementen, *Sitzber. Akad. Wiss. Berlin* (1912), 456–477.
16. F. R. Gantmacher, *Applications of the Theory of Matrices*, Vol. II, Interscience, New York and London, 1959; *MR***21**#6372b.
17. R. L. Graham and H. O. Pollak, On the addressing problem for loop switching, *Bell System Tech. J.* **50** (1971), 2495–2519; *MR***44**#6405.

18. F. Harary, C. King, A. Mowshowitz and R. C. Read, Cospectral graphs and digraphs, *Bull. London Math. Soc.* **3** (1971), 321–328; *MR***45**#3249.
19. W. C. Herndon and M. L. Ellzey, Jr., Isospectral graphs and molecules, *Tetrahedron* **31** (1975), 99–107.
20. M. Hestenes, On the use of graphs in group theory, in *New Directions in the Theory of Graphs* (ed. F. Harary), Academic Press, New York, 1973, pp. 97–128; *MR***50** #12795.
21. A. J. Hoffman, On the polynomial of a graph, *Amer. Math. Monthly* **70** (1963), 30–36; *MR***27**#6257.
22. A. J. Hoffman, The eigenvalues of the adjacency matrix of a graph, in *Combinatorial Mathematics and its Applications* (ed. R. C. Bose and T. A. Dowling), University of North Carolina Press, Chapel Hill, N.C., 1969, pp. 578–584; *MR***42**#110.
23. A. J. Hoffman, On eigenvalues and colorings of graphs, in *Graph Theory and its Applications* (ed. B. Harris), Academic Press, New York, 1970, pp. 79–91; *MR***44**#1601.
24. A. J. Hoffman, On graphs whose least eigenvalue exceeds $-1-\sqrt{2}$, *Lin. Alg. Appl.* **16** (1977), 153–165.
25. A. J. Hoffman and D. K. Ray-Chaudhuri, On the line graph of a symmetric balanced incomplete block design, *Trans. Amer. Math. Soc.* **116** (1965), 238–252; *MR***32**#2342.
26. A. J. Hoffman and D. K. Ray-Chaudhuri, On a spectral characterization of regular line graphs (unpublished).
27. N. Jacobson, *Lie Algebras*, Interscience, New York and London, 1962; *MR***26**#1345.
28. M. Marcus and H. Minc, *A Survey of Matrix Theory and Matrix Inequalities*, Allyn and Bacon, Boston, 1964; *MR***29**#112.
29. B. D. McKay, On the spectral characterization of trees (to appear).
30. A. Mowshowitz, The group of a graph whose adjacency matrix has all distinct eigenvalues, *Proof Techniques in Graph Theory* (ed. F. Harary), Academic Press, New York, 1969, pp. 109–110; *MR***41**#5249.
31. A. Mowshowitz, The characteristic polynomial of a graph, *J. Combinatorial Theory* (*B*) **12** (1972), 177–193; *MR***45**#5011.
32. A. Mowshowitz, The adjacency matrix and the group of a graph, in *New Directions in the Theory of Graphs* (ed. F. Harary), Academic Press, New York, 1973, pp. 129–148; *MR***50**#9679.
33. O. Perron, Zur Theorie der Matrizen, *Math. Ann.* **64** (1907), 248–263.
34. M. Petersdorf and H. Sachs, Spectrum und Automorphismengruppe eines Graphen, in *Combinatorial Theory and its Applications*, III (ed. P. Erdős *et al.*), North-Holland, Amsterdam, 1970, pp. 891–907; *MR***46**#8897.
35. D. K. Ray-Chaudhuri, Characterization of line graphs, *J. Combinatorial Theory* **3** (1967), 201–214; *MR***35**#4119.
36. D. H. Rouvray, Les valeurs propres des molécules qui possèdent un graphe bipartit, *C. R. Acad. Sci. Paris* (*A*) **274** (1972), 1561–1563.
37. H. Sachs, Über selbstkomplementäre Graphen, *Publ. Math. Debrecen* **9** (1962), 270–288; *MR***27**#1934.
38. H. Sachs, Beziehungen zwischen den in einem Graphen enthaltenen Kreisen und seinem charakteristischen Polynom, *Publ. Math Debrecen* **11** (1964), 119–134; *MR***30** #2491.
39. H. Sachs, Über Teiler, Faktoren und charakteristische Polynome von Graphen, II, *Wiss. Z. Techn. Hochsch. Ilmenau* **13** (1967), 405–412; *MR***39**#6774.
40. A. J. Schwenk, Almost all trees are cospectral, in *New Directions in the Theory of Graphs* (ed. F. Harary), Academic Press, New York, 1973, pp. 275–307; *MR***52**#5456.
41. A. J. Schwenk, Computing the characteristic polynomial of a graph, in *Graphs and Combinatorics*, Lecture Notes in Mathematics **406** (ed. R. A. Bari and F. Harary), Springer-Verlag, Berlin, Heidelberg, New York, 1974, pp. 247–261; *MR***52**#7972.
42. A. J. Schwenk, Spectral reconstruction problems, in *Advances in Graph Theory* (ed. F. Harary), Annals N.Y. Academy of Sciences (to appear).
43. J. J. Seidel, Strongly regular graphs with $(-1, 1, 0)$ adjacency matrix having eigenvalue 3, *Lin. Alg. Appl.* **1** (1968), 281–298; *MR***38**#3175.

44. J. J. Seidel, Strongly regular graphs, in *Recent Progress in Combinatorics* (ed. W. T. Tutte), Academic Press, New York, 1969, pp. 185–198; *MR*40#7148.
45. S. S. Shrikhande, On a characterization of the triangular association scheme, *Ann. Math. Statist.* **30** (1959), 39–47; *MR*21#1673.
46. J. H. Smith, Some properties of the spectrum of a graph, in *Combinatorial Structures and their Applications* (ed. R. K. Guy *et al.*), Gordon and Breach, New York, 1970, pp. 403–406; *MR*42#1702.
47. L. Spialter, The atom connectivity matrix and its characteristic polynomial, *J. Chem. Documentation* **4** (1964), 261–274.
48. P. K. Stockmeyer, The falsity of the reconstruction conjecture for tournaments, *J. Graph Theory* **1** (1977), 19–26.
49. W. T. Tutte, All the king's horses, in *Graph Theory and Related Topics* (ed. J. A. Bondy and U. S. R. Murty), Academic Press, New York, 1978 (to appear).
50. R. S. Varga, *Matrix Iterative Analysis*, Prentice-Hall, Engelwood Cliffs, N.J., 1962; *MR*28#1725.
51. D. A. Waller, Regular eigenvalues of graphs and enumeration of spanning trees, in *Teorie Combinatorie*, Tome I (ed. B. Segre), Accademia Nazionale dei Lincei, Rome, 1976, pp. 313–320.
52. H. S. Wilf, The eigenvalues of a graph and its chromatic number, *J. London Math. Soc.* **42** (1967), 330–332; *MR*34#7408.

12
Strongly Regular Graphs

PETER J. CAMERON

1. Introduction

In 1966, Erdős, Rényi and Sós [**25**] proved that, in a finite society in which any two members have exactly one common friend, there is somebody who is everybody else's friend. This result is known as the **friendship theorem**, and can be reformulated as a theorem about graphs:

Theorem 1.1 (Friendship Theorem). *Let G be a graph, and suppose that, for any two distinct vertices v and w, there is a unique vertex u joined to v and w. Then G consists of a number of triangles with a common vertex* (*see Fig.* 1).

There are two parts in the proof of this theorem. First, it is shown by elementary counting that a counter-example to this theorem must be a regular graph. Next, it is shown that the only regular graphs satisfying the hypothesis are K_1 and K_3, neither of which is a counter-example. While so-called "elementary" proofs of this are now known (see, for example, [**38**]), the quickest proof involves computing the eigenvalues of the graph and their multiplicities, using the ideas discussed in Chapter 11.

In 1963, Bose [**4**] introduced the concept of a "strongly regular graph"—namely, a regular graph in which the number of vertices adjacent to two vertices v and w depends only on whether v and w are adjacent or not. (Thus a regular graph satisfying the hypotheses of the friendship theorem would be strongly regular.) In fact, slight variations on this concept occur much earlier in the literature on statistical designs, especially in the works of Bose and his co-authors (for example, **[6]**, [7], **[21]**). Since then, strongly regular graphs have appeared in many contexts, and have been studied in their own right by many people, most notably Seidel **[43]**, **[44]**, **[47]**.

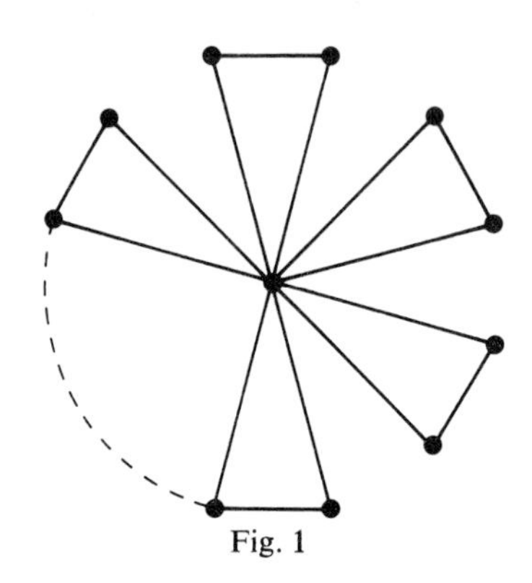

Fig. 1

The parameters of a strongly regular graph are its order, its valency, and the numbers referred to in the above description. Given a set of parameters, we may ask whether there is a graph with these parameters, and (if so) whether it is unique. Simple counting arguments place some restrictions on the parameters of a graph; these restrictions are discussed in Section 2. The next two sections prove deeper necessary conditions. The first of these (the "integrality condition") depends on the fact that the multiplicities of the eigenvalues of the graph must be non-negative integers. The friendship theorem will then appear as a corollary. The second (the "Krein condition") asserts that certain linear combinations of **I** (the unit matrix), **J** (the all-1 matrix), and **A** (the adjacency matrix) are positive semi-definite.

Section 5 gives some examples of strongly regular graphs, with a discussion of the uniqueness question. The remaining sections describe relationships between strongly regular graphs and other combinatorial objects (such as partial geometries, polar spaces, block designs, two-graphs and permutation groups). These are not just "applications", but real interactions, in which both theories are enriched. Strongly regular graphs also have applications in extremal graph theory. For some of these we refer to Section 7 of Chapter 13.

I am grateful to J. J. Seidel for many suggested improvements in the presentation of this chapter.

2. Strongly Regular Graphs

A graph G of order p is **strongly regular** if there exist integers k, λ, μ such that

(*i*) G is regular of valency k;

(*ii*) given any two distinct vertices v and w, the number of vertices adjacent to both v and w is λ if v and w are adjacent, and μ otherwise.

The numbers p, k, λ and μ are called the **parameters** of G. However, we shall be fairly flexible, and admit also as a parameter the number $l = p-k-1$ (the valency of the complement $\bar{G}$). Further parameters will be introduced later. We note that some authors use different symbols for these parameters. The complete and null graphs on any number of vertices are trivially strongly regular. Often these are tacitly excluded.

Theorem 2.1. *If G is a strongly regular graph, then its complement $\bar{G}$ is also strongly regular.*

Proof. It is clear that $\bar{G}$ is regular of valency l. If v and w are adjacent in G, then $k-\lambda-1$ vertices are adjacent to v but not to w, whence $l-k+\lambda+1$ vertices are adjacent to neither v nor w. Similarly, if v and w are not adjacent, then $l-k+\mu$ vertices are adjacent to neither. So $\bar{G}$ is strongly regular, and its parameters are

$$\bar{p} = p, \qquad \bar{k} = l, \qquad \bar{\lambda} = l-k+\mu, \qquad \bar{\mu} = l-k+\lambda+1. \; \|$$

We next show that the parameter l is expressible in terms of k, λ and μ, provided that $\mu \neq 0$.

Theorem 2.2. $k(k-\lambda-1) = l\mu$.

Proof. Take any vertex v, and count edges of the form wz, where w is adjacent to v, and z is not. For each of the k choices for w there are $k-\lambda-1$ choices for z, whereas for each of the l choices for z there are μ choices for w. The result now follows immediately. ||

We conclude this section by investigating the case $\mu = 0$.

Theorem 2.3. *Let G be a strongly regular graph. Then $\mu = 0$ if and only if G is a disjoint union of complete graphs of order $k+1$.*

Proof. From Theorem 2.2, $\mu = 0$ if and only if $\lambda = k-1$—that is, if and only if any two vertices adjacent to a common vertex are adjacent to each other. This is equivalent to the assertion that every component of G is a complete graph. ||

3. The Integrality Condition

In this section we shall prove a very strong necessary condition on the parameters of a strongly regular graph. This result is known as the **integrality condition**, and depends on an analysis of the eigenvalues of the graph. After proving this theorem, we shall discuss its applications and prove the friendship theorem.

Theorem 3.1 (Integrality Condition). *If there is a strongly regular graph with parameters p, k, λ, μ, and $l = p-k-1$, then the numbers*

$$f = \tfrac{1}{2}\left\{k+l+\frac{(k+l)(\mu-\lambda)-2k}{\sqrt{(\mu-\lambda)^2+4(k-\mu)}}\right\}$$

and

$$g = \tfrac{1}{2}\left\{k+l-\frac{(k+l)(\mu-\lambda)-2k}{\sqrt{(\mu-\lambda)^2+4(k-\mu)}}\right\}$$

are non-negative integers.

Proof. Let G be a strongly regular graph with parameters p, k, λ and μ, and let $\mathbf{A}$ be its adjacency matrix. Using some results from Chapter 11, we compute the eigenvalues of $\mathbf{A}$ and their multiplicities.

By the remarks preceding Theorem 2.4 of Chapter 11, the ij-entry of $\mathbf{A}^2$ is the number of paths of length 2 from the ith vertex to the jth; this number is k, λ or μ, according as these vertices are equal, adjacent or non-adjacent. Hence

$$\mathbf{A}^2 = k\mathbf{I}+\lambda\mathbf{A}+\mu(\mathbf{J}-\mathbf{I}-\mathbf{A}),$$

where $\mathbf{I}$ is the identity matrix, and $\mathbf{J}$ is the matrix all of whose entries are 1; so

$$\mathbf{A}^2-(\lambda-\mu)\mathbf{A}-(k-\mu)\mathbf{I} = \mu\mathbf{J}.$$

Since G is regular, the vector $(1, \ldots, 1)^{\mathrm{T}}$ is an eigenvector of $\mathbf{A}$ with eigenvalue k (Chapter 11, Corollary 4.4). It is also an eigenvector of $\mathbf{J}$, with eigenvalue $p = 1+k+l$. Applying the matrix equation to this vector, we get

$$k^2-(\lambda-\mu)k-(k-\mu) = \mu(1+k+l),$$

which reduces to $k(k-\lambda-1) = \mu l$—that is, to Theorem 2.2.

Any other eigenvalue of $\mathbf{J}$ is zero; so, if θ is an eigenvalue of $\mathbf{A}$ different from k, we have

$$\theta^2-(\lambda-\mu)\theta-(k-\mu) = 0.$$

We denote the solutions of this equation by r and s, and we usually make the convention that $r > s$; thus

$$r = \tfrac{1}{2}\{\lambda-\mu+\sqrt{(\lambda-\mu)^2+4(k-\mu)}\} \quad \text{and} \quad s = \tfrac{1}{2}\{\lambda-\mu-\sqrt{(\lambda-\mu)^2+4(k-\mu)}\}.$$

Let f and g be the multiplicities of r and s as eigenvalues of $\mathbf{A}$. Then we have

$$f+g = k+l,$$

and since the trace of $\mathbf{A}$ is zero, we have

$$k+fr+gs = 0.$$

These two linear equations for f and g can be solved, giving the expressions in the statement of the theorem. Since f and g are multiplicities of eigenvalues, they must be non-negative integers! ‖

From Theorem 3.1, it follows that there are two possibilities, according as $(k+l)(\mu-\lambda)-2k$ is equal to, or different from, zero.

Case 1. If $(k+l)(\mu-\lambda)-2k = 0$, then $k+l = 2k/(\mu-\lambda) > k$, so that $0 < \mu-\lambda < 2$. Thus $\mu-\lambda = 1$, and we have $\lambda = \mu-1$, $k = l = 2\mu = f = g$. In this case, the graph G has the same parameters as its complement. A further necessary condition in this case was proved by van Lint and Seidel **[37]**—namely, that the number $p = 4\mu+1$ of vertices must be a sum of two perfect squares. (This condition resembles, in both its statement and its proof, the Bruck–Ryser theorem **[9]** for finite projective planes.)

Case 2. If $(k+l)(\mu-\lambda)-2k \neq 0$, then $(\mu-\lambda)^2+4(k-\mu) = d^2$, for some positive integer d; d divides $(k+l)(\mu-\lambda)-2k$, and the quotient is congruent to $k+l$ (modulo 2). In this case, the eigenvalues r and s are integers.

From the proof of Theorem 3.1, it follows that a strongly regular graph has just three eigenvalues k, r and s. There is a partial converse to this:

Theorem 3.2. *A regular connected graph G is strongly regular if and only if it has just three eigenvalues k, r and s.* ‖

Furthermore, the parameters of G are expressible in terms of its eigenvalues. If $k > r > s$, then the valency of G is k (Chapter 11, Corollary 4.4), and we have

$$\lambda = k+r+s+rs \quad \text{and} \quad \mu = k+rs.$$

As an application of the integrality condition, we prove the friendship theorem (Theorem 1.1).

Proof of the Friendship Theorem. Let G be a graph satisfying the hypothesis of Theorem 1.1, and let v and w be non-adjacent vertices of G. Then there is a unique vertex u adjacent to both v and w. If x is any vertex which is adjacent to v, then there is a unique vertex y adjacent to x and w. A similar statement holds with v and w interchanged. So the valencies of v and w are equal.

Now suppose that G is not regular. Let a and b be vertices whose valencies are unequal, and let c be the unique vertex adjacent to both. From the preceding paragraph, we deduce that a and b are adjacent.

We may suppose, by interchanging a and b if necessary, that the valencies of a and c are unequal. If d is any further vertex, then d is adjacent to at least one of a and b, and to at least one of a and c, but not to both b and c; so d is adjacent to a. It follows that G consists of a number of triangles with common vertex a.

So we may assume that G is regular of valency k. Then, by hypothesis, G is strongly regular, and the parameters λ and μ are both equal to 1. Theorem 3.1 then implies that $f-g = -k/\sqrt{k-1}$ is an integer; so $k-1$ divides k^2. By the remainder theorem, $k-1$ divides 1, so that $k = 0$ or 2, and the graph is a single vertex or a triangle. But neither possibility is a counter-example. This completes the proof. ‖

4. The Krein Condition

Suppose that we attempt to list the parameters of strongly regular graphs of relatively small order. We might check the possible parameter-sets (p, k, λ, μ) to see that Theorem 2.2 and the integrality condition are satisfied, and then try to construct a graph for each of the remaining parameter-sets. In fact, this list will contain some parameter-sets for which no graph can be found; the conditions derived so far are not sufficient for the existence of a strongly regular graph.

The first such parameter-set to arise is $(p, k, \lambda, \mu) = (21, 10, 4, 5)$, which falls in Case 1 of the integrality condition. However, this set is eliminated by the criterion of van Lint and Seidel mentioned in Section 3, since 21 is not a sum of two squares. The next case is (28, 9, 0, 4). There are a variety of *ad hoc* arguments for the non-existence of such a graph; see Biggs [**2**] for one of these. However, there is a general result which covers this particular case—the so-called **Krein condition**:

Theorem 4.1 (Krein Condition). *Let G be a strongly regular graph for which* *both G and its complement $\bar{G}$ are connected, and let k, r and s be the eigenvalues of G. Then*

$$(r+1)(k+r+2rs) \leqslant (k+r)(s+1)^2,$$

and

$$(s+1)(k+s+2sr) \leqslant (k+s)(r+1)^2.$$

Sketch of Proof. We shall outline the idea of the proof of the theorem, without giving the details of the calculations. Let $\mathbf{A}_0 = \mathbf{I}$, $\mathbf{A}_1 = \mathbf{A}$ (the adjacency matrix of G), and $\mathbf{A}_2 = \mathbf{J}-\mathbf{I}-\mathbf{A}$ (the adjacency matrix of $\bar{G}$). By the proof of Theorem 3.1, $\mathbf{A}$ has just three eigenspaces V_0, V_1 and V_2, where V_0 is spanned by the vector $(1, \ldots, 1)^{\mathrm{T}}$, and V_1 and V_2 have dimensions f and g. If $\mathscr{S}$ is the set of linear combinations of $\mathbf{A}_0$, $\mathbf{A}_1$ and $\mathbf{A}_2$, then $\mathscr{S}$ has these eigenspaces, and so $\mathscr{S}$ is precisely the set of matrices with these eigenspaces. For $i = 0, 1, 2$, let $\mathbf{E}_i$ be the matrix with eigenvalue 1 on V_i, and 0 on the other two eigenspaces. Then $\mathbf{E}_0$, $\mathbf{E}_1$ and $\mathbf{E}_2$ also span $\mathscr{S}$.

The set $\mathscr{S}$ is closed under matrix multiplication, but we shall not use this. Instead, we consider the operation of the Hadamard (or pointwise) product of two matrices $\mathbf{A} = (a_{ij})$ and $\mathbf{B} = (b_{ij})$, given by $\mathbf{A} \circ \mathbf{B} = (a_{ij}b_{ij})$. Clearly $\mathbf{A}_i \circ \mathbf{A}_j = \delta_{ij}\mathbf{A}_i$, where δ_{ij} is the Kronecker delta. So $\mathscr{S}$ is closed under Hadamard product, and in particular,

$$\mathbf{E}_i \circ \mathbf{E}_j = \sum_{k=0}^{2} q_{ij}^k \mathbf{E}_k,$$

where q_{ij}^k is the eigenvalue of $\mathbf{E}_i \circ \mathbf{E}_j$ on V_k. The numbers q_{ij}^k can be expressed as algebraic functions of the parameters of the graph.

The Hadamard product $\mathbf{E}_i \circ \mathbf{E}_j$ is a principal submatrix of the Kronecker product $\mathbf{E}_i \otimes \mathbf{E}_j$, and so the eigenvalues q_{ij}^k are bounded above and below by the eigenvalues of $\mathbf{E}_i \otimes \mathbf{E}_j$. But $\mathbf{E}_i \otimes \mathbf{E}_j$ is idempotent, so $0 \leqslant q_{ij}^k \leqslant 1$. It turns out that these inequalities are trivially satisfied, except for the cases $q_{11}^1 \geqslant 0$, $q_{22}^2 \geqslant 0$, which (after some calculation) reduce to the inequalities in the statement of the theorem. ||

In the case mentioned above, we have $k = 9$, $r = 1$ and $s = -5$, and the second inequality of the theorem is violated.

Theorem 4.1 has become known as the Krein condition, because L. L. Scott [**42**] proved a special case of it by applying a result of Krein concerning compact topological groups to finite groups. Subsequent treatments by D. G. Higman [**31**] and P. Delsarte [**23**] have resulted in the simplified proof given here, but the original name has stuck.

As is usual in mathematics, once an inequality has been established, it is important to know what happens if the bound is attained. For the Krein condition, the situation is now fairly well understood.

Theorem 4.2. *Let G be a strongly regular graph, satisfying the hypothesis of Theorem* 4.1, *and attaining either of the bounds of that theorem. Then, for each vertex v, the induced subgraphs on the set of vertices adjacent to v and the set of vertices non-adjacent to v are both strongly regular* (*possibly complete or null*). ‖

This theorem follows from a result of Delsarte, Goethals and Seidel **[24]** on the distribution of finite sets of points on the Euclidean unit sphere. It is perhaps worth noting that the pentagon is the only graph which attains both of the bounds in Theorem 4.1. A near-converse to Theorem 4.2 (using similar techniques) and some applications have been given by Cameron, Goethals and Seidel **[16]**.

5. Some Examples

In this section, we consider a few examples of strongly regular graphs. Others will turn up in various guises throughout the chapter. For an extensive, although not complete, list of known examples, we refer to the survey by X. Hubaut **[36]**.

The **triangular graph** $T(n)$ is the line graph of K_n; thus its vertices can be identified with the 2-element subsets of $\{1, \ldots, n\}$, two vertices being adjacent when the corresponding subsets have non-empty intersection. An easy counting argument shows that $T(n)$ is a strongly regular graph; alternatively, this follows from the equation following Theorem 9.2 of Chapter 11, and Theorem 3.2 of this chapter. The parameters of $T(n)$ are $p = \frac{1}{2}n(n-1)$, $k = 2(n-2)$, $\lambda = n-2$ and $\mu = 4$. The graphs $T(3)$, $T(4)$ and $T(5)$ are respectively the triangle, the octahedron, and the complement of the Petersen graph. We give a characterization of these graphs, due to Chang **[19]**, **[20]**, and Hoffman **[33]**; see also Corollary 6.2, and Chapter 11, Section 9:

Theorem 5.1. *Let G be a strongly regular graph with parameters $p = \frac{1}{2}n(n-1)$, $k = 2(n-2)$, $\lambda = n-2$ and $\mu = 4$ (where $n \geqslant 3$). If $n \neq 8$, then G is isomorphic to $T(n)$. If $n = 8$, then G is isomorphic to one of four graphs, one of which is $T(8)$.* ‖

The three exceptional graphs in the case $n = 8$ were discovered by Chang, and will be defined in Section 9.

A related class consists of the **square lattice graphs** $L_2(n)$ defined as the line graphs of $K_{n,n}$, for $n \geqslant 2$. They are strongly regular with parameters $p = n^2$, $k = 2(n-1)$, $\lambda = n-2$ and $\mu = 2$. A characterization theorem resembling Theorem 5.1 was proved by Shrikhande **[48]**:

Theorem 5.2. *Let G be a strongly regular graph with parameters $p = n^2$, $k = 2(n-1)$, $\lambda = n-2$ and $\mu = 2$ (where $n \geqslant 2$). If $n \neq 4$, then G is isomorphic to $L_2(4)$. If $n = 4$, then G is isomorphic to $L_2(4)$ or the graph in Fig. 2.* ‖

In Fig. 2, the bounding vertices and edges are identified with those on the opposite side of the figure, as indicated by the arrows, so that the graph can be embedded on the torus. We note that 16 is the smallest number of vertices for which there exist non-isomorphic strongly regular graphs with the same parameters.

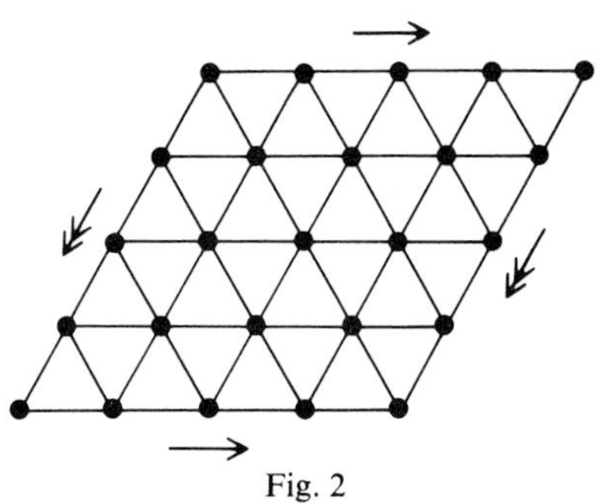

Fig. 2

Next we define an infinite class of graphs satisfying Case 1 of the integrality condition. Let F be the (unique) finite field with p elements, where p is a prime power congruent to 1 (modulo 4). The **Paley graph** $P(p)$ has vertex-set F, with the vertices x and y joined if and only if $x-y$ is a non-zero square in F. (Since -1 is a square in F, this graph is undirected.) The Paley graph is strongly regular with parameters p, $k = \frac{1}{2}(p-1)$, $\lambda = \frac{1}{4}(p-5)$ and $\mu = \frac{1}{4}(p-1)$. The graphs $P(5)$ and $P(9)$ are the pentagon and $L_2(3)$, respectively. In a paper which discusses the existence and uniqueness of strongly regular graphs on fewer than 29 vertices, Seidel [**47**] showed that a strongly regular graph with the parameters of $P(p)$ for $p \leqslant 17$ must be isomorphic to $P(p)$. This is not true in general. An application of the Paley graphs to Ramsey graph theory is given in Chapter 13, following Theorem 7.5.

On the subject of non-uniqueness, we mention a computer result of Bussemaker and Seidel [**12**], which implies the existence of at least 581 pairwise non-isomorphic strongly regular graphs with parameters $p = 35$, $k = 16$, $\lambda = 6$ and $\mu = 8$. In connection with Section 7 of Chapter 11, we note that strongly regular graphs with the same parameters are cospectral.

6. Partial Geometries

We describe in this section a tool introduced by Bose [**4**] for investigating the structure of strongly regular graphs, and for proving the uniqueness of graphs with certain parameters.

A **partial geometry** with parameters (R, K, T), where $R > 1$ and $K > 1$, consists of a set of "points" and a set of "lines", with a relation of "incidence" between points and lines, satisfying the following axioms:

(i) every point is incident with R lines, and every line with K points;

(ii) two points are incident with at most one common line (and two lines with at most one common point);

(iii) if a point p is not incident with a line L, then p is collinear with exactly T points of L. (Formally, there are T incident point-line pairs (q, M) such that q is incident with L, and p with M.)

Conventionally, we identify a line with the set of points incident with it; by (ii), distinct lines are incident with distinct sets of points. A partial geometry has a "dual", obtained by interchanging the labels *point* and *line*, which is a partial geometry with parameters (K, R, T).

The **point-graph** G of a partial geometry is the graph whose vertices correspond to the points of the geometry, and whose edges correspond to collinear point-pairs. It is a routine calculation to verify that G is strongly regular with parameters $p = 1+R(K-1)+(R-1)(K-1)(K-T)T^{-1}$, $k = R(K-1)$, $\lambda = K-2+(R-1)(T-1)$ and $\mu = RT$. Dually, there is a strongly regular "line-graph", namely, the point-graph of the dual geometry (not to be confused with the line graph of G).

There are three important special classes of partial geometries. These and their duals include most known examples:

(i) *Partial geometries with* $T = K$. Any such geometry has the property that every two points are collinear, so that its point-graph is the complete graph. These geometries are the balanced incomplete block designs with $\lambda = 1$, to be discussed in Section 7. If $K = 2$, the line-graph is the triangular graph $T(R+1)$.

(ii) *Partial geometries with* $T = R-1$. In this case, the number of points is K^2, and the lines can be divided into R "parallel classes" with K lines each, satisfying Euclid's postulate: *through each point there passes one line of each parallel class.* These structures are called **nets** of order K and deficiency $K-R+1$, and arise in the theory of affine planes (see Bruck **[8]**). The existence of a net of order K and deficiency $K-R+1$ is equivalent to the existence of $R-2$ mutually orthogonal latin squares of order K. For this reason, the point-graph of a net is often called a *net graph* or a *Latin square graph*. If $R = 2$, the point-graph is isomorphic to $L_2(K)$.

(*iii*) *Partial geometries with* $T = 1$. These are called **generalized quadrangles**, and we shall return to them later.

A strongly regular graph G is called **pseudo-geometric** (R, K, T) if its parameters are $p = 1+R(K-1)+(R-1)(K-1)(K-T)T^{-1}$, $k = R(K-1)$, $\lambda = K-2+(R-1)(T-1)$ and $\mu = RT$; G is called **geometric** if it is the point-graph of a partial geometry. By definition, a geometric graph is pseudo-geometric. An important theorem of Bose [**4**] gives a sufficient condition for the converse to hold:

Theorem 6.1. *Let* G *be a pseudo-geometric graph* (R, K, T), *where* $K > \frac{1}{2}\{R(R-1)+T(R+1)(R^2-2R+2)\}$. *Then* G *is geometric.*

Sketch of Proof. If a graph G is geometric, then the set of points incident with a line L of the geometry corresponds to a complete subgraph of G, which is a clique if $K > T$. Conversely, in order to show that a graph is geometric, we must find a collection of cliques possessing suitable properties (for example, that any clique in the collection has size K, and that any edge lies in a unique clique). Bose showed that, if K is sufficiently large, then the set of all cliques of maximal size has the required properties. ||

In order to illustrate the power of this theorem, we use it to deduce most of Theorems 5.1 and 5.2:

Corollary 6.2. (*i*) *If* $n > 8$, *then* $T(n)$ *is uniquely determined by its parameters;*

(*ii*) *if* $n > 4$, *then* $L_2(n)$ *is uniquely determined by its parameters.*

Proof. (*i*) $T(n)$ is the line graph of K_n, and so is geometric $(2, n-1, 2)$; indeed, it is the unique geometric $(2, n-1, 2)$ graph. If G is strongly regular and has the same parameters as $T(n)$, then G is pseudo-geometric $(2, n-1, 2)$. Bose's theorem shows that, if $n > 8$, then G is geometric, whence G is isomorphic to $T(n)$.

(*ii*) The proof for $L_2(n)$ is similar. ||

The Krein condition (Theorem 4.1) yields the following inequality:

Theorem 6.3. *The parameters* (R, K, T) *of a partial geometry* (*or pseudo-geometric graph*) *satisfy the inequality*

$$(R-1)(K-2T) \leqslant (K-T)^2(K-2). \quad ||$$

This gives a curious postscript to Bose's theorem:

Corollary 6.4. *If a graph* G *satisfies the hypothesis of Theorem* 6.1, *then* $R \leqslant 2T$.

Proof. By Theorem 6.1, G is geometric, and so has a dual (K, R, T). But if $R > 2T$, then the Krein condition (with K and R interchanged) conflicts with the hypothesis of Theorem 6.1. This contradiction establishes the result. ‖

Generalized quadrangles have been studied extensively in their own right. (The reader is warned that another notation is commonly used for their parameters, namely $s = K-1$, $t = R-1$.) Their study is motivated by the existence of classical examples: the geometries of totally isotropic points and lines with respect to polarities of index 2 of finite projective spaces. These have parameters $(R, K, T) = (2, q+1, 1)$, $(q+1, q+1, 1)$, $(q^2+1, q+1, 1)$, $(q+1, q^2+1, 1)$ and $(q^3+1, q^2+1, 1)$, where q is a prime power. Many combinatorial characterizations of the classical quadrangles have been found by Thas and others; a list of theorems appears in **[53]**. A notable example is the following theorem of Buekenhout and Lefèvre **[10]**:

Let S and T be sets of points and lines in a finite projective space, forming a generalized quadrangle Q, and suppose that S contains all the points of every line in T. Then Q is a "classical model".

Theorem 6.3 shows that if a generalized quadrangle has parameters $(R, K, 1)$ with $K > 2$, then $(R-1) \leqslant (K-1)^2$. (This inequality was first proved by D. G. Higman **[30]** by entirely different methods; an elementary proof was found by Cameron **[14]**.) Indeed, the same result holds for any pseudo-geometric graph $(R, K, 1)$, and Theorem 4.2 implies that a pseudo-geometric graph $((K-1)^2+1, K, 1)$ is geometric. However, it is not known whether a pseudo-geometric graph (R, K, T) attaining the bound of Theorem 6.3 with $T > 1$ is necessarily geometric. An interesting test case, suggested by Goethals, is $(R, K, T) = (28, 5, 2)$; there is a unique graph with these parameters (see **[27]**), but it is not known whether or not it is geometric.

Further information on partial geometries and generalized quadrangles can be found in survey papers by Bose **[5]**, Payne **[40]** and Thas **[53]**.

7. Polar Spaces

As a bridge between the last section and this one, we prove the following result **[17]**:

Theorem 7.1. *Let G be a graph in which every edge vw is contained in a triangle vwx with the property that each further vertex is joined to exactly one of v, w and x. Then G is a null graph, a friendship theorem graph, or one of three special graphs on 9, 15 or 27 vertices—namely, $L_2(3)$, the complement of $T(6)$, and the line-graph of the 27 lines in a general cubic surface.*

Sketch of Proof. Arguments resembling those used in the proof of the friendship theorem show that, if G is not of one of the first two types, then G is strongly regular with $k = 2R$, $\lambda = 1$, $\mu = R$, and hence pseudo-geometric $(R, 3, 1)$. If we define a "line" to be the set of vertices of a triangle in G, we see that G is geometric, and must therefore be the point-graph of a generalized quadrangle. The integrality and Krein conditions now show that $R = 2$, 3 or 5, and the uniqueness of the quadrangles is not difficult to establish. ‖

A generalization of Theorem 7.1 was found by Shult **[50]** and Seidel **[45]**. A graph is said to have the **triangle property** if every edge vw is contained in a triangle vwx with the property that each further vertex is joined to just one, or to all, of v, w and x. Shult determined all strongly regular graphs with the triangle property. They are defined by means of bilinear and quadratic forms on vector spaces over the field GF(2), and have $2^{2n-1}+2^{n-1}-1$, $2^{2n}-1$, or $2^{2n+1}-2^n-1$ vertices, for some integer $n \geqslant 2$. Seidel gave another proof of this result; he also showed that if G is a graph with the triangle property which is not strongly regular, then either G is null, or there is one vertex of G which is joined to all of the others.

This theorem was generalized still further by a beautiful result of Buekenhout and Shult **[12]**, which almost characterizes the graphs associated with "polar spaces" (those defined by non-degenerate bilinear or Hermitian forms on finite-dimensional vector spaces). This theorem involves a simple axiom resembling the triangle property. It is interesting to note that one of the exceptional classes of graphs in this theorem consists of the point-graphs of generalized quadrangles.

Shult **[51]** gives a good survey of the Buekenhout–Shult theorem, its implications and applications to finite group theory, and various related results (such as Seidel's determination of graphs with the "cotriangle property"). The details of the axiomatization of polar spaces are not easy, and may be found in lecture notes by Tits **[54]**.

Theorem 7.1 appears in a paper by Cameron, Goethals, Seidel and Shult **[17]**. It is used there as a key lemma in an alternative proof of Hoffman's determination of all "sufficiently large" graphs with least eigenvalue -2 **[34]**. This leads to another proof of Corollary 6.2, since the graphs referred to in that result have least eigenvalue -2. We refer to Section 9 of Chapter 11 for further discussion of these matters.

8. Connection with Block Designs

A well-established area of combinatorial theory which has interacted fruitfully with strongly regular graphs is the theory of designs. A **t-(v, k, λ) design** (or ***t*-design**) is defined to be a set of v "points", and a set of "blocks", with a relation of "incidence" between them, satisfying the conditions:

(*i*) every block is incident with exactly k points;

(*ii*) every t points are incident with exactly λ blocks.

(It is usual to identify a block with the set of points incident with it, and to require that distinct blocks are incident with distinct point sets.)

The most important case is $t = 2$. In this case, the designs are often called *balanced incomplete block designs* defined in Chapter 1, and most combinatorics textbooks give an account of their theory (see, for example, Hall [**28**]). One of the most significant facts about them is "Fisher's inequality",—that a 2-design (with $\lambda > 0$, and $k < v$) has at least as many blocks as points, with equality if and only if any two blocks have exactly λ points in common. A design attaining this bound is called *symmetric*, and its *dual* (obtained by interchanging the names "point" and "block") is again a 2-design.

Block designs originated in the theory of statistical design. The notation betrays this origin—v is the initial letter of "varieties". If we have a number v of varieties to test, it is natural to require that every pair of varieties be tested against each other the same number of times. Fisher's inequality implies that this will require at least v trials. One way of reducing the number of trials needed is to take a graph G whose vertex-set is the set of varieties, and specify that the number of blocks containing a pair $\{x, y\}$ of varieties depends only on whether or not xy is an edge of G. The analysis of variance is simplified if the graph G is strongly regular, and if so, the design is called *partially balanced.* We refer to Bose and Shimamoto [7] for details.

There are many other connections between designs and strongly regular graphs. We describe two of them here; others will occur later. For details, the reader is referred to Cameron and van Lint [**18**]; further results on (v, k, λ)-graphs will be found in Chapter 13.

(*i*) A **(v, k, λ)-graph** is defined to be a strongly regular graph G on v vertices satisfying the condition $\lambda = \mu$; that is, each vertex of G is joined to k other vertices, and any two vertices (adjacent or not) are simultaneously joined to λ vertices. From such a graph we can obtain a 2-(v, k, λ) design by choosing the blocks to be those sets of k vertices adjacent to a fixed vertex x, for each x. This design has equally many points and blocks, and so it is symmetric. However, not every symmetric design can be constructed from a (v, k, λ)-graph. There is a necessary and sufficient condition involving the existence of a certain kind of polarity of the design (an isomorphism between the design and its dual). The friendship theorem includes the assertion that the triangle is the only (v, k, λ)-graph with $\lambda = 1$. However, there exist several interesting examples with $\lambda > 1$, the smallest of which are $L_2(4)$ and the Shrikhande graph of Section 5 (with $(v, k, \lambda) = (16, 6, 2)$). The integrality condition can be used to show that v is bounded by a function of λ ($v \leqslant \lambda^2(\lambda+2)$). It is not known whether such a bound exists for general symmetric designs with $\lambda > 1$.

(*ii*) We saw in Section 6 that a partial geometry with $T = K$ is a 2-$(v, K, 1)$ design, and that its line-graph is strongly regular. Observing that, in such a design, any two blocks have at most one point in common, Goethals and Seidel [**26**] were led to a generalization. A 2-design is called *quasi-symmetric* if there are distinct integers α and β such that every two blocks have α or β points in common. (We assume that both values occur—that is, that the design is not symmetric.) The *block graph* of the design is the graph whose vertices correspond to the blocks, two vertices being adjacent if the corresponding blocks have α points in common. Goethals and Seidel showed that this graph is strongly regular. This observation enabled them to construct several interesting strongly regular graphs. In particular, they obtained strongly regular graphs on 56, 77, 120, 176 and 253 vertices from the well-known 4-(23, 7, 1) design and various other designs related to it. The graph on 253 vertices is referred to in Section 7 of Chapter 13.

9. Two-graphs

A **two-graph** on a set X of points is a collection T of triples of points with the property that every set of four points contains an even number of members of T. Since their original definition by G. Higman in 1971, two-graphs have had important applications to permutation groups, finite geometry, and elliptic geometry, as well as to strongly regular graphs. The theory was developed by J. J. Seidel and D. E. Taylor. Seidel [**46**] has written an excellent survey, so a brief treatment will suffice here.

Given any graph G with vertex-set X, the set $T = T(G)$ of triples of vertices which contain an odd number of edges of G is a two-graph. Every two-graph arises in this way. (To see this, let T be a two-graph on X. Take any point $x \in X$, and let $G(x)$ be the graph in which yz is an edge if and only if $\{x, y, z\} \in T$. Then $T = T(G(x))$.)

However, the correspondence between graphs and two-graphs is not one-to-one. Given a partition of X into two disjoint sets Y and Z, we define the result of **switching** G with respect to this partition to be the graph G' whose edges are those edges of G contained in Y or in Z, together with those pairs $\{y, z\}$, with $y \in Y$, $z \in Z$, for which yz is *not* an edge of G. (Thus, edges between Y and Z are "switched" into non-edges, and *vice versa*.)

It turns out that graphs G and G' with vertex-set X give rise to the same two-graph if and only if one can be switched into the other. It follows that there is a one-to-one correspondence between two-graphs and **switching classes** of graphs (that is, equivalence classes under switching).

Seidel [**47**] showed that any switching class on an odd number of vertices contains a unique Eulerian graph. This is false if the number of vertices is even, but in any case, there are equally many two-graphs and Eulerian graphs on a given number of vertices. This was proved by Mallows and Sloane [**39**] by enumeration (see Section 10 of Chapter 14); Cameron [**15**] has given a short direct proof.

It is convenient to define a "modified adjacency matrix" $\mathbf{B} = (b_{ij})$ of a graph G, where

$$b_{ij} = \begin{cases} -1, & \text{if vertices } i \text{ and } j \text{ are adjacent,} \\ +1, & \text{if vertices } i \text{ and } j \text{ are non-adjacent,} \\ 0, & \text{if } i = j. \end{cases}$$

It follows that if $\mathbf{B}$ and $\mathbf{B}'$ are the matrices of a pair of graphs related by switching, then $\mathbf{B}' = \mathbf{DBD}$, for some diagonal matrix $\mathbf{D}$ with diagonal entries ± 1. In particular, $\mathbf{B}$ and $\mathbf{B}'$ have the same spectrum. We are justified in calling this the **spectrum** of the associated two-graph.

A two-graph T is **regular** if there is an integer a such that every pair of points is contained in exactly a members of T (that is, if it is a 2-design, in the sense of Section 8). Seidel has given the following characterization of regular two-graphs:

Theorem 9.1. *A two-graph is regular if and only if it has exactly two distinct eigenvalues.* ||

It follows easily that if the regular graph G is contained in the switching class of a regular two-graph, then G is strongly regular. There are two possibilities for the parameters of G, depending on which of the two eigenvalues is associated with the eigenvector $(1, \ldots, 1)^{\mathrm{T}}$ (see Theorem 3.2). The switching class of a regular two-graph may contain graphs corresponding to both, one, or neither of these two possibilities. For example, both possibilities occur for the Petersen graph which can be switched into its complement; but for $T(8)$ only one possibility occurs, since the other possibility would give the complement of that graph on 28 vertices whose non-existence was deduced from the Krein condition in Section 4. (Note that a strongly regular graph belongs to the switching class of a regular two-graph if and only if

$$2(\lambda+\mu) = 3k-l-1.)$$

There is another connection with strongly regular graphs. We saw earlier that every switching class contains a graph $G(x)$ in which a given vertex x is isolated. Now we have the following theorem:

Theorem 9.2. *A two-graph is regular if and only if, for some vertex x, the graph $G(x)$ is the disjoint union of $\{x\}$ and a strongly regular graph with $k = 2\mu$. If this holds for one vertex, then it holds for all.* ‖

Examples of strongly regular graphs with $k = 2\mu$ (which give rise to regular two-graphs, by Theorem 9.2) include the following:

(*i*) all strongly regular graphs satisfying Case 1 of the integrality condition (for example, Paley graphs);

(*ii*) all regular graphs with the triangle property of Section 7 (for example, the three regular graphs of Theorem 7.1);

(*iii*) the 581 different graphs on 35 vertices referred to at the end of Section 5 (which produce 91 distinct switching classes).

Corollary 9.3. *If T is a regular two-graph on n points, and if T is not the complete or null two-graph, then n is even.*

Proof. By Theorems 2.2 and 9.2, the parameters of $G(x)$ satisfy $k = 2\mu$ and $l = 2(k-\lambda-1)$. Since the number of points is $k+l+2$, the result follows. ‖

Seidel [**47**] has shown the usefulness of regular two-graphs in the problem of determining all strongly regular graphs with up to 28 vertices. We note in particular that $T(8)$ and the three Chang graphs are equivalent under switching, as are $L_2(4)$ and the Shrikhande graph. In fact, the simplest way of defining the Chang graphs is to give appropriate switching sets of vertices of $T(8)$. Since $T(8) = L(K_8)$, this is equivalent to giving sets of edges of K_8—that is, graphs on 8 vertices. The appropriate graphs are a 1-factor, an octagon, and the disjoint union of a triangle and a pentagon.

A reader familiar with the terminology of algebraic topology will have noticed that a two-graph is a 2-dimensional cocycle (modulo 2) on a simplex, and the assertion that two-graphs are in one-to-one correspondence with switching classes can be translated into the vanishing of cohomology in dimensions 1 and 2. This point of view has provided the short proof mentioned above that there are equally many two-graphs and Eulerian graphs on a given number of vertices. These and other cohomological aspects of two-graphs and their automorphism groups are discussed in [**15**].

Given a set of lines through the origin in Euclidean space $\mathbf{R}^d$, each pair of which are at the same angle α (where $\alpha \neq \frac{1}{2}\pi$), there is a two-graph T whose vertex-set is the given set of lines: T consists of the set of triples of lines on which unit vectors can be chosen whose pairwise inner products are all negative. Any two-graph can be represented in this way (see [**46**]). The smallest regular two-graph (on a set of six points) can be represented by the diagonals of a regular icosahedron. This is shown in Fig. 3, where

$$T = \{013, 024, 035, 041, 052, 123, 234, 345, 451, 512\}.$$

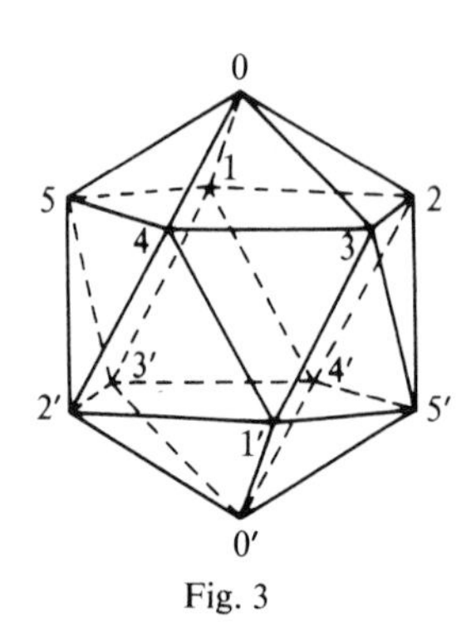

Fig. 3

10. Metrically Regular Graphs

In defining strongly regular graphs, we are aiming for a class of graphs in which any two edges or non-edges "look alike", at least as far as paths of length 2 are concerned. We can obtain a more general class of graphs if we refine the division of pairs of points into edges and non-edges by means of the distance function.

A connected graph of diameter d is said to be **metrically regular** if there are integers p^k_{ij} $(0 \leqslant i, j, k \leqslant d)$ such that, whenever v and w are vertices at distance k, the number of vertices at distance i from v and j from w is equal to p^k_{ij}. (Thus a metrically regular graph of diameter 2 is the same thing as a connected strongly regular graph with $p^0_{11} = k$, $p^1_{11} = \lambda$, $p^2_{11} = \mu$, etc.) It is in fact sufficient to require the defining condition only for those triples i, j, k, for which $j = 1$ and $i = k-1$, k, and $k+1$.

Let G be a metrically regular graph, and let $\mathbf{A}_i$ be the matrix whose (j, k)-entry is 1 if the jth and kth vertices of G are at distance i, and zero otherwise. Then $\mathbf{A}_0 = \mathbf{I}$, and $\mathbf{A}_1 = \mathbf{A}$ (the usual adjacency matrix). From the definition, we can show by induction that $\mathbf{A}_i = \varphi_i(\mathbf{A})$ for $0 \leqslant i \leqslant d$, where φ_i is a polynomial of degree i. Since $\mathbf{A}_0 + \mathbf{A}_1 + \ldots + \mathbf{A}_d = \mathbf{J}$, and $\mathbf{AJ} = k\mathbf{J}$ (where $k = p^0_{11}$ is the valency), $\mathbf{A}$ satisfies the polynomial equation

$$(x-k)(\varphi_0(x)+\varphi_1(x)+\ldots+\varphi_d(x)) = 0.$$

Since **A** has at least $d+1$ distinct eigenvalues (Chapter 11, Theorem 2.5), this is the minimum polynomial of **A**. Conversely, a regular connected graph of diameter d with just $d+1$ distinct eigenvalues is metrically regular.

Following the proof of Theorem 3.1, the eigenvalues of **A** and their multiplicities can be determined, and the integrality and Krein conditions proved for metrically regular graphs. Rather than give the details of this theory (which may be found in Biggs [3], or Delsarte [23]), we shall describe one of its notable applications—namely, the theory of Moore graphs (whose definition is motivated by the next result:

Theorem 10.1. (*i*) *Let G be a connected graph whose diameter is at most d, and whose maximum valency is at most k. Then G has at most* $1+\sum_{i=0}^{d-1} k(k-1)^i$ *vertices, with equality if and only if G is a regular k-valent graph with girth* $2d+1$.

(*ii*) *Let G be a graph whose girth is at least* $2d+1$, *and whose minimum valency is at least k. Then G has at least* $1+\sum_{i=0}^{d-1} k(k-1)^i$ *vertices, with equality if and only if G is a connected regular k-valent graph with diameter d.* ||

A graph attaining the bounds of Theorem 10.1 (with $d > 1$ and $k > 1$) is called a **Moore graph**. Such a graph is metrically regular. The structure of Moore graphs has almost completely been determined, using the integrality condition:

Theorem 10.2. *Let G be a Moore graph with diameter d and valency k.*

(*i*) *If* $d = 2$, *then* $k = 2, 3, 7$, *or* (*possibly*) 57; *in each of the cases* $k = 2$ 3, 7, *G is unique up to isomorphism.*

(*ii*) *If* $d > 2$, *then* $k = 2$, *and G is a circuit of length* $2d+1$. ||

The first part of this theorem was proved by Hoffman and Singleton [**35**]. For $k = 2$, G is the pentagon, and for $k = 3$, G is the Petersen graph; a construction for G when $k = 7$ is given in [**35**]. The existence of a Moore graph of valency 57 is undecided. The second part of the theorem is due to Bannai and Ito [**1**], and Damerell [**22**].

Further generalizations of metrically regular graphs have been made. If the relations of being at given distance in a graph G are replaced by arbitrary symmetric relations, we obtain *association schemes* [**7**]; if we replace undirected edges by directed edges, we obtain *coherent configurations* [**31**], [**23**]. In each case, the integrality and Krein conditions have been shown to hold.

Let us now turn briefly to something less general. In the spirit of the first paragraph of this section, we say that a regular graph G satisfies the ***t*-vertex condition** if any two edges "look alike" in terms of t-vertex subgraphs containing them, with a similar condition for non-edges. The 2-vertex condition is empty, and the 3-vertex condition holds in G if and only if G is strongly regular. Higman [**30**] observed that the point-graph of a generalized quadrangle satisfies the 4-vertex condition, and used this to give the first proof of the inequality $R-1 \leqslant (K-1)^2$ for generalized quadrangles (see Section 6). Sims [**52**] proved a theorem which determines all but finitely many strongly regular graphs which satisfy the 4-vertex condition and have prescribed least eigenvalue s. This result has been discussed by Ray-Chaudhuri [**41**].

11. Permutation Groups

If a graph G admits a group of automorphisms which is transitive on vertices, edges, and non-edges, then G is strongly regular. (In fact, G satisfies the t-vertex condition for all t.) What conditions on a permutation group ensure that it acts in this way on a graph?

Suppose that we are given a group Γ of permutations on a set X, and let us assume that Γ acts transitively on X. Then Γ acts on the Cartesian square X^2 of X. The *rank* of Γ is defined to be the number of orbits on X^2. The rank is at least 2 if $|X| > 1$, since the ordered pairs (x, x) and (x, y) (for $y \neq x$) lie in different orbits. Furthermore, the rank is 2 precisely when Γ is doubly transitive on X.

Suppose now that Γ has rank 3, and has even order. Then Γ contains a permutation of order 2 which interchanges some pair $\{x, y\}$ of vertices. Consider the graph G with vertex-set X, in which two vertices v and w are adjacent if and only if (v, w) lies in the same orbit as (x, y). By assumption, G is undirected. Since it satisfies the condition of the first paragraph, it is strongly regular. The techniques of strongly regular graphs have been very important in the study of permutation groups of rank 3, especially in the hands of D. G. Higman. For a survey of this field, we refer to his expository article [**29**]. We note that there are many known groups of rank 3, giving further examples of strongly regular graphs (see [**36**]).

The theory of coherent configurations, mentioned in the last section, was developed by Higman in order to provide a similar combinatorial framework for studying arbitrary permutation groups, not necessarily transitive. We refer to surveys by Higman [**31**] and Cameron [**13**] for further details.

As a spin-off from this study, several of the so-called "sporadic" finite simple groups have been constructed as groups of automorphisms of strongly regular graphs. The following construction by Higman and Sims [**32**] is one of the simplest of these.

Let P and B be the sets of points and blocks of the (unique) 3-(22, 6, 1) design D. Let G be the graph with vertex-set $\{\infty\} \cup P \cup B$, in which ∞ is joined to every member of P, a vertex in P is joined to a vertex in B if and only if they are incident in D, and two vertices in B are joined if and only if the corresponding blocks are disjoint. Counting arguments based on the properties of D show that, for any vertex v, the vertices adjacent and non-adjacent to v can be identified with the points and blocks of D, so that the same rules for adjacency hold. In particular, G is strongly regular with parameters $p = 100$, $k = 22$, $\lambda = 0$ and $\mu = 6$. It also follows that the automorphism group of G is transitive and has rank 3 on the vertex-set of G. Cameron and van Lint [**18**] have given some explanation of why this construction works. It turns out to be related to the fact that the Krein bound is attained by a graph with these parameters, and also to a question on extensions of designs.

Unfortunately, these methods give no information about doubly transitive groups, since the only graphs admitting a doubly transitive group of automorphisms are the complete and null graphs. But all is not lost. Many of the known doubly transitive groups are groups of automorphisms of two-graphs. Indeed, there is a group-theoretic necessary and sufficient condition for a doubly transitive group to act on a two-graph (see [**15**]). Such a two-graph is necessarily regular, and so the techniques of strongly regular graphs can again be applied. (These matters are discussed in Seidel's survey [**46**].) We close with an example of this connection.

Shult [**49**] has proved the following result, known as the **graph extension theorem**; the proof given here is due to Seidel [**46**]:

Theorem 11.1. *Let G be a graph with vertex-set X, whose automorphism group Γ is vertex-transitive. Let x be a vertex of G, and let Y_1 and Y_2 be the sets of vertices adjacent and non-adjacent to v. Suppose that there are permutations τ_1 and τ_2 of Y_1 and Y_2 respectively, such that*

(*i*) *τ_1 and τ_2 are automorphisms of the induced subgraphs on Y_1 and Y_2;*

(*ii*) *if $y_1 \in Y_1$, $y_2 \in Y_2$, then y_1 and y_2 are adjacent if and only if $\tau_1(y_1)$ and $\tau_2(y_2)$ are non-adjacent.*

Then there is a doubly transitive group Δ of permutations on $X \cup \{\infty\}$, in which the stabilizer of ∞ is equal to Γ.

Sketch of Proof. Let H be the disjoint union of G and the isolated vertex ∞, and let T be the two-graph corresponding to the switching class of H. Let τ be the permutation of $X \cup \{\infty\}$ which interchanges v and ∞, and acts as τ_i on Y_i, for $i = 1, 2$. By hypothesis, the image of H under τ is obtained by switching H with respect to the subset Y_1. So τ is an automorphism of T.

Then the full automorphism group Δ of T has the required properties. Note that T is a regular two-graph, and so, by Theorem 9.2, G must be strongly regular. ||

This theorem can be used to construct many doubly transitive groups as automorphism groups of two-graphs. For example, let G be the pentagon shown in Fig. 4, and take $v = 1$; let τ_1 be the identity permutation on $\{2, 5\}$, and let τ_2 be the transposition on $\{3, 4\}$. The conditions of the graph extension theorem are then satisfied, and we obtain the group $PSL(2, 5)$ of rotations of the regular icosahedron, acting on its set of diagonals (see Section 9).

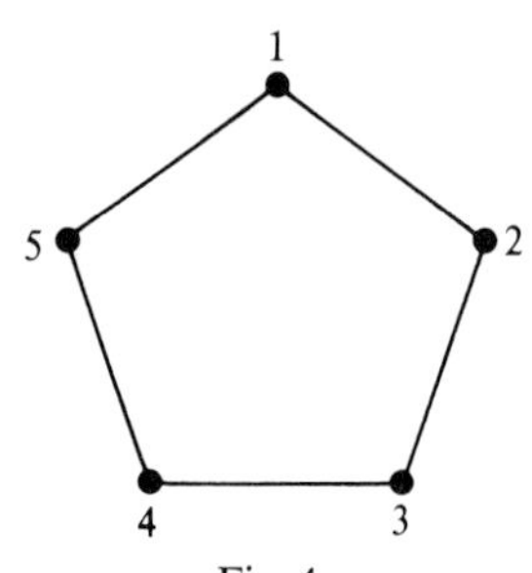

Fig. 4

References

Wherever possible, references have been given to expository articles or survey papers. The reader is especially referred to **[36]**, **[44]** and **[46]**, and (for related topics) to **[3]**, **[5]**, **[13]**, **[18]**, **[31]**, **[41]**, **[51]** and **[53]**.

1. E. Bannai and T. Ito, On finite Moore graphs, *J. Fac. Sci. Univ. Tokyo* **20** (1973), 191–208; *MR*48#1971.
2. N. L. Biggs, *Finite Groups of Automorphisms*, London Math. Soc. Lecture Notes **6**, Cambridge University Press, Cambridge, 1971; *MR*48#5905.
3. N. L. Biggs, *Algebraic Graph Theory*, Cambridge University Press, Cambridge, 1974; *MR*50#151.
4. R. C. Bose, Strongly regular graphs, partial geometries, and partially balanced designs, *Pacific J. Math.* **13** (1963), 389–419; *MR*28#1137.
5. R. C. Bose, Characterization problems of combinatorial graph theory, in *A Survey of Combinatorial Theory* (ed. J. N. Srivastava *et al.*), North-Holland, Amsterdam, 1973, pp. 31–51.
6. R. C. Bose and D. M. Mesner, On linear associative algebras corresponding to association schemes of partially balanced designs, *Ann. Math. Statist.* **30** (1959), 21–38; *MR*21#951.
7. R. C. Bose and T. Shimamoto, Classification and analysis of partially balanced incomplete block designs with two associate classes, *J. Amer. Statist. Assoc.* **47** (1952), 151–184; *MR*14–67.
8. R. H. Bruck, Finite nets, II: Uniqueness and embedding, *Pacific J. Math.* **13** (1963), 421–457; *MR*27#4768.

9. R. H. Bruck and H. J. Ryser, The nonexistence of certain finite projective planes, *Canad. J. Math.* **1** (1949), 88–93; *MR*10–319.
10. F. Buekenhout and C. Lefèvre, Generalized quadrangles in projective spaces, *Arch. Math.* **25** (1974), 540–552.
11. F. Buekenhout and E. E. Shult, On the foundations of polar geometry, *Geometriae Dedicata* **3** (1974), 155–170; *MR*50#3091.
12. F. C. Bussemaker and J. J. Seidel, Symmetric Hadamard matrices of order 36, T.H.-Report 70-WSK-02, Department of Mathematics, Techn. Univ. Eindhoven, 1970; *MR*43#1863.
13. P. J. Cameron, Suborbits in transitive permutation groups, in *Combinatorics* (ed. M. Hall, Jr. and J. H. van Lint), D. Reidel, Dordrecht, 1975, pp. 419–450.
14. P. J. Cameron, Partial quadrangles, *Quart. J. Math.* (*Oxford*) (2) **26** (1975), 61–73.
15. P. J. Cameron, Cohomological aspects of two-graphs, *Math. Z.* **157** (1977), 101–119.
16. P. J. Cameron, J.-M. Goethals and J. J. Seidel, Strongly regular graphs with strongly subconstituents, *J. Algebra* (to appear).
17. P. J. Cameron, J.-M. Goethals, J. J. Seidel and E. E. Shult, Line graphs, root systems, and elliptic geometry, *J. Algebra* **43** (1976), 305–327; *MR*55#175.
18. P. J. Cameron and J. H. van Lint, *Graph Theory, Coding Theory, and Block Designs*, London Math. Soc. Lecture Notes **19**, Cambridge University Press, Cambridge, 1975.
19. Chang Li-Chien, The uniqueness and non-uniqueness of the triangular association scheme, *Sci. Record Peking Math.* (*New Ser.*) **3** (1959), 604–613; *MR*22#7950.
20. Chang Li-Chien, Association schemes of partially balanced designs with parameters $v = 28, n_1 = 12, n_2 = 15$, and $p_{11}{}^2 = 4$, *Sci. Record Peking Math.* (*New Ser.*) **4** (1960), 12–18; *MR*22#7951.
21. W. S. Connor and W. H. Clatworthy, Some theorems for partially balanced designs, *Ann. Math. Statist.* **25** (1954), 100–112; *MR*15–494.
22. R. M. Damerell, On Moore graphs, *Proc. Cambridge. Phil. Soc.* **74** (1973), 227–236; *MR*47#6553.
23. P. Delsarte, An algebraic approach to the association schemes of coding theory, *Philips Res. Repts. Suppl.* **10**, 1973.
24. P. Delsarte, J.-M. Goethals and J. J. Seidel, Spherical codes and designs, *Geometriae Dedicata*, **6** (1977), 363–388.
25. P. Erdős, A. Rényi and V. T. Sós, On a problem of graph theory, *Studia Sci. Math. Hungar.* **1** (1966), 215–235; *MR*36#6310.
26. J.-M. Goethals and J. J. Seidel, Strongly regular graphs derived from combinatorial designs, *Canad. J. Math.* **22** (1970), 597–614; *MR*44#106.
27. J.-M. Goethals and J. J. Seidel, The regular two-graph on 276 vertices, *Discrete Math.* **12** (1975), 143–158.
28. M. Hall, Jr., *Combinatorial Theory*, Blaisdell, Waltham, Mass., 1967; *MR*37#80.
29. D. G. Higman, A survey of some questions and results about rank 3 permutation groups, in *Actes Congrès Intern. Math.*, Tome 1, Gauthier-Villars, Paris, 1971, pp. 361–365.
30. D. G. Higman, Partial geometries, generalized quadrangles, and strongly regular graphs, in *Atti di Conv. Geometria Combinatoria e sue Applicazione* (ed. A. Barlotti), Perugia, 1971, pp. 263–293.
31. D. G. Higman, Invariant relations, coherent configurations, and generalized polygons, in *Combinatorics* (ed. M. Hall, Jr. and J. H. van Lint), D. Reidel, Dordrecht, 1975, pp. 347–363.
32. D. G. Higman and C. C. Sims, A simple group of order 44,352,000, *Math. Z.* **105** (1968), 110–113; *MR*37#2854.
33. A. J. Hoffman, On the uniqueness of the triangular association scheme, *Ann. Math. Statist.* **31** (1960), 492–497; *MR*32#7949.
34. A. J. Hoffman, On graphs whose least eigenvalue exceeds $-1-\sqrt{2}$, *Lin. Alg. Appl.* **16** (1977), 153–165.
35. A. J. Hoffman and R. R. Singleton, On Moore graphs of diameters 2 and 3, *IBM J. Res. Develop.* **4** (1960), 497–504; *MR*25#3857.

36. X. L. Hubaut, Strongly regular graphs, *Discrete Math.* **13** (1975), 357–381.
37. J. H. van Lint and J. J. Seidel, Equilateral point sets in elliptic geometry, *Proc. Nederl. Akad. Wetensch.* (*A*) **69** (= *Indag. Math.* **28**) (1966), 335–348; *MR*34#685.
38. J. Q. Longyear and T. D. Parsons, The friendship theorem, *Proc. Nederl. Akad. Wetensch.* (*A*) **75** (= *Indag. Math.* **34**) (1972), 257–262; *MR*46#5169.
39. C. L. Mallows and N. J. A. Sloane, Two-graphs, switching classes and Euler graphs are equal in number, *SIAM J. Appl. Math.* **28** (1975), 876–880 *MR*55#164.
40. S. E. Payne, Finite generalized quadrangles: a survey, in *Proc. Int. Conf. Projective Planes*, Washington State University Press, Seattle, 1973, 219–261.
41. D. K. Ray-Chaudhuri, Combinatorial characterization theorems for geometric incidence structures, in *Combinatorial Surveys: Proceedings of the Sixth British Combinatorial Conference* (ed. P. J. Cameron), Academic Press, London, 1977, pp. 87–116.
42. L. L. Scott, A condition on Higman's parameters, *Notices Amer. Math. Soc.* (1973), Abstract 701–20–45.
43. J. J. Seidel, Strongly regular graphs with (−1, 1, 0) adjacency matrix having eigenvalue 3, *Lin. Alg. Appl.* **1** (1968), 281–298; *MR*38#3175.
44. J. J. Seidel, Strongly regular graphs, in *Recent Progress in Combinatorics* (ed. W. T. Tutte), Academic Press, New York, 1969, pp. 185–198; *MR*40#7148.
45. J. J. Seidel, On two-graphs and Shult's characterization of symplectic and orthogonal geometries over GF(2), T.H.-Report 73-WSK-02, Department of Mathematics, Techn. Univ. Eindhoven, 1973; *MR*54#10047.
46. J. J. Seidel, A survey of two-graphs, in *Teorie Combinatorie*, Tome I (ed. B. Segre), Accademia Nazionale dei Lincei, Rome, 1976, pp. 481–511.
47. J. J. Seidel, Graphs and two-graphs, in *Proceedings of the Fifth Southeastern Conference on Combinatorics, Graph Theory, Computing*, Congressus Numerantium **X**, Utilitas Math., Winnipeg, 1974, pp. 125–143.
48. S. S. Shrikhande, The uniqueness of the L_2 association scheme, *Ann. Math. Statist.* **30** (1959), 781–798; *MR*22#1048.
49. E. E. Shult, The graph extension theorem, *Proc. Amer. Math. Soc.* **33** (1972), 278–284; *MR*45#3547.
50. E. E. Shult, Characterization of certain classes of graphs, *J. Combinatorial Theory* (*B*) **13** (1972), 142–167; *MR*47#80.
51. E. E. Shult, Groups, polar spaces, and related structures, in *Combinatorics* (ed. M. Hall, Jr. and J. H. van Lint), D. Reidel, Dordrecht, 1975, pp. 451–482.
52. C. C. Sims, On graphs with rank 3 automorphism groups (unpublished).
53. J. A. Thas, Combinatorics of partial geometries and generalized quadrangles, in *Higher Combinatorics* (ed. M. Aigner), D. Reidel, Dordrecht, 1977, pp. 183–199.
54. J. Tits, *Buildings of Spherical Type and Finite BN-Pairs*, Lecture Notes in Mathematics **386**, Springer-Verlag, Berlin, Heidelberg and New York, 1974.

13
Ramsey Graph Theory

T. D. PARSONS

1. Introduction

In any set of six people, there will always be either a subset of three who are mutually acquainted, or a subset of three who are mutually strangers. Similarly, in any sufficiently large (finite) set of people, there will always be either a subset of m mutual acquaintances or a subset of n mutual strangers. These facts may be rephrased in terms of coloring the edges of complete graphs: let each edge of K_p be colored either red or blue. Then if $p \geqslant 6$, there must be either a red triangle or a blue triangle (this is proved in Section 2). Similarly, there is a number $r(m, n)$ such that if $p \geqslant r(m, n)$, then there must be either a "red K_m" (a K_m-subgraph with all its edges colored red) or a "blue K_n".

These assertions are special cases of a celebrated theorem of F. P. Ramsey [**64**]. For simplicity, we state only the version relevant to finite graphs:

Ramsey's Theorem. *Let $m_1, m_2, \ldots, m_k \geqslant 2$ be integers. Then there exists a least positive integer $r(m_1, \ldots, m_k)$ with the property that if $p \geqslant r(m_1, \ldots, m_k)$, and if the edges of K_p are partitioned into k classes $\mathscr{E}_1, \ldots, \mathscr{E}_k$ (some of which may be empty), then for some i there is a complete subgraph K_{m_i}, all of whose edges are in class $\mathscr{E}_i$.* ||

If we think of $\mathscr{E}_i$ as the set of all edges which are assigned the color i (for $1 \leqslant i \leqslant k$), then this theorem asserts that, for edge-colorings involving at most k colors, every sufficiently large complete graph contains a large "monochromatic" complete subgraph. (Note that the term "edge-colorings" has a different meaning here from its meaning in Chapter 5.)

The numbers $r(m_1, \ldots, m_k)$ are called **Ramsey numbers**, in honor of their discoverer. Little is known about them, even in the case of the "two-color" numbers $r(m, n)$. In fact, the only non-trivial numbers known are $r(3, 3) = 6$, $r(3, 4) = 9$, $r(3, 5) = 14$, $r(4, 4) = 18$, $r(3, 6) = 18$, $r(3, 7) = 23$ and $r(3, 3, 3) = 17$. The first four and the last appear in Greenwood and Gleason [**39**], and the other two in Graver and Yackel [**38**] (although $r(3, 6)$ was computed earlier, in less accessible work of Kéry [**56**] and Kalbfleisch [**52**]).

The extreme difficulty encountered in the study of these classical Ramsey numbers has led researchers to pursue various generalized Ramsey numbers, both for their own sake, and for the light they might shed on the classical numbers.

In this chapter we survey both classical and generalized Ramsey numbers, and certain related problems. Other surveys have been given by Burr [**6**], Gardner [**34**] and Harary [**45**].

2. The Classical Numbers

Let us first verify the claim made about K_6 in the Introduction, namely that $r(3, 3) = 6$. Suppose that the edges of K_6 have each been colored red or blue, and consider a vertex v. At least three of the five edges incident to v have the same color, say red. Let the other vertices incident to these three edges be x, y, z. If some edge of the triangle xyz is red, then this edge spans with v a red triangle, and if not, then xyz is a blue triangle. So we have proved that $r(3, 3) \leqslant 6$. On the other hand, the edges of K_5 may be partitioned into two circuits C_5, one red and the other blue, so that $r(3, 3) > 5$. This proves that $r(3, 3) = 6$.

Let us now denote by $(m, n)_p$ a coloring of the edges of K_p with colors red or blue such that there is neither a red K_m nor a blue K_n.

In general, to evaluate a Ramsey number $r(m, n) = N$, we must

(i) establish the upper bound $r(m, n) \leqslant N$;

(ii) provide a "critical coloring" $(m, n)_p$ for $p = N-1$.

In order to obtain the upper bound, we can sometimes use the following result of Greenwood and Gleason [**39**] (see also Erdős and Szekeres [**28**]):

Theorem 2.1. *Let $m, n \geqslant 3$. Then*

$$r(m, n) \leqslant r(m-1, n)+r(m, n-1),$$

with strict inequality if both $r(m-1, n)$ and $r(m, n-1)$ are even.

Proof. Suppose we have an $(m, n)_p$ coloring of the edges of K_p, and let $R(v)$ be the "red valency" and $B(v)$ the "blue valency" of the vertex v. Then $R(v) \leqslant r(m-1, n)-1$, and $B(v) \leqslant r(m, n-1)-1$, so that

$$p-1 = R(v)+B(v) \leqslant r(m-1, n)+r(m, n-1)-2,$$

which proves the inequality. If equality holds, then for each vertex v we must have $R(v) = r(m-1, n)-1$ and $B(v) = r(m, n-1)-1$. Since the number of vertices of odd valency is even, at least one of $r(m-1, n)$ and $r(m, n-1)$ must be odd. ‖

If we start from the obvious facts that $r(m, n) = r(n, m)$, and $r(2, n) = n$, along with the fact that $r(3, 3) = 6$, then Theorem 2.1 gives successively that $r(3, 4) \leqslant 9$, $r(3, 5) \leqslant 14$, and $r(4, 4) \leqslant 18$. We now show that equality holds in these inequalities by presenting appropriate colorings $(m, n)_p$. In order to do this, it suffices in each case to give the subgraph R of K_p spanned by the red edges.

There are exactly three non-isomorphic $(3, 4)_8$ colorings, obtained by taking R to be one of the graphs in Fig. 1. There is a unique $(3, 5)_{13}$ coloring in which R has vertex-set $\mathbf{Z}_{13}$, and whose red edges are those edges xy such that $x-y$ is a cubic residue (modulo 13). There is a unique $(4, 4)_{17}$ coloring in which R has vertex-set $\mathbf{Z}_{17}$, and whose edges are those xy such that $x-y$ is a quadratic residue (modulo 17). It is not difficult to check that these colorings are critical. It is more tedious to show that they are the only critical colorings in each case (see **[52]**, **[54]**, **[56]**).

Using an argument similar to the proof of Theorem 2.1, together with the obvious fact that $r(2, 3, 3) = r(3, 3) = 6$, we may easily prove that $r(3, 3, 3) \leqslant 17$. There are exactly two different colorings $(3, 3, 3)_{16}$ (see **[55]**), where we use the obvious generalization of our previous notation $(m, n)_p$.

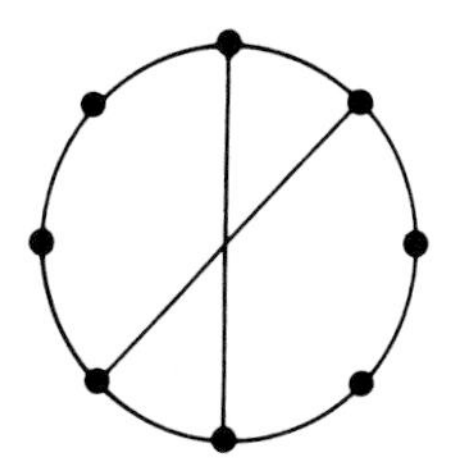
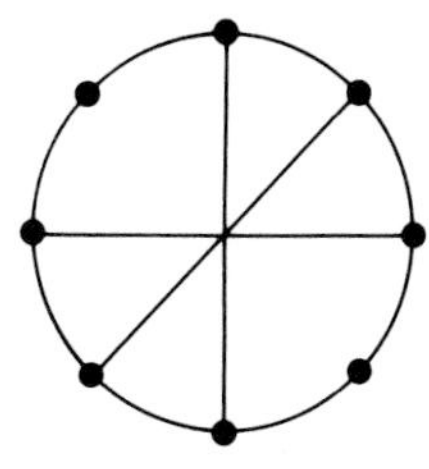
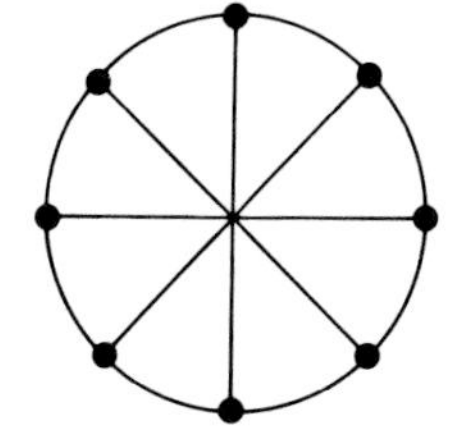

Fig. 1

The first has as its vertex-set the finite field GF(16), where the edge xy is colored red, white or blue, according as $x-y$ is in D_1, D_2 or D_3, the cosets of the subgroup D of cubes in the multiplicative group of non-zero elements of the field. The second coloring has as its vertex-set the additive group $\mathbf{Z}_4 \times \mathbf{Z}_4$, where the edge xy is colored red, white or blue according as $x-y$ is in A_1, A_2 or A_3, where $A_1 = \{(1, 2), (2, 1), (2, 2), (2, 3), (3, 2)\}$, $A_2 = \{(0, 2), (1,0), (1, 1), (3, 0), (3, 3)\}$, and $A_3 = \{(0, 1), (0, 3), (1, 3), (2, 0), (3, 1)\}$ (see[**76**]).

A subset S of an additive group is said to be "sum-free" if $x, y \in S$ implies that $x+y \notin S$. It is easily checked that A_1, A_2, A_3, and D (and hence D_1, D_2 and D_3) are sum-free. If xyz were a monochromatic triangle in either of these colorings, then $x-y$, $y-z$ and $x-z$ would be all in the same set (A_i or D_i). But then $x-z = (x-y)+(y-z)$ would contradict the fact that this set is sum-free. This proves that $r(3, 3, 3) = 17$.

The computations of $r(3, 6)$ and $r(3, 7)$ are long and hard. Unfortunately, the upper bounds provided by Theorem 2.1 are now too slack, and improvement of them requires great effort. To make matters worse, the construction of good lower bound colorings also becomes difficult. Some estimates of other classical numbers are $27 \leqslant r(3, 8) \leqslant 30$, $36 \leqslant r(3, 9) \leqslant 37$, $25 \leqslant r(4, 5) \leqslant 28$, $42 \leqslant r(5, 5) \leqslant 55$, and $51 \leqslant r(3, 3, 3, 3) \leqslant 65$ (see [**12**], [**31**], [**38**], [**42**], [**49**], [**53**], [**73**] and [**75**]).

3. Bounds for $r(n, n)$

In 1947, in one of the first applications of the "probablistic method", now of such basic importance in combinatorics, Erdős [**23**] proved the following result:

Theorem 3.1. *If $n \geqslant 3$, then $2^{\frac{1}{2}n} < r(n, n)$.*

Proof. Let $N \leqslant 2^{\frac{1}{2}n}$. There are $2^{\binom{N}{2}}$ labeled graphs on N vertices, and $2^{\binom{N}{2}-\binom{n}{2}}$ of these contain a given complete subgraph on n vertices. Therefore, the number M of all subgraphs of K_N which contain a subgraph K_n satisfies the inequality

$$M \leqslant \binom{N}{n} 2^{\binom{N}{2}-\binom{n}{2}} < \frac{N^n}{n!} \cdot 2^{\binom{N}{2}-\binom{n}{2}} < \tfrac{1}{2} \cdot 2^{\binom{N}{2}}.$$

(Here the last inequality follows from the fact that $N^n < \frac{1}{2}(n!)2^{\binom{n}{2}}$, which is valid since $n \geqslant 3$ and $N \leqslant 2^{\frac{1}{2}n}$.) If we list all such subgraphs, along with their complements, then there are at most $2M < 2^{\binom{N}{2}}$ graphs in the list. Therefore some subgraph G of K_N contains no K_n subgraph in it or in its complement.

If we now color the edges of G red, and the remaining edges of K_N blue, then the resulting coloring shows that $N < r(n, n)$. ||

Note that this proof uses only the fact that $N^n < \frac{1}{2}(n!)2^{\binom{n}{2}}$. Since, by Stirling's formula, $n!$ is asymptotic to $\sqrt{2\pi n}(\frac{n}{e})^n$, we see that for large n,

$$n \cdot 2^{\frac{1}{2}n}\left\{\frac{1}{e\sqrt{2}}+o(1)\right\} < r(n, n).$$

Spencer [**70**] recently improved this lower bound by a factor of 2.

The upper bound $r(n, n) \leqslant \binom{2n-2}{n-1}$ follows from Theorem 2.1. For large n, Yackel [**78**] has improved this bound by a factor of c loglog $n/\log n$, for some constant c. It is remarkable that the bounds $2^{\frac{1}{2}n} < r(n, n) \leqslant \binom{2n-2}{n-1}$, known since 1947, have been so little improved for large n.

Note that Theorem 3.1 proves the existence of a lower bound coloring, but does not construct one. It is further remarkable that the establishment of a (weaker) lower bound of the form $(1+\varepsilon)^n$, for any fixed $\varepsilon > 0$, and for all large n, has so far resisted all attempts at explicit construction. The best-known construction, due to Frankl [**33**], shows only that $r(n, n) > n^k$, for any k, and for all large n.

Significant progress on the bounds for $r(n, n)$, and the construction of colorings showing that $r(n, n) > (1+\varepsilon)^n$, must be considered to be two of the most important (and difficult) problems of Ramsey graph theory. Also, the known bounds for $r(n, n)$ imply that $2 < (r(n, n))^{1/n} < 4$, for all $n \geqslant 4$. However, it is unknown whether or not $\lim_{n\to\infty} (r(n, n))^{1/n}$ exists. Success in all of these problems will almost surely require major new insights.

4. Definitions and Elementary Results

A k-coloring of a graph G is an assignment $c: E(G) \to \{1, \ldots, k\}$ of a color $c(vw)$ to each edge vw of G. When G and c are given, we let $\mathscr{E}_i$ be the set of edges with color i, $(\mathscr{E}_i)$ be the graph with vertex-set $V(G)$ and edge-set $\mathscr{E}_i$, $N(v, i) = \{w: vw \in \mathscr{E}_i\}$, and $\delta(v, i) = |N(v, i)|$. We write $F \leqslant H$ if F is a subgraph of the graph H. For convenience, the colors 1 and 2 are always taken to be red and blue, and we refer to any $H \leqslant (\mathscr{E}_i)$ for $i = 1$ or 2 as a "red H" or a "blue H". Usually the k-colored graph G will be a complete graph.

Let $\mathscr{F}_1, \ldots, \mathscr{F}_k$ be non-empty (possibly infinite) families of graphs. Then an $(\mathscr{F}_1, \ldots, \mathscr{F}_k)_p$ coloring is a k-coloring of K_p such that, for each i, no member of $\mathscr{F}_i$ is a subgraph of $(\mathscr{E}_i)$. The **Ramsey number** $r(\mathscr{F}_1, \ldots, \mathscr{F}_k)$ is the least integer p such that no $(\mathscr{F}_1, \ldots, \mathscr{F}_k)_p$ coloring exists. If $p = r(\mathscr{F}_1, \ldots, \mathscr{F}_k)-1$, then any $(\mathscr{F}_1, \ldots, \mathscr{F}_k)_p$ coloring is said to be a **critical coloring**.

When $\mathscr{F}_i = \mathscr{F}$, for each i, we write $r_k(\mathscr{F})$ instead of $r(\mathscr{F}_1, \ldots, \mathscr{F}_k)$. If $\mathscr{F}_i = \{G_i\}$ consists of a single graph, we write G_i instead of $\{G_i\}$ in $r(\mathscr{F}_1, \ldots, \mathscr{F}_k)$. Note that if $G_i = K_{m_i}$ for $1 \leqslant i \leqslant k$, then

$$r(G_1, \ldots, G_k) = r(m_1, \ldots, m_k),$$

the classical Ramsey number.

The following statements are clearly true:

Theorem 4.1. *With the above notation,*

(i) $r(\mathscr{F}_1, \ldots, \mathscr{F}_k) = r(\mathscr{G}_1, \ldots, \mathscr{G}_k)$, *for any permutation* $\mathscr{G}_1, \ldots, \mathscr{G}_k$ *of* $\mathscr{F}_1, \ldots, \mathscr{F}_k$;

(ii) if $\mathscr{F}_i \subseteq \mathscr{G}_i$ *for* $1 \leqslant i \leqslant k$, *then* $r(\mathscr{G}_1, \ldots, \mathscr{G}_k) \leqslant r(\mathscr{F}_1, \ldots, \mathscr{F}_k)$;

(iii) if $k > 1$, *then* $r(\mathscr{F}_1, \ldots, \mathscr{F}_k) \geqslant r(\mathscr{F}_1, \ldots, \mathscr{F}_{k-1})$;

(iv) $r(\mathscr{F}_1, \ldots, \mathscr{F}_k) \leqslant \min\{r(G_1, \ldots, G_k) : G_i \in \mathscr{F}_i$ *for* $1 \leqslant i \leqslant k\}$;

(v) if $H_i \leqslant G_i$ *for* $1 \leqslant i \leqslant k$, *then* $r(H_1, \ldots, H_k) \leqslant r(G_1, \ldots, G_k)$. ‖

By Theorem 2.1, and the fact that $r(2, n) = r(n, 2) = n$, the numbers $r(m, n)$ exist for all integers $m, n \geqslant 2$. Also, $r(m_1, \ldots, m_k) \leqslant r(m_1, \ldots, m_{k-2}, r(m_{k-1}, m_k))$, and so, by induction on k, all of the numbers $r(m_1, \ldots, m_k)$ exist. By Theorem 4.1(*v*), $r(G_1, \ldots, G_k) \leqslant r(m_1, \ldots, m_k)$, where $m_i = |V(G_i)|$, so that all of the generalized Ramsey numbers $r(G_1, \ldots, G_k)$ exist. It follows from Theorem 4.1(*iv*) that all Ramsey numbers $r(\mathscr{F}_1, \ldots, \mathscr{F}_k)$ exist.

For any family $\mathscr{F}$ of graphs, we let $\mathscr{F}^*$ be the family of graphs obtained by removing a single vertex (in all possible ways) from each member of $\mathscr{F}$. We can then easily prove the following result:

Theorem 4.2. *In any* $(\mathscr{F}_1, \ldots, \mathscr{F}_k)_p$ *coloring, each induced subgraph* $\langle N(v, i)\rangle$ *with the inherited coloring is an* $(\mathscr{F}_1, \ldots, \mathscr{F}_i^*, \ldots, \mathscr{F}_k)_{p_i}$ *coloring, where* $p_i = \delta(v, i)$ *and* $1 \leqslant i \leqslant k$. ‖

If we now let $N_i = r(\mathscr{F}_1, \ldots, \mathscr{F}_i^*, \ldots, \mathscr{F}_k)$, then the same argument as that used to prove Theorem 2.1 will prove the following generalization of a result of Cockayne [**19**]:

Theorem 4.3.

$$r(\mathscr{F}_1, \ldots, \mathscr{F}_k) \leqslant \left(\sum_{i=1}^{k} N_i\right) - k + 2,$$

with strict inequality if the right-hand side is even and some N_i *is even.* ‖

The next result is sometimes useful in producing lower bounds for

$r(G_1, \ldots, G_k)$. Let $\sigma(G)$ be the minimum size of the smallest part of any partition of $V(G)$ into $\chi(G)$ independent sets of G, where $\chi(G)$ is the chromatic number of G. For example, $\sigma(C_{2m}) = m$.

Theorem 4.4. *Given graphs $G_1, \ldots, G_k$, let $m = \max r(H_1, \ldots, H_{k-1})$, where the maximum is taken over all choices of connected components H_i of G_i, for $1 \leqslant i \leqslant k-1$. Then*

$$r(G_1, \ldots, G_k) \geqslant (\chi(G_k)-1)(m-1)+\min\{m, \sigma(G_k)\}.$$

Proof. Let $s = \min\{m, \sigma(G_k)\}-1$, and let $n = (\chi(G_k)-1)(m-1)+s$. For $1 \leqslant i \leqslant k-1$, let H_i be a component of G_i such that $m = r(H_1, \ldots, H_{k-1})$. Let H be an $(H_1, \ldots, H_{k-1})_{m-1}$ coloring, and let K be an $(H_1, \ldots, H_k)_s$ coloring. Then the coloring of K_n given by $\chi(G_k)-1$ copies of H and one copy of K, with all of the other edges colored k, is easily seen to be a $(G_1, \ldots, G_k)_n$ coloring. ||

In applying this result, we may use the symmetry property of Theorem 4.1(*i*), so that G_k plays no special role. Also we note that for 1-colorings, a degenerate case of our notation is $r(G_1) = |V(G_1)|$, which is used in the lower bound of Theorem 4.4 when $k = 2$.

The following trivial observation reduces the evaluation of numbers $r(G_1, \ldots, G_k)$ to the case where none of the G_i has isolated vertices:

Theorem 4.5. *For $1 \leqslant i \leqslant k$, let $G_i = p_i K_1 \cup H_i$ be the disjoint union of p_i ($\geqslant 0$) isolated vertices and a graph H_i with no isolated vertices, and let $m = \max_i |V(G_i)|$. Then*

$$r(G_1, \ldots, G_k) = \max\{m, r(H_1, \ldots, H_k)\}. \ ||$$

5. Some Generalized Ramsey Numbers

In this section we evaluate certain numbers $r(G_1, \ldots, G_k)$, and we discuss other selected results. For further discussion, see the surveys of Burr **[6]** and Harary **[44]**, **[45]**. The following result is due to Chvátal **[15]**:

Theorem 5.1. *If T is a tree with m vertices, then $r(T, K_n) = (m-1)(n-1)+1$.*

Proof. By Theorem 4.4, $r(T, K_n) \geqslant (n-1)(m-1)+1$. We may assume that $m, n \geqslant 2$. Suppose that H is a $(T, K_n)_p$ coloring, for $p = (m-1)(n-1)+1$. Let v be a vertex of valency 1 in T, and let w be the vertex of T adjacent to v. Using induction on $m+n$, we see that H has a red $T-v$. Let H' be the 2-colored complete graph on $(n-2)(m-1)+1$ vertices, obtained by deleting the red $T-v$ from H. By induction, H' contains a blue K_{n-1}. If some edge joining this blue K_{n-1} to the vertex w (of the deleted $T-v$) is red in H, then H has a red T. Otherwise, w and the K_{n-1} span a blue K_n in H. ||

Burr (private communication) has noted the following result:

Corollary 5.2. *If T is a tree with m vertices, then*

$$r(K_{n_1}, \ldots, K_{n_k}, T) = (m-1)(r(n_1, \ldots, n_k)-1)+1.$$

Proof. Let $p = r(n_1, \ldots, n_k)-1$, and let H be an $(n_1, \ldots, n_k)_p$ coloring. Replace each vertex of H by a K_{m-1} with all of its edges colored $k+1$, and let each edge joining vertices in two different such K_{m-1}'s be the same color as the edge joining the corresponding two vertices of H. The resulting coloring establishes the lower bound.

Now let $n = (m-1)p+1$. By Theorem 5.1 applied to $r(K_{p+1}, T) = n$, in any $(k+1)$-coloring of K_n there is either a K_{p+1} with all of its edges colored with colors from $\{1, 2, \ldots, k\}$ (and hence, for some i, a $K_{n_i} \leqslant (\mathscr{E}_i)$), or else a $T \leqslant (\mathscr{E}_{k+1})$. The result follows. ||

The following result is due to Burr and Roberts [**11**]:

Theorem 5.3. *Let $n_1, \ldots, n_k$ be positive integers, and let $N = (\sum_{i=1}^{k} n_i)-k$. Let $\varepsilon = 1$ if both N and some n_i are even, and $\varepsilon = 2$ otherwise. Then* $r(K_{1,n_1}, \ldots, K_{1,n_k}) = N+\varepsilon$.

Proof. Let $\mathscr{F}_i = \{K_{1,n_i}\}$. Trivially, $r(\mathscr{F}_1, \ldots, \mathscr{F}_i^*, \ldots, \mathscr{F}_k) = n_i$, and so the upper bound follows from Theorem 4.3. Let $f(n_1, \ldots, n_k)$ be the Ramsey number in question. If every n_i is odd, then $N+1$ is odd, so that K_{N+1} can be factored into $\frac{1}{2}N$ Hamiltonian circuits (see Section 11 of Chapter 6). Assigning the color i to the edges of $\frac{1}{2}(n_i-1)$ such circuits, for $1 \leqslant i \leqslant k$, we see that $f(n_1, \ldots, n_k) > N+1$ in this case. Similarly, if $N+1$ is even, then K_{N+1} can be factored into N 1-factors, and if we assign the color i to the edges of n_i-1 such 1-factors, for $1 \leqslant i \leqslant k$, we see that $f(n_1, \ldots, n_k) > N+1$ in this case also. Thus the theorem holds unless both N and some n_i (say, n_1) are even. Then $n_1-1 > 0$ and $(n_1-1)+n_2+\ldots+n_k = N-1$ is odd, and so, by the previous cases and Theorem 4.1(v), we have

$$N+1 \leqslant f(n_1-1, n_2, \ldots, n_k) \leqslant f(n_1, \ldots, n_k). \quad \|$$

An extension of this theorem by Cockayne and Lorimer [**20**] may be stated as

$$r(K_{1,n_1}, \ldots, K_{1,n_k}, nK_2) = \max\{2n, r(K_{1,n_1}, \ldots, K_{1,n_k})+n-1\}.$$

Here $2n$ is an obvious lower bound, whereas $r(K_{1,n_1}, \ldots, K_{1,n_k})+n-1$ is a lower bound, by Theorem 4.4. We omit the upper bound argument. Stahl

[**71**] has generalized Theorem 5.1 by evaluating all numbers $r(F, K_n)$, where F is a forest.

The next theorem (Theorem 5.5) is due to S. L. Lawrence (see [**6**], [**57**]), and depends on the following result, due to Bondy [**5**]:

Lemma 5.4. *If G is a graph of order p in which the sum of the valencies of any two non-adjacent vertices is at least p, then either $p = 2t$ and $G = K_{t,t}$, or else G contains circuits of all lengths k, $3 \leqslant k \leqslant p$.* ‖

Theorem 5.5.

$$r(C_m, K_{1,n}) = \begin{cases} 2n+1, & \text{if } m \text{ is odd, and } m \leqslant 2n+1, \\ m, & \text{if } m \geqslant 2n. \end{cases}$$

Remark. Note that the case of *even* $m < 2n$ is undecided.

Proof. Suppose that we have a $(C_m, K_{1,n})_p$ coloring. Then for any vertex v, $\delta(v, 1)+\delta(v, 2) = p-1$, and $\delta(v, 2) \leqslant n-1$, so that $\delta(v, 1) \geqslant p-n$. Suppose that $p \geqslant 2n$. Then for any two vertices v and w,

$$\delta(v, 1)+\delta(w, 1) \geqslant 2(p-n) = p+(p-2n) \geqslant p.$$

By Lemma 5.4, either $p = 2t$ and $(\mathscr{E}_1) = K_{t,t}$, or else $(\mathscr{E}_1) \geqslant C_k$ for all k satisfying $3 \leqslant k \leqslant p$. If $p = 2n+1 \geqslant m$, the first case is impossible, so that $(\mathscr{E}_1) \geqslant C_m$, which is a contradiction. If $p = m \geqslant 2n$, then since $(\mathscr{E}_1) \ngeqslant C_m$, we must have $p = 2t = m$ and $(\mathscr{E}_1) = K_{t,t}$. But then $(\mathscr{E}_1) \geqslant C_m$, which is also a contradiction. The upper bounds are now established. For *odd* $m \leqslant 2n+1$, let $(\mathscr{E}_1) = K_{n,n}$, and for $m \geqslant 2n$, let $(\mathscr{E}_1) = K_{m-1}$. This gives the required lower bounds. ‖

The proof here shows that $r(C_m, K_{1,n}) \leqslant 2n+1$ for $m \leqslant 2n+1$. For *odd* $m \leqslant 2n+1$, this upper bound is best possible, but for *even* $m < 2n$, complications can arise. For example, if $n > 1$, then $r(C_4, K_{1,n}) \leqslant 1+n+[\sqrt{n}]$, and equality holds here for $n = q^2$ or $n = q^2+1$, where q is any prime power (see [**63**]). Our reason for including the partial results of Theorem 5.5 is to indicate the anomalous behavior possible for Ramsey numbers.

Many evaluations of generalized Ramsey numbers require lengthy case-by-case arguments, and often involve a variety of graph-theoretical tools. The reader should consult the literature for a truer picture of what is involved in such computations.

It is hopeless to expect exact evaluations of most of the numbers $r(G_1, \ldots, G_k)$. Most success must then involve either graphs with a small number of vertices, or else graphs with simple structures.

One way of allowing some arbitrariness in the graphs to be considered, yet still retaining simplicity of structure, is to study $r(m_1G_1, \ldots, m_kG_k)$, for arbitrary fixed graphs $G_1, \ldots, G_k$, as a function of the coefficients $m_1, \ldots, m_k$. (Recall that mG is the disjoint union of m copies of G.) There has been notable success in this direction (see **[10]**, **[21]**, **[59]** and **[60]**). In particular, if $p_i = |V(G_i)|$, then there exists a constant C, depending only on $p_1, \ldots, p_k$, such that $r(m_1G_1, \ldots, m_kG_k) \leqslant \sum_{i=1}^{k} m_ip_i + C$.

Another tactic is to let some simple structure dominate all other parameters. For example, if $G_1, \ldots, G_k$ consists of some mixture of complete graphs, complete bipartite graphs, paths and circuits, then in many cases the numbers $r(C_n, G_1, \ldots, G_k)$ can be evaluated for all sufficiently large n (see **[25]**). Also, if the graph $A(m, s)$ consists of a complete graph K_s with $m-s$ additional edges emanating from one of its vertices, then it can be shown (see **[8]**) that $r(A(m, s), A(n, t)) = (m-1)(n-1)+1$, for fixed s and t, and for all sufficiently large m and n.

Many bounds and estimates have been obtained for generalized Ramsey numbers. In particular, Chvátal and Harary **[16]** have generalized Theorem 3.1 (nearly the same proof applies):

Theorem 5.6. *$r_k(G) > (sk^{q-1})^{1/p}$, where $p = |V(G)|$, $q = |E(G)|$, and s is the order of the automorphism group of G.* ‖

6. The Growth of Generalized Ramsey Numbers

In this section we shall consider the question of which families of graphs exhibit linear growth in their Ramsey numbers. We also discuss the function

$$exr(\mathscr{F}, \mathscr{G}) = \min\{r(F, G) : F \in \mathscr{F}, G \in \mathscr{G}\}.$$

The results here come from Burr and Erdős [7], **[8]**.

The set $\{(G_1, H_1), (G_2, H_2), \ldots\}$ of pairs of graphs is an ***L*-set** if there is a constant c such that for all i, $r(G_i, H_i) \leqslant c(|V(G_i)|+|V(H_i)|)$. We also define $\{G_1, G_2, \ldots\}$ to be an L-set if $\{(G_1, G_1), (G_2, G_2), \ldots\}$ is an L-set. Burr and Erdős have formulated the following conjecture:

Conjecture. *Any set of graphs or pairs of graphs having bounded arboricity is an L-set.*

Theorem 6.1. *Let F, G and H be graphs. Then*

$$r(F, G \cup H) \leqslant \max\{r(F, G)+|V(H)|, r(F, H)\}.$$

Proof. Let n be the indicated maximum, and suppose that we have a 2-coloring of K_n with no $F \leqslant (\mathscr{E}_1)$; then there is an $H \leqslant (\mathscr{E}_2)$. If we delete this H, then there will still be a G with all its edges in $\mathscr{E}_2$. Thus $G \cup H \leqslant (\mathscr{E}_2)$. ‖

The next three results are easy consequences of Theorem 6.1.

Corollary 6.2. *If* $\{(F_1, G_1), (F_2, G_2), \ldots\}$ *and* $\{(F_1, H_1), (F_2, H_2), \ldots\}$ *are L-sets, then so is* $\{(F_1, G_1 \cup H_1), (F_2, G_2 \cup H_2), \ldots\}$. ‖

Corollary 6.3. *If* $\{G_1, G_2, \ldots\}$, $\{H_1, H_2, \ldots\}$ *and* $\{(G_1, H_1), (G_2, H_2), \ldots\}$ *are L-sets, then so is* $\{G_1 \cup H_1, G_2 \cup H_2, \ldots\}$. ‖

Corollary 6.4. *For any graphs G and H, both* $\{G, 2G, 3G, \ldots\}$ *and* $\{(G, H), (2G, 2H), \ldots\}$ *are L-sets.* ‖

Let $G_i = 4^i K_i$, and $H_i = K_{1,n_i}$, where $n_i = i.4^i$. By Theorem 4.4, $r(G_i, H_i) \geqslant (i-1)i.4^i+1$. Thus the sequence $\{(G_1, H_1), (G_2, H_2), \ldots\}$ is not an L-set, even though $\{G_1, G_2, \ldots\}$ and $\{H_1, H_2, \ldots\}$ are L-sets. (The latter is an L-set by Theorem 5.3.) On the other hand, it can be shown that there exists an L-set $\{(G_1, H_1), (G_2, H_2), \ldots\}$ such that neither of the sequences $\{G_1, G_2, \ldots\}$ and $\{H_1, H_2, \ldots\}$ is an L-set.

These examples show that the ratios $r(G, H)/\max\{r_2(G), r_2(H)\}$ and $\min\{r_2(G), r_2(H)\}/r(G, H)$ can (separately) be made arbitrarily large by suitable choice of G and H, since the denominators can grow linearly while the numerators grow faster.

One can test the conjecture on L-sets in many ways. For example, if G has bounded arboricity, then so does $G+K_1$. Also the sequences $\{P_1^m, P_2^m, P_3^m, \ldots\}$ and $\{C_3^m, C_4^m, \ldots\}$, for fixed m, have bounded arboricity. Burr and Erdős have shown that these latter two sequences are L-sets, and that if $\{(G_1, H_1), (G_2, H_2), \ldots\}$ is an L-set, and if the graphs G_i and H_i have bounded arboricity, then $\{(G_1+K_1, H_1+K_1), (G_2+K_1, H_2+K_1), \ldots\}$ is also an L-set.

Further such tests come to mind. Burr and Erdős have asked whether $\{G_1 \times K_2, G_2 \times K_2, \ldots\}$ is an L-set if $\{G_1, G_2, \ldots\}$ is an L-set of bounded arboricity. Also, if $\{G_1, G_2, \ldots\}$ is an L-set of graphs with bounded valency, they asked whether the sequences $\{L(G_1), L(G_2), \ldots\}$ of line graphs, $\{T(G_1), T(G_2), \ldots\}$ of total graphs, and $\{G_1^m, G_2^m, \ldots\}$ of mth-powers (for fixed m), are also L-sets.

Even if the conjecture is true, there will still be other types of L-sets, such as $\{4K_1, 4^2K_2, 4^3K_3, \ldots\}$. This is an L-set since $r_2(K_i) = r(i, i) \leqslant 4^i$, and so, by Theorem 6.1, $r_2(4^iK_i) \leqslant 4^i+2i(4^i-1) \leqslant 3|V(4^iK_i)|$. Indeed, the family of connected graphs $G_i = K_i+4^{2i}K_{i,i}$ has also been shown to be an L-set. Neither of these two families has bounded arboricity.

The study of L-sets is directed at upper bounds. However, the study of $exr(\mathscr{F}, \mathscr{G})$ is concerned with lower bounds. In order to evaluate this function for given families $\mathscr{F}, \mathscr{G}$ of graphs, one must prove that $exr(\mathscr{F}, \mathscr{G}) \geqslant m$, and then exhibit $F \in \mathscr{F}$, and $G \in \mathscr{G}$ with $r(F, G) = m$. For example, let $\mathscr{C}_n$ be the set of connected graphs on n vertices, and let $\mathscr{K}_n$ be the set of graphs with chromatic number n. Then we have

Theorem 6.5. $exr(\mathscr{C}_m, \mathscr{K}_n) = (m-1)(n-1)+1.$

Proof. The lower bound follows from Theorem 4.4, and Theorem 5.1 shows that $r(T, K_n) = (m-1)(n-1)+1$, for a tree $T \in \mathscr{C}_m$, and $K_n \in \mathscr{K}_n$. ‖

An interesting conjecture of Erdős is that, for any $G \in \mathscr{K}_n$, $r_2(G) \geqslant r_2(K_n)$. The first interesting case is for $n = 4$, and here Chvátal and Schwenk **[17]** have shown the conjecture to be true if and only if $r_2(W_6) \geqslant 18$, where W_6 is the wheel of order 6. They have shown only that $17 \leqslant r_2(W_6) \leqslant 20$. For $n > 4$, the problem is difficult, since little is known about $r_2(K_n) = r(n, n)$.

We note in passing that strict inequality can hold in Theorem 4.1(*iv*), since if $\mathscr{F}_1 = \mathscr{F}_2 = \{K_3, P_5\}$, then $r(\mathscr{F}_1, \mathscr{F}_2) < exr(\mathscr{F}_1, \mathscr{F}_2)$.

7. Numbers Related to Combinatorial Designs

Some generalized Ramsey numbers involve combinatorial designs which, when they exist, provide critical or near-critical colorings. In the latter case, improving the estimate by 1 can be difficult, as with the classical numbers. Since the proofs are more instructive than the results, we shall sketch the proofs.

We define the **Turán number** $T(p, G)$ of a graph G to be the maximum number of edges in any graph of order p with no subgraph isomorphic to G. If we have a $(G_1, \ldots, G_k)_p$ coloring in which each $G_i = G$, then $k \,.\, T(p, G) \geqslant \binom{p}{2}$, since there are fewer than $T(p, G)$ edges of any one color. This proves the following results of Irving **[50]**:

Theorem 7.1. *If*

$$T(p, G) < \frac{1}{k}\binom{p}{2},$$

then $r_k(G) \leqslant p$. ‖

Theorem 7.2.

$$r_k(P_4) = \begin{cases} 2k+2, & \text{if } k \equiv 1 \text{ (modulo 3)}, \\ 2k+1, & \text{if } k \equiv 2 \text{ (modulo 3)}, \\ 2k \text{ or } 2k+1, & \text{if } k \equiv 0 \text{ (modulo 3)}. \end{cases}$$

Sketch of Proof. It is easy to see that $T(p, P_4) = p$, if $p \equiv 0$ (modulo 3), and $p-1$ otherwise. Theorem 7.1 gives the required upper bounds. We can construct the lower bound colorings from Kirkman triple systems, by using an existence theorem of Ray-Chaudhuri and Wilson [**65**]. The indeterminacy in the case $k \equiv 0$ (modulo 3) can be resolved when a special kind of triple system exists. ‖

Now let v, k and λ be integers satisfying $v > k > \lambda > 0$. A **(v, k, λ)-graph** is a regular graph with v vertices of valency k, in which any two vertices have exactly λ common neighbors (see [**1**], [**68**], [**74**], and Section 8 of Chapter 12). The following result, due to Lawrence [**57**] and Parsons [**62**], relates (v, k, λ)-graphs to certain Ramsey numbers:

Theorem 7.3. *Let n and λ be positive integers. Then*

$$r(K_{2,\lambda+1}, K_{1,n}) \leqslant 1+n+\tfrac{1}{2}(\lambda+1+[(\lambda-1)^2+4\lambda n]^{\frac{1}{2}}),$$

with equality if and only if there exists an $(n+k, k, \lambda)$-graph.

Remark. Note that k is determined by n and λ, since $k(k-1) = \lambda(n+k-1)$.

Proof. Suppose that we have a $(K_{2,\lambda+1}, K_{1,n})_m$ coloring. Let $\delta_1, \ldots, \delta_m$ be the valencies in $(\mathscr{E}_1)$, and let $\delta = \min(\delta_1, \ldots, \delta_m)$. Then $\delta \geqslant m-n$, and $m\binom{\delta}{2} \leqslant \sum_{i=1}^{n} \binom{\delta_i}{2} \leqslant \lambda\binom{m}{2}$, where the last inequality comes from $(\mathscr{E}_1) \not\supseteq K_{2,\lambda+1}$. The inequalities imply that

$$m^2-(2n+1+\lambda)m+(n^2+n+\lambda) \leqslant 0,$$

and so m cannot exceed the larger root of the quadratic expression. This gives the required inequality. If equality holds, then the right-hand side is an integer, and equality holds in all inequalities of the proof, so we see that $(\mathscr{E}_1)$ is an $(n+k, k, \lambda)$-graph with $k = \delta$. ‖

If $v > k > \lambda > 0$ are integers, then a **(v, k, λ)-design** D is a set V of v points arranged in v k-element blocks $B_1, \ldots, B_v$ such that $i \neq j$ implies $|B_i \cap B_j| = \lambda$ (see Chapter 12). A **polarity** π of D is a bijection from V to $\{B_1, \ldots, B_v\}$ such that $x \in \pi(y)$ implies $y \in \pi(x)$. If $G(D, \pi)$ is the graph with vertex-set V whose edges are those xy such that $x \in \pi(y)$, then letting $(\mathscr{E}_1) = G(D, \pi)$ gives a $(K_{2,\lambda+1}, K_{1,v-k+1})_v$ coloring. This observation, together with the upper bound from Theorem 7.3, gives the following result due to Parsons [**62**]:

Theorem 7.4. *If a (v, k, λ)-design with a polarity exists, then*

$$r(K_{2,\lambda+1}, K_{1,v-k+1}) = 1+v \text{ or } 2+v. \quad \|$$

Such designs are abundant. To show that both of the values $1+v$ and $2+v$ can be attained, consider first the Desarguesian projective planes—that is, the $(q^2+q+1, q+1, 1)$-designs, where q is a prime power; in this case we get $r(K_{2,2}, K_{1,q^2+1}) = q^2+q+2 = 1+v$ (see [**63**]). On the other hand, the graph of the icosahedron provides a critical coloring $(K_{2,3}, K_{1,7})_{12}$, and it follows that for $(v, k, \lambda) = (11, 5, 2)$, we obtain the value

$$r(K_{2,3}, K_{1,7}) = 13 = 2+v.$$

Let $B_m = K_2+mK_1$ be the graph consisting of m triangles with a common base. The Ramsey numbers $r(B_m, B_n)$ have been studied by Rousseau and Sheehan [**67**], and their critical colorings often involve strongly regular graphs. We improve slightly on one of their arguments to obtain an upper bound for these numbers.

Suppose we have a $(B_m, B_n)_p$ coloring. Each edge of $\mathscr{E}_1$ lies in at most $m-1$ triangles of $(\mathscr{E}_1)$, and each edge of $\mathscr{E}_2$ lies in at most $n-1$ triangles of $(\mathscr{E}_2)$; it follows that the number M of monochromatic triangles satisfies

$$M \leqslant M_1 = [\tfrac{1}{3}(m-1)|\mathscr{E}_1|]+[\tfrac{1}{3}(n-1)|\mathscr{E}_2|].$$

Let $\delta_1, \ldots, \delta_p$ be the valencies in $(\mathscr{E}_1)$. From Goodman [**37**], we have $M = \binom{p}{3}-\frac{1}{2}\sum_{i=1}^{p}\delta_i(p-1-\delta_i)$. Treating the δ_i as continuous variables, we next minimize M, subject to the condition that $\sum_{i=1}^{n}\delta_i = 2|\mathscr{E}_1|$. This gives

$$M \geqslant M_0 = \tfrac{1}{24}p(p-1)(p-5)+\frac{2}{p}\left(|\mathscr{E}_1|-\tfrac{1}{2}\binom{p}{2}\right)^2.$$

Now let $x = \frac{1}{2}\binom{p}{2}-|\mathscr{E}_1|$. The requirement $M_0 \leqslant M_1$ implies that

$$\tfrac{1}{24}p(p-1)(p-2m-2n-1) \leqslant \tfrac{1}{3}(n-m)x-\frac{2x^2}{p} \leqslant \tfrac{1}{72}p(n-m)^2,$$

where the last inequality derives from the maximum of the quadratic in x. Simplifying, we get

$$3p^2-6(m+n+1)p+6m+6n+3-n^2-m^2+2mn \leqslant 0.$$

Since p cannot exceed the larger root of this quadratic in p, we get

$$p \leqslant m+n+1+\tfrac{2}{3}\sqrt{3(m^2+mn+n^2)},$$

which gives the following result:

Theorem 7.5. $r(B_m, B_n) \leqslant m+n+2+[\frac{2}{3}\sqrt{3(m^2+mn+n^2)}]$. ||

This bound is best possible since equality holds for $m = n$ whenever $4n+1$ is a prime power, with the critical colorings coming from certain strongly regular graphs (the "Paley graphs"—see Chapter 12 and [**67**]). Suppose that $2m+2n+3 \leqslant r(B_m, B_n)$. Then Theorem 7.5 implies that

$$2m+2n+1 \leqslant \tfrac{1}{3}(n-m)^2.$$

Thus if $\frac{1}{3}(n-m)^2 < 2m+2n+1$, then $r(B_m, B_n) \leqslant 2(m+n+1)$. This is the upper bound obtained by Rousseau and Sheehan, but the same argument gives Theorem 7.5 which is stronger. Taking $(\mathscr{E}_1)$ to be the strongly regular graph with parameters (253, 112, 36, 60) of Goethals and Seidel **[36]**, we may construct a 2-coloring of the edges of K_{253}, thereby showing that $254 \leqslant r(B_{37}, B_{88})$; it follows that, as noted in **[67]**, it is not always true that $r(B_m, B_n) \leqslant 2(m+n+1)$. In fact, Theorem 7.5 then gives

$$254 \leqslant r(B_{37}, B_{88}) \leqslant 255.$$

More careful methods give a better upper bound than Theorem 7.5 in certain cases, and Rousseau and Sheehan have shown that $r(B_1, B_n) = 2n+3$ for $n > 1$, and $r(B_m, B_n) = 4k^2$ for $m = k^2-2$ and $n = k^2+1$, where $k = 3^s2^{s+t-1} \geqslant 2$.

8. Bipartite Ramsey Numbers

Let $G_1, \ldots, G_k$ be bipartite graphs. The **bipartite Ramsey number** $b(G_1, \ldots, G_k)$ is the least integer n such that every k-coloring of $K_{n,n}$ has $G_i \leqslant (\mathscr{E}_i)$, for some i. If $G_i = G$ for each i, we let $b_k(G) = b(G_1, \ldots, G_k)$. Since every bipartite graph is a subgraph of $K_{r,s}$, for large enough r and s, the existence of all numbers $b(G_1, \ldots, G_k)$ follows from the following result:

Theorem 8.1. *Given n, $k > 0$, let $g_k(n)$ be the minimum of $\sum_{j=1}^{n} \binom{\delta_j}{r}$ over all choices of non-negative integers $\delta_1, \ldots, \delta_n$ such that $\sum_{j=1}^{n} \delta_j \geqslant n^2/k$. Then*

$$b_k(K_{r,s}) \leqslant \min\{n : g_k(n) > (s-1)\tbinom{n}{r}\}.$$

Proof. Let A, B be disjoint n-element sets, and suppose that we have a k-coloring of the complete bipartite graph with partite sets A, B. Then, for some i, $|\mathscr{E}_i| \geqslant n^2/k$. Let the vertices of B have valencies $\delta_1, \ldots, \delta_n$ in $(\mathscr{E}_i)$. For each r-element subset S of A, let $\beta(S)$ be the number of vertices of B adjacent in $(\mathscr{E}_i)$ to every vertex of S. Counting in two ways the pairs (S, v) such that $v \in B$ is adjacent in $(\mathscr{E}_i)$ to every vertex of S, we get

$$\sum_S \beta(S) = \sum_j \binom{\delta_j}{r} \geqslant g_k(n).$$

If $g_k(n) > (s-1)\binom{n}{r}$, then the average value of $\beta(S)$ exceeds $s-1$, so that some r-element subset S of A has s vertices of B adjacent in $(\mathscr{E}_i)$ to every vertex of S. It follows that $K_{r,s} \leqslant (\mathscr{E}_i)$. ||

We note that $g_k(n) \geqslant n\binom{n/k}{r} > (s-1)\binom{n}{r}$ for all sufficiently large n, so that $b_k(K_{r,s})$ is finite. This fact was first proved by Erdős and Rado **[27]**. The proof of Theorem 8.1 is due to Chvátal [**14**], and was rediscovered by Beineke and Schwenk [**2**], who used it to prove the following result:

Theorem 8.2. *With the above notation,*

(*i*) $b_2(K_{1,n}) = 2n-1$;

(*ii*) $b_2(K_{2,n}) \leqslant 4n-3$, *with equality holding at least when* $n = 2$, *and when* n *is odd and there exists a Hadamard matrix of order* $2n-2$;

(*iii*) $b_2(K_{2,n}) \geqslant 4n-4$, *if there is a Hadamard matrix of order* $4n-4$.

Proof. (*i*) This part is trivial.

(*ii*) The upper bound follows from Theorem 8.1. If $\mathbf{H}$ is a Hadamard matrix of order $2t$, let

$$\mathbf{M} = \begin{pmatrix} \mathbf{H} & -\mathbf{H} \\ -\mathbf{H} & \mathbf{H} \end{pmatrix}.$$

Let the edge $v_i w_j$ be red or blue according as the ij-entry of $\mathbf{M}$ is 1 or -1, for $1 \leqslant i, j \leqslant 4t$. If $t = n-1$, this coloring of $K_{4t,4t}$ establishes equality.

(*iii*) This part is similar, except that we first normalize the entries of the first row and first column to make them all $+1$, and then delete the first row and first column to get a matrix of order $4n-5$. ‖

(For a discussion of Hadamard matrices, see Ryser [**69**], which also has an elegant chapter on Ramsey's theorem.)

Beineke and Schwenk also showed that $b_2(K_{3,3}) = 17$, $b_2(K_{3,n}) \leqslant 8n-5$, and $b_2(K_{3,n}) \geqslant 8n-7$ if there is a Hadamard matrix of order $4n-4$. Irving [**51**] improved this upper bound to $b_2(K_{3,n}) \leqslant 8n-7$, again with equality if there exists a Hadamard matrix of order $4n-4$. In particular, $b_2(K_{3,4}) = 25$, and $b_2(K_{3,5}) = 33$, thereby solving a problem of Harary [**46**].

A special case of some results of Gyárfás and Lehel [**41**], and Faudree and Schelp [**30**], implies the following results:

Theorem 8.3. *If* P_k *denotes the path graph of order* k, *then*

(*i*) $b(P_{2m}, P_{2n}) = m+n-1$;

(*ii*) $b(P_{2m+1}, P_{2n}) = \begin{cases} m+n, & \text{if } m \geqslant n-1, \\ m+n-1, & \text{otherwise}; \end{cases}$

(*iii*) $b(P_{2m+1}, P_{2n+1}) = \begin{cases} m+n, & \text{if } m \neq n, \\ 2m+1, & \text{if } m = n. \end{cases}$ ‖

Finally, we note the trivial fact that for any bipartite graphs $G_1, \ldots, G_k$, we have $r(G_1, \ldots, G_k) \leqslant 2b(G_1, \ldots, G_k)$.

9. Arrow Relations

Given graphs F, G and H, we write $F \to (G, H)$ ("F arrows G, H") if, for every 2-coloring of F, either $G \leqslant (\mathscr{E}_1)$ or $H \leqslant (\mathscr{E}_2)$. Given r and s, an extension of the problem of evaluating the bipartite Ramsey numbers $b_2(K_{r,s})$ asks for all those integers m and n such that $K_{m,n} \to (K_{r,s}, K_{r,s})$. This is closely related to the well-known problem of Zarankiewicz [79] (see also **[14]**, **[40]** and **[51]**). Gyárfás and Lehel, and Faudree and Schelp, determined, for fixed $s, t \geqslant 2$, all integers m and n such that $K_{m,n} \to (P_s, P_t)$. (This was the result mentioned prior to Theorem 8.3.)

For fixed G and H, it is usually difficult to characterize those graphs F for which $F \to (G, H)$. However, certain properties of such graphs F may be deduced. For example, the parameters $\chi(F)$, $\delta(F)$, $\Delta(F)$ are studied in [9], and we present a theorem from this paper.

Let $\mathscr{G}$, $\mathscr{H}$ be families of graphs. The **chromatic Ramsey number** $chr(\mathscr{G}, \mathscr{H})$ is the minimum value of $\chi(F)$ taken over all F such that $F \to (\mathscr{G}, \mathscr{H})$—that is, every 2-coloring of F has $G \leqslant (\mathscr{E}_1)$ for some $G \in \mathscr{G}$, or $H \leqslant (\mathscr{E}_2)$ for some $H \in \mathscr{H}$.

A **homomorphism** of G into H is a mapping $\phi: V(G) \to V(H)$ such that if v and w are adjacent in G, then $\phi(v)$ and $\phi(w)$ are adjacent in H. We say that H is a **homomorphic image** of G if all vertices and edges of H are images of vertices and edges of G. Finally, we let hom G be the set of all homomorphic images of G, and hom $\mathscr{G} = \bigcup_{G \in \mathscr{G}}$ hom G.

Theorem 9.1. $chr(\mathscr{G}, \mathscr{H}) = r(\text{hom } \mathscr{G}, \text{hom } \mathscr{H})$.

Sketch of Proof. Let $r = r(\text{hom } \mathscr{G}, \text{hom } \mathscr{H})$, and let K be a (hom $\mathscr{G}$, hom $\mathscr{H})_{r-1}$ coloring with vertices $v_1, \ldots, v_{r-1}$. Suppose that $F \to (\mathscr{G}, \mathscr{H})$, and $\chi(F) \leqslant r-1$. Then there is a mapping $\phi: V(F) \to \{v_1, \ldots, v_{r-1}\}$ such that $vw \in E(F)$ implies that $\phi(v) \neq \phi(w)$, and we may regard ϕ as a homomorphism of F into the complete graph K with vertex-set $\{v_1, \ldots, v_{r-1}\}$. Let $vw \in E(F)$ have the color of the edge $\phi(v)\phi(w)$ in K. This is a 2-coloring of F with no red $G \in \mathscr{G}$ and no blue $H \in \mathscr{H}$. Thus $chr(\mathscr{G}, \mathscr{H}) \geqslant r(\text{hom } \mathscr{G}, \text{hom } \mathscr{H})$.

Now let $c_1, \ldots, c_t$ be all the 2-colorings of K_r. For each c_i, we may choose either a $G_i \in \mathscr{G}$ for which some homomorphic image G_i' is red in c_i, or an $H_i \in \mathscr{H}$ for which some homomorphic image H_i' is blue in c_i. Let $\mathscr{G}_1$ and $\mathscr{H}_1$ be respectively the sets of all the G_i and H_i chosen in this way. Then $chr(\mathscr{G}, \mathscr{H}) \leqslant chr(\mathscr{G}_1, \mathscr{H}_1)$, and so it suffices to show that $chr(\mathscr{G}_1, \mathscr{H}_1) \leqslant r$. Let $m = \max\{|V(G)| : G \in \mathscr{G}_1 \cup \mathscr{H}_1\}$. It can be shown that, for large enough n, the complete r-partite graph $K(A_1, \ldots, A_r)$ with partite sets A_i of size n has the property that for any 2-coloring α of its edges, there exist m-element subsets $B_i \subseteq A_i$ such that the complete bipartite subgraphs $K(B_i, B_j)$, for $i \neq j$, are monochromatic with color α_{ij}. Using the definitions of $r = r(\text{hom } \mathscr{G}, \text{hom } \mathscr{H})$, and of $\mathscr{G}_1$ and $\mathscr{H}_1$, and using the 2-coloring of K_r defined by letting $v_i v_j$ have color α_{ij} for $1 \leqslant i < j \leqslant r$, we see that $K(B_1, \ldots, B_r)$ contains either a red $G_i \in \mathscr{G}_1$ or a blue $H_i \in \mathscr{H}_1$. Therefore

$$K(A_1, \ldots, A_r) \rightarrow (\mathscr{G}_1, \mathscr{H}_1),$$

showing that $chr(\mathscr{G}_1, \mathscr{H}_1) \leqslant r$. ∥

Let $F \Rightarrow (G_1, \ldots, G_k)$ signify that any k-coloring of F has, for some i, an induced subgraph G_i of F with $G_i \leqslant (\mathscr{E}_i)$. Deuber **[22]** has proved that for any graphs G and H there exists a graph F such that $F \Rightarrow (G, H)$. Folkman **[32]** has proved that for any integers m and n there exists an F whose largest clique has size $\omega(F) = \max\{m, n\}$, such that $F \rightarrow (K_m, K_n)$. (Here $\Rightarrow$ is the same as $\rightarrow$.) These theorems have been generalized by Nešetřil and Rödl **[61]**:

Theorem 9.2. *Given graphs $G_1, \ldots, G_k$, there exists a graph F with*

$$\omega(F) = \max_i \omega(G_i),$$

such that $F \Rightarrow (G_1, \ldots, G_k)$. ∥

Actually they proved the equivalent theorem where $G_1 = \ldots = G_k = H$; Theorem 9.2 is then obtained by replacing H by the disjoint union

$$G_1 \cup \ldots \cup G_k.$$

These results represent challenging new directions in Ramsey graph theory. Most questions about the properties of those graphs F such that $F \rightarrow$ (or $\Rightarrow$) $(G_1, \ldots, G_k)$ remain open.

10. Variations, Generalizations and Refinements

There are many variants of Ramsey graph problems not covered in the previous sections. We mention briefly a few of them here.

(a) Ramsey Multiplicities

If G is a graph, its **Ramsey multiplicity** $R(G)$ is the minimum number of monochromatic copies of G which must occur in any 2-coloring of K_r, where $r = r_2(G)$. (We might similarly define $R_k(G)$ for k-colorings, but the problem is already hard enough for $k = 2$.) Harary and Prins **[48]** have computed $R(G)$ for some small graphs G, and have shown that $R(B_2) \leqslant 15$, by producing an appropriate coloring of K_{10}. Schwenk (see **[43]**) has shown that equality holds here, using a counting argument based on Goodman's result [**47**] on the number of monochromatic triangles in 2-colorings of K_n. Other than this, little is known, except that $R(K_{1,n})$ is 1 if n is even, and $2n$ if n is odd. Harary has conjectured that $R(G) = 1$ implies $G = K_2$, or $G = K_{1,2m}$ for some m.

(b) Directed Graphs

Given digraphs $D_1, \ldots, D_k$, the **Ramsey number** $r(D_1, \ldots, D_k)$ is the least n such that, for any k-coloring of the arcs of the complete symmetric digraph $\vec{K}_n$, there is, for some i, a copy of D_i all of whose arcs are colored i. This number exists if and only if at most one of $D_1, \ldots, D_k$ contains any directed cycles.

Gyárfás and Lehel [**41**], and Williamson [**77**], have shown that

$$r(\vec{P}_m, \vec{P}_n) = m+n-3$$

for $m, n \geqslant 3$, where $\vec{P}_m$ is the directed path on m vertices. Bermond [**4**] has shown that $r(\vec{P}_{n_1}, \ldots, \vec{P}_{n_k}, D) = (n_1-1)\ldots(n_k-1)(p-1)+1$, for every Hamiltonian digraph D of order p. (It is interesting to note that most of the corresponding numbers for the underlying graphs here have not yet been determined.)

Let TT_n be the transitive tournament of order n. Harary and Hell [**47**] have proved that $r(TT_m, TT_n) = r(m, n)$, the usual classical Ramsey number, and that $r(m, n) \leqslant r(TT_m, \vec{K}_n) \leqslant r(2^{m-1}, n)$. They also showed that

$$r(S_m, S_n) = r(S_m, S_n') = m+n,$$

where S_m is the star with m outgoing arcs, and S_n' is the converse of S_n. Furthermore, they computed $r(D_1, D_2)$ for many small digraphs D_1 and D_2.

(c) The Connected Ramsey Numbers

A graph F is **totally connected** if both F and its complement $\bar{F}$ are connected. The **connected Ramsey number** $cr(G, H)$ is the least integer n such that, for any totally connected graph F with n vertices, either $G \leqslant F$ or $H \leqslant \bar{F}$. Clearly, $cr(G, H) \leqslant r(G, H)$. Since $cr(P_4, H) = 4$, for any H with four or more vertices, strict inequality may hold here.

Sumner [**72**] has shown that if G and H each have at least four vertices, and if neither has a bridge, then $cr(G, H) = r(G, H)$. He also showed that $cr(P_m, P_n) = n+[\frac{1}{2}(m-3)]-1$ for $5 \leqslant m \leqslant n$, so that these numbers differ only slightly from the numbers $r(P_m, P_n) = n+[\frac{1}{2}m]-1$ determined by

Geréncser and Gyárfás [35]. The evaluation of other connected Ramsey numbers $cr(G, H)$, where at least one of G and H has a bridge, remains open. This should first be attempted for those graphs G and H for which $r(G, H)$ is known, such as (P_m, C_n) and (T, K_n). The question of $cr(G, H)$ is primarily of interest for any light it may shed on critical colorings; it differs from $r(G, H)$ only in requiring that $(\mathscr{E}_1)$ and $(\mathscr{E}_2)$ should both be connected in the colorings. One might attempt to study critical colorings by imposing other requirements on the classes $\mathscr{E}_1, \ldots, \mathscr{E}_k$.

(d) Subsets of Colors

Let k and d be integers satisfying $k > d \geqslant 1$, and let $t = \binom{k}{d}$. Given the set $\{1, \ldots, k\}$ of colors, we order the t subsets $A_1, \ldots, A_t$ of d colors. Let $G_1, \ldots, G_t$ be graphs. Then the **d-chromatic Ramsey number** $R_d^k(G_1, \ldots, G_t)$ is the least integer p such that in any k-coloring of K_p, there is, for some i, a copy of G_i all of whose edges are colored with colors from the set A_i. Chung and Liu [13] have defined these numbers, and have studied the case $k = 3$, $d = 2$ in which G_1, G_2 and G_3 are all complete graphs. They were able to prove that $R_2^3(K_4, K_4, K_4) = 10$, and evaluate $R_2^3(K_i, K_j, K_n)$ for $(i, j) = (3, 3)$, $(3, 4)$, $(3, 5)$, and for all n. The numbers R_d^k have obvious relevance to the classical Ramsey numbers.

(e) Endless Variations

In his thesis [3], Benedict gave a framework for Ramsey numbers of a general nature. Let $G_1, G_2, \ldots$ be graphs. If $i \leqslant j$ implies $G_i \leqslant G_j$, then the family $\Phi = \{G_t : t \geqslant 1\}$ is called an **ascending family**. A property π of graphs is an **ascending property** if, whenever G has property π and $G \leqslant H$, then H also has property π. A **factor** F of G is a subgraph of G with $V(F) = V(G)$, and a **factorization** $G = F_1 \oplus \ldots \oplus F_k$ of G is a family of factors $F_1, \ldots, F_k$ whose edge-sets $E(F_1), \ldots, E(F_k)$ partition $E(G)$. (Taking $\mathscr{E}_i = E(F_i)$, $1 \leqslant i \leqslant k$, gives a k-coloring of G.) If Φ is an ascending family of graphs, and if $\pi_1, \ldots, \pi_k$ are ascending properties, then the Ramsey number $r_\Phi(\pi_1, \ldots, \pi_k)$ (when it exists) is the least integer t such that whenever $G_t = F_1 \oplus \ldots \oplus F_k$, then, for some i, F_i has property π_i. The reason for taking Φ and $\pi_1, \ldots, \pi_k$ to be ascending is to ensure that, when $r_\Phi(\pi_1, \ldots, \pi_k)$ exists, then for $t < r_\Phi(\pi_1, \ldots, \pi_k)$ there will always be some factorization of G_t such that no F_i has property π_i, whereas for $t \geqslant r_\Phi(\pi_1, \ldots, \pi_k)$, no such factorization exists. When $\Phi = \{K_t : t \geqslant 1\}$ and π_i is the property $G_i \leqslant G$ of the graph G, then

$r_\Phi(\pi_1, \ldots, \pi_k) = r(G_1, \ldots, G_k)$, the usual generalized Ramsey number. Benedict was concerned mostly with cases in which $\Phi = \{K_{t(i)}: t \geqslant 1\}$, where $K_{t(i)}$ is the complete t-partite graph with partite sets of size i. Of course, the penalty for such generality is that little of much depth can be said.

11. Concluding Remarks

Ramsey discovered his famous theorem in 1928, and published it in 1930 [**64**]. The theorem was rediscovered by Skolem, and by Erdős and Szekeres [**28**] in 1933 (see [**24**, pp. *xix–xxii*] for a delightful account). Since then it has had the most varied offspring, many of which were sired by Erdős, who has maintained a lifelong interest in such problems.

The first published treatment of generalized Ramsey numbers was in 1967 by Geréncser and Gyárfás [**35**], although such numbers were anticipated by Erdős [**23**] in 1947, when he gave the correct upper bound for $r(P_m, K_n)$. Cockayne [**18**] was the first to publish a general statement of the problem. When the time is ripe, a new idea may appear in many places at once, like flowers in springtime. Thus we find that many graph-theorists enjoyed rediscovering the numbers $r(G_1, \ldots, G_k)$.

One standard used to measure mathematics is: does it produce interesting theorems? By this criterion, it appears that Ramsey graph theory has proved its worth.

References

1. R. W. Ahrens and G. Szekeres, On a combinatorial generalization of 27 lines associated with a cubic surface, *J. Austral. Math. Soc.* **10** (1969), 485–492; *MR*42#4419.
2. L. W. Beineke and A. J. Schwenk, On a bipartite form of the Ramsey problem, in *Proceedings of the Fifth British Combinatorial Conference 1975* (ed. C. St. J. A. Nash-Williams and J. Sheehan), Congressus Numerantium **XV**, Utilitas Mathematica, Winnipeg, 1975, pp. 17–22.
3. J. Benedict, On Ramsey numbers defined by factorizations of regular complete multipartite graphs, Ph.D. thesis, Western Michigan University, 1976.
4. J.-C. Bermond, Some Ramsey numbers for directed graphs, *Discrete Math.* **9** (1974), 313–321; *MR*54#5029.
5. J. A. Bondy, Pancyclic graphs, I, *J. Combinatorial Theory* (*B*) **11** (1971), 80–84; *MR*44 #2642.
6. S. A. Burr, Generalized Ramsey theory for graphs—a survey, in *Graphs and Combinatorics*, Lecture Notes in Mathematics **406** (ed. R. A. Bari and F. Harary), Springer-Verlag, Berlin, Heidelberg, New York, 1974, pp. 52–75.
7. S. A. Burr and P. Erdős, On the magnitude of generalized Ramsey numbers for graphs, in *Infinite and Finite Sets, I* (ed. A. Hajnal *et al.*), North-Holland, Amsterdam, 1975, pp. 215–240.
8. S. A. Burr and P. Erdős, Extremal Ramsey theory for graphs, *Utilitas Math.* **9** (1976), 247–258; *MR*55#2633.

9. S. A. Burr, P. Erdős and L. Lovász, On graphs of Ramsey type, *Ars Combinatoria* **1** (1976), 167–190; *MR*54#7308.
10. S. A. Burr, P. Erdős and J. H. Spencer, Ramsey theorems for multiple copies of graphs, *Trans. Amer. Math. Soc.* **209** (1975), 87–99.
11. S. A. Burr and J. A. Roberts, On Ramsey numbers for stars, *Utilitas Math.* **4** (1973), 217–220; *MR*48#3782.
12. F. R. K. Chung, On the Ramsey number $N(3, 3, \ldots, 3; 2)$, *Discrete Math.* **5** (1973), 317–322; *MR*48#4838.
13. K. M. Chung and C. L. Liu, A generalization of Ramsey theory for graphs (to appear).
14. V. Chvátal, On finite polarized partition relations, *Canad. Math. Bull.* **12** (1969), 321–326; *MR*41#5230.
15. V. Chvátal, Tree-complete graph Ramsey numbers, *J. Graph Theory* **1** (1977), 93.
16. V. Chvátal and F. Harary, Generalized Ramsey theory for graphs I. Diagonal numbers, *Periodica Math. Hungar.* **3** (1973), 115–124; *MR*48#10887.
17. V. Chvátal and A. J. Schwenk, On the Ramsey number of the five-spoked wheel, in *Graphs and Combinatorics*, Lecture Notes in Mathematics **406** (ed. R. A. Bari and F. Harary), Springer-Verlag, Berlin, Heidelberg, New York, 1974, pp. 247–261; *MR*51 #234.
18. E. J. Cockayne, An application of Ramsey's theorem, *Canad. Math. Bull.* **13** (1970), 145–146; *MR*41#6721.
19. E. J. Cockayne, Colour classes for r-graphs, *Canad. Math. Bull.* **15** (1972), 349–354; *MR*49#4832.
20. E. J. Cockayne and P. J. Lorimer, On Ramsey graph numbers for stars and stripes, *Canad. Math. Bull.* **18** (1975), 31–34; *MR*54#161.
21. E. J. Cockayne and P. J. Lorimer, The Ramsey graph number for stripes, *J. Austral. Math. Soc.* **19** (1975), 252–256; *MR*51#7950.
22. W. Deuber, Generalizations of Ramsey's theorem, in *Infinite and Finite Sets, I* (ed. A. Hajnal *et al.*), North-Holland, Amsterdam, 1975, pp. 323–332.
23. P. Erdős, Some remarks on the theory of graphs, *Bull. Amer. Math. Soc.* **53** (1947), 292–294; *MR*8–479.
24. P. Erdős, *The Art of Counting* (ed. J. Spencer), M.I.T. Press, Cambridge, Mass., 1975.
25. P. Erdős, R. J. Faudree, C. C. Rousseau and R. H. Schelp, Generalized Ramsey theory for multiple colors, *J. Combinatorial Theory* (*B*) **20** (1976) 250–264.
26. P. Erdős, R. J. Faudree, C. C. Rousseau and R. J. Schelp, On cycle-complete graph Ramsey numbers (to appear).
27. P. Erdős and R. Rado, A partition calculus in set theory, *Bull. Amer. Math. Soc.* **62** (1956), 427–489; *MR*18–458.
28. P. Erdős and G. Szekeres, A combinatorial problem in geometry, *Compositio Math.* **2** (1935), 463–470.
29. R. J. Faudree and R. H. Schelp, All Ramsey numbers for cycles in graphs, *Discrete Math.* **8** (1974), 313–329; *MR*49#10596.
30. R. J. Faudree and R. H. Schelp, Path-path Ramsey type numbers for the complete bipartite graph, *J. Combinatorial Theory* (*B*) **19** (1975), 161–173; *MR*54#153.
31. J. Folkman, Notes on the Ramsey number $N(3, 3, 3, 3)$, *J. Combinatorial Theory* (*A*) **16** (1974), 371–379; *MR*49#111.
32. J. Folkman, Graphs with monochromatic complete subgraphs in every edge coloring, *SIAM J. Appl. Math.* **18** (1970), 19–24; *MR*42#2979.
33. P. Frankl, A constructive lower bound for some Ramsey numbers (to appear).
34. M. Gardner, Mathematical games, *Scientific American* **237**, No. 5 (November 1977), 18–29.
35. L. Gerénсser and A. Gyárfás, On Ramsey-type problems, *Ann. Univ. Sci. Budapest. Eötvös Sect. Math.* **10** (1967), 167–170; *MR*39#1351.
36. J. M. Goethals and J. J. Seidel, Strongly regular graphs derived from combinatorial designs, *Canad. J. Math.* **22** (1970), 597–614; *MR*44#106.
37. A. W. Goodman, On sets of acquaintances and strangers at any party, *Amer. Math. Monthly* **66** (1959), 778–783; *MR*21#6335.

38. J. E. Graver and J. Yackel, Some graph-theoretic results associated with Ramsey's theorem, *J. Combinatorial Theory* **4** (1968), 125–175; *MR***37**#1278.
39. R. E. Greenwood and A. M. Gleason, Combinatorial relations and chromatic graphs, *Canad. J. Math.* **7** (1955), 1–7; *MR***16**–733.
40. R. K. Guy, A many-facetted problem of Zarankiewicz, in *The Many Facets of Graph Theory*, Lecture Notes in Mathematics **110** (ed. G. Chartrand and S. F. Kapoor), Springer-Verlag, Berlin, Heidelberg, New York, 1969, pp. 129–148; *MR***41**#91.
41. A. Gyárfás and J. Lehel, A Ramsey-type problem in directed and bipartite graphs, *Periodica Math. Hungar.* **3** (1973), 299–304; *MR***49**#136.
42. D. Hanson, Sum-free sets and Ramsey numbers, *Discrete Math.* **14** (1976), 57–61.
43. F. Harary, *Graph Theory*, Addison-Wesley, Reading, Mass., 1969; *MR***41**#1566.
44. F. Harary, Recent results in generalized Ramsey theory for graphs, in *Graph Theory and Applications*, Lecture Notes in Mathematics **303** (ed. Y. Alavi *et al.*), Springer-Verlag, Berlin, Heidelberg, New York, 1972, pp. 125–138; *MR***49**#7177.
45. F. Harary, A survey of generalized Ramsey theory, in *Graphs and Combinatorics*, Lecture Notes in Mathematics **406** (ed. R. A. Bari and F. Harary), Springer-Verlag, Berlin, Heidelberg, New York, 1974, pp. 10–17; *MR***50**#12783.
46. F. Harary, The foremost open problems in generalized Ramsey theory, in *Proceedings of the Fifth British Combinatorial Conference 1975* (ed. C. St. J. A. Nash-Williams and J. Sheehan), Congressus Numerantium **XV**, Utilitas Mathematica, Winnipeg, 1975, pp. 269–282.
47. F. Harary and P. Hell, Generalized Ramsey theory for graphs, V. The Ramsey number of a digraph, *Bull. London Math. Soc.* **6** (1974), 175–182; correction in **7** (1975), 87–88.
48. F. Harary and G. Prins, Generalized Ramsey theory for graphs, IV. The Ramsey multiplicity of a graph, *Networks* **4** (1974), 163–173; *MR***49**#4836.
49. R. W. Irving, Contributions to Ramsey theory, Ph.D. thesis, University of Glasgow, 1973.
50. R. W. Irving, Generalized Ramsey numbers for small graphs, *Discrete Math.* **9** (1974), 251–264; *MR***51**#12591.
51. R. W. Irving, A bipartite Ramsey problem and the Zarankiewicz numbers, *Glasgow. Math. J.* (to appear).
52. J. G. Kalbfleisch, Chromatic graphs and Ramsey's theorem, Ph.D. thesis, University of Waterloo, 1966.
53. J. G. Kalbfleisch, Upper bounds for some Ramsey numbers, *J. Combinatorial Theory* **2** (1967), 35–42; *MR***35**#2794.
54. J. G. Kalbfleisch, A uniqueness theorem for edge-chromatic graphs, *Pacific J. Math.* **21** (1967), 503–509; *MR***35**#2789.
55. J. G. Kalbfleisch and R. G. Stanton, On the maximal triangle-free edge-chromatic graphs in three colors, *J. Combinatorial Theory* **5** (1968), 9–20; *MR***37**#3934.
56. G. Kéry, Ramsey egy gráfelméleti tételéröl, *Mat. Lapok* **15** (1964), 204–224; *MR***30** #2484.
57. S. L. Lawrence, Cycle-star Ramsey numbers, *Notices Amer. Math. Soc.* **20** (1973), Abstract A-420.
58. S. L. Lawrence, Bipartite Ramsey theory, *Notices Amer. Math. Soc.* **20** (1973), Abstract A-562.
59. P. J. Lorimer and P. R. Mullins, Ramsey numbers for quadrangles and triangles, *J. Combinatorial Theory* (*B*) **23** (1977), 262–265.
60. P. J. Lorimer and R. J. Segedin, Ramsey numbers for multiple copies of complete graphs (to appear).
61. J. Nešetřil and V. Rödl, The Ramsey property for graphs with forbidden complete subgraphs, *J. Combinatorial Theory* (*B*) **20** (1976), 243–249; *MR***54**#133.
62. T. D. Parsons, Ramsey graphs and block designs, *J. Combinatorial Theory* (*A*) **20** (1976), 12–19.
63. T. D. Parsons, Ramsey graphs and block designs, I, *Trans. Amer. Math. Soc.* **209** (1975), 33–44.

64. F. P. Ramsey, On a problem of formal logic, *Proc. London Math. Soc.* (2) **30** (1930), 264–286.
65. D. K. Ray-Chaudhuri and R. M. Wilson, Solution of Kirkman's schoolgirl problem, *Proc. Sympos. Pure Math.* **19** (1971), 187–204.
66. V. Rosta, On a Ramsey type problem of J. A. Bondy and P. Erdős I, II, *J. Combinatorial Theory* (*B*) **15** (1973), 94–104 and 105–120; *MR* **48**#10894.
67. C. C. Rousseau and J. Sheehan, On Ramsey numbers for $K_2 + \bar{K}_n$ (to appear).
68. A. Rudvalis, (v, k, λ)-graphs and polarities of (v, k, λ)-designs, *Math. Z.* **120** (1971), 224–230; *MR***45**#3220.
69. H. J. Ryser, *Combinatorial Mathematics*, Carus Mathematical Monographs **14**, Mathematical Association of America, Washington D.C., 1963; *MR***27**#51.
70. J. H. Spencer, Ramsey's Theorem—a new lower bound, *J. Combinatorial Theory* (*A*) **18** (1975), 108–115; *MR***51**#2973.
71. S. Stahl, On the Ramsey number $r(F, K_m)$, where F is a forest, *Canad. J. Math.* **27** (1975), 585–589.
72. D. P. Sumner, The connected Ramsey number (to appear).
73. K. Walker, An upper bound for the Ramsey number $M(5, 4)$, *J. Combinatorial Theory* (*A*) **11** (1971), 1–10; *MR***43**#105.
74. W. D. Wallis, A nonexistence theorem for (v, k, λ)-graphs, *J. Austral. Math. Soc.* **11** (1970), 381–383; *MR***42**#1706.
75. E. G. Whitehead, Jr., The Ramsey number $N(3, 3, 3, 3; 2)$, *Discrete Math.* **4** (1973), 389–396; *MR***47**#3229.
76. E. G. Whitehead, Jr., Algebraic structure of chromatic graphs associated with the Ramsey number $N(3, 3, 3; 2)$, *Discrete Math.* **1** (1971), 113–114; *MR***45**#6686.
77. J. E. Williamson, A Ramsey-type problem for paths in digraphs, *Math. Ann.* **203** (1973), 117–118; *MR***48**#3788.
78. J. Yackel, Inequalities and asymptotic bounds for Ramsey numbers, *J. Combinatorial Theory* (*A*) **13** (1972), 56–68; *MR***46**#3370.
79. K. Zarankiewicz, Problem P101, *Colloq. Math.* **2** (1951), 301.

14
The Enumeration of Graphs

EDGAR M. PALMER

1. Introduction

Just as the whole field of graph theory finds its origins in the work of Euler, so also does that part known as graphical enumeration. For example, the first serious study of the so-called "Catalan numbers" $\frac{1}{n}\binom{2n-2}{n-1}$ for the number of triangulations of an $(n+1)$-gon with one oriented edge on its boundary can be traced back to the founding father (see **[3]**, **[11]**). Furthermore, some of the most important and elegant theorems in graphical enumeration have been inspired by the Königsberg bridge problem. In this chapter we shall focus our attention on several of these results as we develop some of the foundations of enumerative analysis.

In such a concise treatment as this, many topics of importance have had to be omitted, or treated very briefly. For example, some readers would have preferred to see a more detailed treatment of Pólya's enumeration theorem, discussed briefly in Section 9. However, fully-developed expositions of it may be found elsewhere (see **[10]**, **[14]**), and one of the aims of the chapter is in fact to show how much can be done without it! Our study of the enumeration of unlabeled trees and of digraphs of various types is also very brief, as is any discussion of the important problems which remain unsolved. Nevertheless, we hope we have been able to give some idea of several of the underlying techniques involved in graphical enumeration, and to present some of the most interesting results in the field. A broader and deeper treatment of graph counting, as well as an extensive bibliography and numerous tables, can be found in the book *Graphical Enumeration* **[14]**.

In the first part of this chapter we derive formulas for the number of labeled graphs of various kinds, such as connected graphs, Eulerian graphs, and trees. In the second half we deal with the more difficult counting problems for unlabeled graphs, again concentrating on connected graphs and Eulerian graphs. This is followed by a discussion of Pólya's theorem, and the chapter concludes with a brief list of some unsolved problems in the area.

The above chapter outline was selected to accomplish three objectives. First, we seek to enable the reader to become quickly acquainted with the elements of the field, and with some of its principal results. Second, we wanted to present the details of the exciting recent discoveries on Eulerian graphs by R. C. Read and R. W. Robinson. At this point, we hope that the reader will be in a position to appreciate the beauty and the difficulty of the unsolved problems in Section 10. Thereby we shall perhaps achieve our third goal—namely, to attract new researchers to this fascinating field.

2. Labeled Graphs

We begin with labeled counting problems because, more often than not, they are substantially easier. A **labeled graph of order p** is a graph of order p, each of whose vertices is assigned a different positive integer from the set $\{1, 2, \ldots, p\}$. Thus the labeling process amounts to arranging the vertices in linear order. The number of distinct ways in which the labels can be associated with the vertices of a given unlabeled graph G is inversely proportional to the order of the automorphism group $\Gamma(G)$, as we now prove:

Theorem 2.1. *The number of ways of labeling the vertices of a given graph G of order p is* $p!/|\Gamma(G)|$.

Proof. The vertices of G can be labeled in $p!$ ways if we ignore isomorphisms of labeled graphs. Any two of these labelings are said to be "equivalent" if one can be obtained from the other by the application of some permutation in $\Gamma(G)$. This is an equivalence relation, and the equivalence classes correspond to different labelings of G. Since there are $\Gamma(G)$ elements in each class, the number of classes must be $p!/|\Gamma(G)|$. (Note that this proof amounts to showing that the different labelings of G correspond exactly to the cosets of the symmetric group S_p with respect to the subgroup $\Gamma(G)$.) ‖

This theorem is illustrated in Fig. 1, which shows the number of ways of labeling all those graphs of order 5 whose vertices are of even valency. Note that the connected graphs in this figure are precisely the Eulerian graphs of order 5. Although simple, Theorem 2.1 is extremely useful for checking results and exploring unfamiliar territory. Furthermore, it can be applied to many different types of graph.

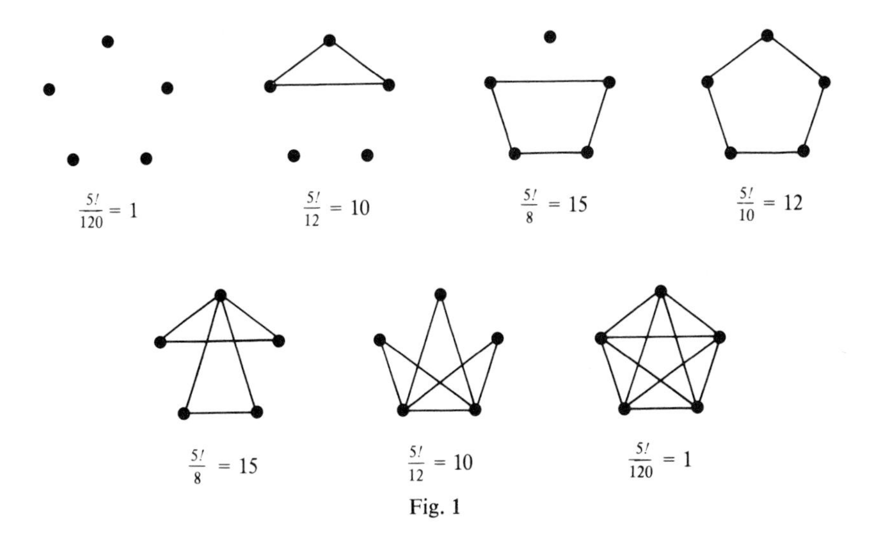

Fig. 1

Many elementary counting results for labeled graphs are immediately accessible because they require little more than a knowledge of the binomial theorem and the multiplication principle. We now give several such formulas for graphs and digraphs.

Theorem 2.2. *The number G_p of labeled graphs of order p is* $2^{\binom{p}{2}}$.

Proof. The result follows from the fact that each of the $\binom{p}{2}$ possible edges can be either present or absent. ‖

Theorem 2.3. *The number $G_{p,q}$ of labeled graphs with p vertices and q edges is*

$$\binom{\binom{p}{2}}{q}.$$

Proof. This follows since each q-element subset of the $\binom{p}{2}$ possible edges determines a different labeled graph. ‖

The results of these theorems may be summarized by introducing the idea of a "generating function". If we consider the polynomial $(1+x)^{\binom{p}{2}}$, then the coefficient of x^q in this polynomial is simply $G_{p,q}$. More generally, if we have a sequence of numbers $s_0, s_1, s_2, \ldots$, then the **generating function** of the sequence is the function S defined by

$$S(x) = s_0 + s_1 x + s_2 x^2 + \ldots .$$

There are numerous advantages to be gained by using generating functions—the notation is often greatly simplified, several different results may be compressed into a single formula, and a substantial number of convenient manipulations can be applied. For example, the separate terms in the expression $1+x$ can be interpreted as standing either for the absence of an edge (1), or for the presence of an edge (x). This suggests that $1+x+x^2$ carries the additional possibility of two edges joining a pair of vertices, so that $(1+x+x^2)^{\binom{p}{2}}$ is the generating function for multigraphs with at most two edges joining each pair of vertices. By using the identity

$$(1-x)^{-1} = 1+x+x^2+\ldots,$$

this idea can be readily extended to give the following result:

Theorem 2.4. *The number $M_{p,q}$ of labeled multigraphs with p vertices and q edges is*

$$(-1)^q \binom{-\binom{p}{2}}{q}.$$

Proof. The number $M_{p,q}$ is the coefficient of x^q in $(1-x)^{-\binom{p}{2}}$, and the result follows by applying the binomial theorem. ||

We conclude this section by stating without proof some similar results for labeled digraphs, tournaments, and bipartite graphs.

Theorem 2.5. (*i*) *The number D_p of labeled digraphs of order p is $2^{p(p-1)}$;*

(*ii*) *the number $D_{p,q}$ of labeled digraphs with p vertices and q arcs is*

$$\binom{p(p-1)}{q};$$

(*iii*) *the number T_p of labeled tournaments of order p is $2^{\binom{p}{2}}$;*

(*iv*) *the number $B_{r,s}$ of labeled bipartite graphs in which the partite sets have sizes r and s is 2^{rs}.* ||

3. Labeled Connected Graphs

In this section we shall show that the number of graphs of a given order can be expressed in terms of the number of connected components of given order. By inverting this relationship, we shall obtain a recurrence relation for the number of connected graphs of given order. This approach is most easily implemented by introducing the concept of a "rooted graph", in which one of the vertices is distinguished as the "root vertex". We may think of the root vertex as being colored black, while the other vertices are colored white. The following recurrence relation for the number of labeled connected graphs is found by counting these rooted graphs in two different ways:

Theorem 3.1. *The number C_p of labeled connected graphs of order p is given by*

$$C_p = 2^{\binom{p}{2}} - \frac{1}{p} \sum_{k=1}^{p-1} k \binom{p}{k} 2^{\binom{p-k}{2}} C_k.$$

Proof. Each labeled graph of order p can be rooted at any one of its vertices, and so the number of rooted graphs of order p is pG_p. On the other hand, the root vertex must lie in some component with k vertices, where $1 \leqslant k \leqslant p$. There are $\binom{p}{k}$ ways of selecting the labels for this component, C_k possible components, and each component can be rooted at any of its k vertices. The remaining $p-k$ vertices can induce G_{p-k} different labeled graphs. On multiplying these factors, and summing over k, we have

$$pG_p = \sum_{k=1}^{p} k \binom{p}{k} G_{p-k} C_k. \tag{1}$$

(Note that we are using $G_0 = 1$ here.) The formula in the theorem now follows by solving this equation for C_p, and substituting the values of G_{p-k} and G_p from Theorem 2.2. ‖

There are several formulas like equation (1) which can be used to enumerate types of connected graph in terms of the corresponding non-connected graphs; for example, one can enumerate Eulerian graphs (see the next section), connected digraphs, oriented graphs, and multigraphs. However, some caution should be exercised because such formulas cannot always be applied (see [**14**, p. 17]).

The relationship between graphs and connected graphs can be neatly expressed using the "exponential generating functions" G and C defined by

$$G(y) = \sum_{p=1}^{\infty} \frac{G_p y^p}{p!}, \qquad C(y) = \sum_{p=1}^{\infty} \frac{C_p y^p}{p!}.$$

Theorem 3.2. *The exponential generating functions G and C satisfy the equation*

$$1+G(y) = \exp C(y). \tag{2}$$

Proof. This result, first found by Riddell **[23]**, can be quickly verified by differentiating both sides of (2) to give

$$G'(y) = C'(y)\exp C(y) = C'(y)(1+G(y)).$$

By comparing the coefficients of y^{p-1} on each side of this equation, we obtain equation (1), which is already known to be true.

However, this routine verification sheds no light on the derivation of (2). Here lies the advantage in using exponential generating functions. One observes that $C^k(y)/k!$ is the exponential generating function for labeled graphs with exactly k components. Summing over all $k \geqslant 0$ gives $\exp C(y)$ on the one hand, and $1+G(y)$ on the other. ‖

Note that the left-hand side of (2) could be simplified slightly to $G(y)$, if $G(y)$ were allowed to include the empty graph (see Fig. 2). This point of view has its advantages, and is found in the literature from time to time (see **[15]**).

Fig. 2

The problem of counting labeled connected graphs with a given number of edges is also most easily approached by using generating functions. Let

$$G_p(x) = (1+x)^{\binom{p}{2}}, \qquad \text{and} \qquad C_p(x) = \sum_{q=p-1}^{\binom{p}{2}} C_{p,q}x^q,$$

where $C_{p,q}$ is the number of labeled connected graphs with p vertices and q edges. Then $G_p(x)$ and $C_p(x)$ are the ordinary generating functions for labeled graphs and connected graphs. We now define the exponential generating functions $G(x, y)$ and $C(x, y)$ by

$$G(x, y) = \sum_{p=1}^{\infty} \frac{G_p(x)\,y^p}{p!}, \qquad \text{and} \qquad C(x, y) = \sum_{p=1}^{\infty} \frac{C_p(x)\,y^p}{p!}.$$

Theorem 3.3. *The exponential generating functions $G(x, y)$ and $C(x, y)$ satisfy the equation*

$$1 + G(x, y) = \exp C(x, y). \tag{3}$$

Proof. As in the proof of Theorem 3.2, we observe that $C^k(x, y)/k!$ counts the labeled graphs with exactly k components. On summing over all $k \geqslant 0$, we obtain $\exp C(x, y)$ which is also equal to $1 + G(x, y)$. ‖

Note that equation (3) reduces to equation (2) if we set $x = 1$. Theorem 3.3 can be used to calculate numerical values for $C_{p,q}$, by deriving a recurrence relation. This is accomplished by taking the partial derivative of (3) with respect to y, and comparing the coefficients of $x^q y^{p-1}$ on both sides. These manipulations are routine, but the resulting formula is a bit more complicated than that of Theorem 3.1.

Corollary 3.4. *The number $C_{p,q}$ of labeled connected graphs with p vertices and q edges is given by*

$$\binom{\binom{p}{2}}{q} - \frac{1}{p} \sum_{k=1}^{p-1} k \binom{p}{k} \sum_{m=0}^{q} \binom{\binom{p-k}{2}}{m} C_{k,q-m}. \; ‖$$

(This formula has been purposely written in such a way as to reveal the alternative direct proof which is similar to that of Theorem 3.1.)

4. Labeled Eulerian Graphs

The reader will recall from Chapter 1 that a connected graph is said to be **Eulerian** if it contains a closed walk which uses each edge exactly once, and that this is the case if and only if every vertex of the graph has even valency. Similarly, a connected digraph is said to be **Eulerian** if it contains a directed closed walk which uses each arc exactly once, and that this is the case if and only if the in-valency of each vertex is equal to its out-valency. Some of the nicest results in graphical enumeration (as well as several difficult unsolved problems) are associated with Eulerian graphs and digraphs, and it is to these that we now turn our attention.

In order to count labeled Eulerian graphs, we first enumerate even graphs—that is, graphs (not necessarily connected) in which every vertex has even valency. Since the Eulerian graphs are precisely the connected even graphs, we can use a formula similar to that given in the proof of Theorem 3.1 to complete the solution.

The first step in this process is given by the following rather surprising result (see [**21**]):

Theorem 4.1. *The number E_p of labeled even graphs of order p is equal to $2^{\binom{p-1}{2}}$, the number of labeled graphs of order $p-1$.*

Proof. We shall describe a one-to-one correspondence between the labeled even graphs of order p and the labeled graphs of order $p-1$. Since the latter number is equal to $2^{\binom{p-1}{2}}$, the result will then follow immediately. In order to establish such a correspondence, we let G be any graph of order $p-1$ whose vertices are labeled with the integers $1, 2, \ldots, p-1$, and we add a new vertex with label p, adjacent to all those vertices of G which have odd valency. As noted by Euler himself (see [**2**, p. 7]), these vertices will be even in number and so the graph constructed will be an even graph of order p, thereby completing the proof. ||

Using this result, together with the method of Theorem 3.1, we immediately obtain the following recurrence relation for the number U_p of labeled Eulerian graphs of order p:

Theorem 4.2. *The number U_p of labeled Eulerian graphs of order p is given by*

$$U_p = 2^{\binom{p-1}{2}} - \frac{1}{p}\sum_{k=1}^{p-1} k\binom{p}{k} 2^{\binom{p-k-1}{2}} U_k. \quad \|$$

Using this recurrence relation, we can easily calculate the first few values of U_p, and the results are displayed in Table 1. The value $U_5 = 38$ can easily be checked by adding together the last four numbers in Fig. 1.

Table 1 Labeled Eulerian graphs

p	1	2	3	4	5	6
U_p	1	0	1	3	38	680

Theorem 4.1 is actually a special case of some other rather interesting and useful results. For example, it provides the initial step for counting labeled graphs with a given number of vertices of odd valency, as we now show:

Theorem 4.3. *The number of labeled graphs of order p with k vertices of odd valency is* $2^{\binom{p-1}{2}}\binom{p}{k}$.

Proof. This result can be proved mechanically by constructing the relevant generating function, and deriving a recurrence relation. Alternatively (and more revealingly), we may observe that the number $2^{\binom{p-1}{2}}\binom{p}{k}$ is the number of even graphs in which k of the vertices are painted white, and the rest are painted black. We can now establish a one-to-one correspondence between these 2-colored even graphs and the graphs with k vertices of odd valency, by replacing the subgraph induced by the white vertices by its complement; the valency of each black vertex remains even, while the white vertices become odd, and the result follows. ||

We can consider Theorem 4.1 as giving the number of labeled even subgraphs of the complete graph K_p. Our next theorem gives a generalization of this result:

Theorem 4.4. *If G is a graph with p vertices and q edges, then there are precisely* 2^{q-p+1} *labeled even subgraphs of order p in G.*

Remark. If $G = K_p$, then $q-p+1 = \binom{p-1}{2}$, and the result reduces to that of Theorem 4.1.

Proof. If H is a labeled even subgraph of order p, then H can be expressed uniquely as the symmetric difference of circuits in a circuit-basis of G. But the number of linearly independent circuits in such a basis is $q-p+1$, and so the total number of even subgraphs of G which can be made up from these circuits is 2^{q-p+1}, as required. ||

Theorem 4.4 has many other interesting applications. For example, it can be used to find the number of even subgraphs of a complete k-partite graph, the n-cube, or an $m \times n$ lattice in the plane or on the torus.

We now turn to the problem of counting labeled Eulerian graphs with a given number of edges as well as vertices. We start with an explicit formula for counting even graphs with q edges, which followed from a bright idea of Read (see [**14**, p. 13], [**21**]):

Theorem 4.5. *The number E(p, q) of labeled even graphs with p vertices and q edges is*

$$\frac{1}{2^p}\sum_{n=0}^{p}\binom{p}{n}\sum_{k=0}^{q}(-1)^k\binom{n(p-n)}{k}\binom{m}{q-k},$$

where

$$m = \binom{n}{2}+\binom{p-n}{2}.$$

Proof. Read's clever idea immediately reveals the proper route to the solution, and so we need only outline the proof. Let S be the set of all 2^p ways in which a positive or negative charge can be assigned to each of the p vertices. A positive or negative charge can then be assigned to each edge by multiplying the charges of its incident vertices. Finally, each "charged graph" G is given the value $\sigma(G) = +1$ or -1, equal to the product of the charges on its edges.

Now let $\mathscr{L}$ be the set of all labeled graphs with p vertices and q edges. Then

$$2^p E(p, q) = \sum_{G \in \mathscr{L}} \sum_{S} \sigma(G), \tag{4}$$

since $\sigma(G) = 1$ whenever G is an even graph, and since all non-even graphs come in oppositely-charged pairs. The proof is completed by reversing the double sum on the right-hand side of (4); the result is the expression in the statement of the theorem. (The following interpretation may be helpful: given an assignment of charges to the vertices, let n be the number of negatively-charged vertices; then $\binom{p}{n}$ is the number of ways of labeling them, and k is the number of edges joining vertices with opposite charges.) ‖

This formula for $E(p, q)$ can also be conveniently expressed in terms of generating functions (see **[21]**):

Corollary 4.6. *The number $E(p, q)$ of labeled even graphs with p vertices and q edges is the coefficient of x^q in*

$$\frac{1}{2^p}(1+x)^{\binom{p}{2}} \sum_{n=0}^{p} \binom{p}{n} \left(\frac{1-x}{1+x}\right)^{n(p-n)}. \; \|$$

In order to obtain the number of labeled Eulerian graphs we imitate Corollary 3.4, giving the following recurrence relation:

Corollary 4.7. *The number $U(p, q)$ of labeled Eulerian graphs with p vertices and q edges is given by*

$$\left(\binom{\binom{p}{2}}{q}\right) - \frac{1}{p} \sum_{k=1}^{p-1} k \binom{p}{k} \sum_{m=0}^{q} \left(\binom{\binom{p-k}{2}}{m}\right) U(k, q-m). \; \|$$

Read **[21]** has also treated similar problems for multigraphs with and without loops.

5. Labeled Trees

In this section we shall digress a little to present a famous classical result of Cayley [**8**]; this was first stated in 1889, although related results had been obtained somewhat earlier. In order to illustrate the variety of techniques that are used in this area, we shall outline three proofs of Cayley's theorem. For further proofs, and a discussion of related ideas, the reader is referred to Moon's survey paper in [**13**], or to his book [**17**].

Theorem 5.1 (Cayley's Theorem). *The number t_p of labeled trees of order p is given by $t_p = p^{p-2}$.*

First Proof (Prüfer [**20**], 1918). We shall establish a one-to-one correspondence between the set of labeled trees of order p, and the set of all ordered symbols $(a_1, a_2, \ldots, a_{p-2})$, where each a_i is an integer satisfying $1 \leqslant a_i \leqslant p$. Since there are precisely p^{p-2} such symbols, the result will then follow immediately. We shall assume that $p \geqslant 3$, since the result is trivial if $p \leqslant 2$.

In order to establish the required correspondence, we first let T be a labeled tree of order p, and show how the symbol can be assigned. If b_1 is the smallest label assigned to any of the end-vertices (vertices of valency 1), we let a_1 be the label of the vertex adjacent to the vertex b_1. We then remove the vertex b_1 and its incident edge, leaving a labeled tree of order $p-1$. If we now let b_2 be the smallest label assigned to any of the end-vertices of our new tree, and let a_2 be the label of the vertex adjacent to the vertex b_2, we can then remove the vertex b_2 and its incident edge, as before. Proceeding in this way until there are only two vertices left gives us the required symbol $(a_1, a_2, \ldots, a_{p-2})$. For example, if T is the labeled tree in Fig. 3, then $b_1 = 2$, $a_1 = 6$; $b_2 = 3$, $a_2 = 5$; $b_3 = 4$, $a_3 = 6$; $b_4 = 6$, $a_4 = 5$; $a_5 = 5$, $b_5 = 1$. The required 5-tuple is therefore (6, 5, 6, 5, 1).

In order to establish the reverse correspondence, we take a symbol $(a_1, \ldots, a_{p-2})$, let b_1 be the smallest number which does *not* appear in it, and join the vertices a_1 and b_1. We then remove a_1 from the symbol, and remove the number b_1 from consideration, and proceed as before. In this way we can

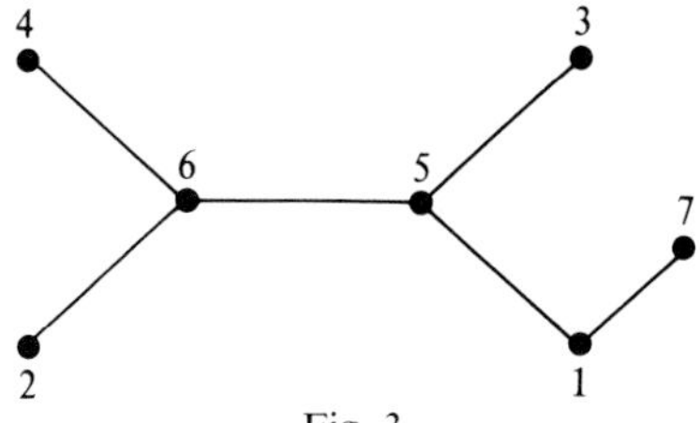

Fig. 3

build up the tree, edge by edge. For example, if we start with the symbol (6, 5, 6, 5, 1), then $b_1 = 2$, $b_2 = 3$, $b_3 = 4$, $b_4 = 6$, $b_5 = 5$, and the corresponding edges are 62, 53, 64, 56, 15; we conclude by joining the last two vertices not yet crossed out—in this case, 1 and 7.

It is a straightforward matter to check that if we start with any labeled tree, find the corresponding symbol, and then find the labeled tree corresponding to that symbol, then we always obtain the tree we started from. The required correspondence is therefore established, and the result follows. ||

Second Proof (Clarke [**9**], 1958). Let $t(p, k)$ be the number of labeled trees of order p, in which a given vertex v has valency k. We shall derive an explicit expression for $t(p, k)$, and the result will follow on summing from $k = 1$ to $k = p-1$.

Let A be any labeled tree in which v has valency $k-1$. The removal from A of any edge wz not incident to v leaves two subtrees, one of which contains v and either w or z (say, w), and the other of which contains z. If we join the vertices v and z, we obtain a labeled tree B in which v has valency k, and we call a pair of trees (A, B) related in this way a **linkage**. Our aim is to count the total number of linkages (A, B).

Since A may be chosen in $t(p, k-1)$ ways, and B is uniquely determined by the edge wz which may be chosen in $(p-1)-(k-1) = p-k$ ways, the total number of linkages (A, B) is clearly $(p-k)t(p, k-1)$. On the other hand, let B be a labeled tree in which v has valency k, and let $T_1, \ldots, T_k$ be the subtrees obtained from B by deleting the vertex v and its incident edges. Then we can obtain a labeled tree A in which v has valency $k-1$ by removing from B just one of these edges (vw_i, say, where w_i lies in T_i), and joining w_i to any vertex u of any other subtree T_j (see Fig. 4). It is clear that the corresponding pair (A, B) of labeled trees is a linkage, and that all linkages may be obtained in this way. Since B may be chosen in $t(p, k)$ ways, and the

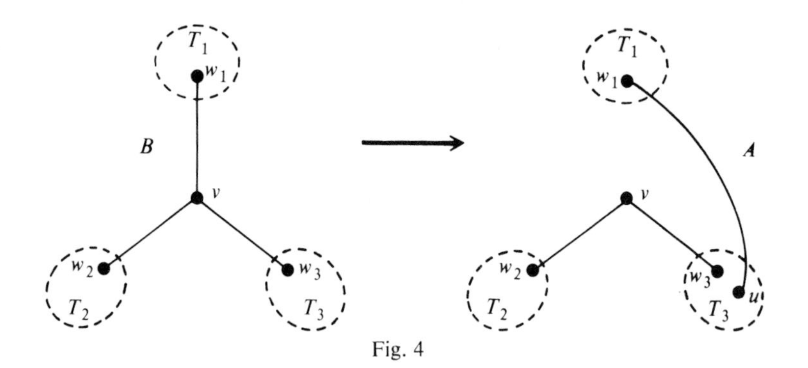

Fig. 4

number of ways in which w_i can be joined to a vertex in any other T_j is $(p-1)-p_i$ (where p_i is the order of T_i), it follows that the total number of linkages (A, B) is

$$t(p, k)\{(p-1-p_1)+\ldots+(p-1-p_k)\} = (p-1)(k-1)t(p, k),$$

since $p_1+\ldots+p_k = p-1$. We have thus shown that

$$(p-k)t(p, k-1) = (p-1)(k-1)t(p, k).$$

On iterating this result, and using the obvious fact that $t(p, p-1) = 1$, we deduce that

$$t(p, k) = \binom{p-2}{k-1}(p-1)^{p-k-1}.$$

On summing over all possible values of k, we deduce that

$$\begin{aligned} t_p &= \sum_{k=1}^{p-1} t(p, k) = \sum_{k=1}^{p-1} \binom{p-2}{k-1}(p-1)^{p-k-1} \\ &= \{(p-1)+1\}^{p-2} = p^{p-2}. \, \| \end{aligned}$$

Third Proof. In this proof, we shall use the multinomial cofficients, defined by

$$\binom{n}{r_1, \ldots, r_k} = \frac{n!}{r_1!r_2!\ldots r_k!} \quad \text{if} \quad r_1+\ldots+r_k = n, \text{ and } 0 \text{ otherwise.}$$

We first prove that if $t(p; \rho_1, \ldots, \rho_p)$ is the number of labeled trees of order p, in which the vertex-valencies are $\rho_1, \rho_2, \ldots, \rho_p$, then

$$t(p; \rho_1, \ldots, \rho_p) = \binom{p-2}{\rho_1-1, \rho_2-1, \ldots, \rho_p-1}.$$

We may assume that $\rho_1 \geqslant \rho_2 \geqslant \ldots \geqslant \rho_p$, so that $\rho_p = 1$, and we proceed by induction on p.

Since the vertex corresponding to ρ_p $(=1)$ can be joined to any vertex of valency 2 or more, it is clear that

$$t(p; \rho_1, \ldots, \rho_p) = \sum t(p-1; \rho_1, \rho_2, \ldots, \rho_i-1, \ldots, \rho_{p-1}),$$

where the summation extends over all i for which $\rho_i \geqslant 2$. But by induction,

$$\begin{aligned} &t(p-1; \rho_1, \rho_2, \ldots, \rho_i-1, \ldots, \rho_{p-1}) \\ &\qquad = \binom{p-3}{\rho_1-1, \rho_2-1, \ldots, \rho_i-2, \ldots, \rho_{p-1}-1}, \end{aligned}$$

so that

$$t(p;\rho_1,\ldots,\rho_p) = \sum_{i;\rho_i\geq 2}\binom{p-3}{\rho_1-1,\rho_2-1,\ldots,\rho_i-2,\ldots,\rho_{p-1}-1}$$

$$= \binom{p-2}{\rho_1-1,\rho_2-1,\ldots,\rho_{p-1}-1},$$

by a well-known identity involving multinomial coefficients (see [**1**, p. 33]). But this is equal to

$$\binom{p-2}{\rho_1-1,\rho_2-1,\ldots,\rho_p-1},$$

since $\rho_p = 1$, and the formula is established.

To conclude the proof of Cayley's theorem, we use the multinomial theorem

$$(a_1+\ldots+a_p)^n = \sum \binom{n}{n_1,\ldots,n_p} a_1^{n_1}a_2^{n_2}\ldots a_p^{n_p},$$

where the summation extends over all non-negative integers n_i satisfying $n_1+\ldots+n_p = n$. In order to apply this theorem, we put $n = p-2$, $a_i = 1$, and $n_i = \rho_i-1$. The right-hand side enumerates all labeled trees (note that $\sum n_i = p-2$, since $\sum \rho_i = 2p-2$), and the left-hand side is simply

$$(1+\ldots+1)^{p-2} = p^{p-2},$$

as required. ‖

6. Unlabeled Graphs

We now turn our attention to unlabeled problems. A variety of such problems can be solved by interpreting the answer as the number of equivalence classes determined by the action of an appropriate permutation group. For example, if X is the set of all $2^{\binom{p}{2}}$ labeled graphs of order p, then the symmetric group S_p of order $p!$ can be represented on X in a natural way—namely, each permutation α acts on the graph by permuting the labels of the individual vertices.

Let A be a permutation group on a finite object set X. For each element x in X, let $A(x)$ consist of all those permutations α in A which leave x fixed (the "stabilizer" of x). The *orbit* (or *transitivity class*) of x, denoted by $\mathcal{O}(x)$, consists of all elements y in X such that $\alpha(x) = y$ for some permutation α in A. In the example of the previous paragraph, each orbit consists of the different labelings of some unlabeled graph.

The number of orbits of a permutation group can be expressed as the average number of objects fixed by the permutations; this is an old result of Cauchy and Frobenius, sometimes called "Burnside's lemma" [**4**, p. 191], and it permits the efficient solution of many enumeration problems:

Theorem 6.1. *If A is a permutation group with finite object set X, then the number $N(A, X)$ of orbits of A is given by*

$$N(A, X) = \frac{1}{|A|} \sum_{\alpha \in A} N(\alpha),$$

where $N(\alpha)$ is the number of objects of X fixed by α.

Proof. We first express A in terms of its cosets modulo the subgroup $A(x)$:

$$A = A(x) \cup \alpha_2 A(x) \cup \ldots \cup \alpha_n A(x).$$

But these cosets correspond precisely to the elements of $\mathcal{O}(x)$—that is,

$$\mathcal{O}(x) = \{x, \alpha_2(x), \ldots, \alpha_n(x)\}.$$

Hence, for any x in X,

$$|A| = |A(x)||\mathcal{O}(x)|. \tag{5}$$

We now select one element of X from each of the orbits of A to obtain the set $\{x_1, \ldots, x_m\}$, where $m = N(A, X)$. On substituting each x_i for x in (5), and adding, we find

$$m|A| = \sum_{i=1}^{m} |A(x_i)||\mathcal{O}(x_i)|. \tag{6}$$

Since $|A(x)| = |A(y)|$ whenever x and y are in the same orbit of A, we get

$$\sum_{i=1}^{m} |A(x_i)||\mathcal{O}(x_i)| = \sum_{x \in X} |A(x)|. \tag{7}$$

But clearly

$$\sum_{x \in X} |A(x)| = \sum_{\alpha \in A} N(\alpha), \tag{8}$$

and the result follows by combining (6), (7) and (8), and dividing by $|A|$. ‖

In order to count graphs of order p, we take X to be the set of labeled graphs of order p, and A to be the representation of S_p which permutes these labeled graphs by permuting their labels. So the number of orbits of A is precisely the number of unlabeled graphs of order p. On applying Theorem 6.1, we can express this number g_p in the form

$$g_p = \frac{1}{p!} \sum_{\alpha \in S_p} N(\alpha),$$

where $N(\alpha)$ is the number of labeled graphs which are fixed or unaffected by the change of labels under the action of α. These are listed in Table 2 for all five types of permutations of four vertices. The middle row shows the number $h(\alpha)$ of different permutations with the same disjoint cycle decomposition as α. On applying Burnside's lemma, we get

$$g_4 = \frac{1}{4!}(2^6+6.2^4+8.2^2+3.2^4+6.2^2) = 11.$$

Table 2

α	(1)(2)(3)(4)	(1)(2)(34)	(1)(234)	(12)(34)	(1234)
$h(\alpha)$	1	6	8	3	6
$N(\alpha)$	2^6	2^4	2^2	2^4	2^2

The key idea used in calculating $N(\alpha)$ is that the vertices labeled v and w are adjacent in a graph fixed by α if and only if those labeled $\alpha^k(v)$ and $\alpha^k(w)$ are adjacent for every value of k. One can then select v and w from the same cycle of α, or from different cycles, and pursue the various possibilities.

In order to illustrate this procedure, we shall verify the entries in the last row of Table 2.

Case 1. $\alpha = (1)(2)(3)(4)$.

Since α is the identity permutation, each labeled graph is left fixed by α, and hence $N(\alpha) = G_4 = 2^6$.

Case 2. $\alpha = (1)(2)(34)$.

The vertices labeled 3 and 4 may or may not be adjacent in a labeled graph fixed by α, giving rise to two possibilities. Similarly, the vertices labeled 1 and 2 may or may not be adjacent, giving rise to two further possibilities. If the vertices labeled 1 and 3 are adjacent, then $\alpha(1)$ and $\alpha(3)$ (that is, 1 and 4) must also be adjacent, giving two more possibilities, namely that 1 is adjacent to both 3 and 4, or to neither. Finally, a similar argument applies to the vertices labeled 2 and 3, giving another two possibilities. So

$$N(\alpha) = 2.2.2.2 = 2^4.$$

Case 3. $\alpha = (1)(234)$.

If the vertices labeled 2 and 3 are adjacent in a labeled graph fixed by α, then $\alpha(2)$ and $\alpha(3)$ (that is, 3 and 4) must also be adjacent, and so must $\alpha^2(2)$ and $\alpha^2(3)$ (that is, 4 and 2). Hence the vertices 2, 3 and 4 must either form a triangle, or must be mutually non-adjacent, giving rise to two possibilities. If the vertices labeled 1 and 2 are adjacent, then so are $\alpha(1)$ and $\alpha(2)$ (that is, 1 and 3), and $\alpha^2(1)$ and $\alpha^2(2)$ (that is, 1 and 4). Hence the vertex 1 is adjacent to all, or to none, of the vertices 2, 3 and 4, giving two more possibilities. So $N(\alpha) = 2.2 = 2^2$.

Case 4. $\alpha = (12)(34)$.

The vertices labeled 1 and 2 may or may not be adjacent, and similarly for the vertices labeled 3 and 4. These give rise to 2.2 possibilities. If the vertices 1 and 3 are adjacent, then the vertices $\alpha(1)$ and $\alpha(3)$ (that is, 2 and 4) must also be adjacent. Hence edges joining the vertices 1 and 3, and the vertices 2 and 4, are either both present or both absent, yielding two more possibilities. A similar situation holds for edges joining the vertices 1 and 4, and the vertices 2 and 3. So $N(\alpha) = 2.2.2.2 = 2^4$.

Case 5. $\alpha = (1234)$.

If the vertices labeled 1 and 2 are adjacent in a labeled graph fixed by α, then $\alpha(1)$ and $\alpha(2)$ (that is, 2 and 3) must also be adjacent, and so must $\alpha^2(1)$ and $\alpha^2(2)$ (that is, 3 and 4), and $\alpha^3(1)$ and $\alpha^3(2)$ (that is, 4 and 1). So the vertices 1, 2, 3 and 4 must either form a circuit (in that order), or all four edges must be absent, giving rise to two possibilities. Also the edges joining 1 and 3, and $\alpha(1)$ and $\alpha(3)$ (that is, 2 and 4), must either be both present or both absent, yielding two more possibilities. So $N(\alpha) = 2.2 = 2^2$.

Note, for future reference, that in *Case 5*, the latter cycle of two edges is exactly half the length of the other cycle. This is a characteristic of all even cycles, and furthermore these cycles of "half length" contribute exactly 1 to the valency of each of their vertices, when they are included in a fixed graph. Note also that each permutation α induces a permutation α' on the unordered pairs of vertices, and that in order to obtain $N(\alpha)$, we simply raise the number 2 to a power equal to the number of cycles of unordered pairs in α'. The base 2 is used because each of these cycles of pairs can either be included in, or excluded from, a graph fixed by α.

In fact, a general formula for $N(\alpha)$ can be derived, which depends only on the cycle decomposition of α. We let the vector $\mathbf{j} = (j_1, \ldots, j_p)$ represent a partition of p with j_k cycles of length k, for $k = 1, \ldots, p$. Then it is easy to see that there are altogether

$$h(\mathbf{j}) = \frac{p!}{\prod_k k^{j_k} j_k!}$$

permutations in S_p with j_k cycles of length k; each term of the form k^{j_k} arises from the fact that a k-cycle can start with any of the k elements, and each term of the form $j_k!$ arises from the fact that the k-cycles may be permuted in $j_k!$ ways. It follows that the number $N(\mathbf{j})$ of labeled graphs fixed by a permutation α with j_k cycles of length k, for each k, takes the form

$$N(\mathbf{j}) = 2^{g(\mathbf{j})},$$

where $g(\mathbf{j})$ is the number of cycles of unordered pairs in the permutation α' induced by α. This accounts for the form of the expression for the number of graphs in the following theorem of Pólya (see [**13**, p. 34]):

Theorem 6.2. *The number g_p of graphs of order p is given by*

$$g_p = \frac{1}{p!}\sum_{\mathbf{j}} h(\mathbf{j})2^{g(\mathbf{j})},$$

where

$$g(\mathbf{j}) = \sum_k \left(kj_{2k+1} + kj_{2k} + k\binom{j_k}{2}\right) + \sum \gcd(r, t)j_r j_t,$$

and the last summation extends over all pairs r, t with $r < t$.

Proof. The main formula of the theorem follows from an application of Burnside's lemma, and our remarks immediately preceding the statement of the theorem. It remains to show that $g(\mathbf{j})$ is the correct number of cycles of unordered pairs.

Let α be a permutation with partition $\mathbf{j} = (j_1, \ldots, j_p)$, so that, for each k, α has j_k cycles of length k. The permutation α' is induced by α on the unordered pairs $\{v, w\}$ of vertices labeled by the set $\{1, \ldots, p\}$. There are two types of cycles to consider—those cycles in α' consisting of unordered pairs of vertices which are all in the same cycle of α (*Case 1*), and those cycles in α' consisting of unordered pairs of vertices which are in different cycles of α (*Case 2*); we obtain the desired expression for $g(\mathbf{j})$ by considering these cases in turn:

Case 1. Let z be a cycle of length m in α; without loss of generality, we can assume that $z = (1, \ldots, m)$. There are $\binom{m}{2}$ unordered pairs of vertices in this cycle. If m is odd, these unordered pairs will be permuted in cycles of length m, and so there will be exactly $\frac{1}{2}(m-1)$ of them. On summing over all cycles of odd length in α, we obtain the first term $\sum_k kj_{2k+1}$ of $g(\mathbf{j})$. If m is even, the unordered pair $\{1, 1+\frac{1}{2}m\}$ is part of a special cycle of length $\frac{1}{2}m$, whereas all the other cycles of pairs have length m. (Compare this with *Case 5* for $\alpha = (1234)$, above.) It follows that there are $\frac{1}{2}m$ cycles altogether, including the one of "half length". If we now sum over all cycles of even length in α, we obtain the second term $\sum_k kj_{2k}$ of $g(\mathbf{j})$.

Case 2. Let z_r and z_t be cycles of length r and t in α. There are rt unordered pairs of vertices with one vertex in each of these cycles. Each unordered pair is permuted by α' in a cycle of length $\operatorname{lcm}(r, t)$, and so the number of pairs here is simply $\gcd(r, t)$. When the two cycles z_r and z_t have the same length (k, say), there are $\binom{j_k}{2}$ sets of different cycles, and each produces k cycles of unordered pairs. Summing over all possible values of k gives the third term $\sum_k k\binom{j_k}{2}$ of $g(\mathbf{j})$. Finally, when $r < t$, there are $j_r j_t$ pairs of such cycles in α, giving rise to $\gcd(r, t)\, j_r j_t$ cycles of unordered pairs in α', and this accounts for the final contribution $\sum_{r<t} \gcd(r, t)\, j_r j_t$. ‖

The corresponding formula for digraphs is very similar, and was found by Harary [**12**]. We need only determine the number of cycles of *ordered* pairs in a permutation α'' induced by α. Each cycle in α' (of unordered pairs) splits into two cycles of ordered pairs, except in one instance—the special cycle of pairs joining opposite vertices in an even cycle of α yields just one self-converse cycle of arcs. It follows that the exponent $d(\mathbf{j})$ in the next theorem is obtained by doubling $g(\mathbf{j})$ and subtracting the number of even cycles in α, and the result takes the following form:

Theorem 6.3. *The number d_p of digraphs of order p is given by*

$$d_p = \frac{1}{p!} \sum_{\mathbf{j}} h(\mathbf{j}) 2^{d(\mathbf{j})},$$

where $d(\mathbf{j}) = 2g(\mathbf{j}) - \sum_k j_{2k}$. ‖

The values of g_p (for $p \leqslant 24$) and d_p (for $p \leqslant 8$), computed from the formulas of this section are tabulated in [**14**, pp. 241–242]. The first few of these values are given in Table 3.

Table 3 Graphs and digraphs

p	1	2	3	4	5	6	7
g_p	1	2	4	11	34	156	1044
d_p	1	3	16	218	9608	1 540 944	882 033 440

7. Unlabeled Connected Graphs

Just as in the labeled case, the relationship between the number of unlabeled graphs and the number of unlabeled connected graphs can be concisely expressed using generating functions. However, products of ordinary generating functions for unlabeled graphs cannot be interpreted in exactly the same manner as was used in the labeled case. Because of this, the statement of Theorem 7.1, due to Riddell ([**23**]; see also [**14**]), bears only a superficial resemblance to that of Theorem 3.2 for labeled connected graphs.

Let c_p be the number of connected graphs of order p, and let g and c be the ordinary generating functions defined by

$$g(y) = \sum_{p=1}^{\infty} g_p y^p, \qquad c(y) = \sum_{p=1}^{\infty} c_p y^p.$$

We then have the following result:

Theorem 7.1. *The ordinary generating functions g and c satisfy the equation*

$$1+g(y) = \exp\left(\sum_{k=1}^{\infty} \frac{c(y^k)}{k}\right).$$

Proof. We first observe that the generating function for graphs in which each component is of order n is simply

$$(1+y^n+y^{2n}+\ldots)^{c_n} = (1-y^n)^{-c_n}.$$

It follows that the product of this expression over all n enumerates graphs with components of arbitrary size, and hence that

$$\begin{aligned}1+g(y) &= \prod_{n=1}^{\infty} (1-y^n)^{-c_n} \\ &= \exp\left(\sum_{n=1}^{\infty} -c_n \log(1-y^n)\right) \\ &= \exp\left(\sum_{n=1}^{\infty} \sum_{k=1}^{\infty} \frac{c_n y^{nk}}{k}\right).\end{aligned}$$

The result follows by interchanging the order of summation inside the parentheses, and observing that, for each k,

$$\sum_{n=1}^{\infty} \frac{c_n (y^k)^n}{k} = \frac{c(y^k)}{k}. \quad \|$$

Implicit in the statement of Theorem 7.1 is an effective method for calculating the values of c_p, if the numbers g_p are known:

Corollary 7.2. *If c_p and g_p are as defined above, and if $g_0 = 1$, then*

$$g_p = \frac{1}{p} \sum_{k=1}^{p} \left(\sum_{d|k} d c_d\right) g_{p-k}.$$

Proof. The result follows by differentiating the result of Theorem 7.1, and comparing the coefficients of y^{p-1}. ‖

It is not difficult to obtain an explicit formula for c_p. It was shown by Cadogan [5] that if a_k is the coefficient of y^k in $\log(1+g(y))$, then

$$c_p = \sum_{d|p} \frac{\mu(d)}{d} a_{p/d},$$

where μ is the classical Möbius function. As one might expect, the proof involves Möbius inversion, and the details are given in [**14**, pp. 91–92]. The values obtained are listed in Table 4.

Table 4 Graphs and connected graphs

p	1	2	3	4	5	6	7
g_p	1	2	4	11	34	156	1044
c_p	1	1	2	6	21	112	853

Just as in Section 3, one can imitate the above treatment, using generating functions of two variables, to obtain expressions for the number of connected graphs with p vertices and q edges. For details, the reader is referred to [**5**] or [**14**, pp. 92–93].

8. Unlabeled Eulerian Graphs

Just as in the labeled case, the enumeration of unlabeled Eulerian graphs is most easily accomplished by first counting even graphs. This was first completed by Robinson (see [**24**], [**14**, p. 113]), using a clever modification of the formula in Theorem 6.2 for graphs.

Theorem 8.1. *The number e_p of even graphs of order p is given by*

$$e_p = \frac{1}{p!} \sum_{\mathbf{j}} h(\mathbf{j}) 2^{e(\mathbf{j})},$$

where $e(\mathbf{j}) = g(\mathbf{j}) - \sum_k j_k + \operatorname{sgn}(\sum_k j_{2k+1})$.

Proof. If X is the set of all labeled even graphs, and if A is the symmetric group S_p, then we can use Theorem 6.1 to deduce that

$$e_p = \frac{1}{p!} \sum_{\mathbf{j}} h(\mathbf{j}) N(\mathbf{j}),$$

where $N(\mathbf{j})$ is the number of even graphs which are fixed by a permutation α with j_k cycles of length k, for $k = 1, \ldots, p$. As before, we can write $N(\mathbf{j}) = 2^{e(\mathbf{j})}$; the proof splits into three cases, according as α has no odd cycles, exactly one odd cycle, or more than one odd cycle:

Case 1. If α has no odd cycles, then the "troublesome cycles" of pairs of opposite vertices induced by all even cycles in α must be excluded, since otherwise each vertex in these cycles would have odd valency in any graph fixed by α. It follows that

$$e(\mathbf{j}) = g(\mathbf{j}) - \sum_k j_{2k}.$$

Case 2. If α has one odd cycle, then the troublesome cycles are no longer so troublesome, since they can be included or excluded as necessary in order to preserve the even valency of each vertex in a graph fixed by α. As in *Case 1*, they must be excluded from $g(\mathbf{j})$, and so

$$e(\mathbf{j}) = g(\mathbf{j}) - \sum_k j_{2k}.$$

Case 3. If α has m odd cycles, where $m > 1$, then we consider a particular odd cycle z, and $m-1$ cycles of pairs each of which has one vertex in z and the other vertex in one of the other odd cycles. Excluding these, and the troublesome cycles, leaves

$$e(\mathbf{j}) = g(\mathbf{j}) - \sum_k j_{2k} - (m-1),$$

for inclusion or exclusion. Once these selections have been made, then the troublesome cycles and (independently) the $m-1$ other odd cycles can be included or excluded in order to correct the parity. Note that after this has been done, every vertex in the graph fixed by α will have even valency, including the vertices of z, because there are an odd number of them. ||

Just as in Theorem 4.3, it is quite straightforward to modify the above result to handle the enumeration of graphs with a given number of vertices of odd valency. The calculation essentially involves counting even graphs with an even number of black vertices and the rest white.

In order to go from counting even graphs to counting Eulerian graphs, we use the procedures adopted to prove Theorem 7.1 and Corollary 7.2. If e and u are the ordinary generating functions defined by

$$e(y) = \sum_{p=1}^{\infty} e_p y^p, \qquad u(y) = \sum_{p=1}^{\infty} u_p y^p,$$

we obtain the following results:

Theorem 8.2. *The ordinary generating functions e and u satisfy the equation*

$$1+e(y) = \exp\left(\sum_{k=1}^{\infty} \frac{u(y^k)}{k}\right). \;\|$$

Corollary 8.3. *If u_p is the number of Eulerian graphs of order p, and e_p is as above, then*

$$e_p = \frac{1}{p} \sum_{k=1}^{p} \left(\sum_{d|k} du_d\right) e_{p-k}. \;\|$$

Using Möbius inversion, we can invert this formula to give

$$u_p = \sum_{d|p} \frac{\mu(d)}{d} b_{p/d},$$

where b_k is the coefficient of y^k in $\log(1+e(y))$. The first few values of e_p and u_p found by Robinson [23] are listed in Table 5.

Table 5 Even and Eulerian graphs

p	1	2	3	4	5	6	7
e_p	1	1	2	3	7	16	54
u_p	1	0	1	1	4	8	37

9. Pólya's Enumeration Theorem

We come now to one of the most important theorems in the subject—Pólya's enumeration theorem [19]. While Theorem 6.1 can be used to good effect in a variety of enumeration problems, as we have seen in Sections 6–8, the extra power and flexibility of Pólya's theorem enables far more enumeration problems to be solved quickly and efficiently. In particular, the use of Pólya's theorem facilitates the enumeration of graphs and multigraphs with a given number of edges, digraphs with a given number of arcs, trees (both rooted and unrooted), and tournaments, to name but a few (see [12]). Further details of these applications may be found in [14].

The central ideas involved in the use of Pólya's theorem for graphical enumeration are the cycle index of a permutation group, the concept of a figure-counting series, and the pair groups $S^{(2)}$ and $S^{[2]}$. We shall deal with these in turn.

The Cycle Index

Let A be a group of permutations on a finite set X, and for each $\alpha \in A$, let $j_k(\alpha)$ be the number of cycles of length k in the cycle decomposition of α. Let $s_1, \ldots, s_n$ be undefined variables, where n is the number of elements of X. Then the **cycle index** of A, denoted by $Z(A)$, is defined by

$$Z(A) = \frac{1}{|A|} \sum_{\alpha \in A} s_1^{j_1(\alpha)} s_2^{j_2(\alpha)} \ldots s_n^{j_n(\alpha)}.$$

Example. If $A = S_3$, then there are six permutations—namely,

$$(1)(2)(3),\ (12)(3),\ (13)(2),\ (1)(23),\ (123) \text{ and } (132),$$

and the cycle index is therefore

$$\begin{aligned} Z(S_3) &= \tfrac{1}{6}(s_1^3 + s_1 s_2 + s_1 s_2 + s_1 s_2 + s_3 + s_3). \\ &= \tfrac{1}{6}(s_1^3 + 3s_1 s_2 + 2s_3). \end{aligned}$$

Similarly, the cycle indices of S_4 and S_5 are given by

$$Z(S_4) = \tfrac{1}{24}(s_1^4 + 8s_1 s_3 + 6s_1^2 s_2 + 3s_2^2 + 6s_4),$$

and

$$Z(S_5) = \tfrac{1}{120}(s_1^5 + 10s_1^3 s_2 + 20s_1^2 s_3 + 15s_1 s_2^2 + 30s_1 s_4 + 20s_2 s_3 + 24s_5),$$

and more generally we can write

$$Z(S_n) = \frac{1}{n!} \sum_{\mathbf{j}} h(\mathbf{j}) s_1^{j_1} \ldots s_n^{j_n},$$

where $h(\mathbf{j})$ is as defined in Section 6.

The Figure-counting Series

When counting graphs or multigraphs with a given number of edges, we wish to be able to distinguish between those pairs of vertices which are joined by no edges, those which are joined by one edge, and so on. In order to do this, we imitate our ideas in Section 2, and indicate the absence of an edge by 1,the presence of one edge by x, the presence of two edges by x^2, and so on. The idea of this is that any term of the form kx^q which appears in the result of our enumeration problem will tell us that precisely k of the objects we are counting have q edges. In order to push this idea through, we use the idea of a figure-counting series, some examples of which are as follows:

(i) for graphs and digraphs, the figure-counting series is $1+x$; the terms 1 and x refer respectively to the absence, and presence, of an edge;

(*ii*) for multigraphs with at most two edges joining any pair of vertices, the figure-counting series is $1+x+x^2$;

(*iii*) for multigraphs in general, the figure-counting series is $1+x+x^2+\ldots$, or simply $1/(1-x)$.

If $c(x)$ is a figure-counting series, and if $Z(A)$ is the cycle index of some permutation group A, then we define $Z(A, c(x))$ to be the expression we obtain when we replace each variable s_k by $c(x^k)$. For example, if $A = S_3$, and $c(x) = 1+x$, then we have

$$\begin{aligned} Z(S_3, 1+x) &= \tfrac{1}{6}\{(1+x)^3+3(1+x)(1+x^2)+2(1+x^3)\} \\ &= 1+x+x^2+x^3. \end{aligned}$$

More generally, it will turn out that if A is any permutation group, then the coefficient of x^k in $Z(A, 1+x)$ is the number of equivalence classes of k-sets in X, where two k-sets S and S' are called *equivalent* if $\alpha(S) = S'$ for some $\alpha \in A$. Thus if $A = S_n$, then every two k-sets are equivalent, and we get

$$Z(S_n, 1+x) = 1+x+\ldots+x^n.$$

When we come to the enumeration of graphs, we want to choose our group A in such a way that we get the same graph whenever there is a permutation α of the vertices which preserves adjacency. This leads to the idea of a "pair group".

The Pair Group $S_p^{(2)}$

If A is any permutation group on a set X with n elements, the **pair group** $A^{(2)}$ is the permutation group on $X^{(2)}$ (the set of 2-subsets of X) induced by A. This means that each permutation $\alpha \in A$ induces a permutation $\alpha^{(2)}$ in $A^{(2)}$, defined by

$$\alpha^{(2)}(\{x_1, x_2\}) = \{\alpha(x_1), \alpha(x_2)\}.$$

In particular, if $A = S_n$, we get the pair group $S_n^{(2)}$. We can also define the **ordered pair group** $A^{[2]}$ induced by A to be the permutation group on the subset of $X \times X$ consisting of those pairs (x_1, x_2) for which $x_1 \neq x_2$. If $A = S_n$, we get the ordered pair group $S_n^{[2]}$.

Since the groups $S_n^{(2)}$ and $S_n^{[2]}$ are permutation groups, we can calculate their cycle indices. Before stating the results for a general value of n, we shall indicate what is involved by looking at the case $n = 4$. The method we use is to take the terms of $Z(S_4)$, one at a time, and investigate what happens to them as we replace S_4 by $S_4^{(2)}$. As an example, we observe that each term of the form s_2^2 is transformed into one of the form $s_1^2 s_2^2$, as we indicate in Fig. 5.

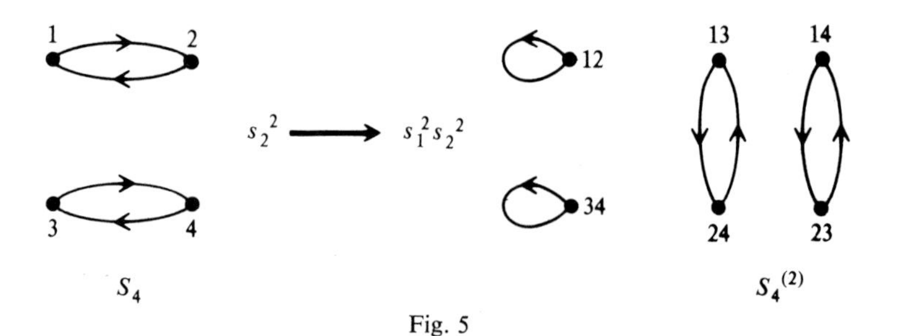

Fig. 5

In the same way, s_1^4 becomes s_1^6, $s_1^2 s_2$ becomes $s_1^2 s_2^2$, $s_1 s_3$ becomes s_3^2, and s_4 becomes $s_2 s_4$. The required cycle index is therefore

$$Z(S_4^{(2)}) = \tfrac{1}{24}(s_1^6 + 8s_3^2 + 9s_1^2 s_2^2 + 6s_2 s_4).$$

Similarly,

$$Z(S_4^{[2]}) = \tfrac{1}{24}(s_1^{12} + 6s_1^2 s_2^5 + 8s_3^4 + 3s_2^6 + 6s_4^3).$$

In investigating what happens for $n > 4$, it is necessary to find out what happens to a typical cycle. In Fig. 6 we show what happens to cycles of lengths $k = 2, 3, 4, 5$ and 6, as we go from S_n to $S_n^{(2)}$; note that if the cycle-length is odd ($2k+1$, say), then we get k cycles of length $2k+1$, and if the cycle-length is even ($2k$, say), then we get $k-1$ cycles of length $2k$ and one cycle of length k. Using arguments of this type one can then prove the following results; the details of the proofs are given in [**14**, pp. 84–87, 121–122].

Theorem 9.1. *The cycle indices of* $S_n^{(2)}$ *and* $S_n^{[2]}$ *are given by*

$$Z(S_n^{(2)}) = \frac{1}{n!}\sum_{\mathbf{j}} h(\mathbf{j}) \prod_k s_{2k+1}^{kj_{2k+1}} \prod_k (s_k s_{2k}^{k-1})^{j_{2k}} s_k^{k\binom{j_k}{2}} \cdot \prod_{r<t} s_{\mathrm{lcm}(r,t)}^{\gcd(r,t) j_r j_t},$$

and

$$Z(S_n^{[2]}) = \frac{1}{n!}\sum_{\mathbf{j}} h(\mathbf{j}) \prod_k s_k^{(k-1)j_k + 2k\binom{j_k}{2}} \cdot \prod_{r<t} s_{\mathrm{lcm}(r,t)}^{2\gcd(r,t) j_r j_t}. \quad \|$$

Pólya's Theorem

Before stating the main results of this section, let us calculate $Z(S_4^{(2)}, 1+x)$, which is obtained by replacing each s_k in $Z(S_4^{(2)})$ by $1+x^k$. We then have

$$\begin{aligned} Z(S_4^{(2)}, 1+x) &= \tfrac{1}{24}\{(1+x)^6 + 8(1+x^3)^2 + 9(1+x)^2(1+x^2)^2 + 6(1+x^2)(1+x^4)\} \\ &= 1 + x + 2x^2 + 3x^3 + 2x^4 + x^5 + x^6. \end{aligned}$$

The reason this is important is that the coefficient of each power x^k is equal to the number of graphs with four vertices and k edges; for example, there are

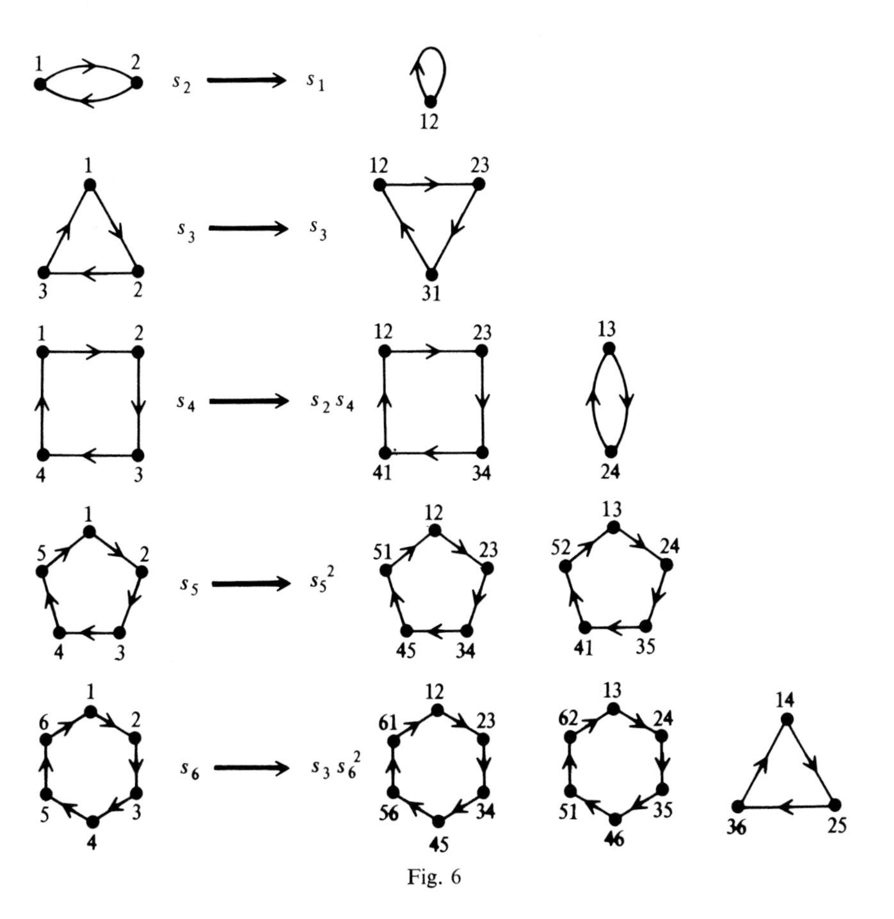

Fig. 6

three graphs with four vertices and three edges. We say that $Z(S_4^{(2)}, 1+x)$ is the **generating function** for graphs of order 4, in the sense that the number of such graphs with q edges is just the coefficient of x^q in this expression. In fact, results of this kind hold much more generally, and in the following theorem we state some typical results (see [**12**]) which give the number of graphs, digraphs and multigraphs of order p; in each case, the number of these with a given number of edges or arcs can be calculated by looking at the corresponding coefficient:

Theorem 9.2. (*i*) *The generating function for graphs of order p is* $Z(S_p^{(2)}, 1+x)$;

(*ii*) *the generating function for digraphs of order p is* $Z(S_p^{[2]}, 1+x)$;

(*iii*) *the generating function for multigraphs of order p is* $Z(S_p^{(2)}, (1-x)^{-1})$. ‖

In fact, the usual version of Pólya's theorem is far more general than this, but for the purposes of this chapter the more general version is unnecessary. For a general treatment of Pólya's theorem, see [**10**], [**14**] or [**19**].

We conclude this section by stating without proof some results on the enumeration of trees, which may (but need not) be proved using Pólya's theorem. It turns out that one needs to enumerate rooted trees first, and with this in mind, we let T_p and t_p denote, respectively, the number of rooted and unrooted trees of order p, and define the ordinary generating functions T and t by

$$T(x) = \sum_{p=1}^{\infty} T_p x^p, \qquad t(x) = \sum_{p=1}^{\infty} t_p x^p.$$

We can now state our main results for trees:

Theorem 9.3. *With the above notation,*

(*i*) $T(x) = x \prod_{k=1}^{\infty} (1-x^k)^{-T_k}$;

(*ii*) $T(x) = x \exp\left\{\sum_{k=1}^{\infty} \dfrac{T(x^k)}{k}\right\}$;

(*iii*) $t(x) = T(x) - \frac{1}{2}\{T^2(x) - T(x^2)\}$. ‖

Remarks. Part (*i*) of Theorem 9.3 was proved by Cayley in 1857 (see [**7**] or [**2**, pp. 39–43]), and gives one (in principle, at least) a way of calculating $T_1, T_2, \ldots$ in turn. The result of part (*ii*) is due to Pólya [**19**] and may be derived directly from part (*i*). Alternatively, it can be proved by deriving a Pólya-type result involving the cycle index $Z(S_n, T(x))$, and then summing over n; the details are given in [**14**, pp. 52–53]. Finally, the result of part (*iii*) is a celebrated result of Otter [**18**], which can be obtained from part (*i*) by building up unrooted trees from their center or bicenter; an alternative proof appears in Chapter 3 of [**14**].

The values of T_p and t_p (for $p \leqslant 26$) are tabulated in [**14**, p. 244]. The first few of these values are given in Table 6.

Table 6 Rooted and unrooted trees

p	1	2	3	4	5	6	7	8	9	10
T_p	1	1	2	4	9	20	48	115	286	719
t_p	1	1	1	2	3	6	11	23	47	106

10. Unsolved Problems

Although many enumeration problems have been solved, a great many remain open. In this final section, we present a few of the most important outstanding problems. Further problems are discussed in Chapter 10 of **[14]**.

(1) Labeled Graphs

As we have observed above, labeled enumeration problems are usually easier than the corresponding unlabeled problems, because there is less symmetry involved. However, there are a few notable exceptions to this, where the unlabeled problem has been solved, but not the labeled one. In particular, the number of self-complementary graphs and digraphs has been obtained by Read **[22]**, but the corresponding labeled problem is unsolved. An analogous situation holds for self-converse digraphs (digraphs isomorphic to their converse), where the unlabeled problem has been solved using a result known as the "power group enumeration theorem" (see **[10]**, or Chapter 6 of **[14]**).

(2) Even Graphs

Although even graphs have been enumerated (see Section 8), it is not known how many even graphs of order p there are with a specified number of edges. The usual methods for counting graphs with a given number of edges—in particular, Pólya's theorem—fail in this instance. As a consequence of this, no progress has been made on the problem of enumerating Eulerian graphs with a given number of edges. Of possible relevance to this problem is the idea of "switching". A graph is **switched** with respect to a vertex v if the edges incident to v are deleted, and the edges absent from v are introduced; two graphs are then said to be "equivalent" if they can be obtained from each other by a succession of switchings. Using a very neat application of Theorem 6.1, Mallows and Sloane **[16]** showed that the number of equivalence classes is equal to the number of even graphs of the same order, and a natural correspondence between the two has been described by Cameron **[6]**. (For the context in which Cameron's work arose, see Section 9 of Chapter 12.)

(3) Eulerian Digraphs

A digraph is **even** if the in-valency of each vertex is equal to its out-valency, and is **Eulerian** if it is even and connected. Labeled even digraphs have not been enumerated, and in consequence neither have Eulerian digraphs. One possible approach to this problem is to consider orientations of even graphs. If we remove all symmetric pairs of arcs from an even digraph D, the result is an "even orientation" of an even graph. So the number $ED_{p,q}$ of labeled even digraphs of order p with q arcs can be expressed in terms of the number $H_{p,k}$ of even orientations of all labeled even graphs with p vertices and k edges, as follows:

$$ED_{p,q} = \sum_{k+2m=q} \binom{\binom{p}{2}-k}{m} H_{p,k}.$$

The problem then reduces to the determination of the values of $H_{p,k}$.

A special case of the above problem is the enumeration of labeled Eulerian tournaments—that is, the regular tournaments; this problem amounts to determining $H_{p,k}$ when p is odd, and $k = \binom{p}{2}$.

(4) Eulerian Walks in Graphs

Although there is an explicit expression for the number of Eulerian walks in a given labeled Eulerian digraph (the so-called "BEST result", due to van Aardenne-Ehrenfest and de Bruijn [**26**], a special case of which appeared earlier in Smith and Tutte [**25**]), there is no known corresponding formula for graphs. In particular, the number of Eulerian walks in the complete graph K_{2s+1} is unknown. A solution of this problem may depend on the determination of all the even orientations of K_{2s+1}.

(5) Various Types of Graph

Although graphs, connected graphs, even graphs, Eulerian graphs, and trees of a given order have been enumerated, there are many classes of graphs for which no method has yet been found. These include the planar graphs, Hamiltonian graphs, k-chromatic graphs ($k \geqslant 3$), line graphs, and graphs with identity automorphism group, and there are many others. In addition, the enumeration of graphs with given girth, maximum valency, connectivity, genus, radius, or any of a whole variety of different parameters, is unsolved except in a few special cases.

References

1. C. Berge, *Principles of Combinatorics*, Academic Press, New York, 1971; *MR*38#5635 and **42**#5805.
2. N. L. Biggs, E. K. Lloyd and R. J. Wilson, *Graph Theory 1736–1936*, Clarendon Press, Oxford, 1976.

3. W. G. Brown, Historical note on a recurrent combinatorial problem, *Amer. Math. Monthly* **72** (1965), 973–977.
4. W. Burnside, *Theory of Groups of Finite Order*, 2nd ed., Cambridge University Press, London, 1911; reprinted by Dover, New York, 1955; *MR***16**–1086.
5. C. C. Cadogan, The Möbius function and connected graphs, *J. Combinatorial Theory* (*B*) **11** (1971), 193–200; *MR***44**#1605.
6. P. J. Cameron, Cohomological aspects of two-graphs, *Math. Z.* **157** (1977), 101–119.
7. A. Cayley, On the theory of the analytical forms called trees, *Phil. Mag.* (4) **13** (1857), 172–176 = *Math. Papers, Vol. 3*, Cambridge University Press, 1890, pp. 242–246.
8. A. Cayley, A theorem on trees, *Quart. J. Pure Appl. Math.* **23** (1889), 376–378 = *Math. Papers, Vol. 13*, Cambridge University Press, 1897, pp. 26–28.
9. L. E. Clarke, On Cayley's formula for counting trees, *J. London Math. Soc.* **33** (1958), 471–474; *MR***20**#7282.
10. N. G. de Bruijn, Pólya's theory of counting, in *Applied Combinatorial Mathematics* (ed. E. F. Beckenbach), Wiley, New York, 1964, pp. 144–184.
11. L. Euler, [Unsigned summary of "Enumeratio modorum, quibus figurae planne rectiliniae per diagonales dividuntur in triangula" by J. A. von Segner], *Novi Comm. Acad. Sci. Imp. Petropol.* **7** (1758–9), 13–15 = *Opera Omnia* (*1*), *Vol. 26*, Soc. Scient. Nat. Helveticae, Lausanne, 1953, pp. xvi–xviii.
12. F. Harary, The number of linear, directed, rooted, and connected graphs, *Trans. Amer. Math. Soc.* **78** (1955), 445–463; *MR***16**–844.
13. F. Harary, *A Seminar on Graph Theory*, Holt, Rinehart and Winston, New York, 1967; *MR***35**#5348.
14. F. Harary and E. M. Palmer, *Graphical Enumeration*, Academic Press, New York, 1973; *MR***50**#9682.
15. F. Harary and R. C. Read, Is the null-graph a pointless concept?, in *Graphs and Combinatorics*, Lecture Notes in Mathematics **406** (ed. R. A. Bari and F. Harary), Springer-Verlag, Berlin, Heidelberg, New York, 1974, pp. 37–44; *MR***50**#12819.
16. C. L. Mallows and N. J. A. Sloane, Two-graphs, switching classes and Euler graphs are equal in number, *SIAM J. Appl. Math.* **28** (1975), 876–880; *MR***55**#164.
17. J. W. Moon, *Counting Labelled Trees*, Canad. Math. Monographs, No. 1, Canad. Math. Congress, Montreal, 1970; *MR***43**#98.
18. R. Otter, The number of trees, *Ann. of Math.* (2) **49** (1948), 583–599; *MR***10**–53.
19. G. Pólya, Kombinatorische Anzahlbestimmungen für Gruppen, Graphen und chemische Verbindungen, *Acta Math.* **68** (1937), 145–254.
20. H. Prüfer, Neuer Beweis eines Satzes über Permutationen, *Arch. Math. Phys.* (*3*) **27** (1918), 142–144.
21. R. C. Read, Euler graphs on labelled nodes, *Canad. J. Math.* **14** (1962), 482–486; *MR***26**#4338.
22. R. C. Read, On the number of self-complementary graphs and digraphs, *J. London Math. Soc.* **38** (1963), 99–104; *MR***26**#4339.
23. R. J. Riddell, Contributions to the theory of condensation, Ph.D. thesis, University of Michigan, Ann Arbor, 1951.
24. R. W. Robinson, Enumeration of Euler graphs, in *Proof Techniques in Graph Theory* (ed. F. Harary), Academic Press, New York, 1969, pp. 147–153; *MR***40**#7154.
25. C. A. B. Smith and W. T. Tutte, On unicursal paths in a network of degree 4, *Amer. Math. Monthly* **48** (1941), 233–237.
26. T. van Aardenne-Ehrenfest and N. G. de Bruijn, Circuits and trees in oriented graphs, *Simon Stevin* **28** (1951), 203–217; *MR***13**–857.

15
Some Applications of Computers in Graph Theory

RONALD C. READ

1. Introduction

The ways in which the theory of graphs interacts with the theory and practice of computing are many and various. There is the vast and rapidly growing area of graph theory algorithms—the study of algorithms that enable one to determine properties of graphs. For example, there are algorithms for determining the connectedness or chromatic number of a graph, or for deciding whether or not a graph is planar, or whether or not two graphs are isomorphic, and so on. This area becomes even larger if one includes algorithms that operate on networks (that is, graphs with weighted edges), since much of operations research relates to work of this kind.

Several books have been written covering greater or lesser expanses of this large area, in addition to the multitude of papers on specific topics. However, we shall not be concerned in this chapter with graph theory algorithms of this kind, except perhaps incidentally.

A flow-chart for a computer program is a graph of a sort, and, more generally, many of the structures that arise in the organization of computer programs, scheduling, handling of peripherals, and many other aspects of the practice of computing, can be represented as graphs and studied using graph-theoretical techniques. Thus graph theory can be usefully applied in many ways to the theory and practice of computation. However, we shall not be discussing these applications either.

Instead, in this chapter, we shall look at a small but interesting area in the interaction between the two disciplines, and consider ways in which computations have led to new theoretical results in graph theory, and where new theorems have been obtained that would not have been obtained—sometimes, indeed, would not even have been envisaged—had it not been for the use of an electronic computer somewhere along the way.

2. The Four-Color Theorem

The example, *par excellence*, of the kind of computer application that we shall be studying in this chapter is provided by the recent proof, by Appel and Haken in 1976, of the four-color theorem—a proof which depended in a vital manner on the use of a computer. This theorem is the subject of Chapter 4, but it will be of interest here to take a brief look at the role that the computer played in the proof since, as we shall see, it was not just a question of mere number-crunching or routine verification of results.

We shall start by quickly reviewing the general method of the proof, referring the reader to Chapter 4 for a fuller treatment. It has been known for many years that in order to prove that any map can be colored with four colors one need only consider "normal" maps—that is, maps in which each region has at least five neighboring regions, and where at most three regions can have a single point in common. If normal maps exist that are not 4-colorable, then there is at least one that is "minimal", in the sense that the number of regions is the smallest possible.

Suppose that we have a configuration C of regions in a map M, such that their union is a connected subset of the plane. We say that C is **reducible** if we can modify M by replacing C by some configuration C' having fewer regions in such a way that if the modified map M′ is 4-colorable, then from a 4-coloring of it we can deduce a 4-coloring of M. Clearly a minimal 5-chromatic map cannot contain any reducible configurations.

Any configuration can be regarded as a ring, or annulus, consisting of a number (r, say) of regions, with the other regions in its interior. Such a configuration is described as an ***r*-ring configuration**, r being called the **ring-size**. It is fairly easy to show that 2-ring, 3-ring and 4-ring configurations are reducible; but 5-ring and 6-ring configurations are not always reducible.

Suppose now that we have found a set U of configurations with the following two properties:

(*i*) the set U is **unavoidable**—that is, every minimal 5-chromatic map must contain at least one member of U;

(*ii*) every member of U is reducible.

Then the four-color theorem follows. For, a minimal 5-chromatic map cannot contain any reducible configuration, and this contradicts property (*i*).

This then is the method of proof: *find an unavoidable set of configurations, each of which is reducible.*

Discharging Procedures

A powerful method of studying unavoidable sets of configurations is provided by "discharging procedures", of which we now give an example.

Associate with every k-sided region a "charge" of $6-k$. Thus pentagons alone have positive charge, hexagons are uncharged, and the major regions (those with more than six sides) have negative charge. It can be shown by Euler's formula (see Chapter 4) that the total charge over all regions is positive. The positively-charged pentagons are then "discharged" by distributing their charges among the major regions which are either adjacent to them, or are connected to them by a chain of hexagons.

Under the discharging procedure (and under a second discharging process if the first should lead to a major region accumulating a positive charge), most configurations end up having no residual positive charge, whereas a few are left with a positive charge. It follows that every map must contain at least one of these latter, since the total charge is positive. This brief and greatly oversimplified description will perhaps give an idea of the sort of method by which unavoidable sets can be tracked down.

It was known previously that an unavoidable set would have to include at least one (and probably very many) configurations of ring-size 12 or more. Thus it was clear that one could not expect a simple proof along these lines. Nevertheless, refinements of the discharging procedure and other improvements brought the problem to the stage where it was reasonably amenable to a computer attack. An initial computer program was written and the search for an unavoidable set was started. The progress of this program was constantly monitored, and modifications were made to it whenever necessary. Thus, for example, it was felt that it would be worthwhile to limit the ring-size of the configurations, even at the risk of increasing the number of configurations eventually produced. In consequence, whenever the program produced a configuration of ring-size 15 or more, the discharging procedure was adjusted so that this configuration would not occur.

One of the most interesting things about this procedure was that during the course of making these manual changes to the program, Appel and Haken become more and more aware of what sort of adjustments were necessary, and why. Eventually a point was reached where enough insight into this aspect of the problem had been gained to make it more worthwhile to carry out this part of the proof by hand; and this is what they eventually did.

The end result was a set of 1936 configurations forming an unavoidable set. This was not the smallest possible such set since, in the finding of it, economy of configurations was often subordinated to ease of construction. The size of this set therefore depends on a compromise between the extra time needed to test for reducibility every configuration of an unnecessarily large set, and the extra time required to find a smaller set.

Reducibility

It is convenient to think of the proof of the four-color theorem as consisting of two more or less independent parts, the second of which is to show that every configuration in the set U produced by the first part is reducible. In point of historic fact, the two parts were interdependent. Thus, for example, if a candidate for inclusion in U could not be shown to be reducible in a reasonable length of time, then modifications were made to the method of constructing U so that this configuration did not have to be considered. Nevertheless, once an unavoidable set has been found, the matter of checking the configurations for reducibility can be regarded as a separate operation.

With the exception of one particularly recalcitrant configuration, every configuration with ring-size 11 or more was tested by computer programs written for this purpose. Configurations of smaller ring-size had already been extensively studied, and could be accounted for.

There is not too much that can be said about these programs without going into excessive detail (which can be found in **[2]**, **[3]**). In any event, this part of the proof was more in the nature of a routine computer application; the interesting man–machine interaction that characterized the first part of the proof had no counterpart in the second. The outcome was, of course, that every one of the 1936 configurations in the set U was shown to be reducible, and, as a result, Appel and Haken were able to announce that the four-color conjecture had at last become the four-color theorem.

3. Characteristic Polynomials and Spectra of Graphs

A concept that lies (explicitly or implicitly) behind almost any discussion of graphs, to the extent that it can be regarded as the fundamental concept of graph theory, is that of "graph isomorphism". The formal definition is as follows: two graphs G and H are **isomorphic** if there exists a one-to-one correspondence between their vertex-sets which preserves adjacency—that is, such that two vertices of G are adjacent if and only if the corresponding vertices of H are adjacent. Isomorphism expresses what, in less formal language, we mean when we say that two graphs are "the same" graph. Thus the two graphs of Fig. 1 are isomorphic, even though, as depicted, they may look very different.

The question of isomorphism is one that is of particular interest to chemists. The structural formula of an organic chemical compound is a diagram which represents the bondings between the atoms in a molecule of the compound. An example is given in Fig. 2, in which C stands for a carbon atom and H stands for a hydrogen atom. From this example it will be seen that a structural

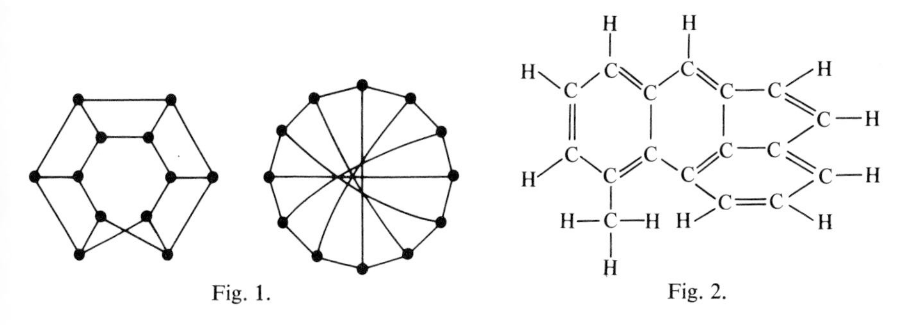

Fig. 1. Fig. 2.

formula is a special kind of graph having several different kinds of vertices, according to how many different kinds of atoms occur in the compound. A chemist who has isolated some compound from a natural product will naturally want to know whether this compound is already known, or whether it is something quite new. If he determines the various bondings in the molecule, he can draw the structural formula. But, as we have seen, two drawings of the same graph may look very different, and the problem that confronts the chemist is therefore that of determining whether his structural formula is isomorphic to that of a known compound. Chemists have, over the years, developed their own methods for coping with this problem, largely (although not entirely) independently of the efforts of graph-theorists in the same direction.

It is often not realized to what extent, in thinking about graphs, we think in terms of *labeled graphs* (that is, graphs whose vertices have been distinguished from each other by ascribing to them distinct labels, or in some other way). Thus to form an adjacency matrix $\mathbf{A} = (a_{ij})$ of a graph, we must first label it, as is immediately seen from the definition of an adjacency matrix—namely, $a_{ij} = 1$ if the vertices i and j are adjacent, and 0 otherwise. If we draw a graph, or even just picture one in our mind's eye, we implicitly label its vertices; for the vertices are distinguished from each other by their different spatial positions. In this sense, the two graphs of Fig. 1 are differently labeled, and it is for this reason that they "look different".

One very convincing way of demonstrating that two graphs are not isomorphic is to find a property which one graph has and the other does not, provided that this property is one which the graphs would both have if they were isomorphic. Thus two graphs with different valency-sequences cannot be isomorphic; a graph with a Hamiltonian circuit cannot be isomorphic to a graph that does not have one, and so on. A property which is such that two isomorphic graphs must either both have the property, or both lack it, is said to be a **graph invariant**; so also is a parameter which is such that two isomorphic graphs must have the same value of the parameter. Thus the property of having a Hamiltonian circuit is a graph invariant; the numbers of vertices and edges of a graph are graph invariants, as is also the valency-sequence. Put another way, a graph invariant is a property that does not depend on the way in which a graph has been labeled.

If two graphs are found to have a large number of graph invariants in common, and no property has been found for which they differ, one might suspect that they are isomorphic: but this would not constitute a proof. What would be useful would be a "complete graph invariant"—namely, a property or parameter which two graphs both have if and only if they are isomorphic. Unfortunately no such invariants have been found which are not either equivalent to the straight definition of isomorphism, or at least as difficult to apply. Since the discovery of an easily-computable complete graph invariant would solve the basic problem of chemical documentation described above, as well as many other problems, the search for such an invariant has been assiduous and still continues (see, for example, [**46**]).

The requirement that a graph invariant should be independent of the labeling of a graph automatically focuses attention on properties that clearly have this independence. We have already noted that the adjacency matrix $\mathbf{A}$ of a graph does not have this property; but clearly its characteristic polynomial $|\lambda\mathbf{I}-\mathbf{A}|$ is independent of the labeling. Hence the sequence of coefficients in this polynomial is a graph invariant. So is the spectrum of the graph—the set of eigenvalues, the roots of the corresponding polynomial equation. If the characteristic polynomials of small connected graphs (up to 5 vertices, say) are computed, they turn out to be different for non-isomorphic graphs, and this has prompted the suggestion, from time to time, that the characteristic polynomial is a complete graph invariant. In point of fact, some isolated examples of "cospectral graphs" (non-isomorphic graphs with the same spectrum) have been known for many years, as mentioned in Section 7 of Chapter 11. In 1968, a systematic computer calculation of the characteristic polynomials for all graphs with up to 7 vertices, using a list of 7-vertex graphs compiled by B. R. Heap of the National Physical Laboratory, revealed the existence of many other examples (see **[24]**).

Although the idea that the characteristic polynomial is a complete graph invariant is thus discredited, it is still tempting to suppose that it might be true if we restrict our attention to special kinds of graphs, or if we supplement the information provided by the characteristic polynomial with other facts. The conjecture that the characteristic polynomial is complete for the graphs of chemical structures was refuted by Balaban and Harary **[5]**. It is not difficult to formulate more elaborate invariants related to the characteristic polynomial, and to conjecture that they are complete, but naturally as the conjectures become more complicated, the finding of counter-examples becomes more difficult. There are no very obvious ways of tailoring a graph so as to make it have a given spectrum, so that a search for cospectral graphs is largely a hit-or-miss affair. Systematic computer scrutiny of all graphs is possible only to the extent that lists of graphs are available (see Section 4), and is therefore not applicable to the more complicated conjectures.

Such evidence as is available does not encourage one to believe that the spectrum of a graph, even for graphs of a restricted type, can be made into a useful complete graph invariant. Thus Schwenk **[48]** has shown that almost all trees are cospectral, in the sense that the property of being distinguishable from other graphs on the basis of the spectrum alone is the exception rather than the rule, at least for trees (Theorem 7.2 of Chapter 11). Whether this last statement is true of graphs in general is still not known, but an indication that it might be so is provided by some recent work by Godsil and McKay **[19]**. They computed the characteristic polynomials and spectra of all graphs on 9 vertices (using a catalogue produced by Baker, Dewdney and Szilard **[4]**) and found that about 18.6% of them were not determined by their spectra. It looks as if this percentage increases as the number of vertices increases, although this conclusion must be regarded as tentative.

The computer work of Godsil and McKay has shed some light on many other problems concerning spectra of graphs, including some that do not seem to have been studied before. Thus they were able to characterize those cospectral pairs of graphs whose complements were also cospectral, and those whose line graphs are cospectral. They also investigated graphs which are cospectral to their complements, and obtained many similar results relating to cospectral trees and bipartite graphs. Further results on cospectral graphs and related topics will be found in Chapter 11.

4. Roots of Chromatic Polynomials

The many attempts that have been made to prove the four-color theorem have provided a very strong impetus to progress in many branches of graph theory, and in particular to the study of chromatic polynomials and their properties. The concepts of graph coloring, and of the chromatic polynomial of a graph, have already been given in Chapter 1, and so it will not be necessary to give them here. For a general survey of chromatic polynomials and their properties, see [**39**].

The early investigations of chromatic polynomials, usually in the context of planar graphs, were carried out mainly by Whitney [**53**], [**54**] and Birkhoff [**8**], [**9**]. One method of computing these polynomials, that of using "broken circuits", was given by Whitney [**54**], but later researchers have tended to use a somewhat different method which we shall briefly consider. It was known to Whitney, who mentioned it in [**53**] as having been used by R. M. Foster.

We recall from Chapter 1 that if G is a graph containing the edge $e = vw$, and if G' and G'' are the graphs obtained by respectively deleting and contracting the edge e, then

$$P_G(k) = P_{G'}(k) - P_{G''}(k).$$

If, in the process of contracting the edge e, we create a double edge (which will happen if v and w have a common adjacent vertex), we can replace the double edge by a single edge, since a double edge implies exactly the same restriction on the coloring as does the single edge—namely, that v and w must not be given the same color. Hence the chromatic polynomial of any graph can be expressed in terms of the chromatic polynomial of a graph with one fewer edge and that of a graph with one fewer vertex. If this operation is repeated as often as necessary (that is, until all the graphs that have been derived from the original graph are without edges, at which stage it is clearly impossible to go any further), then the chromatic polynomial of the original graph will have been expressed in terms of the chromatic polynomials of graphs with no edges, and these are simply powers of k.

Although this method, among others, provides a method—and a very straightforward one at that—for calculating chromatic polynomials, it is of little use except for comparatively small graphs. For the computation of chromatic polynomials is one of the class of "*NP-complete problems*", to which many graph-theoretical problems belong. These problems have the property that all known algorithms for solving them require exponential time. By this is meant that the time taken for a computer program to solve any one of them is an exponential function (or worse) of the number p of vertices in the graph. This property imposes severe limitations on the size of graph that can be handled by such a program. For example, advances in computer technology and in the speed of computers do not increase very much the size of problem that can be handled, whereas programs for which the running time is some power of p (the so-called "*polynomial programs*") can be expected to handle appreciably larger graphs, given (say) a twofold increase in computer speed. Moreover, these NP-complete problems have the property that if any one of them can be shown to be solved by a polynomial algorithm, then a polynomial algorithm exists for each of them, whereas if it could be proved that no polynomial algorithm was possible for any one of them, then the same would be true of all of them. Since there are many quite important problems in this set—problems on which much time and ingenuity has been spent—this suggests intuitively that it is unlikely that there is a polynomial algorithm for any of them. For further information on NP-complete problems, see, for example, [**1**].

Thus the computation of chromatic polynomials, even using a fast computer, becomes difficult for more than about 15 to 20 vertices. For special types of graph it may be possible to devise special methods to speed up the computation, and if the graph has a lot of symmetry it may be possible to make use of this for the same purpose. This is achieved at the cost of increasingly complex programming, and it can often happen that the eye can detect symmetries, yielding valuable short cuts, which a computer program can perceive only with great difficulty. In this connection it is relevant to mention a remarkable *tour de force* carried out by D. W. Hall and his colleagues [**21**] who, over the space of several years, computed by hand the chromatic polynomial of the graph formed by the faces of the truncated icosahedron—a graph with 32 vertices!

The four-color theorem can be formulated in terms of chromatic polynomials as follows:

$$\textit{for every planar graph } G,\ P_G(4) > 0.$$

The fact that the four-color theorem can be stated in this way has led to a great deal of interest in the properties of the chromatic polynomials of planar graphs, and most of the work referred to above by Whitney and Birkhoff related to planar graphs. A wealth of information was obtained in these early papers concerning such chromatic polynomials and their properties.

More recently, as a direct result of a computer investigation, properties of a new and different kind came to light. I refer to work carried out by Berman and Tutte in 1968. Properties of polynomials can be expressed in terms of their roots, so it is reasonable to ask questions such as:

what complex numbers can be roots of a polynomial equation of the form $P_G(x) = 0$?,

Berman and Tutte were interested in planar graphs, particularly triangulations (planar graphs in which every region is a triangle). In order to gain some insight into the behavior of their chromatic polynomials, they took two lists of such graphs, one prepared by R. A. Bari **[6]**, and the other by G. Faulkner **[17]**, computed the chromatic polynomials, and hence found the roots. These roots, when plotted in the complex plane, turn out to have a markedly non-random distribution, and to have several other interesting properties (see **[7]**). For one thing, the roots show a pronounced tendency to cluster along a curve having the general shape of a cardioid (see **[32**, p. 197]).

Perhaps the most remarkable property, however, is that these polynomials, generally have a non-integral real root which is very close to the value $\frac{1}{2}(3+\sqrt{5}) = 2.618\,034\ldots$. They called this number the "golden root", after the classical golden section number from which it differs by 1. Agreement with this value was seen to be better for the larger graphs, and for the graph related to the truncated icosahedron (mentioned above), the root agreed with the golden root to 7 places of decimals. Clearly there was some significance in all this—but what?

An explanation of this interesting behavior came a little later when Tutte **[51]** produced the following theorem.

Theorem 4.1. *If G is a plane triangulation with p vertices, and if* τ *is the golden section number* (*so that* $1+\tau$ *is the golden root*), *then*

$$|P_G(1+\tau)| \leqslant \tau^{5-p}. \quad \|$$

This theorem shows that the chromatic polynomial takes a small value at the golden root (especially small if the number of vertices is large), and hence is almost certain to be zero somewhere in the immediate neighborhood.

This computer investigation, and more recent work, has disclosed other interesting properties of the roots of these polynomials, including some other numbers at or near which roots are likely to occur. One of these is the number $2+\tau$, and Tutte **[52]** has proved the following remarkable result which he calls the "golden identity":

$$P_G(2+\tau) = (2+\tau)\tau^{3p-10}P_G^2(1+\tau).$$

Another number is $2+2\cos(2\pi/7) = 3.246\,98\ldots$—the "silver root", and S. Beraha has suggested that all the numbers of the form $2+2\cos(2\pi/n)$, for $n \geqslant 3$, are significant in the theory of chromatic polynomials of triangulations. More recent work by Tutte and others shows that such numbers do, in fact, turn up in the study of these graphs.

It is natural to ask whether something similar happens if we consider the chromatic polynomials of all graphs, instead of restricting ourselves to a special type of planar graph. An attempt to answer this question was made in 1974 by Farrell **[16]**, who found the chromatic polynomials and their roots for all graphs on 8 vertices; these graphs were taken from a list of such graphs prepared by Heap **[25]**. When these roots are plotted in the complex plane, an interesting pattern emerges—one that could also be detected (although less easily) on similar plots for graphs with 6 and 7 vertices. Around certain "popular" roots (that is, complex numbers that are roots of many chromatic polynomials) there appear roughly circular areas in which no other roots occur at all. For example, centered on the root $\frac{1}{2}(5+\sqrt{3}i)$ there is a roughly circular area of radius about 0.1, inside which no other roots are found. The actual plot obtained from this investigation is not easy to reproduce here, but the sketch in Fig. 3 shows its general appearance.

One naturally asks why the roots should show this sort of behavior, but to this question no answer has yet been found. Tutte, in a personal communication, has suggested the possibility that those chromatic polynomials that have

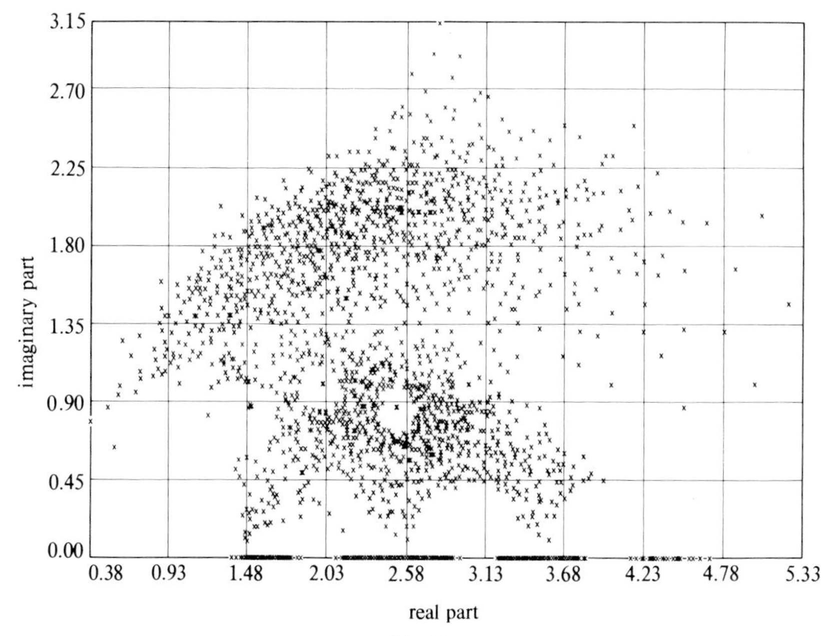

Fig. 3.

a root at one of these values may also have a large derivative there. This would explain why such polynomials do not have another root in the immediate neighborhood. If, in addition, the polynomials that do not have this number as a root assume a large value there, this would explain why they do not have a root nearby. However, this is all speculation. Possibly a better idea of the behavior of these roots could be obtained from a similar plot for the graphs on 9 vertices, but this has yet to be done.

5. Catalogs of Graphs and Other Configurations

Mathematics is often portrayed as a purely deductive discipline which progresses inexorably by logical steps from premises to conclusions, from definitions to theorems; a discipline in which scientific induction (as distinct from the mathematical variety)—the non-logical intuitive leap from a number of special cases to a universal assertion—plays no part. But this is true only on the surface, and as seen in the final versions of published work. In the course of mathematical discovery, intuition and induction play as important a role as in any of the sciences. We saw, in the last two sections, how what could quite properly be called "experimental evidence" suggested conjectures which led to proven theoretical results, or which have the potentiality of doing so. We also saw that in order to obtain this experimental evidence it was necessary to have lists or catalogs of graphs of certain kinds.

Thus the production of catalogs of all graphs of a particular kind is a potentially useful pursuit. Several such catalogs have been produced from time to time, and we shall now discuss some of them and the methods used to compile them. Because one is more likely to be able to "induce" valid theorems if one has a lot of experimental data, and because the numbers of graphs tend to become quite large even for comparatively small numbers of vertices and edges, these catalogs need to be produced on fast electronic computers used to their utmost capacity. Clearly, also, the method used to produce the catalogs must be the best that can be devised.

The first sizable catalog of graphs was produced in 1946 by Kagno [**27**]. This was generated by hand, and in addition to giving all the 156 graphs on 6 vertices, it also gave their automorphism groups. This catalog had some errors that were subsequently corrected (see [**28**]).

The 1044 graphs on 7 vertices are sufficiently numerous to rule out hand-computation as a reliable method of generation, and the first catalog giving these graphs was produced by Heap in 1965, using the computer at the National Physical Laboratory. Later, in 1969, Heap produced a catalog of the 12 346 graphs on 8 vertices (see [**25**]). The 274 668 graphs on 9 vertices were cataloged in 1974 by Baker, Dewdney and Szilard [**4**] at the University of Western Ontario, as the end result of a lengthy computer program. This catalog (it is stored on a single reel of magnetic tape) remains the most extensive catalog of graphs so far produced.

The graphs on 10 vertices are very numerous—there are 12 005 168 of them altogether—but there would be no great problem in storing these graphs; they could be quite comfortably accommodated on magnetic tape. Unfortunately, the methods at present available for generating graphs (even the orderly algorithms that we shall consider shortly) appear to be inadequate to produce these graphs in a reasonable length of time, unless some substantial improvements or effective short cuts can be found.

The digraphs on 4 and 5 vertices were generated by the present author on an IBM 1620 computer in 1966 [**40**] (there are 218 and 9608 of these, respectively). Quite recently, using improved techniques, R. Cameron and I have produced a catalog of the 1 540 744 digraphs on 6 vertices, together with a selection of properties for each one.

Unlabeled unrooted trees were listed up to 13 vertices by Morris [**31**], and up to 18 vertices by Frazer [**18**]; I am not aware of any more extensive list than this. Tournaments (complete oriented graphs) on up to 7 vertices have been cataloged by McWha [**30**].

Trivalent connected graphs are comparatively few in number, but their construction is not easy. They were constructed up to 14 vertices by Bussemaker and others [**11**] in 1975, and up to 18 vertices by Faradžev [**15**] in 1976. The numbers of such graphs are as follows:

Table 1 Trivalent connected graphs

p	4	6	8	10	12	14	16	18
Number	1	2	5	19	85	509	4060	41301

(The number 4060 appears incorrectly as 4006 in Faradžev [**15**].) Faradžev also gave the numbers of 4-regular graphs up to 14 vertices, and of 5-regular graphs up to 12 vertices.

Planar graphs of various kinds have been generated from time to time. Bouwkamp and others [**10**] listed 3-connected planar graphs up to 19 vertices, and Faulkner in 1971 generated lists of cyclically k-connected planar trivalent graphs [**17**].

An interesting point here is that for graphs, digraphs, trees, tournaments, and many other kinds of graph, the numbers of these configurations with a given number of vertices and edges is known from theoretical formulas, whereas for the planar graphs just mentioned, no theoretical formula is known.

The same is true of another kind of combinatorial configuration, sufficiently graph-like in nature to be included here, namely the "polyominoes". Polyominoes are connected figures formed by fitting together unit squares in the plane so that adjacent squares abut along the whole of a common edge. (An example is shown in Fig. 4; for general information on polyominoes, see **[20]**.) The "*polyomino problem*" (also called the "cell-growth problem") is

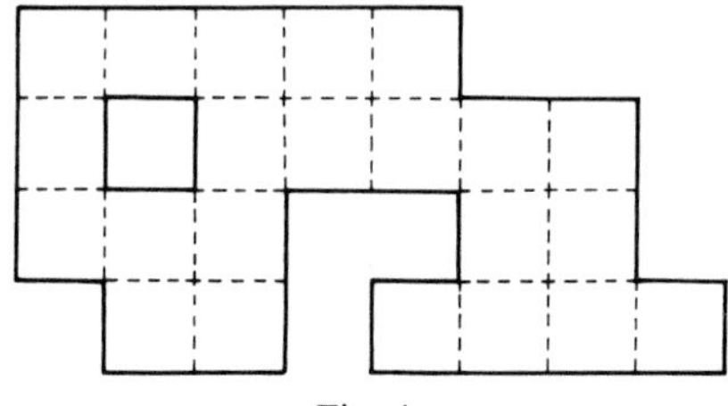

Fig. 4.

that of determining the number $P(n)$ of polyominoes with exactly n squares (or cells). Here we regard two polyominoes as distinct if neither can be obtained from the other by translation, rotation, or reflection in the plane. No theoretical formula for $P(n)$ is known, and it seems unlikely that one ever will be. Some formulas are known for special kinds of polyominoes (see **[42]**), and the asymptotic behavior of $P(n)$ has been studied, but the only knowledge that we have of the exact values of $P(n)$ comes from computer programs that have actually constructed all of the polyominoes. Thus the lists of polyominoes that have been produced give values of $P(n)$ as a rather expensive by-product, in addition to whatever intrinsic use the lists themselves might have. (One might well wonder what possible practical use a list of polyominoes could have—I did myself; yet I have been consulted by students of architecture who apparently found in polyominoes something of practical interest.)

Polyominoes for $n \leqslant 8$ were enumerated on the MANIAC II computer at Los Alamos by Harary in 1958 (see **[22]**). The value of $P(9)$ was found by me in 1960 **[42]** by a hand calculation that did not give the polyominoes themselves (the value of $P(10)$ given in that paper is wrong). Parkins and others generated polyominoes on a computer, and gave the values up to $P(15)$ in 1966 (see **[34]**), but the last value turned out to be incorrect. Around 1970, Lunnon used an Atlas computer in background mode, over a period of a year or two, to generate the polyominoes up to $n = 19$, thus correcting and extending Parkin's results (see [**29**]). In 1974, an undergraduate student of mine, Hugh Redelmeier, using an improved generating technique of his own invention (which he has so far not published), duplicated Lunnon's results on a PDP11 computer at the University of Waterloo in the course of a day or two, and incidentally detected a small error in one of Lunnon's figures. He has since used the computer at the University of Toronto to investigate polyominoes on as many as 23 cells. Redelmeier's program does not keep the polyominoes, since they are too numerous; it merely counts them as they are produced, after which they are discarded. The following table gives the most up-to-date information, as far as I am aware, of what is known about the exact values of $P(n)$.

Table 2 Polyominoes

n	$P(n)$	n	$P(n)$
1	1	13	238 591
2	1	14	901 971
3	2	15	3 426 576
4	5	16	13 079 255
5	12	17	50 107 909
6	35	18	192 622 052
7	108	19	742 624 232
8	369	20	2 870 671 950
9	1285	21	11 123 060 678
10	4655	22	43 191 857 688
11	17 073	23	168 047 007 728
12	63 600		

Despite the diversity of graphs just mentioned, the general methods used to generate them are fairly similar; they are recursive, and they generate graphs of a certain size from those of smaller size already generated.

For graphs with a given number p of vertices, we can take the graphs on q edges (which we suppose to have been already listed), and add an extra edge to each one in all possible ways. Clearly every graph with $q+1$ edges will be obtained at least once like this, and most will be obtained many times over. If we eliminate all duplicate graphs, we shall have a list of all the graphs with $q+1$ edges. Thus by starting with $q = 0$ (for which there is just one graph), we can generate all the graphs on p vertices. The same method applies for directed graphs.

For trees, we can go from a list of trees on p vertices to a list of those with $p+1$ vertices, as follows. We take each tree on p vertices (supposing them already listed), and add a new edge joining an existing vertex to a new vertex. This will give all the trees with $p+1$ vertices, again with many duplicates which we must eliminate.

As an example of a similar process for plane graphs, consider the 3-connected plane trivalent graphs. Suppose we have listed those of order $2m$ (the number of vertices must be even). In order to construct those of order $2m+2$, we take each of the listed graphs, and in each one, and in all possible ways, we choose two edges which lie on the boundary of a common region of the plane and join their mid-points, as shown in Fig. 5. Again, this process gives rise to many duplicates which have to be eliminated.

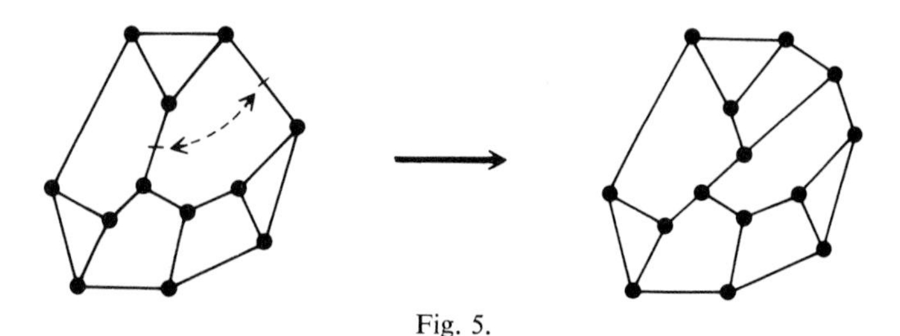

Fig. 5.

For polyominoes the method is similar. If to every polyomino with n cells we add an extra cell, attaching it to one (or more) existing cells along an edge (edges) in all possible ways, we obtain all of the polyominoes with $n+1$ cells, again with many duplications.

The above descriptions are considerably over-simplified—in practice there are many refinements and short cuts that can be introduced—but the basic idea behind most of these enumerations is as given, and we can see that the manner of going from a graph to a larger graph (which we shall call the "augmenting operation") is usually quite simple. The difficulty in implementing these procedures lies in the elimination of duplicates from the resulting new list. We have to be able to tell whether two graphs are the same, and we are therefore faced with the isomorphism problem, which, as we saw earlier, is very intractable for graphs in general. For some kinds of graphs, however, recognition of isomorphism can be very simple. Thus two polyominoes can easily be tested to see whether they are the same or not, and for trees there are good algorithms which can test for isomorphism, in time proportional to the number of vertices (see **[1]**).

Even for graphs and digraphs the situation is not all that bad, since the size of graph being dealt with is not all that large. A graph may have as many as $p!$ different adjacency matrices, according to the way in which the vertices are labeled. Hence isomorphism can be tested in a brute force manner by carrying out $p!$ operations, and although this number increases worse than exponentially with p, it is still manageably small ($=720$) when $p = 6$ (as required for the 6-vertex digraphs).

In practice, each graph that is produced is converted to some standard description (its "code"), and the list of graphs so far produced is searched to see if that code is already included. This simplifies the search procedure, but the formation of the code still involves (at least potentially) $p!$ operations. This, however, is not really the problem; the cataloguing of the 9-vertex graphs was carried out despite the fact that $9! = 362\,880$. The real difficulty with these programs is the sheer amount of data to be handled, and if, as in the above procedure, the whole list of graphs so far produced has to be searched every time a new candidate is generated by the augmenting operation, then the whole thing becomes impossibly time-consuming. This is especially the case if the list is stored on auxiliary storage (tape or disk), as it eventually must be.

Recently, in [**43**], I described a class of algorithms (called "orderly algorithms", applicable to a fairly wide variety of cataloging problems) which do away with the need for all this searching of lists. By keeping the lists in a particular order, and by a careful choice of augmenting operation, it becomes possible to tell whether a graph just generated is a new one or not. Determining this fact may still require $O(p!)$ operations, but the list-searching, which is what really takes up the time, is avoided altogether. It was by the use of such an algorithm that the catalog of 6-vertex digraphs was produced.

A bonus from the use of orderly algorithms is that the augmenting operation may turn out to be simpler than before. Thus it is shown in [**43**] that rooted trees on $p+1$ vertices can be formed from those on p vertices by adding a new edge, not at every vertex of the smaller tree (as given above for unrooted trees), but only to those at the extreme right-hand side of a standard drawing of the tree. This makes for a faster algorithm. Nevertheless, this algorithm for generating rooted trees is only of academic interest. It is not as good as an algorithm given by Scoins [**49**], and is completely outclassed by a recent algorithm due to T. Beyer of the University of Oregon (personal communication). This latter algorithm simply churns out from scratch the rooted trees on p vertices, one after the other, with no need for any input list and without any list-searching. Indeed, to produce each tree requires a maximum of $2p$ computer operations—either "compare" or "transmit". Moreover, a trivial modification will cause it to produce only trees of a given height, or of a given range of heights. What is more, I understand that a modification of this algorithm will give the unrooted unlabeled trees with only a small increase in the time taken to produce each tree. This makes catalogs of trees largely unnecessary, since for most purposes the trees required could be generated by the program requiring them as quickly as they could be read from magnetic tape.

6. The Use of Computers in Graphical Enumeration

We have already remarked, when discussing the cataloging of graphs, how it sometimes happens that the production of a catalog is the only way in which we can determine the numbers of graphs of a given kind, whereas for other types of graph there are theoretical formulas which give their numbers. Graphical enumeration—the branch of graph theory which is concerned with the determination of such formulas—has quite a respectable history, going back at least to the time of Arthur Cayley, and those who practice it (known amongst themselves as "the graph-theorists who count"!) have developed many powerful techniques for deriving formulas, or (more usually) generating functions. Further results in this area may be found in Chapter 14.

As examples of these two ways of enumerating graphs, consider two results due to Cayley. In **[12]** he gave the formula p^{p-2} for the number of labeled trees on p vertices (see Section 5 of Chapter 14). This is a simple formula from which the number of trees on any given number of vertices can be immediately found. Cayley also showed **[13]** that the number T_p of (unlabeled) rooted trees on p vertices is given by the equation

$$T_1x + T_2x^2 + T_3x^3 + \ldots = x(1-x)^{-T_1}(1-x^2)^{-T_2}\ldots, \tag{1}$$

where the left-hand side is the generating function for the trees in question. In this case, the determination of specific values of T_p is much more complicated. It is easy to verify that if we know the values of $T_1, T_2, \ldots, T_p$, and substitute these values into the right-hand side of (1), then the value of T_{p+1} on the left-hand side will be determined. Thus we can calculate the sequence $\{T_p\}$ as far as we wish, but we can find a given value of T_p only by calculating all the previous values in the sequence. This situation is quite common in graphical enumeration; it is rarely that the solution to a problem is in a form from which the required numbers can be computed easily. For that reason it is frequently necessary to use a computer to evaluate these numbers, in order to be reasonably certain that no arithmetical errors have been made, since the computation of quite small numbers (of the kind needed to check the accuracy of a catalog) may well be sufficiently complicated that hand calculations cannot be trusted. We shall now look at some examples of such computer calculations.

The structural formula for a chemical compound is a graph of a special kind, as we saw in Section 2. Figure 6 gives the structural formula of a hydrocarbon (a compound of hydrogen and carbon only) for which the graph is a tree, and which has no multiple edges. Such a compound is known as an "alkane". Since carbon atoms have valency 4, and hydrogen atoms have valency 1, it is easy to deduce that if the number of carbons is n then the number of hydrogens is $2n+2$ (chemists would therefore describe this by the symbolism C_nH_{2n+2}). A question of interest to chemists is, "how many different alkanes are there with n carbon atoms?". If we interpret this to mean, "how many different structural formulas (like Fig. 6) are there for a given value of n?", then we have a problem in enumerative graph theory.

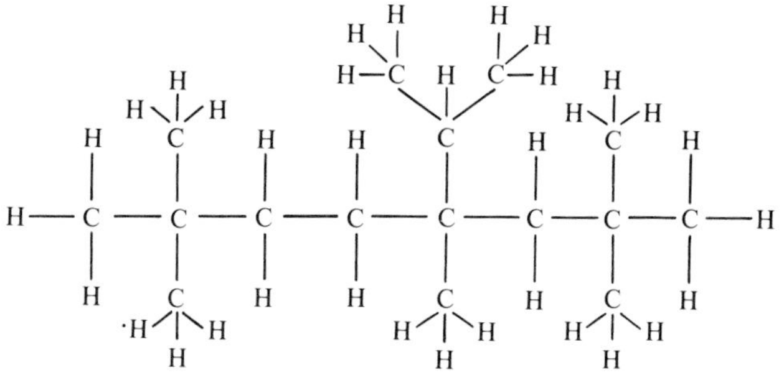

Fig. 6.

Let us look, for convenience, at a slightly simpler problem, that of finding the number of different "monosubstituted alkanes" (alkanes in which one hydrogen atom has been replaced by something else, called X). We can simplify the problem by leaving out the hydrogen atoms (they can be put back in a unique way), thus obtaining a sort of carbon skeleton of the compound. One carbon atom is distinguished from the rest—it is the one to which X is attached—and the problem is therefore equivalent to that of enumerating rooted trees in which no vertex has valency greater than 4, and where the root has valency at most 3.

Cayley solved this problem [**14**], and Pólya [**36**] embedded it in a much more general type of problem. Let a_p be the number of these trees with p vertices, and write

$$A(x) = \sum_{p=0}^{\infty} a_p x^p$$

for the generating function of the sequence $\{a_p\}$. Then we have, following Pólya, the result

$$A(x) = 1+\frac{x}{6}\{A^3(x)+3A(x)A(x^2)+2A(x^3)\}. \tag{2}$$

Again it is easy to verify that if the numbers a_p are known up to a certain value $p = m$, and are substituted into the right-hand side of (2), then the left-hand side is now determined up to and including the term in x^{m+1}, so that the value of a_{m+1} is now known. Thus the numbers a_p can be calculated recursively one by one, but the calculation is tedious and prone to error if carried out by hand. Cayley made some errors in this calculation, which were corrected by Henze and Blair [**26**] and Perry [**35**], who continued the calculation up to $p = 30$, by which time the numbers had become extremely large (for example, $a_{30} = 95\,991\,365\,288$). However, these authors also made a few minor errors—typographical, it would seem, rather than arithmetical.

Recently, in [**44**], I looked at a generalization of this and other similar problems, namely to determine the numbers of hydrocarbons in which double and triple bonds are allowed (see Fig. 7 for an example). Since this requires

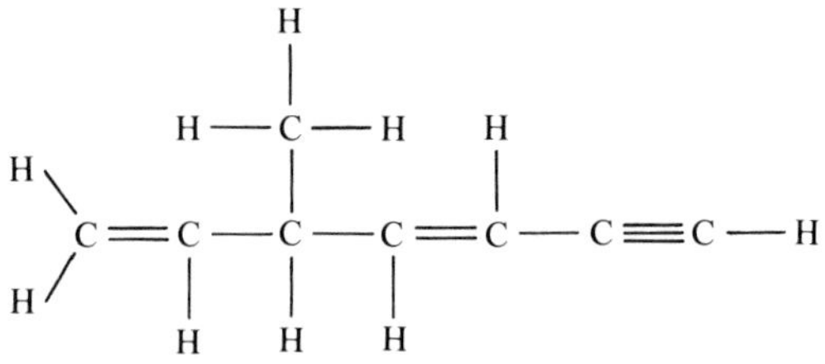

Fig. 7.

the simultaneous consideration of three parameters (the number of carbon atoms, the number of double bonds, and the number of triple bonds), the resulting formulas and equations are for generating functions of three variables. Under these conditions a hand computation is virtually out of the question for all but very small values of the parameters, and a computer was used to obtain the results given in this paper. This sort of calculation gives rise to some interesting programming problems on how to manipulate series in several variables, especially when (as is the case here) many of the series are quite sparse.

As another example of an enumeration formula from which numerical results can be obtained, although not easily, consider the problem of determining the number of graphs with p vertices and q edges. Harary [23] showed that this number is the coefficient of x^q in

$$\frac{1}{p!}\sum_{\mathbf{j}} h(\mathbf{j})\left\{\prod_{k}\left((1+x^{2k+1})^{kj_{2k+1}}(1+x^k)^{j_{2k}+k\binom{j_k}{2}}(1+x^{2k})^{(k-1)j_{2k}}\right).\prod_{r<t}(1+x^{\operatorname{lcm}(r,t)})^{\gcd(r,t)j_r j_t}\right\}, \tag{3}$$

where

$$h(\mathbf{j}) = \frac{p!}{\prod_k k^{j_k}.j_k!},$$

and where the summation extends over all partitions $\mathbf{j}$ of the integer p—that is, over all solutions in non-negative integers of the equation

$$j_1+2j_2+3j_3+\ldots+pj_p = p,$$

where j_k denotes the number of parts equal to k occurring in the partition $\mathbf{j}$. The reader should compare the expression in (3) with the expression for $Z(S_n^{(2)})$ in Theorem 9.1 of Chapter 14.

Even for $p = 4$, the expansion of (3) as a polynomial in x (giving the required number of graphs for each value of q) is quite time-consuming. Writing a computer program to compute these numbers is not easy either, but once it is done, it can be used to provide reliable values. These numbers have been calculated by Stein and Stein [50] up to $p = 18$. The total numbers of graphs (all values of q) have been found by Palmer and King [33] up to $p = 24$, using (3) with $x = 1$.

Some formulas that have been given as solutions of enumeration problems in graph theory are so intractable, and yield numerical results so reluctantly, that it is a moot point whether they can really be regarded as solutions. Thus it was shown in [37] that the number of trivalent general graphs (in which loops and multiple edges are allowed) on $2n$ vertices is given by the formula

$$N(h_{2n}[h_3]*h_{3n}[h_2]) \tag{4}$$

in the improved notation of [41]. Although this is an elegant and concise formula, it is useless in practice. For to obtain results from it requires the manipulation of symmetric functions or, equivalently, of polynomials in an indefinite number of variables. Even the simplest case, $n = 1$, is far from easy; it requires about a page of working, and after the mountain has thus labored it is delivered of a tiny mouse in the form of the answer "2". This is all the more ridiculous since with only a moment's thought we can see that this is the answer, the two graphs being those of Fig. 8. In addition to all this, the

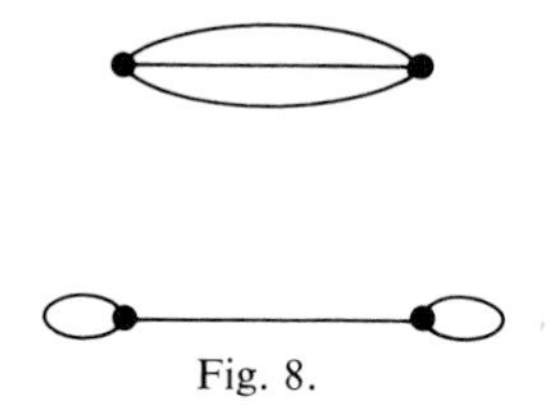

Fig. 8.

computations that have to be performed are not at all suitable for implementation on a computer.

One should not, however, be too hasty in condemning formulas such as (4) as being without value. Improved techniques may be able to transmute them into other formulas from which numerical results can be more readily obtained, and this is indeed the case with many problems of this kind. Consider, for example, the enumeration of labeled graphs with a given valency-sequence. This enumeration was originally given in [**38**] as a formula every bit as obscure as (4), and yet, by the use of what are known as *S*-functions, numerical results for this problem can be obtained without too much effort, and by methods that are quite suitable for computer implementation. Although we cannot go into details (they can be found in [**41**], I should like to mention a rather curious kind of multiplication which arises in this application.

An *S*-function is a symmetric function related to a partition of a positive integer (m, say). If the partition is denoted by (λ), and consists of parts λ_i such that

$$\lambda_1 \geqslant \lambda_2 \geqslant \lambda_3 \geqslant \ldots \geqslant \lambda_k,$$

then the corresponding *S*-function is denoted by $\{\lambda\}$. In the special case where (λ) is the partition of m into exactly one part m, the corresponding *S*-function is the homogeneous product sum h_m (see [**41**]). The basic operation to be performed in finding the numbers of graphs with a given valency-sequence is that of multiplying a general *S*-function $\{\lambda\}$ by one of these homogeneous product sums, and expressing the result as a linear combination of *S*-functions. This is done by the following rather bizarre procedure:

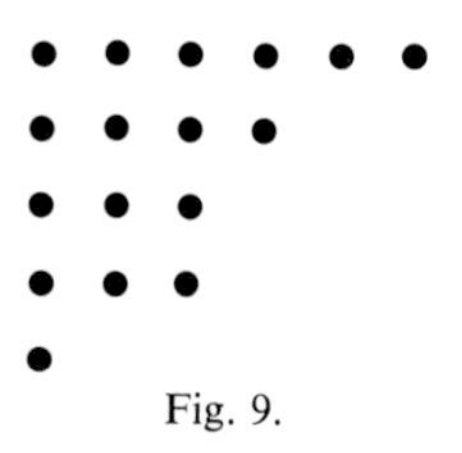

Fig. 9.

We represent the partition (λ) by a Young diagram, of which Fig. 9 is an example. There are k rows of dots, the ith row containing λ_i dots, and the rows are left-justified. The partitions of the S-functions that occur in the product of $\{\lambda\}$ by h_n are those whose Young diagrams can be obtained from that of (λ) by adding n further "new" dots to its rows (and also, possibly, starting one new row), subject to the following two conditions:

(*i*) the resulting Young diagrams must be proper—that is, no row contains more dots than the one above;

(*ii*) no new dot may lie in the same vertical line as another new dot—that is, each new dot must have an old dot immediately above it.

In order to find the number of labeled general graphs with valency-sequence $m_1, m_2, \ldots, m_k$, we need to expand in the above way the product

$$h_{m_1} h_{m_2} h_{m_3} \ldots h_{m_k},$$

and sum the coefficients of those S-functions whose partitions have only even parts. By summing coefficients for partitions of other special kinds, we can find the number of graphs for which loops are not allowed (but multiple edges are), for which loops (but not multiple edges) are allowed, and also those for which neither loops nor multiple edges are allowed; for details, see [**41**].

Some of the sequences that arise from graphical enumeration problems, such as the sequence $\{g_p\}$ of the number of graphs on p vertices, have been computed to quite extraordinary lengths. Thus g_{24} is a 60-digit number! The motivation for producing such enormous numbers could be attributed to that same *furor arithmeticus* that goaded the pre-computer calculators of π, but there is more to it than that. The difficulty or ease of calculating numerical results can give an indication of the merit of the theoretical formula from which they are obtained. Thus if a sequence has been laboriously calculated up to 10 terms at the cost of vast amounts of computer time, and then a new formula is found which gives these results in a few seconds, the chances are that the new result represents a breakthrough from which new theoretical results can be expected. Moreover, the fact that a new method duplicates known results up to large numbers encourages us to believe that it is valid, even though it does not prove it. An apparently valid method may have a flaw that does not become apparent until quite large numbers are reached, and thus the agreement of only a few early terms is not always very convincing.

7. Visual Display and Handling of Graphs

Paper and pencil are the traditional tools of the mathematician, and the number of theorems in graph theory that were conjectured as the result of mathematical "doodling"—the construction, manipulation, and examination of hand-drawn graphs—must surely be extremely large. Yet there is a danger in making conjectures from hand-drawn figures; small graphs may well be atypical in their properties, while the drawing and manipulation of large graphs is more difficult and more prone to error. Hence a very useful adjunct to a graph-theorist's tool-kit is a computer program that will display graphs on a screen and permit their manipulation in various ways. There are, I believe, several such programs in existence; I shall describe briefly one that I have produced myself, this choice being made simply because it is the one with which I am most familiar.

This *Graph Manipulation Package*, as it is called, runs on a PDP11/20 computer, driving a Tektronix 4001A visual display terminal. Text and diagrams are displayed by means of a 1023×760 rectangular array of points on the screen, which gives sufficient resolution for quite large graphs to be displayed with reasonable clarity. Except for a strip along the bottom of the screen (the "scratch pad"), used for messages and other input and output, the displayed material is not refreshed, and can be erased only by erasing the whole screen. The visual display terminal has its own keyboard, and apart from the initial signing on, all operations are performed from this keyboard.

Communication between the operator and the program is mainly via a repertoire of two-character commands. When the program is in "command mode" (that is, ready to accept a command), the typing of one of these two-character commands causes the appropriate portion of the program to be entered. Thus the drawing of a graph on the screen will usually start with the command "AV" (=add a vertex). This causes two perpendicular cross-hairs to appear on the screen, the positions of which (and hence of their point of intersection) are controlled by a joy-stick at the side of the terminal. Pressing the space bar of the keyboard causes a small circle to appear at the point of intersection of the cross-hairs. This command, like many others, is repetitive, and one can go on drawing little circles to represent vertices of a graph as long as one likes (up to the limit of the program—255 vertices). When all the required vertices have been placed, pressing the ESCAPE key returns the program to command mode.

The command "AE" (=add an edge) enables edges to be added to the display. These are indicated by positioning the cross-hairs first on one end-vertex and then on the other, whereupon the program draws an edge connecting the two vertices. Other commands will produce edges made up of dotted lines, dashed lines or extra-thick lines. Others enable specified vertices and edges to be deleted (which, for the reason given above, entails erasing the whole screen and redrawing the graph in its modified form—but this is very quick).

Manipulation of the shape of the graph is effected via commands which move vertices from one position to another, shrink or expand selected portions of the display, rotate the display about a given point through a given angle, and so on. The instruction "LB" causes the labels 1, 2, 3, . . ., p to be displayed against the vertices on the screen. The particular labeling thus given to the graph is of no special importance—it depends on the order in which the vertices were taken when the graph was originally constructed, and on the manipulations subsequently performed. If a particular labeling is required, the command "RL" is available to relabel the graph in any desired way. Other commands are provided to enable the displayed graph to be stored in disk storage for later use, to display on the screen a graph so stored, and to display such a graph on the screen side by side with the currently displayed graph. Hard copies of the display can be obtained at any time.

Another set of commands serves to display commonly-used classes of graphs. Complete graphs, complete bipartite graphs, wheels and circuits can be obtained merely by entering the appropriate numbers of vertices. Other commands produce the complement and the line graph of the displayed graph. Provision is also made for the construction of a random graph on entering the required number of vertices and edges.

Incorporated into the Graph Manipulation Package are a few graph-theoretical algorithms for determining some property of the displayed graph. Thus the command "SP" produces a spanning tree of the graph (identified by thickened edges); "HC" finds a Hamiltonian circuit, if there is one; "HD" does the same, but then redraws the graph so that the Hamiltonian circuit is a regular polygon; "BL" redraws the graph so as to separate its components and display the blocks and cut-vertices of the graph.

There are some other commands that have not been mentioned, but the above are the most important ones. A complete list is given in the instruction booklet [**45**] for the package. The program contains a complete description of itself, which can be consulted at any time by the operator if he has forgotten what commands are available, or if he wants operating instructions for some particular command. This information is displayed on the screen, and when it has served its purpose, the graph that the operator was working on is displayed again.

The program is written mainly in FORTRAN, but the subroutines for performing the frequent operations of drawing vertices, edges, and so on, are in the assembler language. The program is divided into six parts which are kept as overlays. When a command is given, the program checks to see whether the overlay containing that command is currently in storage. If it is, it executes the relevant part of that overlay; if it is not, it first calls in the appropriate overlay from disk storage (this takes a second or so). Commands that are likely to be used a lot together are kept in the same overlay. This makes possible a program which, taken in its entirety, is too large to fit into the core storage of the computer. It also makes it quite easy to hang other programs on to the package. Thus if you want to be able to construct graphs, and then compute your favorite graph invariant, all you need is an algorithm and a program to compute that invariant. By making a few trivial changes to the source programs, recompiling and linking together with the new program, the computation of the invariant is included as one of the commands under the system. As an example of this, and also of how visual display programs can be useful in suggesting results and theorems, I can cite the study of the principal tripartition of a graph. This concept—due to P. Rosenstiehl—concerns a partition of the edges of a graph into three sets, according to how the edges are related to the cycle and cocycle spaces of the graph (the details and precise definition need not concern us here). Rosenstiehl and I have recently written a paper [**47**] on this topic, setting out a number of properties of this tripartition. A great many of the results contained in it were suggested and verified by the use of the Graph Manipulation Package just described, to which had been appended an algorithm for calculating the tripartition. This calculation is not all that easy to do by hand, so that by using this program we were able, by trial and error, to gain an insight into the way in which certain manipulations of the graph (such as the addition and deletion of edges, to mention just two) altered the nature of the tripartition—an insight which it would have been quite impossible to obtain by pencil and paper methods.

Acknowledgement

This work was supported by grant A8142 of the National Research Council of Canada.

References

1. A. V. Aho, J. E. Hopcroft and J. D. Ullman, *The Design and Analysis of Computer Algorithms*, Addison-Wesley, Reading, Mass., 1974.
2. K. Appel and W. Haken, Every planar map is four colorable: Part 1, Discharging, *Illinois J. Math.* **21** (1977), 429–490.
3. K. Appel, W. Haken and J. Koch, Every planar map is four colorable: Part 2, Reducibility, *Illinois J. Math.* **21** (1977), 491–567.

4. H. H. Baker, A. K. Dewdney and A. L. Szilard, Generating the nine point graphs, *Math. Comp.* **28** (1974), 833–838.
5. A. T. Balaban and F. Harary, The characteristic polynomial does not uniquely determine the topology of a molecule, *J. Chem. Documentation* **11** (1971), 258–259.
6. R. A. Bari, Absolute reducibility of maps of at most 19 regions, Ph.D. thesis, Johns Hopkins University, 1966.
7. G. Berman and W. T. Tutte, The golden root of a chromatic polynomial, *J. Combinatorial Theory* **6** (1969), 301–302; *MR***39**#98.
8. G. D. Birkhoff, A determinant formula for the number of ways of coloring a map, *Ann. of Math.* (*2*) **14** (1912–3), 42–46 = *Collected Mathematical Papers, Vol. 3*, American Mathematical Society, New York, 1950, pp. 1–5.
9. G. D. Birkhoff, On the polynomial expressions for the number of ways of coloring a map, *Ann. Scuola Norm. Sup.* (*Pisa*) (*2*) **3** (1934), 85–104 = *Collected Mathematical Papers, Vol. 3*, American Mathematical Society, New York, 1950, pp. 29–47.
10. C. J. Bouwkamp, A. J. Duijvestijn and P. Medena, Table of c-nets of orders 8 to 19 inclusive, Phillips Res. Lab., Eindhoven, 1960.
11. F. C. Bussemaker, S. Cobeljic, D. M. Cvetković and J. J. Seidel, Computer investigation of cubic graphs, T.H.-Report 76-WSK-01, Department of Mathematics, Techn. Univ., Eindhoven, 1976.
12. A. Cayley, A theorem on trees, *Quart. J. Pure Appl. Math.* **23** (1889), 376–378 = *Math. Papers, Vol. 13*, Cambridge University Press, Cambridge, 1897, pp. 26–28.
13. A. Cayley, On the theory of the analytical forms called trees, *Phil. Mag.* (*4*) **13** (1857), 172–176 = *Math. Papers, Vol. 3*, Cambridge University Press, Cambridge, 1890, pp. 242–246.
14. A. Cayley, On the analytical forms called trees, with applications to the theory of chemical combinations, *Rep. Brit. Assoc. Advance. Sci.* **45** (1875), 257–305 = *Math. Papers, Vol. 9*, Cambridge University Press, Cambridge, 1896, pp. 427–460.
15. I. A. Faradžev, Constructive enumeration of regular graphs (Russian), *Uspekhi Mat. Nauk* **31** (1976), 246; *MR***54**#12566.
16. E. J. Farrell, On a general class of graph polynomials, Ph.D. thesis, University of Waterloo, 1973.
17. G. B. Faulkner, The recursive generation of cyclically k-connected cubic planar graphs, Ph.D. thesis, University of Waterloo, 1971.
18. R. J. Frazer, Graduate course project (unpublished), Department of Combinatorics and Optimization, University of Waterloo, 1973.
19. C. Godsil and B. McKay, Some computational results on the spectra of graphs, *Combinatorial Mathematics IV*, Lecture Notes in Mathematics **560** (ed. L. R. A. Casse and W. D. Wallis), Springer-Verlag, Berlin, Heidelberg, New York, 1976, pp. 73–92; *MR***55**#2662.
20. S. W. Golomb, *Polyominoes*, Allen and Unwin, London, 1966.
21. D. W. Hall, J. W. Siry and B. R. Vanderslice, The chromatic polynomial of the truncated icosahedron, *Proc. Amer. Math. Soc.* **16** (1965), 620–628; *MR***31**#3361.
22. F. Harary, Unsolved problems in the enumeration of graphs, *Magyar Tud. Akad. Mat. Kutató Int. Közl.* **5** (1960), 63–95; *MR***26**#4340.
23. F. Harary, The number of linear, directed, rooted and connected graphs, *Trans. Amer. Math. Soc.* **78** (1955), 445–463; *MR***16**-844.
24. F. Harary, C. King, A. Mowshowitz and R. C. Read, Cospectral graphs and digraphs, *Bull. London Math. Soc.* **3** (1971), 321–328; *MR***45**#3249.
25. B. R. Heap, The production of graphs by computer, in *Graph Theory and Computing* (ed. R. C. Read), Academic Press, New York, 1972, pp. 47–62.
26. H. R. Henze and C. M. Blair, The number of isomeric alcohols of the methane series, *J. Amer. Chem. Soc.* **53** (1931), 3077–3085.
27. I. N. Kagno, Linear graphs of degree $\leqslant 6$ and their groups, *Amer. J. Math.* **68** (1946), 505–520, and **69** (1947), 872; *MR***8**-46.
28. I. N. Kagno, Corrections to the paper "Linear graphs of degree $\leqslant 6$ and their groups", *Amer. J. Math.* **77** (1955), 392; *MR***16**-733.

29. W. F. Lunnon, Counting polyominoes, in *Computers in Number Theory* (ed. A. O. L. Atkin and B. J. Birch), Academic Press, London, 1971, pp. 347–352.
30. P. McWha, Graduate course project (unpublished), Department of Combinatorics and Optimization, University of Waterloo, 1973.
31. P. A. Morris, A catalogue of trees, *Math. Comp.* **27** (1973), 1006.
32. C. St. J. A. Nash-Williams, Possible directions in graph theory, in *Combinatorial Mathematics and its Applications* (ed. D. J. A. Welsh), Academic Press, London, 1971, pp. 191–200.
33. E. M. Palmer and C. King, Calculation of the number of graphs of order $p = 1(1)24$ (to appear).
34. T. R. Parkin, L. J. Lander and D. R. Parkin, Polyomino enumeration results, SIAM Fall Meeting, Santa Barbara, 1967.
35. D. Perry, The number of structural isomers of certain homologues of methane and methanol, *J. Amer. Chem. Soc.* **54** (1932), 2918–2920.
36. G. Pólya, Kombinatorische Anzahlbestimmungen für Gruppen, Graphen und chemische Verbindungen, *Acta Math.* **68** (1937), 145–254.
37. R. C. Read, The enumeration of locally restricted graphs, I, *J. London Math. Soc.* **34** (1959), 417–436; *MR***21**#7162.
38. R. C. Read, The enumeration of locally restricted graphs, II, *J. London Math. Soc.* **35** (1960), 344–351; *MR***25**#3863.
39. R. C. Read, An introduction to chromatic polynomials, *J. Combinatorial Theory* **4** (1968), 52–71; *MR***37**#104.
40. R. C. Read, The production of a catalogue of digraphs on 5 nodes, Scientific Report No. UWI/CC1, University of the West Indies, 1966.
41. R. C. Read, The use of *S*-functions in combinatorial analysis, *Canad. J. Math.* **20** (1968), 808–841; *MR***37**#5108.
42. R. C. Read, Contributions to the cell-growth problem, *Canad. J. Math.* **14** (1962), 1–20; *MR***24**#A1219.
43. R. C. Read, Every one a winner, or How to avoid isomorphism search when cataloguing combinatorial configurations, in *Algorithmic Aspects of Combinatorics* (ed. B. Alspach *et al.*), North-Holland, Amsterdam, 1978, pp. 107–120.
44. R. C. Read, The enumeration of acyclic chemical compounds, in *Chemical Applications of Graph Theory* (ed. A. T. Balaban), Academic Press, London, 1976, pp. 25–61.
45. R. C. Read, A visual graph manipulation package for general use, Research Report CORR 75-9, Department of Combinatorics and Optimization, University of Waterloo, 1975.
46. R. C. Read and D. G. Corneil, The graph isomorphism disease, *J. Graph Theory* **1** (1977), 339–363.
47. P. Rosenstiehl and R. C. Read, On the principal tripartition of a graph, in *Advances in Graph Theory* (ed. B. Bollobás), North-Holland, Amsterdam, 1978.
48. A. J. Schwenk, Almost all trees are cospectral, in *New Directions in the Theory of Graphs* (ed. F. Harary), Academic Press, New York, 1973, pp. 275–307; *MR***52**#5456.
49. H. I. Scoins, Placing trees in lexicographic order, in *Machine Intelligence*, Vol. 3 (ed. D. Michie and N. L. Collins), Oliver and Boyd, Edinburgh, 1969, pp. 43–60.
50. M. L. Stein and P. R. Stein, Enumeration of linear graphs and connected linear graphs up to $p = 18$ points Report LA-3775 UC-32, Mathematics and Computers, TID-4500, Los Alamos Scientific Laboratory of the University of California, 1967.
51. W. T. Tutte, On chromatic polynomials and the golden ratio, *J. Combinatorial Theory* **9** (1970), 289–296; *MR***42**#7557.
52. W. T. Tutte, More about chromatic polynomials and the golden ratio, in *Combinatorial Structures and their Applications* (ed. R. K. Guy *et al.*), Gordon and Breach, New York, 1970, pp. 439–453.
53. H. Whitney, A logical expansion in mathematics, *Bull. Amer. Math. Soc.* **38** (1932), 572–579.
54. H. Whitney, The coloring of graphs, *Ann. of Math.* (2) **33** (1932), 688–718.

Index of Definitions

This index gives the main reference for each of the terms included. When a given term features in an important way in more than one place, a reference is given for each occurrence.